Organic Agriculture

Edited by Shaon Kumar Das

Published in London, United Kingdom

IntechOpen

Supporting open minds since 2005

Organic Agriculture
http://dx.doi.org/10.5772/intechopen.89775
Edited by Shaon Kumar Das

Contributors
Rita Pupalienė, Francisco Daniel Hernandez Castillo, Francisco Castillo, Marco Antonio Tucuch Pérez, Roberto Arredondo Valdes, Shaon Kumar Das, Tamanreet Kaur, Haluk Gedikoglu, Nisa Sansel Tandogan, Md. Motaher Hossain, Farjana Sultana, Muthuraman Yuvaraj, Siddhartha Das, Sudeepta Pattanayak, Peyandi Paraman Mahendran, Eman Tawfik Hussien, Ravikant Avasthe, Aušra Marcinkevičienė, Marina Keidan, Rimantas Velička, Zita Kriaučiūnienė, Lina Marija Butkevičienė, Robertas Kosteckas

First published in London, United Kingdom, 2020 by IntechOpen
IntechOpen is the global imprint of INTECHOPEN LIMITED, registered in England and Wales, registration number: 11086078, 5 Princes Gate Court, London, SW7 2QJ, United Kingdom
Printed in Croatia

British Library Cataloguing-in-Publication Data
A catalogue record for this book is available from the British Library

Additional hard and PDF copies can be obtained from orders@intechopen.com

Organic Agriculture
Edited by Shaon Kumar Das
p. cm.
Print ISBN 978-1-78984-668-3
Online ISBN 978-1-78984-669-0
eBook (PDF) ISBN 978-1-78984-720-8

For EU product safety concerns:
IN TECH d.o.o., Prolaz Marije Krucifikse Kozulić 3, 51000 Rijeka, Croatia, info@intechopen.com or visit our website at intechopen.com.

Meet the editor

Shaon Kumar Das is working as a Scientist SS (Agricultural Chemistry) at ICAR RC for NEH Region, Sikkim Centre, Tadong, Gangtok under the Indian Council of Agricultural Research, New Delhi. He received his M.Sc. in 2010 from IARI, New Delhi along with a gold medal for his outstanding contribution. He joined the Agricultural Research Service (ARS) in 2011 as a scientist. He completed his Ph.D in 2020 from Visva Bharati University in Soil Science and Agricultural Chemistry and worked on integrated organic nutrient management. He has published 27 national and 40 international research articles, 10 review articles in national and international journals, 30 popular articles, 28 extension folders, 2 books, and 25 book chapters in edited books. He received a DST-INSPIRE fellowship in 2010 from DST, Govt. of India. He has been awarded the best scientist award by the ICAR RC for the NEH Region, Umiam, Meghalaya in 2018. He received six young scientist awards and 3 best oral presentation awards from different reputable societies. He was the Principal Investigator (PI) of different externally funded projects like DST, DBT, NMHS, ICAR etc. He has been involved in the research on carbon sequestration, soil fertility management, soil acidity reclamation, and organic nutrient standardization of major hill crops.

Contents

Preface

Organic crop production is the science and art of growing field crops, fruits, vegetables, and flowers by adopting the essential principles of organic agriculture in soil building and conservation, pest management, and heirloom variety conservation. This book provides detailed insights into organic farming in agriculture, biological efficacy in the management of plant diseases, organic nutrient management, socio-economic dimensions of adoption of conservation practices, nonchemical weed control, plant growth promoting fungi for phytostimulation, nanotechnological approaches, and finally vermicomposting. The book primarily focuses on research and development based organic agriculture and horticulture production technologies, and has attempted to abridge information on organic crop production of the major food grain crops. The book also contains comprehensive information on the various related dimensions of organic crop production.

It is a matter of great privilege and honor for me to express my esteemed gratitude to all my senior colleagues who conceived, detailed, and shaped the problem. Their sagacious guidance, persistent involvement, scholarly suggestions, relentless encouragement, prudent admonitions, constant and ungrudging help, and affectionate behavior has been a fountain of great inspiration to me. I implore my feeling of profound gratitude to the researchers for their keen interest, cordial attitude, generous help, and valuable suggestions during the entire course of work. I owe a great deal to all the scientific community for their wise counsel and steadfast help during this investigation. I avail myself of this opportunity to convey my heartfelt thanks to contributors of the book chapters for providing necessary facilities and valuable help during the entire span of work. I express my heartfelt thankfulness to my brother, mother, father, wife, and son for their encouragement, cordial attitude, and help during the entire span of my study. Candid thanks are due to all my seniors, juniors, and friends for their helping nature, company, inspiration, and moral support, which made my passage smooth over the rough terrain. Finally, I express my special thanks to all the staff of the IntechOpen from whom I received ever-willing help and moral support.

Shaon Kumar Das
Indian Council of Agricultural Research (ICAR),
India

Section 1

Organic Farming

Chapter 1

Role of Organic Farming in Agriculture

Muthuraman Yuvaraj, Peyandi Paraman Mahendran and Eman Tawfik Hussien

Abstract

Organic farming could be an all-encompassing generation administration framework that empowers and improves agroecosystem wellbeing, counting biodiversity, natural cycles, and soil biological activity. It stresses the role of management activities in preference to the use of off-farm data, considering that regional conditions require locally adapted systems. This can be achieved using agronomic, biological, and mechanical methods, in equal share to synthetic materials, to carry out any specific role inside the organization. Organic farming is still only a small industry, which represents only 2% of global food sales. However, it is growing in importance in the world. It is hard to get information due to lack of official statistics and the level of confidentiality of systems of organic produce. Soil practices such as crop rotations, organic fertilizers, symbiotic associations, cover crops, inter-cropping, and minimum tillage are central to organic practices. The static arrangements of soil are achieved by soil fauna and vegetation. Besides, cycling of nutrients and energy is enhanced by increasing the retentive abilities of the soil for nutrients and water.

Keywords: organic farming, vermicompost, soil health

1. Introduction

Organic farming is defined as a production system that avoids or largely eliminates the usage of synthetic compounded fertilizers, growth regulators, pesticides, and farm animal feed additives. To maximize feasible extent, the organic farming system depends on crop rotation, green manures, legumes, animal manures, crop residues, off-farming natural squanders, and aspect of biological pest control to preserve soil fertility and productivity to sustain plant and thereby curbing pests, diseases, and unwanted weeds. Organic farming methods are internationally regulated and legally executed by many countries, based in great part on the standards set by the International Federation of Organic Agriculture Movements (IFOAM), an international umbrella organization for organics established in 1972 [1].

After the launch of green revolution in India, substantial growth in the output of food grains was achieved. This was achieved through the utilization of improved crop varieties and higher levels of inputs of plant foods and plant protection chemicals. Merely there has been a rise in production, which was accomplished by the monetary value of soil health. The organic farming and ecological agriculture are one of the alternative agriculture systems to overcome the problems of soil degradation and declining soil fertility [2].

Soil organic matter is an essential soil component. These are residuals of dead plants, animals, and microbial tissues. Examples of organic manures are farmyard manure, compost, and green manure. They are added to the soil to the stock of organic matter. These added organics undergo a series of microbial decompositions, and finally, humus is formed. Humus is a light bulky amorphous material of dark brown to black color composed of organic compounds. Tropical soils characterized by low organic and loam less than of clay soils. The low organic matter is primarily due to climate, especially for high temperature and cultural practices, while organic matter contents increase with rainfall. In tropical and subtropical areas, there is much organic matter created, and it decays very quickly. Any organic matter added to the grounds will be decomposed (over 90% in a year), and thus, it is a Herculean job to produce the organic matter; content ranges from less than 1 to 15% [3].

2. Guidelines on production of organic produce

2.1 Duration of conversion period

Firstly, the formation of an organic controlling system and construction of soil fertility needs a temporal time known as change period. This change period may not continuously be the identical period to advance soil fertility and re-establish the equilibrium of the environment in all cases. It is defined as the time in which all the farm, including the livestock, is converted into organics. This was confirmed by the norms laid in the National Standards for organic products. The entire farm unit should be converted to organic in a phased manner, and the grower should present an alternative plan to the certification body when applying for accreditation.

Second step isolation belts are maintained all around the organic unit. This would drastically bring down the net area under organic cultivation. In view of this, a community approach is suggested to a group of continuous farms.

Thirdly, for existing conventional plantations, a changeover point for organic culture required a minimum of 3 years.

Finally, for a newly planted area, the first yield itself can be considered as organic, as the organic produce has a pre-bearing period of 1–2 years.

2.2 Farm designing

Firstly, an organic farm has to be a self-sustaining unit. Also, farm designing plays a very important role in optimizing the utilization of resources within the farm topography of the land and varieties of crops to be cultivated. Aside from these, border trees, compost yard, bounds, storage home, cattle shed, and farmhouse can be suitably incorporated.

Secondly, the topography of the location of the cattle shed, compost yard, etc., should be decided. All the structures are better if at comparatively higher elevation than the cropped areas as it prevents water logging inside the cattle shed, store house, and office.

Finally, border trees: on the boundary of the farm, multipurpose border like Neem, Karanj or any other local trees of importance are planted 10 m apart. These border trees provide useful as wind break, abundant biomass for green manuring and composting, and preparation of pest management aids like Neem seed kernel extract to control insect pests.

2.3 Cattle shed

The cattle shed should be near the compost yard. The site for the cattle shed is better if it is at higher elevation. The shed should provide a comfortable and hygienic habitat to keep the cattle healthy and free of diseases. The floor may slope toward the dung channel to provide satisfactory drainage and facilitate collection of urine. Managers can be 0.75 m wide with all corners rounded off in cement. The cattle shed should be preferably oriented in east to west direction to have proper public discussion and with neem or peepal trees around it [4].

2.4 Store house

A storehouse should be maintained to stock the farm implements and the produce after harvest. The storehouse can be at least half a meter above the farm level and should throw a projection of 0.5 m at that height to protect the produce from rats and other rodents. The storehouse should be damp proof. The windows and doors could be lined with fine wire mesh [5].

2.5 Border plants

The planting of leguminous trees is suggested to be grown as border crops for biomass. The trees mentioned earlier would also enrich the soil fertility and help in the improvement of the soil structure [6, 7].

3. Components of organic farming system

The significant research in an organic farming system has been difficult to design due to the nature of organic philosophy. There is no shortage of essential component of research, which could be done to increase both nitrogen fixation (in legumes on component crops) and on weed and insect control measures. This research would touch only marginally on the basic questions inherent in organic philosophy [8, 9].

Principles of organic agriculture

- Use renewable resource in locally organized production system
- Increase balance between crop production and animal husbandry
- Give all livestock condition of life with due consideration for the basic aspects of their intake behavior
- Decrease all forms of pollution
- Processing organic products with renewable resources
- Production of fully biodegradable organic products
- Progress toward an entire production processing and distribution chain, which is both socially and ecologically responsible

Benefits of organic farming

- Organic agriculture helps to prevent environmental degradation and avoid a chain reaction in the environment from chemical sprays and dusts
- Organically grown crops are believed to be healthier and nutritional food for man and animals
- Organic fertilizer is considered as complete plant food. Organic matter restores the pH of the soil, which may become acidic due to the continuous application of chemical fertilizer.
- Organic manure is the principal component of organic farming to produce optimal conditions in the soil for high yields and good quality.
- Most of the organic manures are wastes or byproducts, which on accumulation may contribute to contamination. In this method of organic farming pollution is reduced.
- Organic farming is labor intensive in the nation which will also help in generating more employment in rural areas that will help in reducing economic inputs.
- As a whole, adoption of organic farming provides a better and balanced environment and better products and living conditions to the human beings.

4. Different available organic inputs

4.1 Organic droppings (manure)

The farm-yard manure (FYM) and vermicomposting, etc., are typically little in nutrient contented. So, great use rates are difficult to meet crop nutrient supplies. However, in numerous emerging nations (like India), the obtainability of organic manures is not sufficient for crop demands partly due to its wide use of cattle dung in energy production. Green manuring with Sesbania plant, cowpea, and green gram is effective to improve the soil content of organic matter. However, employment of inexperienced manuring has declined in the previous couple of decades because of intensive cropping and socioeconomic reasons. Taking these constraints, International Federation of Organic Agriculture Movement (IFOAM) and Codex Alimentarius have approved the use of some inorganic sources of plant nutrients like rock phosphate, basic slag, rock potash, etc., in organic farming systems [10].

4.2 Bacterial and fungal biofertilizers

Nitrogen fixation on the surface of the earth is mainly by microorganisms, representing 67.3% among all the bases of N fixation. Subsequent microorganism and plant life biofertilizers will be utilized as an ingredient of organic farming in numerous crops.

4.2.1 Rhizobium

The efficiency of nitrogen-fixing microorganisms viz. *Rhizobia* for legume crops e.g. *Rhizobium*, *Bradyrhizobium*, *Azorhizobium*, *Mesorhizobium,* and *Sinorhizobium*.

Legumes are infected by these bacteria all over the world. These rhizobia have a N_2-fixing capability up to 450 metric weight units N ha^{-1} counting on host-plant species and microorganism strains. Carrier based inoculants will primarily be coated with seeds for the introduction of microorganism strains into soil [11].

4.2.2 *Azotobacter*

Nitrogen will be fixed by independent microorganisms in cereal crops with no interdependency. Such free-living bacteria are: *Azotobacter sp*. for dissimilar cereal crops; *Herbaspirillum spp* and *Acetobacter diazotrophicus*.

4.2.3 *Azospirillum*

The gram-positive bacteria *Azospirillum* colonizes in a remarkably kind of yearly and perennial florae. Studies indicate that *Azospirillum* will proliferate the development of crops like flowers, cotton, oak, tomato, sugar beet, pepper, carrot, wheat, and rice. The crop yield can upsurge from 5 to 30%. Inoculum of *Azotobacter* and *Azospirillum* will be created and applied as in humate origination finished seed coating.

5. Plant growth-promoting rhizobacteria

Numerous microorganisms promote plant growth area. The unit is jointly known as plant growth-promoting rhizobacteria (PGPR). PGPR improves plant growth by colonizing the root system. Huge inhabitants of microorganisms recognized in implanting material and roots develop an incomplete sink for nutrients in the rhizosphere [12].

6. Phosphorus-solubilizing microorganism (PSB)

Phosphorus is also an important nutrient similar to nitrogen for plants. This part is important for the nodulation by bacteria genus and even to nitrogen fixers, *Azolla* and BGA. Phosphorus-solubilizing microorganism (PSB) fungi create on the market are insoluble phosphorus to the plants. It will increase crop weight up to 200–500 metric unit/ha and so 30–50 kg Super Phosphate will be preserved. Most predominant phosphorus-solubilizing microorganism (PSB) belongs to the genera *Bacilli* genus.

6.1 Mycorrhizal fungi

Mycorrhizal fungi which cause root-colonizing increase tolerance to many sever metal contamination and drought. Mycorrhizal fungi improve soil quality additionally by having an on-the-spot influence on soil aggregation and also aeration and water dynamics. An interesting potential of these fungi is their ability to permit plant access to nutrient sources [13–15].

6.2 Blue Chlorophyta (BGA)

The BGA represents the most important, most numerous and cosmopolitan cluster of microscopic organisms that perform an oxygenic chemical process. These are as well-known as Cyanophyceae and cyanobacteria [15–18].

6.3 Azolla

A floating fern *'Azolla'* hosts element fixing BGA *Anabaena azollae*. *Azolla* contains 3.4% nitrogen (on dry weight basis) and adds organic matter in soil. This biofertilizer is used for rice cultivation. There are six species of *Azolla* viz. *A. pinnata*, *A. microphylla*, *A. mexicana*, *A. filiculoides*, *A. nilotica,* and *A. caroliniana* [19–20].

7. Soil management in organic farming

It is essential to take care to build up comparison of average yields, standard deviations, and coefficients of variation over five successive years. This includes crops and soil fertility, based on three inextricably interrelated components of soil management. These components are physical (water-holding capacity, structure, etc.), chemical (nutrient dynamics, pH), and biological (soil biota). The soil fertility of organic farming, can be defined as: well-managed soil organic matter in the soil, good soil structure, diverse soil biota and a high nutrient and water-holding capacity by using compost and stable manure. The organic soil is one that can build up over time buffering capacity and resistance to an imbalance in growing conditions as part of the strategy to enhance the self-regulatory capacity of the farm ecosystem [21, 22].

8. Weed management in organic farming

Weeds are often cited as the most significant problem in organic farming systems, and they are certainly the problem that most concerns the farmers, who are looking at changing over their farm from a standard one into an associate-in-nursing organic one.

9. Botanical pesticide use in organic farming

9.1 Nicotine

Nicotine is obtained from tobacco or related Nicotiana species and is one of the oldest botanical insecticides in use today. It is also one of the most toxic to warm-blooded animals, and it is readily absorbed through the skin.

9.2 Sabadilla

Sabadilla is another botanical insecticide. It is extracted from the seeds of the sabadilla lilly. The veratrine alkaloid is the active ingredient. Sabadilla is a botanical insecticide with low toxicity. However, its dust can be extremely irritating to the eyes and can produce sneezing if inhaled.

9.3 Neem

Neem is a botanical pesticide that comes from the neem tree, which is native of India. This tree supplies at least two compounds, salannin and azadirachtin, that have insecticidal activity and other unknown compounds with fungicidal activity. Neem pesticide controls gypsy moths, western flower thrips, sweet potato white flies, leaf miners, loopers, caterpillars, and mealybugs.

10. Conclusions

The contemporary form of natural farming is being popular in the world in a timely fashion, in particular in developed countries. Organic farming device is a choice, and an appropriate management system would help to enhance the soil health environment and hence expand the productive ranges and the enhance quality of crops. The natural farming system utilizes extremely multifaceted and combined residing classifications to accomplish their cease of maintainable harvest and inventory output. Organic agriculture is a potential choice due to the fact that it enlivens the soil, strengthens the natural resource base, and sustains organic production at degrees to commensurate the carrying potential of the managed agro eco-system. In addition to this, export market can also be tapped by group initiatives in organic farming. In a country like India, food production has to grow steadily.

Author details

Muthuraman Yuvaraj[1]*, Peyandi Paraman Mahendran[2] and Eman Tawfik Hussien[3]

1 Agricultural College and Research Institute, Vazhavachanur, Tiruvannamalai, Tamil Nadu, India

2 Department of Crop Management, Agriculture College and Research Institute, Kudumiyanmalai, Tamil Nadu Agricultural University, Tamil Nadu, India

3 Faculty of Science, Helwan University, Egypt

*Address all correspondence to: yuvasoil@gmail.com

References

[1] Nanda SK, Bhattacharya B, Chandra PS. Promoting organic farming in India: Extension implications. Management Extension Research Review. 2003;**4**(2):9-20

[2] Palaniappan SP, Annadurai K. Organic Farming: Theory and Practice. Vol. 9. India: Scientific Publishers; 1999. pp. 1-257

[3] Hole DG, Perkins AJ, Wilson JD, Alexander IH, Grice PV, Evans AD. Does organic farming benefits biodiversity. Biological Conservation. 2005;**122**:113-130

[4] Jayanthi C. Sustainable Farming System and Lowland Farming of Tamil Nadu. IFS Adhoc Scheme. Completion Report. 2002

[5] Mader P, Fliebach A, Dubois D, Gunst L, Fried PM, Niggli U. Soil fertility and biodiversity in organic farming. Science. 2002;**296**:1694-1697

[6] Oehl F, Sieverding E, Mader P, Dubois D, Ineichen K, Boller T, et al. Impact of long-term conventional and organic farming on the diversity of arbuscular mycorrhizal fungi. Oecologia. 2004;**138**:574-583

[7] Parr JF, Stewart BA, Hornick SB, Singh RP. Improving the sustainability of dryland farming systems: A global perspective. Advanced in Soil Science. 1990;**13**:1-8

[8] Rundlof M, Smith HG. The effect of organic farming on butterfly diversity depends on landscape context. Journal of Applied Ecology. 2006;**43**:1121-1127

[9] Verbruggen E, Rolling WFM, Gamper HA, Kowalchuk GA, Verhoef HA, Van Der Heijden MGA. Positive effects of organic farming on below ground mutualists: Large scale comparison of mycorrhizal fungal communities in agricultural soils. New Phytologist. 2010;**186**:968-979

[10] Dobbs TL, Smolik JD. Productivity and profitability of conventional and alternative farming systems: A long-term on-farm paired comparison. Journal of Sustainable Agriculture. 1996;**91**(1):63-79

[11] Drinkwater LE, Wagoner P, Sarrantonio M. Legume-based cropping systems have reduced carbon and nitrogen losses. Nature. 1998;**396**:262-265

[12] Edwards S. The impact of compost use on crop yields in Tigray, Ethiopia. Institute for Sustainable Development (ISD). In: Proceedings of the International Conference on Organic Agriculture and Food Security. Rome: FAO; 2007

[13] Finamore A, Britti MS, Roselli M, Bellovino D, Gaetani S, Mengheri E. Novel approach for food safety evaluation. Results of a pilot experiment to evaluate organic and conventional foods. Journal of Agricultural and Food Chemistry. 2004;**52**:7425-7431

[14] Reicosky DC. Long-term effect of moldboard plowing on tillage-induced CO2 loss. In: Kimble JM, Lal R, editors. Agricultural Practices and Policies for Carbon Sequestration in Soil. Boca Raton: CRC Press; 2002. pp. 87-97

[15] Roschewitz I, Gabriel D, Tscharntke T, Thies C. The effects of landscape complexity on arable weed species diversity in organic and conventional farming. Journal of Applied Ecology. 2005;**42**:873-882

[16] Kustermann B, Kainz M, Hulsbergen KJ. Modeling carbon cycles and estimation of greenhouse gas emissions from organic and conventional farming systems.

Renewable Agriculture and Food Systems. 2008;**23**:38-52

[17] Olsson ME, Andersson CS, Oredsson S, Berglund RH, Gustavsson KE. Antioxidant levels and inhibition of cancer cell proliferation in vitro by extracts from organically and conventionally cultivated strawberries. Journal of Agricultural and Food Chemistry. 2006;**54**:1248-1255

[18] Pimentel D, Harvey C, Resosudarmo P, Sinclair K, Kurz D, McNair M, et al. Environmental and economic costs of soil erosion and conservation benefits. Science. 1995;**267**:1117-1123

[19] Reganold JP, Palmer AS, Lockhart JC, Macgregor AN. Soil quality and financial performance on biodynamic and conventional farms in New Zealand. Science. 1993;**260**:344-349

[20] Staiger D. The nutritional value of foods from conventional and biodynamic agriculture. IFOAM Bulletin No. 4. 1988. pp. 9-12

[21] Bengtsson DJ, Ahnstrom J. Effects of organic agriculture on biodiversity and abundance: A meta-analysis. Journal of Applied Ecology. 2005;**42**:261-269

[22] Clark MS, Horwath WR, Shennan C, Scow KM, Lantini WT, Ferris H. Nitrogen, weeds and water as yield-limiting factors in conventional, low-input, and organic tomato systems. Agriculture Ecosystems & Environment. 1999;**73**:257-270

Chapter 2

Nanotechnological Approaches in Sustainable Agriculture and Plant Disease Management

Siddhartha Das and Sudeepta Pattanayak

Abstract

Every year approximately 30–50% of crops suffer with different kinds of biotic stresses. Rapidly growing agrochemical industries and their diverse products make the environment more toxic and simultaneously hazardous for plant heath and soil health. Such types of agrochemicals are toxic, hazardous, carcinogenic, non-eco-friendly. Therefore, this is the ideal time to think about some more effective alternatives against those problems. Nanotechnological approaches bring the alternatives in the form of decreasing toxicity, improving shelf-life, increasing solubility for poorly water-soluble agrochemicals, minimum use with maximum effect, slow leaching efficiency with long-term effect with coupling of eco-friendly naturalistic way. The way of nanoparticle application in agriculture, specifically disease management, is unique, where it can be used singly or by coupling with fungicidal, herbicidal, insecticidal, RNA-interference molecules. Though it has such a positive impact, very few products will be commercially available in our market due to high price of particular products and well-established long field trial efficacy detection among insect, pest-pathogen, and environment. Application of nanomolecules in other progressive fields has been emerging, whereas advancement in agricultural applications needs to be boosted up through skilled knowledge transfer and basic understanding of its fundamental aspect.

Keywords: nanotechnology, IDM, sustainable, eco-friendly

1. Introduction

The population of the world is increasing in an alarming rate. The global demand for food production will need to double by 2050. But, the climate change such as prolonged drought, sudden increase of temperature, unpredictable rains, and floods are the main barrier to achieve this global food demand [1]. Moreover, significant crop losses, that is, 20–40% per year due to some biotic causes is another major problem in worldwide basis [2]. The biotic causes mostly include insects, nematodes, pathogens, weeds, human, and animal interaction. Earlier studies revealed that approximately 20 and 26% yield loss occurred due to pathogen and insect infestation [3]. Most developed and developing countries depend upon chemical pesticides to control the diseases and pest incidence as these are easily available and show the result quickly. Industrial business and marketing policy also sometimes cross their barrier to gain some profit. Therefore human made behavioristic alteration of our environment also considered as a crucial responsible factor.

But they do not think those chemicals are harmful to the ecosystem, poisonous to beneficial insects, create pest resurgence, 90% loss of chemical during or post application in field leading to more economical loss for farmers. By keeping all the aforesaid reality based problem in mind, a new potential concept and technique known as nanotechnology is developed by the scientists, which is cost-effective, reliable, eco-friendly, and very effective. Although, this concept has not gained more focus in agriculture compared to pharmacology and medicine, but still it takes part a major role in plant breeding, nano sensors, plant hormone delivery, limited application of chemicals, plant health management etc. [4–6].

Nanotechnology refers to the technology related to application and manipulation of nanoparticles, that is, very small particles or materials having one or more dimensions, that is, 1–100 nm and fashioned with exclusive physical, chemical, and biological properties [7]. Several scientists have worked on the desired characteristics of nanoparticles such as pore shape, size etc. for accurate and specific application through adsorption or encapsulation of effective pesticide [8]. Application of nanoparticles can be functionalized in two ways, that is, first, nanoparticles directly involves as a plant protectant and second, nanoparticles used as carrier. On the first way researchers have tested the viability of the nanomolecules and mode of action against a particular pathogen. On the second way nanoparticles used with the existing fungicides or pesticide, herbicides, RNAi-mediated coupled component, to boost its activity. Nanoparticles used as a carrier have some positive sites like it can increase solubility of the coupled component, increase shelf-life, boosting site- and target-specific activity, reduce toxicity level, and maintain environmental safety. The implementation and utilization of fruits of nanotechnological applications, in terms of plant disease detection and management, gene editing and transformation is at infancy stage till now due to insufficiency of knowledge and skills. In the present context, the use of nanotechnology in plant disease management and allied sectors of agriculture is a real challenging task, which is objectified and framed with the present chapter.

Context of sustainable agriculture and nanotechnology

1. **Application of nanotechnology to increase crop productivity and quality:** A variety of nanoparticles were tested previously to study their efficacy for boosting the yield as well as quality of the crop. It was observed that the nanoparticles modify enzymatic action, electron transport system, and influence the nutrition uptake. Carbon nanoparticles are reported to increase the yield of bitter gourd by enhancing the biomass, water content, fruit weight, length, and number. Lentil seeds show high germination rate and growth when treated with silica-based nanoparticles. Titanium dioxide nanoparticles can boost the water uptake, breakdown of organic compounds, photosynthesis capacity, etc. [9].

2. **Increase of photosynthetic efficiency through nanotechnological application:** Seeds incorporated with nanoparticles can promote the photosynthetic activity in plants. The nanoparticles respond to a specific wavelength of light and increase the optical potential of leaves which in result effect the hills reaction. Titanium dioxide-based nanoparticles can enhance the water and nutrient uptake, light absorbance and activity of Rubisco activase enzyme resulting more photosynthetic activity in plants. These nanoparticles induce the plants to be photosynthetically more active, grow faster, resulting in quality food production in less period of time [9, 10].

3. **Improvement in water retention and management:** Nanotechniques can be imposed in agriculture to reduce the loss through evaporation, irrigation, and

to stabilize soil horizons in addition to lessening the ecotoxicity. Nanoparticles or nanotubes are found to be effective in retaining the water inside the hollow core for longer period by modifying the xylem vessel mechanism and metabolism of plant cells. Due to the partial solubility and dispersible characteristics, the water is retained inside plant cells for more time. In 2009, Corredor et al. demonstrated carbon nanotube treatment in pumpkin plant to analyze the water retention capacity and observed the accumulated carbon nanotubes inside the plant cell, which act as a water transporter. Due to the cohesive force, the water moves continuously through the nanotubes in plant cell forming a nanosized water stream [11].

4. **Significance in the field of plant disease management:** Recent days, nanotechnology is taking part a major role in plant disease management due to its eco-friendly nature and potentiality. Among all nanobased particles, silver nanoparticles stand out in the frontline in plant defense. These nanoparticles disturb the cell DNA, metabolic activity, electron transport chain, nutrient uptake of micro-organisms leading to death. The pathogenic fungi which can be controlled by using silver nanoparticles are *Colletotrichum gloeosporioides, Alternaria solani, Fusarium oxysporum, F. solani, Macrophomina phaseolina, Rhizoctonia solani, Aspergillus niger* etc. Silica-silver based nanoparticles are reported to inhibit the growth of bacteria *Pseudomonas syringae, Xanthomonas campestris* pv. *vesicatoria* up to 100%. Copper-based nanoparticles were found to be effective against bacteria *Xanthomonas oryzae*pv. *Oryzae*, *Xanthomonas campestris* pv. *phaseoli* and fungi *Fusarium solani, Alternaria solani, Aspergillus flavus* etc. [12, 13].

   A. **Limitations of different applied disease management strategies:** Drawbacks or limitation of all applied disease management practices are depicted under **Table 1**.

   B. **Types of nanoparticles for plant disease management:** Nanoparticles can be used to manage the plant disease through two different mechanisms: as protectants where the nanoparticles only protect the plants and as carrier where the nanoparticles contain potential pesticides or some other active compounds [2]. Schematic representations of mode of action of these two modes are shown under **Figure 1**.

      I. **Nanoparticles used as protectants:** Potential nanoparticles used alone to protect the plants from pathogenic micro-organisms and applied directly to plant and plant parts. Several metal nanoparticles like copper, silver, zinc oxide, titanium dioxide is experimented for their antagonistic effects against all pathogenic bacteria, virus, fungus, and concluded as successful potential protectants [14–16]. These nanoparticles have more shelf life, easily soluble in water, and show site specific uptake as compared to conventional chemicals.

         a. **Silver nanoparticle:** It is in the frontline due to its "green synthesis" production mechanism in bacteria, yeast, fungi, and plants [17]. From previous studies, it is reported that silver nanoparticles have antifungal effect against *Alternaria alternata, Sclerotinia sclerotiorum, Macrophomina phaseolina, Botrytis cinerea, Rhizoctonia solani,* etc. by well diffusion assay [18] while antiviral effect on sun hemp rosette virus when sprayed on leaves and bean yellow mosaic virus when applied after infection [19].

Techniques	Physical Method	Chemical method	Biological method
Strategies applied	Soil solarization	Fungicide, pesticide or related agrochemicals	Resistant varieties, Use of biocontrol agents
Limitations	• Longer duration • Selective towards thermophilic and thermotolerant pathogens • Climate dependent • Availability of crop-free fields for long period • Survival of pathogen in deeper soil zone • Pollution from plastic residue after treatment	• Resistance to fungicide or pesticide • Harmful effect on non-target organisms • Non ecofriendly • Carcinogenic • Non-economic	• Wide host range of pathogen • Survival of pathogen for long period • Costly • Lack of proper awareness and training

Table 1.
Limitations of different disease management prcatices.

b. **Nano carbon:** The carbon nanoparticles not only show antimicrobial activity in plant health management but also positively affect the plant growth. These nanoparticles from graphene oxide sheets have shown antimicrobial activity against *Aspergillus niger, A. oryzae, and Fusarium oxysporum in vitro* [20]. Many researchers have anticipated about the mechanism behind the inhibition of microbial growth by carbon nanomaterials and explained that the nano edges of graphene oxide come in direct contact with the chemical compounds present in cell wall of fungi and bacteria making them inactive [21–23].

c. **Titanium dioxide nanoparticle:** When this nanoparticle is used as fertilizer was known to show antibacterial, antiviral and insecticidal characteristics.

d. **Poly dispersed gold nanoparticles**: It was reported to inactive the Barley yellow mosaic virus and develop plant resistance [24, 25].

e. **Chitosan nanoparticles** control alfalfa mosaic virus, *Fusarium* spp., *Botrytis* spp., *Pyriularia grisea* while show limited effect against bacteria [26]. The mechanisms behind the antimicrobial properties of chitosan are inhibition of the growth of pathogen, protein synthesis, and ATPase activity, agglutination, interruption of cell membrane, disruption of nutrient flow etc. Chitosan nanoparticles have also been reported to control aphids, root knot nematode, and cotton leaf worm.

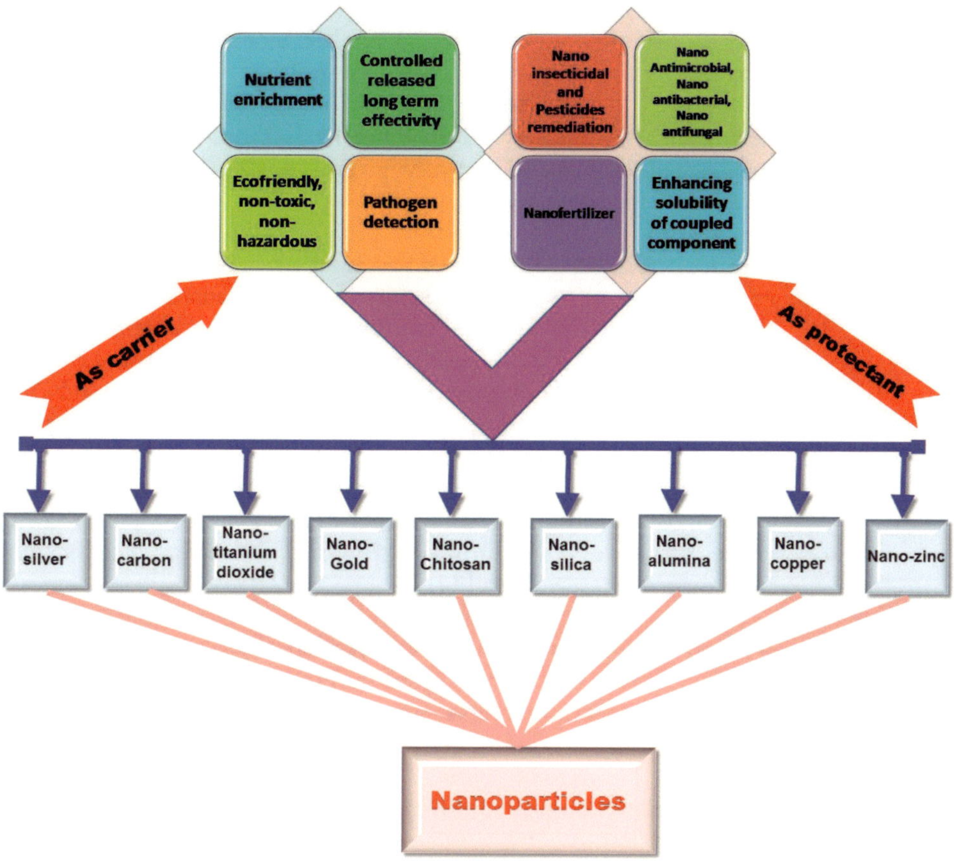

Figure 1.
Schematic representation of different nanoparticles and their mode of action in term of carrier and protectant.

II. **Nanoparticles that act as carriers:** Nanoparticles are engineered in such a way that these can carry, encapsulate, or absorb active potential compounds to develop effective agricultural pesticides. These carrier nanoparticle-based pesticides have less toxicity, slow leaching capacity, and highly target specific. The below-mentioned nanoparticles are mainly used as carrier:

a. **Silica nanoparticles:** These nanoparticles have specific size, shape, and structure making them able to deliver the active compounds in the target site [27]. They are mainly round or circular like structure having a pore like holes, for example, mesoporous silica nanoparticles (MSNs) or porous hollow silica nanoparticles (PHSNs). The active compounds are mainly loaded in the inner hole to keep them protect and proper release in target sites. The outer shell guards the active compounds against degradation by UV light.

b. **Nano alumino-silicate**: The major advantages for the use of nano alumino-silicate are that it is biologically active, eco-friendly, and more effective than other nanoparticles. The alumino-silicate nanotube containing active compounds when sprayed on plants, the insects present on plants intake the active compounds contained nanotubes [28].

c. **Chitosan nanoparticles:** Due to the low solubility in aqueous media, chitosan nanoparticles are applied by mixing with some organic or inorganic or copolymer compounds to increase its solubility [29]. These compounds stick to the stem and leaf epidermis facilitating them for prolonged effect in target sites [27]. Presence of amines and hydroxyl groups, enable the nanoparticles to enhance its properties [30].

d. **Nanocopper:** Copper is one of the most effective and potential chemicals in combating wide range of plant diseases. For the first time in 2013, Giannousi et al. studied the potential antimicrobial characteristics of nanocopper particles and concluded that the nanocopper particles are more promising than other copper-based chemicals [31]. From past studies, it was reported to control or minimizes the growth of the pathogen of Fusarium wilt of tomato, *Verticillium* wilt of eggplant, Leaf spot, leaf blight diseases etc. [32, 33]. Biosynthesized nanocopper derived from *Streptomyces griseus* was observed to limit the growth of *Poria hypolateritia* causing root rot disease of tea [34].

e. **Nanozinc:** The characteristic antimicrobial properties of nanozinc particles are mostly studied in vitro condition. It was reported to control bacteria, fungi, for example, *Alternaria alternata, Botrytis cinerea, Sclerotinia sclerotiorum, Rhizopus stolonifera, Rhizoctonia solani, Mucor* spp., *Fusarium oxysporum,* and *Penicillium* spp. [35, 36]. Disease suppression in protected cultivation and green house condition through nanozinc is also possible.

f. **Solid lipid nanoparticles (SLNs):** These nanoparticles contain lipids, which remain in solid state at room temperature. Therefore, the lipophilic compounds show its prolonged effect by slow releasing of active compounds without adding any extra organic solvents [37]. The limitations are that the active compound may escape during storage [38].

g. **Layered double hydroxides (LDHs):** These nanoparticles break down when come in contact with water or carbon dioxide [39]. The positively charged LDH nanoparticles make the active compounds easier to move through the plant cell wall.

h. **Nanoemulsions:** Nanoemulsions are combination of more than one liquid that does not mix easily. These nanoparticles contain active ingredients in the droplets of diameter 500 nm or less and lessen the degradation of chemicals or active ingredients inside [35].

i. **Dendrimers:** These nanoparticles are branched, similar to tree like structure, having a central core, which is occupied by functional groups [1]. Very few reports and research have done on dendrimers related to plant pathology. It can be referred as a great delivery vehicle as it can transport chemicals or pesticides in basipetal manner after a foliar spray. It can enhance the permeability in diseased plant cells.

III. Application of nanoparticles to boost/develop commercial agro-protectants (herbicide/fungicide/insecticide)

a. **Nanoparticles as carriers for herbicides:** Nanocarrier-based herbicide studied mainly confined to reduce the environmental toxicity, which occurs due to conventional herbicides. The herbicides, carrier-based nanoparticles with target pest and toxicity is listed in the below **Table 2**.

b. **Nanoparticles as carriers for fungicides:** The mostly used nanoparticle carriers for fungicidal use are silica, chitosan, polymer mixes etc. Nanoparticles for carrier-based fungicides are much popular as they can reduce the loss through volatilization, increase solubility and release chemicals at target sites slowly. The list of nanocarrier-based fungicide, its target pest with crop details, is given in below **Table 3**.

c. **Nanoparticles as carriers for insecticides:** Nanoparticles can be used to lower the toxicity level while increasing its solubility in water. Additionally, these particles lessen the volatilization loss resulting effective protection on targeted sites. The nanoparticles as carrier of insecticide, its target and crop details given in the below-mentioned **Table 4**.

IV. **Application of nanoparticles in gene delivery system and to moderate gene expression:** The nanoparticles can be referred as molecular cargo as they can easily transfer the genetic materials such as nucleic

Nanoparticle	Herbicide	Target pest	Toxicity or soil leaching	Reference
Chitosan	Paraquat (L)	Maize, Mustard	Chinese hamster ovary cells *A. cepa*, Soil sorption	[40]
SLN	Simazine (C)	*R. raphanistrum*	Mouse fibroblast cells Maize	[41]
Chitosan/tripolyphosphate	Imazapic (B) and Imazapyr (B)	Black-jack (*B. pilosa*)	Chinese hamster ovary cells *A. cepa* Soil biota	[42]
Chitosan	Diuron (C)	*E. crusgalli*	Maize	[43]
Nanosized rice husk	2,4-D (I)	Maize	*Brassica Sp.* Soil sorption	[44]
Polymer	Metolachlor	*D. sanguinalis*	Preosteoblast cell line (mammal)	[45]
Polymeric	Simazine (C)	-	*C. elegans*	[46]

Table 2.
Nanocarrier-based herbicidal characterization and their target pests [40–46].

Nanoparticle	Fungicide	Target pest	Crop	Reference
Silver	Tebuconazole Propineb Fludioxonil	*Bipolaris maydis*	-	[47]
PVP and PVP copolymer	Chlorothalonil, Tebuconazole	*Gloeophyllum trabeum*	Southern pine sapwood	[48]
Gold	Ferbam	-	Tea	[49]
PVC	Chlorothalonil, Tebuconazole	Turkey tail (*T. versicolor)* *G. trabeum*	Southern yellow pine and Birch pine	[50]
Bacterial ghosts	Tebuconazole	*L. nodorum, E. graminis, S. fuliginea* and *P. teres*	Wheat, Barley and cucumber	[51]
Chitosan	Carbendazim	*F. oxysporum*and *A. parasiticus*	Cucumber, Maize, Tomato	[52]

Table 3.
List of nanocarrier-based fungicides and their application in plant disease management [47–52].

Nanoparticle	Insecticide	Target pest	Crop	Reference
Silica	Chlorfenapyr	*H. armigera, P. xylostella*	*Brassica chinese*	[53]
Sodium alginate	Imidacloprid	Leafhopper (Jassids)	-	[54]
Dendrimers	Thiamethoxam	Cotton bollworm cells and larvae (*H. armigera*)	-	[55]
Chitosan	Azadirachtin	Tobacco cutworm (*S. litura*) culture ovarian cell lines	-	[56]
PHSN	Avermectin	*P. xylostella* larvae	*Brassica oleracea*	[57]
Zinc oxide and chitosan	Azadirachtin	-	Groundnut bruchid (*C. serratus*)	[58]
Polydopamine	Avermectin	-	Cotton leaves and corn leaves	[59]

Table 4.
Nano-protectant with insecticidal activity [53–59].

acids to the target site and induce to express the new genetic material in the recipient. Surface-functionalized vertically arranged carbon nanofibers act as molecular cargo and carry the plasmid DNA to the desired site, expressing the characters in the recipient similar to other conventional methods. Researchers had developed a fluorescent labeled starch nanoparticle-based transgene vector that is able to enter the plant cell and nuclear membrane to deliver the genetic materials. In this system, ultrasound and fluorescent label help in gene delivery by producing transient pores and visual tracking of transgene respectively. Some nanoparticles carry the genetic material by holding it tightly so that it will not get detached from the nanoparticle thereby expressing for short time without integrating to the genetic material of the recipient. Apart from this, sometimes nanoparticle-based gene transfer show negative effect by modifying some genes.

V. **Nanoparticles used for abiotic stress tolerance:** Nanoparticles are reported to reduce or tolerate abiotic stress like salt stress, heavy metal toxicity, biotic stress etc. In addition to this, nanoparticles can provide mechanical strength to plants, promote germination and seedling growth, help in nutrient and water uptake, etc. Zinc, copper, and iron nanoparticles are observed up to 40 times less lethal than their salts. Hence, these nanoparticles widen the scope for more productivity than other salt-based applications [60]. In 2014, Sabaghnia and Janmohammadi reported that SiO_2nanoparticles enhanced germination percentage, shoot and root length, seedling fresh and dry weight in lentil plant [61]. In 2017, Taran et al. concluded that the drought tolerance in winter wheat can be alleviated by inducing higher antioxidant enzyme activity when treated with the colloidal solution of Cu and Zn nanoparticles [62].

C. Preparation of silver nanoemulsion

a. **Wet chemistry method:** This method is one of the best methods in preparation of silver nanoemulsion as it combines in molecular level and versatile in nature. This method is first time carried out by Guzman et al. in 2009. In this method, silver nitrate, hydrazine hydrate, and sodium dodecyl sulfate act as metal precursor, reducing agent, and stabilizing agent respectively. The creation of silver nanoparticles was observed with the help of UV-Vis absorption spectroscopy [63].

b. **Ion implantation method:** This method was carried out by Popok and his coworkers in 2005. In this method, silver ions were implanted to synthesize the metal nanoparticles in SiO_2 by using 30 KeV energy and ion current density of 4–15 $\mu A/ cm^2$. The analysis of this new compound was done by using optical spectroscopy and atomic force microscopy [53].

c. **Physical vapor deposition:** This method was explained by Lin and his coworkers in 2003. In this method, the overall size of silver trifluoroacetate reduced up to 7–11 nm under isoamyl ether and oleic acid solution by properly maintaining the temperature variation. This process is widely accepted and very smooth to manage [54].

5. **Nanotechnological applications to reduce postharvest loss:** Nanotechnology can be used to limit the post-harvest loss of solid, liquid or processed food

items and beverages. A sensory coating is applied on the product through wrapping or encapsulation method, which slightly modifies the color of the product by improving the taste and shelf-life of the food item. Thin layer of nanoparticles protects the food products from spoilage by holding moisture. Some of the nanoparticles proven promising in lessening the post-harvest loss are carbon nanoparticles, zinc oxide etc., which are mainly antibacterial [9].

6. **Nanotechnological applications in agriculture and animal husbandry:** In livestock management and animal husbandry, nanochips are installed in poultry and livestock to check their health and behavior through sensory technique. Quality of the poultry and livestock products can also be tested by using this method. The quality of meat or any contaminated particles present in it can be checked to avoid microbial spoilage [64]. In case of milk and milk products, **pasteurization** and improved nutrition to increase its quality are the possible ways through nanotechnological system. If nano carrier-based particles are employed in egg production, it will lessen pathogenic infection and transport the nutrients properly resulting quality egg production [65].

7. **Nanotechnological applications in fishery and aquaculture:** Nanotechnology is found to be one of the effective methods in aquaculture in order to increase the nutritional quality like protein and oil and also to promote the health of the fish and other sea foods [9]. Sensory-based drug delivery system was proved to be promising in detection of pathogen present in fish thereby promoting good health and quality nutrition. Nanoparticles mixed in water can also detect the presence of algae or any other micro-organisms and control eutrophication by reducing the phosphorus compounds [66]. Additionally, the encapsulated nanoparticles can effectively be used for fish feed, promoting the fish growth, increasing water quality etc. [67].

8. **Waste management approaches through nanotechnological applications for sustainable agriculture:** Researchers have done enough study on sustainable management of both solid and liquid waste in such a way that it can reduce the soil and water contamination. Scientists observed that implementation of nanotechnique, for example, bionanomaterials and nanobiomimics in waste management is a promising solution. Nanofibers can effectively be used to prepare absorbable substances that can degrade the waste materials in less time without leaving any toxic substances in addition to improve the production and quality of biomass [9].

9. Concerns about nanotechnology in terms of environmental protection as well as human development:

    a. **Toxicological and environmental safety concerns:** The more use of nanoparticles can lead to negativity on agriculture as many nanosubstances have proven toxic to some beneficial micro-organisms. The uncontrolled application of nanobased agrochemicals can usher to environmental pollution and disturbances in food chain. Before mass release of any nanoparticles, it should be checked properly for safety protocols of toxic composition and its residual effect [9]. A previous study has reported that the pure aluminum-based nanoparticles inhibit root growth of many plants while aluminum oxide-based nanoparticles act as a pollutant and hamper crop growth [68]. Similarly, nanozinc oxide-based products improve the food quality of soybean while nano-CeO_2 has reported to reduce the soybean yield and nitrogen fixing potential [69]. Therefore, the nanobased products should be in controlled use with prior precaution measures and existing regulatory frameworks to lessen the negative effects on agriculture.

b. **Legal and regulatory action:** Nonhomologous legal protocols refer to nano entities. In some developing countries like India, some compounds or particles with known application characteristics, if modified to nanoparticles, it will not be registered under patent. Section 3d of Indian patent act limited or restricted the common or already known materials from patent registration thereby opening the ways to do more research on it [70]. These types of laws have widened the paths to find out the scope of nanotechnological application, which will lead to sustainable agriculture.

c. **Socioeconomic perspective:** The nanotechnological application not only used to get more benefits in case of farmers as well as entrepreneur but also to take our agricultural system toward sustainability. All the nanoproducts should be properly labeled, which will make the farmers to choose the required products. Nanoparticles and nanofibers can improve the efficacy of agrochemicals in less period of time. The use of natural occurring nano formulations is very old. The potentiality of these formulations is very large, which can be known from the patents registered from last year's [9].

2. Summary and future scope

Nanotechnology is one of the emerging methods in agriculture that can bring revolution to bring sustainability. Nanotechnology is not only the potential solution for agriculture but also in other allied sectors like animal husbandry, dairy, poultry, fishery etc. The use of nano-components is one of the potentials, eco-friendly and economical method by using the entire proper regulatory framework. This method should be followed properly by keeping in view of all ethical, regulatory, toxic, and policy concerns. The nanoparticles should come in to market after repeated experiment both in vitro and lab condition.

Nanotechnology helps to developed multiple new methods for green house to field disease management as well as new area of molecular editing and manipulation for plants and pathogenic stress. Therefore diverse nanoparticles like nano-Ag, nano-Cu, nano-Zn used as potential arsenal against various destructive diseases which causes severe yield loss. One of the most important or striking feature is that considerable amount of reduction of metals compared to inorganic agrochemicals. Using of carbon nanoparticles to control disease is also under trial. Every time one thing must be remembered whenever any metal or metallic ions we are applying for nanomaterial preparation, that must be consumable through food chain additionally nontoxic and nonhazardous. It is also found that mechanism of action for each disease, different diseases on the same host and different diseases on various hosts are not the same way of treatment. Because the mode of action is different for each cases. Application of nano-globule in organic farming system is slowly upgrading and matter under consideration of developing countries agricultural policy. Some countries used copper oxychloride, copper hydroxide and copper oxide in their organic farming system in very lesser amount. Application of nanocopper in organic farming system found to be very effective for its sustainable, environment friendly and long term approach. Researchers are trying to develop micro and macro nutrient enriched plant growth nano promoter. This type of nano-complex not only gives protection against various destructive diseases but also enhancing the plant growth with its slow leaching compatibility mode. Application of different metallic ions and carbon-nanomer under genetic delivery system is the modern topic of research. Introduction of modern tolls like quantum dots and biosensor (with combination of nanotechnology) in plant pathology can greatly replace conventional ELISA technique for plant viral detection.

The use of nanobased pesticide, herbicide and fungicide, is an attractive advancement in plant health management keeping in focus of the healthy environment. It has the ability to deliver the pesticide to the specified target with low evaporation loss, improve the solubility, overcome resistance of pesticide in insects and pests, and have better shelf-life. Due to lacking of long-term based in vivo experimental trial, observation or monitoring, more research should be carried out on the particular field. To brand this method as a successful one, material scientists and biologists should cooperate with each other to gain more deep knowledge on complex nanosystems. The reliable, efficient nanoparticles should be chosen that can be biocompatible and no harmful to the plant system. Before releasing, any nanoparticles-based products in to the market, their toxicity, impact on environment, and other risk assessment factors should be analyzed properly. This method is yet to reach to the farmers mostly in developing countries. Therefore, government should work on this to create awareness among the farmers. The journey of nanotechnology in plant disease management and different other sectors of agriculture is under progressive research and wide adaptation through farmers level. The multidisciplinary nanotechnological approaches needs support from government and government-induced schemes or policies and also support from funding agencies, which will lead to a sustainable agriculture system by lessening the global food production as a challenge for future.

Conflict of interest

The authors solemnly and confidently declared that they have no conflict of interest. They are not attached with any other academic or research institutes, except the mentioned affiliation in terms of contribution. Additionally, they also declared that they are not attached with any kind of financial interests or non-financial (or personal interests) interests with any other organization or person. The subject matter of this chapter is totally original and unique or if taken from any other literature properly cited under the references of this manuscript.

Author details

Siddhartha Das* and Sudeepta Pattanayak
Department of Plant Pathology, M.S. Swaminathan School of Agriculture, Centurion University of Technology and Management, Paralakhemundi Odisha, India

*Address all correspondence to: siddhartha.das10@gmail.com

References

[1] Elmer W, White JC. The future of nanotechnology in plant pathology. Annual Review of Phytopathology. 2018;**56**:111-133

[2] Worrall EA, Hamid A, Mody KT, Mitter N, Pappu HR. Nanotechnology for plant disease management. Agronomy. 2018;**8**(285):1-24

[3] Bhattacharjee R, Dey U. An overview of fungal and bacterial biopesticides to control plant pathogens/diseases. African Journal of Microbiology Research. 2014;**8**(17):1749-1762

[4] Sinha K, Ghosh J, Sil PC. New pesticides: A cutting-edge view of contributions from nanotechnology for the development of sustainable agricultural pest control A2—Grumezescu, AlexandruMihai. In: New Pesticides and Soil Sensors. Cambridge, MA, USA: Academic Press; 2017. p. 5

[5] Balaure PC, Gudovan D, Gudovan I. Nanopesticides: A new paradigm in crop protection. In: New Pesticides and Soil Sensors. Academic Press; 2017. pp. 129-192

[6] Hayles J, Johnson L, Worthley C, Losic D. Nanopesticides: A review of current research and perspectives. In: New Pesticides and Soil Sensors. Academic Press; 2017. pp. 193-225

[7] Sekhon BS. Nanotechnology in Agri-food production: An overview. Nanotechnology, Science and Applications. 2014;7:31-53

[8] Yang W, Peters JI, Williams RO. Inhaled nanoparticles—A current review. International Journal of Pharmaceutics. 2008;**356**:239-247

[9] Podder K, Vijayan J, Ray S, Adak T. Chapter 10: Nanotechnology for sustainable agriculture. In: Singh RI, Mondal S, editors. Biotechnology for Sustainable Agriculture: Emerging Approaches and Strategies. United Kingdom: Woodhead Publishing; 2018. pp. 281-303

[10] Noji T, Kamidaki C, Kawakami K, Shen JR, Kajino T, Fukushima Y, et al. Photosynthetic oxygen evolution in mesoporous silica material: Adsorption of photosystem II reaction Centre complex into 23 nm nanopores in SBA. Langmuir. 2011;**27**(2):705-713

[11] Corredor E, Testillano PS, Coronado MJ, Gonzalez-Melendi P, Fernandez-Pacheco R, Risueno MC. Nanoparticle penetration and transport in living pumpkin plants: In situ subcellular identification. BMC Plant Biology. 2009;**9**(1):45

[12] Gogoi R, Dureja P, Singh PK. Nano formulations: A safer and effective option for agrochemicals. Indian Farming. 2009;**59**(8):7-12

[13] Essa AMM, Khallaf MK. Antimicrobial potential of consolidation polymers loaded with biological copper nanoparticles. BMC Microbiology. 2016;**16**:144

[14] Kah M, Hofmann T. Nanopesticide research: Current trends and future priorities. Environment International. 2014;**63**:224-235

[15] Gogos A, Knauer K, Bucheli TD. Nanomaterials in plant protection and fertilization: Current state, foreseen applications, and research priorities. Journal of Agricultural and Food Chemistry. 2012;**60**:9781-9792

[16] Kim DY, Kadam A, Shinde S, SarataleRG PJ, Ghodake G. Recent developments in nanotechnology transforming the agricultural sector: A transition replete with opportunities. Journal of the Science of Food and Agriculture. 2018;**98**:849-864

[17] Rafique M, Sadaf I, Rafique MS, Tahir MB. A review on green synthesis of silver nanoparticles and their applications. Artificial Cells, Nanomedicine, and Biotechnology: An International Journal. 2017;**45**:1272-1291

[18] Krishnaraj C, Ramachandran R, Mohan K, Kalaichelvan P. Optimization for rapid synthesis of silver nanoparticles and its effect on phytopathogenic ungi. Spectrochimica Acta Part A: Molecular and Biomolecular Spectroscopy. 2012;**93**:95-99

[19] Jain D, Kothari S. Green synthesis of silver nanoparticles and their application in plant virus inhibition. Journal of Mycology and Plant Pathology. 2014;**44**:21-24

[20] Wang X, Liu X, Chen J, Han H, Yuan Z. Evaluation and mechanism of antifungal effects of carbon nanomaterials in controlling plant fungal pathogen. Carbon. 2014;**68**:798-806

[21] Berry TD, Filley TR, Blanchette RA. Oxidative enzymatic response of white-rot fungi to single walled carbon nanotubes. Environmental Pollution. 2014;**193**:197-204

[22] Chen J, Wang X, Han H. A new function of graphene oxide emerges: inactivating phytopathogenic bacterium, *Xanthomonas oryzaepv*. oryzae. Journal of Nanoparticle Research. 2013;**15**(5):1658

[23] Liu S, Wei L, Hao L, Fang N, Chang MW. Sharper and faster "nano darts" kill more bacteria: A study of antibacterial activity of individually dispersed pristine single-walled carbon nanotube. ACS Nano. 2009;**3**(12):3891-3902

[24] Elbeshehy EKF, Elazzazy AM, Aggelis G, et al. Frontiers in Microbiology. 2015;**6**:453

[25] Alkubaisi NAO, Aref NMMA, Hendi AA. Method of inhibiting plant virus using gold nanoparticles. US Patents US9198434B1, 1 December 2015; 2015

[26] Kochkina Z, Pospeshny G, Chirkov S. Inhibition by chitosan of productive infection of T-series bacteriophages in the *Escherichia coli* culture. Mikrobiologiia. 1994;**64**:211-215

[27] Kashyap PL, Xiang X, Heiden P. Chitosan nanoparticle-based delivery systems for sustainable agriculture. International Journal of Biological Macromolecules. 2015;**77**:36-51

[28] Patel N, Desai P, Patel N, Jha A, Gautam HK. Agro nanotechnology for plant fungal disease management: A review. International Journal of Current Microbiology and Applied Sciences. 2014;**3**(10):71-84

[29] Mody VV, Cox A, Shah S, Singh A, Bevins W, Parihar H. Magnetic nanoparticle drug delivery systems for targeting tumor. Applied Nanoscience. 2014;**4**:385-392

[30] Malerba M, Cerana R. Chitosan effects on plant systems. International Journal of Molecular Sciences. 2016;**17**:996

[31] Giannousi K, Avramidis I, Dendrinou-Samara C. Synthesis, characterization and evaluation of copper-based nanoparticles as agrochemicals against *Phytophthora infestans*. RSC Advances. 2013;**3**(44):21743-21752

[32] Elmer WH, De La Torre-Roche R, Pagano L, Majumdar S, Zuverza-Mena N, et al. Effect of metalloid and metallic oxide nanoparticles on *Fusarium* wilt of watermelon. Plant Disease. 2018;**102**(7):1394-1401

[33] Evans I, Solberg E, Huber DM. Copper and plant diseases. In: Datnoff LE, Elmer WH, Huber DM,

editors. Mineral Nutrition and Plant Disease. St. Paul: American Phytopathological Society; 2007. pp 177-188

[34] Ponmurugan P, Manjukarunambika K, Elango V, Gnanamangai BM. Antifungal activity of biosynthesised copper nanoparticles evaluated against red root-rot disease in tea plants. Journal of Experimental Nanoscience. 2016;**11**(13):1019-1031

[35] He L, Liu Y, Mustapha A, Lin M. Antifungal activity of zinc oxide nanoparticles against *Botrytis cinerea* and *Penicillium expansum*. Microbiological Research. 2011;**166**(3):207-215

[36] Wani AH, Shah MA. A unique and profound effect of MgO and ZnO nanoparticles on some plant pathogenic fungi. Journal of Applied Pharmaceutical Science. 2012;**2**(3):40-44

[37] Li M, Huang Q, Wu Y. A novel chitosan-poly (lactide) copolymer and its submicron particles as imidacloprid carriers. Pest Management Science. 2011;**67**:831-836

[38] Ekambaram P, Sathali AAH, Priyanka K. Solid lipid nanoparticles: A review. Scientific Reviews and Chemical Communications. 2012;**2**:80-102

[39] Tamjidi F, Shahedi M, Varshosaz J, Nasirpour A. Nanostructured lipid carriers (NLC): A potential delivery system for bioactive food molecules. Innovative Food Science and Emerging Technologies. 2013;**19**:29-43

[40] Grillo R, Pereira AE, Nishisaka CS, de Lima R, Oehlke K, Greiner R, et al. Chitosan/tripolyphosphate nanoparticles loaded with paraquat herbicide: An environmentally safer alternative for weed control. Journal of Hazardous Materials. 2014;**278**:163-171

[41] De Oliveira JL, EVR C, Gonçalves da Silva CM, Pasquoto T, Lima R, Fraceto LF. Solid lipid nanoparticles co-loaded with simazine and atrazine: Preparation, characterization, and evaluation of herbicidal activity. Journal of Agricultural and Food Chemistry. 2015;**63**:422-432

[42] Maruyama CR, Guilger M, Pascoli M, Bileshy-José N, Abhilash P, Fraceto LF, et al. Nanoparticles based on chitosan as carriers for the combined herbicides imazapic and imazapyr. Scientific Reports. 2016;**6**:19768. DOI: 10.1038/srep19768

[43] Chidambaram R. Application of rice husk nano sorbents containing 2, 4-dichlorophenoxyacetic acid herbicide to control weeds and reduce leaching from soil. Journal of the Taiwan Institute of Chemical Engineers. 2016;**63**:318-326

[44] Tong Y, Wu Y, Zhao C, Xu Y, Lu J, Xiang S, et al. Polymeric nanoparticles as a metolachlor carrier: Water-based formulation for hydrophobic pesticides and absorption by plants. Journal of Agricultural and Food Chemistry. 2017;**65**:7371-7378

[45] Popok V, Stepanov A, Odzhaev V. Synthesis of silver nanoparticles by the ion implantation method and investigation of their optical properties. Journal of Applied Spectroscopy. 2005;**72**(2):229-234

[46] Lin X, Teng X, Yang H. Direct synthesis of narrowly dispersed silver nanoparticles using a single source precursor. Langmuir. 2003;**19**:10081-10085

[47] Huang W, Wang C, Duan H, Bi Y, Wu D, Du J, et al. Synergistic antifungal effect of biosynthesized silver nanoparticles combined with fungicides. International Journal of Agriculture and Biology. 2018;**20**:1225-1229

[48] Liu Y, Yan L, Heiden P, Laks P. Use of nanoparticles for controlled

release of biocides in solid wood. Journal of Applied Polymer Science. 2001;**79**:458-465

[49] Hou R, Zhang Z, Pang S, Yang T, Clark JM, He L. Alteration of the non systemic behaviour of the pesticide ferbam on tea leaves by engineered gold nanoparticles. Environmental Science & Technology. 2016;**50**:6216-6223

[50] Liu Y, Laks P, Heiden P. Controlled release of biocides in solid wood. II. Efficacy against *Trametes versicolor* and *Gloeophyllum trabeum* wood decay fungi. Journal of Applied Polymer Science. 2002;**86**:608-614

[51] Hatfaludi T, Liska M, Zellinger D, Ousman JP, Szostak M, Jalava K, et al. Bacterial ghost technology for pesticide delivery. Journal of Agricultural and Food Chemistry. 2004;**52**:5627-5634

[52] Kumar S, Kumar D, Dilbaghi N. Preparation, characterization, and bio-efficacy evaluation of controlled release carbendazim-loaded polymeric nanoparticles. Environmental Science and Pollution Research. 2017;**24**:926-937

[53] Song MR, Cui SM, Gao F, Liu YR, Fan CL, Lei TQ, et al. Dispersible silica nanoparticles as carrier for enhanced bioactivity of chlorfenapyr. Journal of Pesticide Science. 2012;**37**: 258-260

[54] Kumar S, Bhanjana G, Sharma A, Sidhu M, Dilbaghi N. Synthesis, characterization and on field evaluation of pesticide loaded sodium alginate nanoparticles. Carbohydrate Polymers. 2014;**101**:1061-1067

[55] Liu X, He B, Xu Z, Yin M, Yang W, Zhang H, et al. A functionalized fluorescent dendrimer as a pesticide nanocarrier: Application in pest control. Nanoscale. 2015;**7**:445-449

[56] Lu W, Lu ML, Zhang QP, Tian YQ, Zhang ZX, Xu HH. Octahydrogenated retinoic acid-conjugated glycol chitosan nanoparticles as a novel carrier of azadirachtin: Synthesis, characterization, and in vitro evaluation. Journal of Polymer Science, Part A: Polymer Chemistry. 2013;**51**:3932-3940

[57] Kaziem AE, Gao Y, Zhang Y, Qin X, Xiao Y, Zhang Y, et al. Amylase triggered carriers based on cyclodextrin anchored hollow mesoporous silica for enhancing insecticidal activity of avermectin against *Plutella xylostella*. Journal of Hazardous Materials. 2018;**359**:213-221

[58] Jenne M, Kambham M, Tollamadugu NP, Karanam HP, Tirupati MK, Balam RR, et al. The use of slow releasing nanoparticle encapsulated Azadirachtin formulations for the management of *Caryedon serratus* O. (groundnut bruchid). IET Nanobiotechnology. 2018;**12**:963-967

[59] Jia X, Sheng WB, Li W, Tong YB, Liu ZY, Zhou F. Adhesive polydopamine coated avermectin microcapsules for prolonging foliar pesticide retention. ACS Applied Materials & Interfaces. 2014;**6**:19552-19558

[60] Sytar O, Novicka N, Taran N, Kalenska S, Ganchurin V. Nanotechnology in modern agriculture. Physics—Alive. 2010;**9**:113-116

[61] Sabaghnia N, Janmohammadi M. Effect of nano-silicon particles application on salinity tolerance in early growth of some lentil genotypes. Annales UMCS. 2014;**69**:39-55

[62] Taran N, Storozhenko V, Svietlova N, Batsmanova L, Shvartau V, Kovalenko M. Effect of zinc and copper nanoparticles on drought resistance of wheat seedlings. Nanoscale Research Letters. 2017;**12**:60

[63] Guzman M, Dile J, Godet S. Synthesis of silver nanoparticles by

chemical reduction method and their antibacterial activity. International Journal of Chemical Engineering and Applications. 2009;**2**:3

[64] Rajkumar RS, Kandeepan G, Prejit Susitha K. Poultry research priorities to 2020. In: Sasidhar PVK, editor. Proceedings of National Seminar. Contributory Papers. Izatnagar, India: Central Avian Research Institute; 2006

[65] Kannaki TRV, Verma PC. Poultry research priorities to (2020). In: Sasidhar PVK, editor. Proceedings of National Seminar. Keynote Addresses and Lead Papers. Izatnagar, India: Central Avian Research Institute; 2006b. pp. 273-277

[66] Copetti D, Finsterle K, Marziali L, Stefani F, Tartari G, Lurling M. Eutrophication management in surface waters using lanthanum modified bentonite: A review. Water Research. 2016;**97**:162-174

[67] Van Oosterhout F, Lurling M. The effect of phosphorus binding clay (Phoslocks) in mitigating cyanobacterial nuisance: A laboratory study on the effects on water quality variables and plankton. Hydrobiologia. 2013;**710**:265-277

[68] Burklew CE, Ashlock J, Winfrey WB, Zhang B. Effects of aluminum oxide nanoparticles on the growth, development, and microRNA expression of tobacco (*Nicotiana tabacum*). PLoS One. 2012;7(5):e34783

[69] Priester JH, Ge Y, Mielke RE, et al. Soybean susceptibility to manufactured nanomaterials with evidence for food quality and soil fertility interruption. Proceedings of the National Academy of Sciences. 2012;**109**:E2451-E2456

[70] Reddy SBAP. The "efficacy" of Indian patent law: Ironing out the creases in section 3(d). SCRIPTed. 2008;**5**:1-35

Section 2

Nutrient Management

Chapter 3

Packages of Organic Nutrient Management as Soil Policy for Upgrading Cropping System to Restore Soil Productivity

Shaon Kumar Das and Ravikant Avasthe

Abstract

The indigenous farming systems are, by and large, organically practiced. Organic farming systems facilitate the buildup of soil organic matter, reducing risk of erosion and runoff and enhancing nutrient storehouse in soils for plants. Rapid developments in organic farming promotion necessitated continuous flow of technology to meet day-to-day challenges. Farmyard manure (FYM), compost, and green manure are the most important and widely used bulky organic manures. Manuring with different short-duration legumes is suitable for maintenance of soil quality in terms of adding nitrogen to soil. Sustainable quantity of potassium can be maintained by vegetative mulching with crop residues. The use of balanced dosages of mixed compost at 5–10 t/ha along with 2 t/ha dolomite increases yield of maize, rice, mustard, and soybean. This article briefly describes about the integrated organic nutrient management as soil policy for upgrading cropping system to restore soil productivity.

Keywords: organic farming, balanced dose, soil, policy, manure, cropping system

1. Introduction

Sikkim enjoys a wide range of climate, physiographic, geology, and vegetation that influence formation of different kinds of soils. Hills of Sikkim mainly consist of gneissose and half-schistose rocks, producing generally poor and shallow brown soils [1, 2]. The soil is coarse, with large concentrations of iron oxide, and ranges from neutral to acidic, making it lacking in mineral nutrients. This type of soil tends to support evergreen and deciduous forests [3]. Rock consists of phyllites and schists, which is much younger in age and is highly susceptible to weathering and erosion [4–8]. This combined with the state's heavy rainfall causes extensive soil erosion and the loss of soil nutrients through leaching. Soils of Sikkim belong to 3 orders, 7 suborders, 12 great groups, and 26 subgroups. It is observed that inceptisols are dominant (42.84%) followed by entisols and mollisols occupying 42.52 and 14.64%, respectively. Percentage area under Zn deficiency (<0.6 mg kg^{-1}) in Sikkim is 15.69% (202.35 sq. km) of the geographic area having highest Zn deficiency in South Sikkim district (82.07 sq. km, 19.1% of TGAD) followed by East (56.84 sq. km, 13.3% of TGAD), West (48.91 sq. km, 15.7 of TGAD), and North (14.53 sq. km,

11.8% of TGAD). Percentage area under Mn deficiency (<3.5 mg kg^{-1}) in Sikkim is 10.16% (131.02 sq. km) of the geographic area having highest Mn deficiency in South Sikkim (48.72 sq. km, 11.3 of TGAD) followed by East (34.52 sq. km, 8.1% of TGAD), North (28.82 sq. km, 23.13% of TGAD), and West (18.96 sq. km, 6.1% of TGAD). Total degraded area in Sikkim is 60,000 ha (9% of TGA), of which West Sikkim is highly degraded, followed by South Sikkim and North Sikkim [9–13]. Erosional hazard has affected about 2000 ha (0.28% of TGA of the state). South Sikkim is the worst affected district, followed by West Sikkim and North Sikkim [14, 15]. Sikkim being hilly state practicing terraced agriculture on an extensive scale could successfully control soil erosion [16].

2. Integrated organic nutrient management practices

The major challenge in organic agriculture is the availability of huge quantities of organic inputs for satisfying the farm demand. The use of animal excreta-based manure alone is not sufficient for meeting the nutrient needs of the crops. It is, therefore, necessary to utilize all the sources available on and off farm effectively [17]. The resource components available for nutrient management in organic horticulture are the following: farmyard manure (FYM), crop residue, weed biomass, green manures, biofertilizers, composts/phospho-compost, vermicomposting, oil cakes, mulching/cover crop, liquid manures, biodynamic preparation, botanicals, legumes in cropping sequence, and certified commercial products. Maintenance of soil fertility may be achieved through organic matter recycling, enrichment of compost, vermicomposting, animal manures, urine, farmyard manure, litter composting, use of botanicals, green manuring, etc. Biofertilizers like *Azolla*, *Azospirillum*, *Azotobacter*, *Rhizobium* culture, PSB, etc. can be used. Sawdust from untreated wood, calcified seaweed, limestone, gypsum, chalk, magnesium rock, and rock phosphate can be used [18]. Various sprays like vermiwash, liquid manure, etc. can be used in crops for nourishing the soil and plant. Farmyard manure, compost, and green manure are the most important and widely used bulky organic manures. Partially decomposed FYM has to be applied 3–4 weeks before sowing, while well-decomposed FYM should be applied immediately before sowing [19]. Manuring with different short-duration legumes is suitable for maintenance of soil quality in terms of adding nitrogen to soil. Nitrogen addition by sun hemp (150–200 kg/ha N) and dhaincha (125–175 kg/ha N) is highly beneficial for the succeeding crops and even for the subsequent crops too. Crop residue can also produce 2.47 kg N, 0.53 kg P, and 8.87 kg K per ha. Edible oil cakes of mustard and nonedible oil cakes from neem, karanj, and castor can serve the dual purpose of manure and bio-pest control [20–25]. Vermicompost can be used for a wide variety of horticultural, ornamental, and vegetable crops at any stage. Generally vermicompost is applied at 3–5 t/ha in row zones for field crops, whereas, for fruit crops, it is preferred to use the same mixing with equal amount of FYM in periodic interval. The general recommendation dose of vermicompost is 6–8 t/ha for field crop and 3–5 t/ha for subtropical fruits [26, 27]. In case of soil application, desired strain of biofertilizer is normally mixed with 20 times well-decomposed FYM to maintain uniformity of mixture and applied in furrows. However, for seedling treatments, biofertilizer slurry is made (1:10 ratio) in water, and roots are emerged in suspension for about 30 minutes. For cereals like, maize, baby corn, buckwheat, upland rice, and finger millet, it was suggested to apply 10–20 t/ha FYM along with 5.0 t/ha vermicompost, whereas, for low P and low K, the dosages are 6–12 t/ha FYM and 3–4 t/ha vermicompost. It is suggested that goat/pig/poultry at 3.0 t/ha along with FYM at 5.0 t/ha is a good source of organic zinc supplement in zinc-deficient soils. For spices like ginger, turmeric, and large cardamom, it is suggested to apply well-decomposed FYM along with neem cake at 3.0 t/ha and biofertilizer slurry in rows at planting time in variable dosages under low NPK situations [28]. The temperate climate with high organic matter is highly suitable for fruits like mandarin, chayote, strawberry, pear, etc. in the state. Application of well-decomposed FYM along with neem cake and vermicompost at variable dosages during land preparation and biofertilizer treatment before transplanting can be beneficial for improving fruit quality even under the stress of NPK in soils [29].

3. Year-round cropping systems of major crops for lower and mid hills (300: 2000 m amsl)

For rainfed areas, the predominant cropping systems are maize + beans-vegetable pea; maize + beans-barley; maize + beans-rajma; maize + beans-buckwheat; maize + beans-toria; soybean-buckwheat; and soybean-toria. For irrigated areas, the predominant cropping systems are maize (green cobs)-pahenlo dal-buckwheat; maize-vegetable pea; rice-vegetable pea-maize (green cobs); rice-fenugreek (leafy vegetable)-baby corn; rice-sunflower-*dhaincha* (green manuring); and rice-vegetable pea. Important vegetable cropping systems under low-cost plastic tunnels are broccoli-spinach-coriander-broccoli-coriander system; broccoli-coriander-cabbage-radish-coriander system; coriander-radish-fenugreek-spinach-coriander system; cabbage-local rayo sag-broccoli-coriander system; cabbage-spinach-broccoli-coriander system; and coriander-radish-fenugreek-cauliflower-pak choi system [30, 31]. Important vegetable cropping sequences for low-cost plastic rain shelter are tomato-pea-tomato system; bitter gourd-pea-tomato system; bottle gourd-capsicum-pea system; and sponge gourd-pea-tomato system. Important vegetable cropping sequences for low-cost polyhouse are cucumber-cabbage-tomato system; capsicum-broccoli-tomato system; and cucumber-cauliflower-tomato system. Important vegetable cropping sequences for open condition are okra-pea-cole crops system; okra-cole crops-local rayo sag/leafy vegetables system; dalley chili + local rayo sag/leafy vegetables as intercrop; okra-garlic-local rayo sag/leafy vegetables system; ginger-pea system; and okra-potato-local rayo sag/leafy vegetables system. **Table 1** represents the organic nutrient available in Sikkim from all possible sources.

4. Nutrient management in major crops of Sikkim

1. Maize (*Zea mays* L.): Application of dolomite at 2 t/ha + mixed compost at 2.5 t/ha + neem cake at 0.5 t/ha + vermicompost at 2.5 t/ha (ICAR Sikkim, 2011). Apply FYM at 15 t/ha 20 days before planting along with 150 kg rock phosphate. Neem cake at 150 kg/ha for nutrient supply and control of soil-borne insect pests. Green manuring: sun hemp and dhaincha another alternative. Seed inoculant: *Azospirillum*, *Azotobacter*, and PSB at 20 g/kg seed.

2. Rice (*Oryza sativa* L.): Apply FYM at 10 t/ha to supplement recommended dose of N + P + K for maintaining soil fertility. Practice of raising a pre-kharif crop like green gram, cowpea, sun hemp, or Sesbania for use as green manure. Biofertilizers (blue-green algae or Azolla) capable of providing 20–25 kg N/ha. Neem cake at 150 kg/ha provides protection against soilborne diseases and improves nutrition of rice crops. 5 t FYM + 2 t vermicompost + green manures/weed biomass before 20 days transplanting and 250–300 kg neem cakes during transplanting of rice crop are best nutrient management options. Mixed

Sl. no.	Animal	Livestock population of Sikkim (19th livestock census 2012)	Manure production rate per animal per day in kilograms	Amount of manure produced per day (in kilograms)	Manure/year (tons)	Manure/year On dry weight basis (in tons)
1	Cattle	140,467 CB—126,519 Ind.—13,948	25–30	3,511,675–4,214,010	1281761.375–1538113.65	384528.4125–461434.095
2	Buffalo	703	25–30	17,575–21,090	6414.875–7697.85	1924.4625–2309.355
3	Sheep	2634	2–3	5268–7903	1922.82–2884.595	576.846–865.3785
4	Goat	113,364	2–3	226,728–340,092	82755.72–124133.58	24826.716–37240.074
5	Pig	29,907	5	149,535	54580.275	16374.0825
6	Yak	4036	25	100,900	36828.5	11048.55
7	Poultry	451,966	0.6	271179.6	98980.554	29694.1662
Total				4282860.6–5104709.6	1563243.9–1863219	468973.17–558965.7

Table 1.
Organic nutrient available in Sikkim from all possible sources.

compost at 15 t/ha + neem cake at 1 t/ha. Green manure crops like dhaincha, sun hemp, and cowpea capable of accumulation of 4–5 t/ha of dry biomass and 100 kg N2/ha in 50–60 days.

3. Rapeseed and mustard (Brassica sp.): Apply FYM at 10 t/ha or vermicompost at 5 t/ha during last field preparation. Vermicompost along with Azotobacter and PSB considerably enhances mustard yield. Apply different oil cakes at 0.5 to 1.0 t /ha to meet demand of micronutrient and S demand of the crop. Mixed compost at 5 t/ha + vermicompost at 1.0 t/ha + neem cake at 1.0 t/ha + dolomite at 1.0 t/ha was recommended (ICAR Sikkim, 2011).

4. Soybean (Glycine max) (L.) Merr.: Being a leguminous crop, require less N than other crops. Apply FYM at 5–10 t/ha and incorporate into soil during final land preparation. Apply neem cake at 1 t/ha + mixed compost at 2.5 t/ha + dolomite at 1 t/ha (ICAR Sikkim, 2011). Seed inoculation with Bradyrhizobium japonicum culture (500 g/75 kg seed) + PSB/PSM (6.5 g/ kg seed).

   Buckwheat (*Fagopyrum esculentum* Moench.): Application of vermicompost at 1.5 t/ha recorded the higher grain yield of buckwheat. Efficient crop in extracting phosphorus of low availability from the soil. Azophos seed treatment (APST) + mixed compost at 5 t/ha + neem cake at 0.5 t/ha (ICAR Sikkim, 2011).

5. Baby corn (Zea mays L.): Well-decomposed FYM at 10 t/ha should be applied 20 days before sowing of crop. Baby corn should be inoculated with N-fixing nonsymbiotic microorganism like Azospirillum, Azotobacter, etc. and PSB at 20 g/kg seed.

6. Finger millet (*Eleusine coracana*): Apply 5 t FYM/ha 15 days prior to sowing of the crop. Biofertilizers like Azospirillum brasilense (N-fixing) and Aspergillus awamori (P-solubilizing) apply at 25 g/kg seed.

7. Black gram (*Vigna mungo* L.): FYM or mixed compost at 5 tons/ha enhances the yield. Seed inoculation with Rhizobium strains increases seed yield and uptake of nutrients. Additional nutrient may be supplied through water-soluble organic granules at 5 kg/acre mixed with FYM, vermicompost, or mixed compost.

8. Large cardamom (Amomum subulatum Roxb.): If the land is not terraced, soil base may be made by cutting topsoil from upper half and placed on lower half followed by mulching. At plant base, mulching with easily degradable organic materials is good for conserving both moisture and soil. Mulching improves soil physical condition and fertility. Dried organic matter, leaves, weeds, etc. can be used as mulch. During planting, pits are filled with topsoil mixed with FYM at 1–2 kg/pit. FYM/compost at 5 kg/plant at least twice a year in April to May and August to September is beneficial.

9. Ginger (*Zingiber officinale* L.): Well-decomposed and dried cattle manure or compost at 25–30 t/ha + neem cake at 2 t/ha + biofertilizer (Azospirillum + PSB) at 5–6 kg/ha helps in reducing incidence of rhizome rot of ginger and increases yield. Two months after planting, vermicompost at 5 t/ha should also be applied for better growth and production. Since edible part is rhizome, prior to planting of seed rhizome in soil, a half foot layer (6″) of leaf increases production of ginger by loosening soil texture around seed rhizome at later stages.

10. Turmeric (*Curcuma longa* L.): Needs heavy manuring. Apply FYM at 15–20 t/ha along with 250 kg neem cake or vermicompost at 10 t/ha. Integrated application of FYM at 10 t/ha and vermicompost at 5 t/ha along with 250 kg neem cake. O.M. along with biofertilizers like Azospirillum and Bacillus for better nutrition. Dolomite at 2 t/ ha to ameliorate soil acidity.

11. Mandarin (*Citrus reticulata* Blanco.): Young plants manured once/year, bearing plants twice/year (June to July and after harvesting in December to January) at 10–20 kg FYM/tree or 2–2.5 kg vermicompost/tree. Micronutrients through foliar sprays of water-soluble organic sources or nano-fertilizers at 0.2%. Dolomite at 100–200 g/plant for every second year. Neem cake at 2 t/ha during active growth stage in July to August.

12. Kiwifruit (*Actinidia chinensis*): Plants are heavy nitrogen feeders. Apply well-decomposed FYM at 25–30 t/ha and neem cake at 2 t/ha after vines have several inches of new growth during early spring. During active fruit growth stage, vermicompost at 2 kg/plant should also be given for better growth, production, and fruit quality.

    Cole crops (Brassica spp.): Well-decomposed FYM or compost should be applied at 5.0 kg/m2 along with neem cake at 200 g/m2 at the time of final land preparation. Root dipping of seedlings in Azospirillum + PSB (20%) for 15 minutes at the time of planting. Additional application of vermicompost in cole crops at 1 kg/m2 further improves production.

13. Potato (*Solanum tuberosum* L.): Proper soil fertility management alone accounts for 20.7% of all yield contributing factors. Well-decomposed and dried cattle manure or compost at 25–30 t/ha and neem cake at 2 t/ha should be applied.

5. Identified crops for marketing outside state from Sikkim

The most important crops which have been identified in Sikkim as commercial crop for marketing outside state are large cardamom, ginger, turmeric, buckwheat, cymbidium (flower), and tea. **Table 2** represents the marketing of organic produce in Sikkim.

Agency	Jurisdiction of marketing	Products
Sikkim Marketing Federation (SIMFED)	Within and outside Sikkim	Sikkim mandarin, kiwi, ginger, turmeric, buckwheat, rajma, and vegetables
Farmers Producers Organizations	Within Sikkim	Vegetables
Nature's gift (private entity)	Outside Sikkim	Ginger, turmeric, buckwheat

Table 2.
Marketing of organic produce in Sikkim.

6. Strategies for increasing organic farm productivity in Sikkim

Single-cropping should be avoided and preferably 2–3 crops should be grown together. If for any reason it is not possible to grow mixed or intercrops, then grow different crops in adjacent plots to maintain diversity. At any given time, legumes must occupy at least 30% of total cropping area. The legumes are nitrogen-fixing and can also be good source of mulching from the crop residues. High-yielding varieties require high nutrient inputs; they should be replaced with improved varieties suitable for organic management [31–35]. The same crop or same cropping sequence should not be repeated in the same field in two consecutive seasons/years (except for some legume crops such as mung bean or cowpea), and the field must be rotated every 2–3 years. Adoption of conservation tillage practices for improving soil quality and conserving soil moisture. Cover cropping, in situ residue management and restoration of degraded lands for soil moisture conservation, and improved C-sequestration should be practiced [36, 37]. Integrated farming systems and watershed development with animal, fishery, and suitable cropping for soil and moisture conservation and nutrient recycling should be practiced. The use of water-saving and nutrient-saving technologies, viz., system of rice intensification (SRI) and aerobic rice, should be popularized [38]. Rainwater harvesting: in situ (land configuration, mulching with locally available biomass, etc.) and ex situ (ponds, micro-water harvesting structures like *jalkund*, etc.) for ensuring year-round high-value crop production [38]. Adoption of conservation irrigation practices like drip, sprinklers, etc. in situ biomass management in shifting cultivation instead of biomass burning for improving soil carbon economy and hydrology should be practiced. Adoption of low-cost plastic tunnels, low-cost plastic rain shelters, and greenhouse (low cost) for year-round production of high-value low-volume vegetable crops should be promoted. Sufficient application of organic matter is crucial for soil fertility management especially for achieving satisfactory yields with good-quality product. Integration of integrated farming system is a necessity for organic farming [39]. It is also important to strengthen the animal husbandry section with main emphasis on poultry and piggery because majority of the population consume meat [40]. Besides, both are more profitable ventures. Composting of locally available biomass and construction of vermibeds for vermicomposting is also essential. Need-based crop diversification which allows more crops per unit area per unit time and per drop of water with due consideration of market demand should be enhanced. Introduction of new oilseeds and pulse crops which have yield potentials to meet the pulses and oilseed requirement of the region should be promoted. Recycling of all kinds of biomass and crop residues for minimizing the dependence of nutrient requirement from outside should be practiced [24, 41–45]. Adoption of soil conservation measures and careful soil cultivation that does not lead to soil erosion and conserves the soil moisture should be practiced. Integrated organic nutrient management strategies should be adopted. Uses of biofertilizers, green manuring, and concentrated organic manures like neem cake should be used for proper nutrition. Preventive measures should be adopted to manage pests, diseases, and weeds [46]. Awareness should be created for off-season vegetable production on scientific lines. There should also be an adoption of cool transport chain, pre-cooling units, packing houses, short- and long-term cold stores, etc. for minimizing the postharvest losses. Extension network for dissemination should be strengthened and the adoption of appropriate knowledge/technologies monitored [47, 48]. Agri Export Zone should be identified by the government for export of organic products, and contract cultivation/cooperative farming should be encouraged.

Author details

Shaon Kumar Das* and Ravikant Avasthe
ICAR Research Complex for NEH Region, Sikkim Centre, Gangtok, Sikkim, India

*Address all correspondence to: shaon.iari@gmail.com

References

[1] Avasthe RK, Das SK, Reza SK. Integrated Nutrient Management through Organic Sources. In: Handbook on Organic Crop Production in Sikkim. (Eds. R.K. Avasthe, Yashoda Pradhan and Khorlo Bhutia). Published by Sikkim Organic Mission. Tadong, Gangtok, Sikkim: Govt. of Sikkim and ICAR Research Complex, Sikkim Centre; 2014. pp. 317-328

[2] Benke MB, Hao XO, Donovan JT, Clayton GW, Lupwayi NZ, Caffyn P, et al. Livestock manure improves acid soil productivity under a cold northern Alberta climate. Canadian Journal of Soil Science. 2009;**90**:685-697

[3] Crawford TW, Singh U, Breman H. Solving agricultural problems related to soil acidity in central Africa great lake region. CATALIST Project Report; 2008. 64 p

[4] Das SK. Role of micronutrient in rice cultivation and management strategy in organic agriculture-a reappraisal. Agricultural Sciences. 2014;**5**(09):765

[5] IARI. Crop Residues Management with Conservation Agriculture: Potential, Constraints and Policy Needs. New Delhi: Indian Agricultural Research Institute; 2012. p. 32

[6] Schoenau JJ, Davis JG. Optimizing soil and plant responses to land applied manure nutrients in the Great Plains of North America. Canadian Journal of Soil Science. 2006;**86**:587-595

[7] Soils of Sikkim for Optimising Land Use, NBSS Publ. 60b, 1996, ISBN: 81-85460-39-6. NBSS&LUP; 2016

[8] Soil Series of Sikkim. NBSS&LUP Publ.105, (2004) ISBN: 81-85460-83-3

[9] Das SK, Avasthe RK, Gopi R. Vermiwash: Use in organic agriculture for improved crop production. Popular Kheti. 2014;**2**(4):45-46

[10] Roy A, Das SK, Tripathi AK, Singh NU. Biodiversity in north East India and their conservation. Progressive Agriculture. 2015;**15**(2):182-189

[11] Das SK, Mukherjee I, Kumar A. Effect of soil type and organic manure on adsorption–desorption of flubendiamide. Environmental Monitoring and Assessment. 2015;**187**(7):403

[12] Das SK, Avasthe RK, Singh M, Sharma K. Biobeds: On-farm biopurification for environmental protection. Current Science. 2015;**109**(9):1521-1521

[13] Das SK, Avasthe RK. Carbon farming and credit for mitigating greenhouse gases. Current Science. 2015;**109**(7):1223

[14] Das SK, Avasthe RK, Singh R, Babu S. Biochar as carbon negative in carbon credit under changing climate. Current Science. 2014;**107**(7):1090-1091

[15] Mukherjee I, Das SK, Kumar A. Degradation of flubendiamide as affected by elevated CO2, temperature, and carbon mineralization rate in soil. Environmental Science and Pollution Research. 2016;**23**(19):19931-19939

[16] Das SK, Avasthe RK, Singh M. Buckwheat: The natural enhancer in rhizosphere phosphorus. Current Science. 2015;**109**(10):1763

[17] Das SK, Mondal T. Mode of action of herbicides and recent trends in development: A reappraisal. International Journal of Agricultural and Soil Science. 2014;**2**:27-32

[18] Mukherjee I, Das SK, Kumar A. Atmospheric CO2 level and temperature affect degradation of pretilachlor and butachlor in

Indian soil. Bulletin of Environmental Contamination and Toxicology. 2018;**100**(6):856-861

[19] Das SK. Rice cultivation under changing climate with mitigation practices: A mini review. Universal Journal of Agricultural Research. 2017;5(6):333-337

[20] Das SK, Das SK. Acid sulphate soil: Management strategy for soil health and productivity. Popular Kheti. 2015;**3**(2):2-7

[21] Barman H, Das SK, Roy A. Zinc in soil environment for plant health and management strategy. Universal Journal of Agricultural Research. 2018;**6**:149-154

[22] Das SK, Avasthe RK, Yadav A. Secondary and micronutrients: Deficiency symptoms and management in organic farming. Innovative Farming. 2017;**2**(4):209-211

[23] Das SK, Ghosh GK. Soil hydro-physical environment as influenced by different biochar amendments. International Journal of Bio-resource and Stress Management. 2017;**8**(5):668-673

[24] Das SK. Soil carbon sequestration strategies under organic production system: A policy decision. Ağrı. 2019;**8**(1):1-6

[25] Das SK. Qualitative evaluation of fodder trees and grasses in hill region. Journal of Krishi Vigyan. 2019;**7**(2):276-279

[26] Das SK, Avasthe RK. Soil organic nutrients management through integrated approach: A policy for environment & ecology. Environmental Analysis & Ecology Studies. 2018;**4**(1):1-8

[27] Roy A, Das A, Sas SK, Datta M, Datta J, Tripathi AK, et al. Impact analysis of National Agricultural Innovation Project (NAIP): A paradigm shift in income and consumption in Tripura. Green Farming. 2018;**9**(3):559-564

[28] Das SK, Avasthe RK, Singh M, Roy A. Managing soil fertility under organic production system through integrated approach. Green Farming. 2018;**9**(3):449-454

[29] Das SK, Mukherjee I. Propesticides and their implications. In: Insecticides: Agriculture and Toxicology. Intech Publication; 2018. p. 107

[30] Das SK, Avasthe RK. Plant nutrition management strategy: A policy for optimum yield. Acta Scientific Agriculture. 2018;**2**(5):65-70

[31] Das SK, Avasthe RK, Singh M, Yadav A. Soil health improvement using biochar application in Sikkim: A success story. Innovative Farming. 2018;**3**(1):48-50

[32] Barman H, Das SK, Roy A. Future of nano science in technology for prosperity: A policy paper. NanoScience and Technology. 2018;**5**(1):1-5

[33] Das SK, Avasthe RK, Singh M, Dutta SK, Roy A. Zinc in plant-soil system and management strategy. Ağrı. 2018;**7**(1):1-6

[34] Das SK, Avasthe RK. Development of innovative low cost biochar production technology. Journal of Krishi Vigyan. 2018;**7**(1):223-225

[35] Das SK, Avasthe RK, Sharma P, Sharma K. Rainfall characteristics pattern and distribution analysis at Tadong East Sikkim. Indian Journal of Hill Farming. 2017;**30**(2):326-330

[36] Roy A, Singh NU, Tripathi AK, Yumnam A, Sinha PK, Kumar B, et al. Dynamics of pulse production in north-east region of India- a

state-wise analysis. Economic Affairs. 2017;**62**(4):655-662

[37] Das SK, Avasthe RK. Livelihood improvement of rural tribal farmers through soil health management, input support system and training-a success story. Innovative Farming. 2017;**2**(3):171-173

[38] Das SK, Avasthe RK, Sharma K, Singh M, Sharma P. Soil fertility assessment in different villages of East Sikkim District. Indian Journal of Hill Farming. 2017;**30**(1):14-16

[39] Das SK, Ghosh GK, Mukherjee I, Avasthe RK. Nano-science for agrochemicals in plant protection. Popular Kheti. 2017;**5**(4):173-175

[40] Das SK, Ghosh GK, Avasthe RK. Biochar amendments on physico-chemical and biological properties of soils. Ağrı. 2017;**6**(2):79-87

[41] Sharma M, Rana M, Sharma P, Das SK. Effect of different organic substrates and plant botanicals on growth and flowering of chincherinchee (Ornithogalum thyrsoides jacq). Indian Journal of Hill Farming. 2016;**29**(2):72-74

[42] Sharma P, Sharma K, Das SK. Ethno medicinal plants uses in health care by the Himalayan tribal people in India. Popular Kheti. 2016;**4**(3):41.45

[43] Roy A, Dkhar DS, Tripathi AK, Singh NU, Kumar D, Das SK, et al. Growth performance of agriculture and allied sectors in the north East India. Economic Affairs. 2014;**59**(Special):783-795

[44] Das SK. Nanoparticles advanced characterization techniques: A view point. Journal Atoms and Molecules. 2017;**7**(4):1091-1098

[45] Das SK, Avasthe RK, Ghosh GK, Dutta SK. Pseudocereal buckwheat with potential anticancer activity. Bulletin of Pure and Applied Sciences Section B-Botany. 2019;**38**(2):94-95

[46] Singh M, Das SK, Avasthe RK. Effect of multipurpose trees on production and soil fertility on large cardamom based agroforestry system in Sikkim Himalaya. Indian Journal of Agroforestry. 2018;**20**(2):25-29

[47] Singh M, Gupta B, Das SK. Soil organic carbon (SOC) density under different agroforestry systems along an elevation gradient in North-Western Himalaya. Range management and Agroforestry. 2018;**39**(1):8-13

[48] Das SK, Ghosh GK, Avasthe RK. Preparation and characterization of biochars for their application as a soil amendment. Indian Journal of Hill Farming. 2018;**31**(1):141-145

Chapter 4

Vermicomposting: An Effective Option for Recycling Organic Wastes

Tamanreet Kaur

Abstract

Urbanization and industrialization resulted in rapid increase in volume of solid waste; its management has become one of the biggest problems today. Solid wastes can be disposed off by methods like land filling, incineration, conversion into biogas, recycling, and composting, but its overproduction has led to inappropriate disposal practices such as their indiscriminate and inappropriately timed application to agricultural fields that ultimately leads to water and soil pollution. However, if handled properly, these organic wastes can be used for vermicomposting; it is an effective recycling technology that improves the quality of the products which is disinfected, detoxified, and highly nutritive. It is a low cost, eco-biotechnological process of waste management in which earthworms are used to cooperate with microorganisms in order to convert biodegradable wastes into organic fertilizer. Earthworms excreta (vermicast) is a nutritive organic fertilizer rich in humus, NPK, micronutrients, beneficial soil microbes; nitrogen-fixing, phosphate solubilizing bacteria, actinomycets, and growth hormones auxins, gibberlins and cytokinins, is a suitable alternative to chemical fertilizers, being an excellent growth promoter and protector for crop plants. Thus, vermiculture not only results in management of soild waste but also produces excellent nutrient enriched vermicompost. Vermicompost is beneficial for sustainable organic agriculture and maintaining balanced ecosystem.

Keywords: organic waste, earthworms, vermicompost, nutrients, sustainable organic agriculture

1. Introduction

Increasing world population has resulted in higher consumption of goods and services that has driven a substantial increase of organic wastes originating from households, industry, and agriculture [1]. Much of the organic wastes are highly infectious as they contain a variety of pathogenic microorganisms. Dumping of organic wastes in open areas generates serious environmental issues such as the accumulation of heavy metals in soil, pollution of ground and surface waters due to leaching and run-off of nutrients. These organic wastes when applied directly to agricultural fields cause soil environment-related problems including phytotoxicity [2]. These wastes represent a valuable organic resource, which could be recycled and transformed into nutrient rich fertilizer and/or soil conditioner [3–5]. Moreover growing awareness about adverse effects of agricultural chemicals on human health has increased interest in organic agriculture [6]. Organic agriculture also promotes ecological conservation due to judicious use of natural resources [7–9]. In demand for safe and sustainable strategies to treat organic wastes includes best known practices of composting and vermicomposting for biological stabilization of solid organic wastes by transforming them into a safer and more stabilized material that can be used as a source of nutrients and soil conditioner in agricultural applications [10–12]. Vermicomposting is one of the most efficient means to mitigate and manage environmental pollution problems [13]. Recently, many studies are being done to establish vermicompost as one of the preferred organic substitutes to chemical fertilizers [14, 15]. Vermicompost is more rich in NPK, micronutrients and beneficial soil microbes (nitrogen fixing and phosphate solubilizing bacteria and actinomycetes), an excellent growth promoter and protector for crop plants [16, 17] than compost [18, 19].

1.1 Vermicomposting- a preferred approach in organic farming

Vermicomposting (vermis from the Latin for worm) is a mesophilic process [20] which involves a joint action of earthworms (active at 10–32°C) and mesophilic microbes [21] for the conversion of organic wastes into a valuable end product known as vermicompost. Whereas, composting involves the degradation of organic waste by microorganisms under controlled conditions, in which the organic material undergoes a characteristic thermophilic stage that allows sanitization of the waste by elimination of pathogenic microorganisms [22]. Composting is also used to treat manures, green wastes or municipal solid wastes [23]. However, vermicomposting gives a higher-quality end product than composting due to joint action of enzymatic and microbial activities that occur during the process [24]. This process is faster than traditional composting as the material passes through the earthworm gut, whereby the resulting earthworm castings are rich in microbial activity and plant growth regulators, and fortified with pest repellence attributes as well [25, 26]. Compared to traditional composting method, vermicomposting also results in mass reduction, shorter processing time, and high levels of humus with reduced phytotoxicity [27]. Thus, vermicompost is considered an ideal manure for organic agriculture as it is nutrient rich and contains high quality humus, plant growth hormones, enzymes, and substances that are able to protect crops against pests and diseases [28, 29]. Moreover, vermicompost has high porosity, aeration, drainage, and water-holding capacity [20]. In addition to increased N availability, C, P, K, Ca and Mg plant nutrient availability in the earthworm casts are also found [30]. Plant growth hormones namely cytokinins and auxins are found in organic wastes processed by earthworms [31]. They also release certain metabolites, such as vitamin B, vitamin D and similar substances into the compost [32]. Thus, earthworms accelerate the mineralization rate and convert the manures into casts with higher nutritional value and degree of humification than traditional method of composting [33]. The composition of commonly available nutrients in vermicompost is as follows: Organic carbon 9.5–17.98%, Nitrogen 0.5–1.50%, Phosphorous 0.1–0.30%, Potassium 0.15–0.56%, Sodium 0.06–0.30%, Calcium and Magnesium 22.67–47.60 meq/100 g, Copper 2–9.50 mg/kg, Iron 2–9.30 mg/kg, Zinc 5.70–11.50 mg/kg, Sulfur 128–548 mg/kg [34]. Hence, vermicomposting enables biological transformation of wastes into a valuable organic fertilizer [35, 36]. Vermicompost is popularly called as black gold and has become one of the major components of organic farming system [26].

2. Role of earthworm in vermicomposting

2.1 Biology of earthworm

Earthworms are invertebrates belonging to the phylum Annelida, class Oligochaeta and family Lumbricidae. The earthworms are long, elongated, cylindrical, soft bodied animals with uniform ring like structures consisting of segments along the length of their body outwardly highlighted by circular grooves called annuli. On the ventral surface of sides of the body each segment bears four pairs of short, stubby bristles, or setae used for its movement. Earthworms have an opening at the anterior end is mouth and the one at the posterior is anus. Earthworms possess both male and female gonads, so are called as hermaphrodites. They deposit their eggs in a cocoon without any larval stage. At the time of egg laying, the sexually mature worms contain a distinctive epidermal ring just beneath the anterior segments called, clitellum, which has gland cells to form a viscid, girdle like structure known as cocoon. The number of fertilized ova in each cocoon has 1–20 lumbricid worms.

2.2 Classification of earthworm

There are about 3320 species of earthworms all over the world [37], but hardly 8–10 species are suitable for vermicompost preparation. Earthworms have been extensively utilized for the recycling of a variety of organic wastes like municipal solid wastes [38] wheat straw [39], sewage sludge [40], forestry waste [41], vegetable waste [42], farmyard manure [43], sorghum stalk, wheat straw, paddy straw [44], coir pith [45]. Renowned scientists, Charles Darwin called earthworms as the 'unheralded soldiers of mankind', and Aristotle described them as the 'intestine of earth', as they could digest a wide range of organic materials [46, 47]. On the basis of morpho-ecological characteristics, earthworms have been classified into three categories [48]; Anecic (Greek word "out of the earth") – these are burrowing worms that only come to the surface at night to drag food down into their permanent burrows deep within the mineral layers of the soil. Endogeic (Greek word "within the earth") – these are also burrowing worms but their burrows are typically more shallow and they feed on the organic matter inside the soil, so they come to the surface only rarely. Epigeic (Greek word "upon the earth") – these worms live on the surface litter and feed on decaying organic matter. They do not have any permanent burrows. These "decomposers" are the type of worm used in vermicomposting. Two tropical species, African night crawler, *Eudrilus eugeniae* (Kinberg) and Oriental earthworm, *Perionyx excavates* (Perrier) and two temperate ones, red earthworm, *Eisenia andrei* (Bouche) and tiger earthworm, *Eisenia fetida* (Savigny) are extensively used in vermicomposting [49–51]. Most vermicomposting facilities and studies are using the worms *E. andrei* and *E. fetida* due to their high rate of consumption, digestion, and assimilation of organic matter, tolerance to a wide range of environmental factors, short life cycles, high reproductive rates and endurance and resistance during handling [52]. A few other species *Drawida nepalensis*, *Lampito mauritrr. Dichogaster* spp., *Polypheretima elongate, Amynthas* spp. *Dendrobaena octaedra, Eisenia hortensis* [53] have also been used for composting under specific conditions.

2.3 How does earthworm facilitate vermicomposting?

Earthworms promote the growth of "beneficial decomposer aerobic bacteria" in organic waste material and also act as a grinder, crusher, chemical degrader and a biological stimulator of waste material [54, 55]. Earthworm hosts millions of decomposer (biodegrader) microbes [56], hydrolytic enzymes and hormones that helps in rapid decomposition of complex organic matter into vermicompost in a relatively smaller duration of 1–2 months [57] as compared to traditional composting method which takes nearly 5 months [58]. The mechanism of formation of vermicompost by earthworms occurs in following steps; organic material consumed by earthworm is softened by the saliva in the mouth of the earthworms. Food in esophagus is further softening and neutralization by calcium and physical breakdown in muscular gizzard results in particles of size <2 μ, thereby giving an enhanced surface area for microbial processing. This finally ground material is exposed to various enzymes such as protease, amylase, lipase, cellulase and chitinase secreted in lumen by stomach and small intestine [12]. Moreover, microbes associated with intestine facilitate breaking down of complex biomolecules into simple compounds. Only 5–10% of the ingested material is absorbed into the tissues of worms for its growth and the rest is excreted as vermicast. The vermicast is a good organic fertilizer and soil conditioner. High-quality vermicast can be produced by worms such as the red wrigglers (*E. fetida*) as it contains humus with high levels of nutrients that has good potential for the production of organic fertilizer. Vermiwash is a liquid fertilizer and used as a foliar spray produced by passing water through columns of vermiculture beds [59].

2.4 Suitable environmental conditions for earthworms

Optimum conditions of temperature 15–20°C (limits 4–30°C), Moisture content 80–90% (limits 60–90%), Oxygen – Aerobicity, Ammonia content of the waste Low: <1 $mg{\cdot}g^{-1}$ (0.016 oz.1b–1), Salt content Low < 0.5% and pH of 5–9 are preferred for stable life cycle of earthworm.

2.5 Starter food for multiplication of earthworms

To attain the desired earthworm population their starter food includes 1:1 mixture of cow dung and decaying leaves in a cement tank/wooden box/plastic bucket with proper drainage facilities and on attaining sufficient number of earthworms, subsequently other sources of organic wastes can be provided. Compost worms being voracious feeders, consume in excess of their body weight each day but they prefer some foods to others. Manures are the most commonly used worm feedstock, with dairy and beef manures generally considered the best natural food for *E. fetida* [60]. The unit should be kept in shade. Sufficient moisture level should be maintained by occasional sprinkling of water. Within 1–2 months, the worms multiply 300 times, which can be used for large scale vermicomposting.

3. Methods of vermicomposting

Earthworms are used to convert organic waste material into dark brown nutrient rich humus that is a good source of manure for plants. Worms can also degrade specific pollutants and might allow community formation of useful microorganisms. Three commonly used methods for vermicomposting are discussed below:

1. Bin composting: The most common method for small scale composting is bin composting method. The bin can be constructed of several materials such as wooden/plastic/recycled containers like bathtubs and barrels. A vermicompost bin may be in different sizes and shapes, but its average dimensions are 45 × 30 × 45 cm. Around 10 holes with 1–1.5 cm in diameter holes in bottom, sides and cap of bin is useful for aeration and drainage.

2. Pit composting: For large scale composting, pits of sizes 2.5 m × 1 m × 0.3 m under thatched sheds with sides left open are advisable. The bottom and sides of the pit should be made hard with a wooden mallet.

3. Pile composting: Pile method is mostly used for vermicomposting in larger scale. The piles can be made in porch place like greenhouse or in a floor with some facilities for drainage in warm climate. The pile size may vary in length and width, however, its height is average height of bin used for bin composting.

4. Conventional steps involved in vermicomposting

Prior to the vermicomposting process, it is preferred to assign pre-composting of organic waste (thermophilic composting), which comprises a short period of high temperature for facilitating mass reduction, waste stabilization, and pathogen reduction [61, 62]. Thermophilic composting results in sanitization of organic wastes and elimination of toxic compounds [63]. Although pathogen removal occurs during transit in the worm gut [64] but thermophilic composting prior to vermicomposting is advisable to avoid the earthworm mortality. Then, after some days of high temperature, pre-mature compost is cooled by spreading it as thin layers on vermicomposting beds. Vermicomposting can be done either in containers, pits or piles.

1. **Materials required for vermicomposting:** Carbon and nitrogen-rich organic materials, spade, ground space, stakes, hollow blocks, plastic sheets or used sacks, water (according to the season) and water sprinklers, shading materials, nylon net or any substitute to cover the beds, and composting earthworms.

2. **Site Selection:** Vermicompost production can be done at any place which is having shades, cool and has high humidity. For instance, abandoned cattle shed, or poultry shed or unused buildings or artificial shading could also be provided.

3. **Shredding of organic waste material:** The collected organic waste material should be processed for shredding along with mechanical separation of the metal, glass and ceramics that should be kept aside.

4. **Pre-digestion of organic waste material:** Pre-digestion of organic waste should be done for at least 20–25 days prior by mixing the waste material along with raw material (e.g., cattle dung slurry). Regular watering is required for partially digesting it and making it fit for earthworm consumption. Raw material to be used includes for composting – cow dung, crop residues, farm wastes, vegetable market wastes and fruit wastes. Cow dung should be at least 20–25 days old to avoid excess heat generation during the composting process. Moreover addition of higher quantities of acid-rich substances such as citrus wastes should be avoided. It is important to mix carbonaceous with nitrogenous organic materials at the right proportions to obtain a C: N ratio of about 30:1, as it results in product of highest stability, the best fertilizer-value and with lowest potential for environmental pollution. For example, rice straw and fresh manure are mixed at about 25:75 ratio by weight. When the material with higher carbon content is used with C:N ratio exceeding 40:1, it is advisable to add nitrogen supplements to ensure its effective decomposition.

5. **Earthworm bed preparation:** An hospitable living environment for worms called bedding is prepared. Bedding is a material that provides the worms with a relatively stable habitat with following characteristics:

   i. **High absorbency**: As earthworms breathes through their skins and therefore bedding must be able to absorb and retain water fairly well. Worms dies if its skin dries out.

   ii. **Good bulking potential**: Worms respire aerobically and different bedding materials affect the overall porosity of the bedding, including the range of particle size and shape, the texture, and the strength and rigidity of its structure. If bedding material is too dense or packs too tightly, then the flow of air is reduced or eliminated. This overall effect is referred as the material's bulking potential.

   iii. **Low protein and/or nitrogen content/high Carbon:** Earthworms consume their bedding as it breaks down and it is very important for this process to be slow. High protein/nitrogen levels can result in rapid degradation of bedding and its associated heating, creating inhospitable or fatal conditions. High carbon content is required as earthworms and microbes in the feed mixtures activate microbial respiration and degradation of organic wastes, thereby increasing the loss of organic carbon during the vermicomposting process [65, 66]. Various bedding material according to absorbency, bulking potential and C:N are enlisted in **Table 1**.

6. **Vermiculture bed:** Vermiculture bed can be prepared by placing a first layer of saw dust, newspaper, straw, coir waste, sugarcane trash etc. at the bottom of tub/container. Newspaper is one of bedding material that high in absorbency whereas for the sawdust the level of absorbency is poor to medium. A second layer of moistened fine sand of 3 cm thick should be spread over the culture bed followed by a layer of garden soil (3 cm). The floor of the unit should be compacted to prevent earthworm's migration into the soil.

7. **Loading of organic waste mixture in bed:** Third layer of the pre-digested organic waste prepared is added. Thereafter a thin layer of cow dung mixture is placed on the surface of waste material as starter food for compost worms. Then compost worms are to be added without spreading them out. Earthworms consume various organic wastes and reduce the volume by 40–60%. Earthworm eats waste equivalent to its body weight, and produce cast about 50% of the wastes, it consumes in a day.

8. **Composting process:** After addition of compost worms wait for at least 15 days for the thermophillic process to end. During this process there is a rapid increase in temperature followed by a gradual decrease. During this period turning to the material 2–3 times at 4–5 days interval is required. Its temperature should be maintained at 30°C, when temperature approaches ambient temperature (<35°C) covering is to be removed and for temperature maintenance, upturning and regular sprinkling of water is advisable. Prominent precautionary measures include; Composting pit should be covered with nylon net or any substitute material to serve as barrier against predators like ants, birds, lizards as it may disturb the activity of earthworm, Blockage of side air vents should be avoided as it can quickly lead to putrefaction and extreme

DOI: http://dx.doi.org/10.5772/intechopen.91892

weather conditions such as frost, heavy rainfall, drought and overheating should be avoided. No smell comes out of composting site if the right products or bedding and feed are used. The vermicompost once formed completely will give the smell of moist soil. Maturity could be judged visually also by observing the formation of granular structure of the compost at the surface of the tank. Next step is to make a heap in sunlight on a plastic sheet and keep it for 1-2 hours. The worms will gather at the bottom of heap. After removing vermicompost on top, the worms settled down at the bottom can be carefully collected for use in the next batch of vermicomposting.

Bedding Material	Absorbency	Bulking Pot.	C:N Ratio
Horse manure	Medium-good	Good	22–56
Peat moss	Good	Medium	58
Corn silage	Medium-Good	Medium	38–43
Hay–general	Poor	Medium	15–32
Straw–general	Poor	Medium-Good	48–150
Straw–oat	Poor	Medium	48–98
Straw–wheat	Poor	Medium-Good	100–150
Paper from municipal waste stream	Medium-Good	Medium	127–178
Newspaper	Good	Medium	170
Bark–hardwoods	Poor	Good	116–436
Bark–softwoods	Poor	Good	131–1285
Corrugated cardboard	Good	Medium	563
Lumber mill waste–chipped	Poor	Good	170
Paper fibre sludge	Medium-Good	Medium	250
Paper mill sludge	Good	Medium	54
Sawdust	Poor-Medium	Poor-Medium	142–750
Shrub trimmings	Poor	Good	53
Hardwood chips, shavings	Poor	Good	451–819
Softwood chips, shavings	Poor	Good	212–1313
Leaves (dry, loose)	Poor-Medium	Poor-Medium	40–80
Corn stalks	Poor	Good	60–73
Corn cobs	Poor-Medium	Good	56–123
Paper mill sludge	Good	Medium	54
Sawdust	Poor-Medium	Poor-Medium	142–750
Shrub trimmings	Poor	Good	53
Hardwood chips, shavings	Poor	Good	451–819
Softwood chips, shavings	Poor	Good	212–1313
Leaves (dry, loose)	Poor-Medium	Poor-Medium	40–80
Corn stalks	Poor	Good	60–73
Corn cobs	Poor-Medium	Good	56–123

Table 1.
List of some of the commonly used earthworm bedding material.

5. Effect of abiotic factors on vermicomposting

The most important abiotic factors which affect vermicomposting process include moisture, pH, temperature, aeration, pH value, ammonia and salt content.

1. **Moisture:** A strong relationship exists between the moisture content of organic wastes and the growth rate of earthworms. In a comparative study on vermicomposting process and earthworm's growth at different temperature and moisture ranges showed that 65–75% is most suitable range of moisture at all ranges of vermicomposting temperature [67]. The bedding used for vermicomposting must be able to hold sufficient moisture as earthworms respire through their skins and moisture content in the bedding of less than of 45% can be fatal to the worms. Although epigenic species, *E. fetida* and *E. andrei* can survive moisture ranges between 50% and 90%, but they grow more rapidly between 80% and 90% [20, 68]. The bacteria also plays vital role in vermicomposting. Its activity decreases in moisture content lower than 40% and it almost stops in lower than 10% [69].

2. **Temperature:** Earthworm's activity, metabolism, growth, respiration and reproduction are greatly influenced by temperature [70]. The temperature for the stable development of earthworm population should not exceed 25°C [71]. Although *E. fetida* cocoons survive extended periods of deep freezing and remain viable [72] but they do not reproduce and do not consume sufficient food at single digit temperatures. It is generally considered necessary to keep the temperatures preferably 15°C for vermicomposting efficiency and 20°C for effective reproductive vermiculture operations. Temperatures above 35°C will cause the worms to leave the area or if they cannot leave, they will quickly die. Bacterial activity is also greatly depended on temperature as it multiplies by two per each 10°C increase in temperature and is quite active around 15–30°C.

3. **Aeration:** Earthworms are oxygen breathers and cannot survive in anaerobic conditions. They operate best when compost material is porous and well aerated. Earthworms also help themselves by aerating their bedding by their movement through it. *E. fetida* have been reported to migrate in high numbers from oxygen depleted water saturated substrate, or in which carbon dioxide or hydrogen sulfide has accumulated.

4. **pH value:** The pH value is also one of the important factors affecting the vermicomposting process [73]. Epigenic worms can survive in a pH range of 5–9 [74]. The pH of worm beds tends to drop over time. If the food source/bedding is alkaline, than pH of bed drop to neutral or slightly alkaline and if the food source is acidic than the pH of the beds can drop well below 7. The pH can be adjusted upwards by adding calcium carbonate or peat moss for adjusting pH downward can be introduced into the mix. Although microorganisms which are active in vermicomposting which can maintain their activity even in lower pH of around 4 but recommended pH range for compost is around 6.5–7.5.

5. **Ammonia and salt content:** Earthworms cannot survive in organic wastes containing high levels of ammonia. Worms are also very sensitive to salts and they prefer salt contents less than 0.5% [75]. However, many types of manures

have high salt contents and if they are to be used as bedding, they should be leached first to reduce the salt content, it is done by simply running water through the material for a period of time [60].

6. Common methods of vermicompost harvesting

The vermicompost is ready within 60–90 days and ultimately the material becomes black, granular, lightweight, moderately loose, crumbly and humus-rich. Watering must be avoided two to three days before emptying the beds to facilitate the separation of worms from the compost. Common procedures for harvesting the vermicompost are briefly described below. Any method may be adopted exclusively by preference. Moreover, two or more methods may be applied on the same pile. Except for the first method, the rest are intended for bulk harvesting.

6.1 Manual harvesting of Vermicompost

This method is practiced if one wants to collect small amounts of vermicast just a few days after the compost pile is stocked with composting worms. In this case only top layer is covered with a thin layer of vermicast and rest of pile has not fully decomposed. The vermicast on top of the pile are simply gathered by hand/trowel and transferred directly into a container. This method is recommended if there is need of organic soil amendment in preparing a fertile potting mix. With time, as vermicompost is collected at the bottom of the pile it is further collected by hand.

6.2 Vermicompost harvesting by pyramidal heap

The vermicompost is first gathered to form a pyramid like heap within the composting enclosure provided that the heap is exposed to light or it is transferred on to a flat surface elsewhere in open sun on a plastic sheet or a sack. This method of harvesting vermicompost takes the advantage of the earthworm's sensitivity towards light as they will tend to move deep into the pyramid. Vermicompost from the bottom, sides, and top surface of the heap is then collected by hand or with a trowel. After the first cycle of vermicompost collection, a few minutes are passed to provide sufficient time for the earthworms to move deeper and another cycle is commenced. For faster rate of harvesting vermicompost, the original heap is divided into several smaller heaps.

6.3 Screening or sieving of vermicompost

This method of vermicompost harvesting is done manually with tool consists of mesh wire nailed on wood called sieve. A small portion from vermicompost pile spread on flat floor is transferred into a sieve and it is shaken so that fine vermicompost falls on the ground. Any undecomposed subtrates and earthworms are retained in the screener and the worms are separated manually.

6.4 Vermicompost harvesting by inducing the migration of earthworms

This method of harvesting vermicompost is based on earthworms' ability to detect sources of food. Earthworms have the habit of abandoning the pile exhausted of food and moving towards fresher palatable source. Despite many modifications in this technique, but the basic principle is the same to provide fresh or more palatable food to cause the migration of earthworms from the exhausted pile to the new food source.

7. Storing and packing of vermicompost

The harvested vermicompost should be stored in dark and cool place as sunlight will lead to loss of moisture and nutrient content. Moreover, harvested vermicompost material should be stored in open rather than packed in sacs. Packing should be done at the time of selling and laminated sac is always advisable. During compost storage in open place, periodical sprinkling of water should be done to maintain moisture level and beneficial microbial population. Vermicompost can be stored for longer periods of one year without loss of its quality, if its moisture is maintained at 40% level.

8. Role of vermicompost on soil fertility

The key role of vermicompost is change in physical, chemical and biological properties of soil by earthworm activities and they thus called as soil managers [59]. It substantially improves soil structure, texture, aeration and prevents soil erosion. It increases the macropore space ranging from 50 to 500 μm, resulting in improved air-water relationship in the soil thereby favorably affecting plant growth [76]. It also favorably affects soil pH, its microbial population and soil enzyme activities [77]. Moreover, vermicompost is rich source of nutrients such as nitrates, phosphates and exchangeable calcium and soluble potassium [30]. Apart from adding mineralogical nutrients, vermicompost is also rich in beneficial micro flora such as N-fixers, P-solubilizers, cellulose decomposing micro-flora, etc. It also reduces the proportion of water soluble chemical, which causes possible environmental contamination [78]. Mucus excreted by earthworm's digestive canal produces some antibiotics and hormone-like biochemicals thereby boosting plant growth [70] and enhancing the decomposition of organic matter in soil [79]. Vermicompost has been reported to have favorable influence on the growth and yield parameters of several crops like paddy, sugarcane, brinjal, tomato, and okra [59]. Thus, vermicompost acts a soil conditioner [80] and a slow-release fertilizer [81] that ultimately improves soil structure, soil fertility, plant growth and suppresses diseases caused by soil-borne plant pathogens, increases crop yield [82–84].

9. Conclusion

Chemical fertilizers are produced from "vanishing resources" of earth and crops grown on chemical fertilizers have low and contaminated nutrient value in comparison to grown naturally or organic way. To preserve the agro-ecosystem and protect human health from the harmful chemical fertilizers 'Ecological Agriculture and Organic Farming' has to be promoted as the new emerging concept of "Organic Farming" focuses mainly on production of chemical free foods. Organic farming with use of organic fertilizers like "vermicompost" could substitute the chemical fertilizers and can reduce the economic cost and may also lead to organic products which fetches higher price in the market.

Conflict of interest

No conflict of interest is indulged.

Author details

Tamanreet Kaur
Department of Zoology, Kanya Maha Vidyalaya, Jalandhar, India

*Address all correspondence to: tamanreetkaur@gmail.com

References

[1] Hoornweg D, Bhada-Tata P, Kennedy C. Waste production must peak this century. Nature. 2013;**502**:615-617. DOI: 10.1038/502615a

[2] Hsu JH, Lo SL. Chemical and spectroscopic analysis of organic matter transformations during composting of pig manure. Environmental Pollution. 1999;**104**:189-196. DOI: 10.1016/S0269-7491(98)00193-6

[3] Marshall RE, Farahbakhsh K. Systems approaches to integrated solid waste management in developing countries. Waste Management. 2013;**33**:988-1003. DOI: 10.1016/j.wasman.2012.12.023

[4] Bernstad A, Schott S, Wenzel H, la Cour JJ. Identification of decisive factors for greenhouse gas emissions in comparative life cycle assessments of food waste management—An analytical review. Journal of Cleaner Production. 2016;**119**:13-24. DOI: 10.1016/j.jclepro.2016.01.079

[5] Calabi-Floody M, Medina J, Rumpel C, Condron LM, Hernandez M, Dumont M, et al. Smart fertilizers as a strategy for sustainable agriculture. Advances in Agronomy. 2018;**147**:119-157. DOI: 10.1016/bs.agron.2017.10.003

[6] Sinha RK, Chan A. Study of emission of greenhouse gases by Brisbane households practicing different methods of composting of food & garden wastes: aerobic, anaerobic and vermicomposting [project report]. South East Queensland: Griffith University; 2009

[7] Reganold JP, Palmer AS, Lockhart JC, Macgregor AN. Soil quality and financial performance on biodynamic and conventional farms in New Zealand. Science. 1993;**260**:344-349. DOI: 10.1126/science.260.5106.344

[8] Letourneau DK, Goldstein B. Pest damage and arthropod community structure in organic vs. conventional tomato production in California. Journal of Applied Ecology. 2001;**38**:557-570. DOI: 10.1046/j.1365-2664.2001.00611.x

[9] Mader P, Fliessbach A, Dubois D, Gunst L, Fried P, Niggli U. Soil fertility and biodiversity in organic farming. Science. 2002;**296**:1694-1697. DOI: 10.1126/science.1071148

[10] Lazcano C, Gomez-Brandon M, Dominguez J. Comparison of the effectiveness of composting and vermicomposting for the biological stabilization of cattle manure. Chemosphere. 2008;**72**:1013-1019. DOI: 10.1016/j.chemosphere.2008.04.016

[11] Bernal MP, Alburquerque JA, Moral R. Composting of animal manures and chemical criteria for compost maturity assessment: A review. Bioresource Technology. 2009;**100**:5444-5453. DOI: 10.1016/j.biortech.2008.11.027

[12] Dominguez J, Edwards CA. Relationships between composting and vermicomposting: Relative values of the products. In: Edwards CA, Arancon NQ, Sherman RL, editors. Vermiculture Technology: Earthworms, Organic Waste and Environmental Management. Boca Raton, Florida; 2010. pp. 1-14

[13] Waleed SA. Cow manure composting by microbial treatment for using as potting material: An overview. Pakistan Journal of Biological Sciences. 2016;**19**:1-10. DOI: 10.3923/pjbs.2016.1.10

[14] Sinha RK, Herat S, Valani D, Chauhan K. Vermiculture and Sustainable Agriculture. American-Eurasian Journal of Agricultural & Environmental Sciences. 2009;**5**(special issue):01-55

[15] Adhikary S. Vermicompost, the story of organic gold: A review. Agricultural Sciences. 2012;**3**:905-917. DOI: 10.4236/as.2012.37110

[16] Sinha K, Valani D, Soni B, Chandran V. Earthworm Vermicompost: A Sustainable Alternative to Chemical Fertilizers for Organic Farming. Agriculture Issues and Policies. New York: Nova Science Publishers Inc; 2011. p. 71

[17] Chauhan HK, Singh K. Potancy of Vermiwash with Neem plant parts on the Infestation of Earias vittella (Fabricius) and Productivity of Okra (*Abelmoschus esculentus*) (L.) Moench. Asian Journal of Research in Pharmaceutical Sciences. 2015;**5**:36-40. DOI: 10.5958/2231-5659.2015.00006.5

[18] Maulini-Duran C, Artola A, Font X, Sánchez A. Gaseous emissions in municipal wastes composting: Effect of the bulking agent. Bioresource Technology. 2014;**172**:260-268. DOI: 10.1016/j.biortech.2014.09.041

[19] Cerda A, Artola A, Font X, Barrena R, Gea T. Sanchez. A composting of food wastes: Status and challenges. Bioresource Technology. 2018;**248A**:57-67. DOI: 10.1016/j.biortech.2017.06.133

[20] Edwards CA, Burrows I. The potential of earthworm composts as plant growth media. In: Edwards CA, Neuhauser E, editors. Earthworms in Waste and Environmental Management. Netherlands: SPB Academic Press; 1988. pp. 21-32

[21] Benitez E, Nogales R, Elvira C, Masciandaro G, Ceccanti B. Enzymes activities as indicators of the stabilization of sewage sludges composting by Eisenia foetida. Bioresource Technology. 1999;**67**:297-303. DOI: 10.1016/S0960-8524(98)00117-5

[22] Lung AJ, Lin CM, Kim JM, Marshall MR, Nordstedt R, Thompson NP, et al. Destruction of Escherichia coli O157:H7 and *Salmonella enteritidis* in cow manure composting. Journal of Food Protection. 2001;**64**:1309-1314. DOI: 10.4315/0362-028x-64.9.1309

[23] Goyal S, Dhull SK, Kapoor KK. Chemical and biological changes during composting of different organic wastes and assessment of compost maturity. Bioresource Technology. 2005;**96**:1584-1591. DOI: 10.1016/j.biortech.2004.12.012

[24] Bajsa O, Nair J, Mathew K, Ho GE. Vermiculture as a tool for domestic wastewater management. Water Science and Technology. 2003;**48**:125-132. DOI: 10.2166/wst.2004.0821

[25] Vermi Co. Vermicomposting technology for waste management and agriculture: An executive summary [Internet]. 2001. Avaliable from: http://www.vermico.com/summary.htm [Accessed: 12 February 2020]

[26] Crescent T. Vermicomposting. Development alternatives (DA) sustainable livelihoods [Internet]. 2003. Available from: http://www.dainet.org/livelihoods/default.htm [Accessed: 12 February 2020]

[27] Atiyeh RM, Edwards CA, Subler S, Metzger JD. Pig manure vermicompost as a component of a horticultural bedding plant medium: Effects on physicochemical properties and plant growth. Bioresource Technology. 2001;**78**:11-20. DOI: 10.1016/s0960-8524(00)00172-3

[28] Sinha RK, Agarwal S, Chauhan K, Valani D. The wonders of earthworms and its vermicompost in farm production: Charles Darwin's friends of farmers, with potential to replace destructive chemical fertilizers from agriculture. The Journal of Agricultural Science. 2010;**1**:76-94. DOI: 10.4236/as.2010.12011

[29] Sinha RK, Herat S, Bharambe G. Brahambhatt a Vermistabilization of sewage sludge (biosolids) by earthworms: Converting a potential biohazard destined for landfill disposal into a pathogen-free, nutritive and safe biofertilizer for farms. Waste Management & Research. 2010;**28**:872-881. DOI: 10.1177/0734242X09342147

[30] Orozco FH, Cegarra J, Trujillo LM, Roig A. Vermicomposting of coffee pulp using the earthworm *Eisenia foetida*: Effects on C and N contents and the availability of nutrients. Biology and Fertility of Soils. 1996;**22**:162-166. DOI: 10.1007/BF00384449

[31] Krishnamoorthy RV, Vajrabhiah SN. Biological activity of earthworm casts: An assessment of plant growth promoter levels in casts. Proceedings of the Indiana Academy of Sciences. 1986;**95**:341-351

[32] Nielson RL. Presence of plant growth substances in earthworms demonstrated by paper chromatography and the went pea test. Nature. 1965;**208**:1113-1114. DOI: 10.1038/2081113a0

[33] Jeyabal A, Kuppuswamy G. Recycling of organic wastes for the production of vermicompost and its response in rice–legume cropping system and soil fertility. European Journal of Agronomy. 2001;**15**:153-170. DOI: 10.1016/S1161-0301(00)00100-3

[34] Kale RD. Vermicomposting has a bright scope. Indian Silk. 1995;**34**:6-9

[35] Bozym M. Biologiczne przetwarzanie biodegradowalnej frakcji odpadow komunalnych i osadów ściekowych w wermikulturze. Prace Instytutu Ceramiki i Materialow Budowlanych. 2012;5:335-369

[36] Kostecka J. Badania nad wermikompostowaniem odpadów organicznych. Zeszyty Naukowe Akademii Rolniczej w Krakowie. Rozprawy. 2000;**68**:1-88

[37] Bhatnagar RK, Palta R. Earthworm Vermiculture and Vermicomposting. Ludhiana: Kalyani Publishers; 1996

[38] Ciavatta C, Govi M, Pasotti L, Sequi P. Changes in organic matter during stabilization of compost from municipal solid waste. Bioresource Technology. 1993;**43**:141-145. DOI: 10.1016/0960-8524(93)90173-9

[39] Bannik CG, Joergensen RG. Change in N fractions during composting of wheat straw. Biology and Fertility of Soils. 1993;**16**:269-274. DOI: 10.1007/BF00369303

[40] Govi M, Ciavatta C, Gessa C. Evolution of organic matter in sewage sludge a study based on the use of humification parameters and analytical elecuofocasing. Bioresource Technology. 1993;**44**:175-180. DOI: 10.1016/j.wasman.2010.04.030

[41] Marttinez-Inigo MJ, Almedros G. Kinetic study of the composting of evergreen cak forestry waste. Waste Management and Research. 1994;**12**:305-314. DOI: 10.1177/0734242X9401200403

[42] Vallini G, Pera A. Green compost production from vegetable waste separately collected in metropolitan garden produce markets. Biological Wastes. 1989;**29**:33-41. DOI: 10.1016/0269-7483(89)90101-8

[43] Jakobsen ST, Gade N, Moller I, Franzen E. Experiments on decomposition of farmyard manure at Egtved field station. Egtved Avlsstationer for Kod Produkter. 1988-1989;**1990**:1-48

[44] Gaur AC, Singh G. Recycling of rural and urban wastes through conventional composting and vermicomposting. In: Tandon HLS, editor. Recycling of Crop, Animal, Human and Industrial Wastes in Agriculture. New Delhi: Fertilizer

development and consultation organization; 1995. pp. 1-49

[45] Jothimani S. Organic farming in coconut. Indian Coconut Journal. 1994:48-49

[46] Darwin F, Seward AC. More letters of Charles Darwin. In: John M, editor. A record of his work in series of hitherto unpublished letters. London; 1903. p. 508. DOI: 10.5962/bhl.title.1477

[47] Martin JP. Darwin on Earthworms: The Formation of Vegetable Moulds. Ontario: Bookworm Publishing; 1976

[48] Bouche MB. Strategies lombriciennes. In: Lohm V, Persson T, editors. Soil Organisms as Components of Ecosystems. Proccedings on Sixth International Soil Zoology Colloquium Ecological Bulletin. Vol. 25. Stockholm; 1977. pp. 122-132

[49] Graff O. Preliminary experiment of vermicomposting of different waste materials using Eudrilus eugeniae Kingberg. In: Appelhof M, editor. Proccedings of the Workshop on 'Role of Earthworms in the Stabilization of Organic Residues'. Michigan, USA: Malanazoo Publication; 1981. pp. 179-191

[50] Beetz A. Worms for composting (vermicomposting). ATTRA-National Sustainable Agriculture Information Service [Livestock Technical Note]. June 1999

[51] Sinha RK, Heart AS, Asadi R, Carretero E. Vermiculture technology for environmental management: Study of action of earthworms *Eisenia foetida, Eudrilus euginae and Parionyx excavatus* on biodegradation of some community wastes in India and Australia. The Environmentalist. 2002;**22**:261-268. DOI: 10.1023/A:1016583929723

[52] Lavelle P, Brussaard L, Hendrix P, editors. Earthworm Management in Tropical Agroecosystems. Wallingford: CAB International; 1999

[53] Dash MC. Role of earthworms in the decomposer system. In: Singh JS, Gopal B, editors. Glimpses of ecology. New Delhi: India International Scientific Publication; 1978. pp. 399-406

[54] Binet F, Fayolle L, Pussard M. Significance of earthworms in stimulating soil microbial activity. Biology and Fertility of Soils. 1998;**27**: 79-84. DOI: 10.1007/s003740050403

[55] Singleton DR, Hendrixb PF, Colemanb DC, Whitmana WB. Identification of uncultured bacteria tightly associated with the intestine of the earthworm *Lumbricus rubellus* (Lumbricidae; Oligochaeta). Soil Biology and Biochemistry. 2003;**35**:1547-1555. DOI: 10.1016/ S0038-0717(03)00244-X

[56] Nagavallemma KP, Wani SP, Stephane L, Padmaja VV, Vineela C, Babu Rao M, et al. Vermicomposting: Recycling wastes into valuable organic fertilizer. Global Theme on Agrecosystems [Report]. Patancheru,Andhra Pradesh: International Crops Research Institute for the Semi-Arid Tropics; 2004

[57] Sanchez-Monedero MA, Roig A, Paredes C, Bernal MP. Nitrogen transformation during organic waste composting by the Rutgers system and its effects on pH, EC and maturity of the composting mixtures. Bioresource Technology. 2001;**78**:301-308. DOI: 10.1016/s0960-8524(01)00031-1

[58] Munnoli PM. Teixeira da Silva JA, Bhosle, S. dynamics of the soil - earthworm – Plant relationship: A review. Dynamic soil. Dynamic Plant. 2010;**4**:1-21

[59] Ismail SA. Vermicology: The Biology of Earthworms. Orient Longman Limited: Chennai; 1997

[60] Gaddie RE, Donald ED. Earthworms for Ecology and Profit. 1st Vol. Scientific Earthworm Farming. California: Bookworm Publishing Company; 1975. p. 180

[61] Nair J, Sekiozoic V, Anda M. Effect of pre-composting on vermicomposting of kitchen waste. Bioresource Technology. 2006;**97**:2091-2095. DOI: 10.1016/j.biortech.2005.09.020

[62] Frederickson J, Howell G, Hobson AM. Effect of precomposting and vermicomposting on compost characteristics. European Journal of Soil Biology. 2007;**43**:S320-S326. DOI: 10.1016/j.ejsobi.2007.08.032

[63] Mupondi L. Improving Sanitization and Fertiliser Value of Dairy Manure and Waste Paper Mixtures Enriched with Rock Phosphate through Combined Thermophilic Composting and Vermicomposting [Thesis]. South Africa: University of Fort Hare; 2010

[64] Soobhany N, Mohee R, Garg VK. Inactivation of bacterial pathogenic load in compost against vermicompost of organic solid waste aiming to achieve sanitation goals: A review. Waste Management. 2017;**64**:51-62. DOI: 10.1016/j.wasman.2017.03.003

[65] Rostami R, Najafi SH, Nabaee A, Eslami A. Survey of *E. foetida* population on pH, C/N ratio process rate in vermicompost. Journal of Environmental Studies. 2010;**35**:93-98

[66] Garg VK, Kaushik P. Vermistabilization of textile mill sludge spiked with poultry droppings by epigeic earthworm *Eisenia fetida*. Bioresource Technology. 2005;**96**:1063-1071. DOI: 10.1016/j.biortech.2004.09.003

[67] Suthar S. Potential utilization of guargum industrial waste in vermicompost production. Bioresource Technology. 2006;**97**:2474-2477. DOI: 10.1016/j.biortech.2005.10.018

[68] Dominguez J, Edwards CA. Effects of stocking rate and moisture content on the growth and maturation of *Eisenia andrei* (Oligochaeta) in pig manure. Soil Biology and Biochemistry. 1997;**29**:743-746. DOI: 10.1016/S0038-0717(96)00276-3

[69] Tchobanoglous G, Theisen H, Vigil S. Integrated Solid Waste Management. New York: McGraw-Hill; 1993. p. 4

[70] Edwards CA, Bohlen PJ. Biology and Ecology of Earthworm. 3rd ed. London: Chapman and Hall; 1996. p. 426

[71] Sinha RK, Nair J, Bharambe B, Patil S, Bapat P. Vermiculture revolution: A low-cost and sustainable technology for management of municipal and industrial organic wastes (solid and liquid) by earthworms with significantly low greenhouse gas emissions. In: Daven JI, Klein RN, editors. Progress in Waste Management Research. New York: Nova Science Publishers; 2008. pp. 158-229

[72] Georg. Feasibility of Developing the Organic and Transitional Farm Market for Processing Municipal and Farm Organic Wastes Using Large-Scale Vermicomposting. Good Earth Organic Resources Group, Halifax, Nova Scotia [Internet]. 2004. Available from: http://www.alternativeor-ganic.com [Accessed: 15 February 2020]

[73] Gajalakshmi S, Abbasi SA. Earthworms and vermicomposting. Indian Journal of Biotechnology. 2004;**3**:486-494

[74] Edwards CA, editor. Earthworm Ecology. 2nd ed. Boca Raton: CRC Press; 1998. p. 389

[75] Gunadi B, Blount C, Edwards CA. The growth and fecundity of *Eisenia fetida* (Savigny) in cattle solids pre-composted for different periods. Pedobiologia. 2002;**46**:15-23. DOI: 10.1078/0031-4056-00109

[76] Marinari S, Masciandaro G, Ceccanti B, Grego S. Influence of organic and mineral fertilisers on soil biological and physical properties. Bioresource Technology. 2000;**72**:9-17. DOI: 10.1016/S0960-8524(99)00094-2

[77] Maheswarappa HP, Nanjappa HV, Hegde MR. Influence of organic manures on yield of arrowroot, soil physico-chemical and biological properties when grown as intercrop in coconut garden. Annals of Agricultural Research. 1999;**20**:318-323

[78] Mitchell A. Production of *Eisenia fetida* and vermicompost from feed-lot cattle manure. Soil Biology and Biochemistry. 1997;**29**:763-766. DOI: 10.1016/S0038-0717(96)00022-3

[79] Edwards CA, Arancon NQ. Vermicomposts suppress plant pest and disease attacks. Biocycle. 2004;**45**:51-53

[80] Albanell E, Plaixats J, Cabrero T. Chemical changes during vermicomposting (*Eisenia fetida*) of sheep manure mixed with cotton industrial wastes. Biology and Fertility of Soils. 1988;**6**:266-269. DOI: 10.1007/bf00260823

[81] Atiyeh RM, Arancon NQ, Edwards CA, Metzger JD. The influence of earthworm-processed pig manure on the growth and productivity of marigolds. Bioresource Technology. 2001;**81**:103-108. DOI: 10.1016/s0960-8524(02)00017-2

[82] Chaoui H, Edwards CA, Brickner M, Lee S, Arancon N. Suppression of the plant diseases, Pythium (damping off), Rhizoctonia (root rot) and Verticillum (wilt) by vermicomposts. Proceedings of Brighton Crop Protection Conference – Pests and Diseases II. 2002;(8B-3):711-716

[83] Scheuerell SJ, Sullivan DM, Mahaffee WF. Suppression of seedling damping-off caused by *Pythium ultimum*, and *Rhizoctonia solani* in container media amended with a diverse range of Pacific northwest compost sources. Phytopathology. 2005;**95**: 306-315. DOI: 10.1094/PHYTO-95-0306

[84] Singh R, Sharma RR, Kumar S, Gupta RK, Patil RT. Vermicompost substitution influences growth, physiological disorders, fruit yield and quality of strawberry (Fragaria x ananassa Duch.). Bioresource Technology. 2008;**99**:8507-8511. DOI: 10.1016/j.biortech.2008.03.034

Section 3

Biological Aspects

Chapter 5

Application and Mechanisms of Plant Growth Promoting Fungi (PGPF) for Phytostimulation

Md. Motaher Hossain and Farjana Sultana

Abstract

Plant growth-promoting fungi (PGPF) constitute diverse genera of nonpathogenic fungi that provide a variety of benefits to their host plants. PGPF show an effective role in sustainable agriculture. Meeting increasing demand for crop production without damage to the environment is the biggest challenge nowadays. The use of PGPF has been recognized as an environmentally friendly way of increasing crop production. These fungi have proven to increase crop yields by improving germination, seedling vigor, plant growth, root morphogenesis, photosynthesis, and flowering through either a direct or indirect mechanism. The mechanisms of PGPF involve solubilizing and mineralizing nutrients for easy uptake by plants, regulating hormonal balance, producing volatile organic compounds and microbial enzyme, suppressing plant pathogens and ameliorating abiotic stresses. Successful colonization is an intrinsic factor for most PGPF to exert their beneficial effects on plants. A certain level of specificity exists in the interactions between plant species and PGPF for root colonization and growth promoting effects. There is a gap between the number of reported efficacious PGPF and the number of PGPF as biofertilizer. Efforts should be strengthened to improve the efficacy and commercialization of PGPF. Hence, this chapter summarizes valuable information regarding the application and mechanisms of PGPF in sustainable agriculture.

Keywords: seed germination, seedling vigor, root morphogenesis, yield, root colonization, formulation

1. Introduction

The world's population exceeded ~7 billion just after 2010, and still continues to grow fast. Roughly, 83 million people are added to the world's population every year and with this pace of growth, the global population is projected to reach around 9.7 billion by 2050, ~24% higher than today [1]. In order to feed this large population, crop production must increase by approximately 25–70% above current production levels [2]. Intensification of agriculture is considered a potential solution. By relying on intensive use of fertilizers, pesticides and other inputs, agricultural intensification increases the productivity of existing farmland and delivers more food to the added population. However, the chemical-based crop

IntechOpen

intensification produces more food in a way that the future production potential of farmland is being undermined and the environment is being affected. An increasingly degraded soil, overwhelming health hazards from soil and water pollution, disturbed natural microbial populations are a few of the direct implications in chemical-intensive agriculture. To avoid these potentially harmful effects of agrochemicals in agriculture, alternative approaches must be persuaded. An ecocentric approach that provides both environmental and economic benefits is increasingly needed. Organic farming is one of many such approaches that promote agroecosystem health, ensuring sustainable intensification in agriculture.

The uniqueness of microorganisms and the dynamic part played by them in sustaining agricultural ecosystems have made them likely candidates for playing a central role in organic-based modern agriculture. Fortunately, plant roots harbor an abundant association of beneficial microorganisms. Root exudates are the largest source of carbon that attracts the microbial populations and allow them to forge an intimate association with host plants [3]. In response, the rhizosphere microbial populations play versatile roles in transforming, mobilizing and solubilizing soil nutrients, which are crucial for plant growth and development. Among the diverse rhizosphere microbial population, fungi known as plant growth promoting fungi (PGPF) are receiving a growing attention in recent days. Over the decades, varieties of PGPF have been studied including those belong to genera *Trichoderma, Penicillium, Phoma* and *Fusarium* [4]. Studies have shown that PGPF modulate plant growth and enhance resilience to plant pathogens without environmental contamination [5]. The positive effects of PGPF on plant and environment make them well fitted to organic agriculture.

The course of plant growth promotion by PGPF is a complex process and often cannot be attributed to a single mechanism. A variety of direct and indirect mechanisms, including solubilization of minerals, synthesis of phytohormones, production of volatile organic compounds, exploitation of microbial enzymes, increases in nutrient uptake, amelioration of abiotic stresses and suppression of deleterious phytopathogens are involved. These wide arrays of interconnected mechanisms help PGPF maintaining rhizosphere competence and stability in host performance. Compared to the large number of PGPF identified in the laboratory, only a small fraction of them is in agricultural practice worldwide. Inconsistent performance of the inoculated PGPF under field conditions limits the commercial application of them. Development of appropriate formulation could improve the performance in the field and pave the way for commercialization of the PGPF. An ideal formulation of PGPF should fit with existing application technologies, protect biological actives from stress, ensure viability, remains unaffected after storage under ambient conditions, ensure microbial actives in the field and be cost effective [6].

Considering the aspects discussed above, the need for superior PGPF to supplement inorganic chemical fertilizers as one of the crucial steps of moving toward organic farming practices has been highlighted. Inclusion of new techniques in these processes has been vital to the development of novel PGPF applications. This review will therefore attempt to shed light on the recent findings related to the impact of PGPF on plant growth and yield, duration of their effects, host specificity of the cooperation, root colonization mechanisms, their modes of action and commercial formulation for enhancement of plant growth and yield. The knowledge produced from this review could be very useful to those who are apprehensive about environmental protection and agricultural sustainability.

2. Plant growth promoting fungi (PGPF)

Plants have intricate relationships with an array of microorganisms, particularly rhizosphere fungi and bacteria, which can lead to an increase in plant vigor, growth and development as well as changes in plant metabolism [7]. The group of rhizosphere fungi that colonize plant roots and enhance plant growth is referred to as PGPF [4]. PGPF are heterogeneous group of nonpathogenic saprotroph fungi. They can be separated into endophytic, whereby they live inside roots and exchange metabolites with plants directly, and epiphytic, whereby they live freely on the root

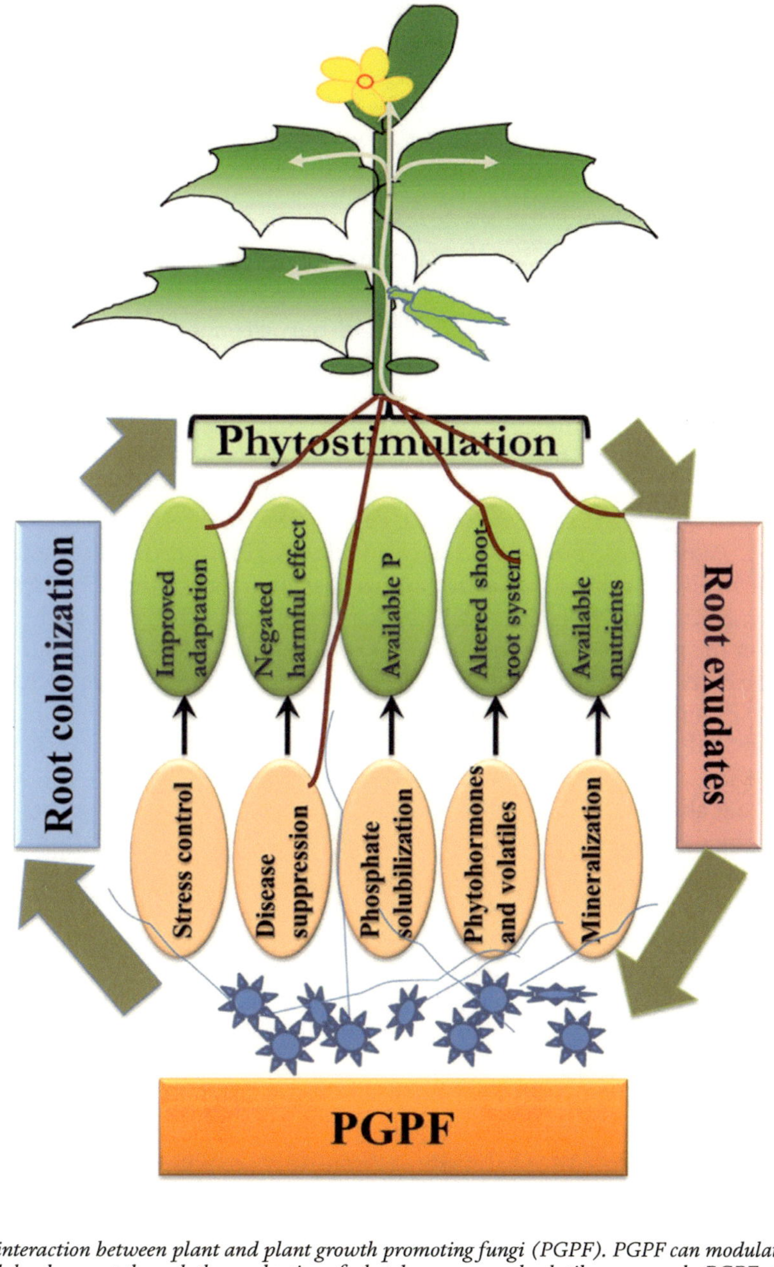

Figure 1.
Beneficial interaction between plant and plant growth promoting fungi (PGPF). PGPF can modulate plant growth and development through the production of phytohormones and volatile compounds. PGPF also influence plant nutrition via solubilization of phosphorus and mineralization of organic substrates. PGPF modify plant functioning against biotic and abiotic stresses by negating their harmful effects.

surface and free-living PGPF, which live outside plant cells, i.e., in the rhizosphere [5]. PGPF establish a non-obligate mutualism with a broader range of host plants. That is why symbiotic mycorrhizal fungi are not considered as PGPF, although they are known to improve growth of the plants [8]. Moreover, PGPF encompass a diverse taxonomic group in comparison to mycorrhiza. They are often involved in a range of complex interactions with plants and develop distinct strategies to mediate improvements in seed germination, seedling vigor, plant growth, flowering and productivity of host plants (**Figure 1**). PGPF are not only associated with the root to mediate positive effects on plant growth and development but also have beneficial effects on suppressing phytopathogenic microorganisms [9]. Not every organism identified as PGPF will improve plant growth under all conditions or in association with all plant hosts [10]. Some PGPF biocontrol inoculants usually contain necrotrophic mycoparasites such as *Trichoderma* spp. [11], while a limited number such as *Sphaerodes mycoparasitica* is biotrophic mycoparasitic agent [12]. Therefore, PGPF are considered one of the potential active ingredients in both biofertilizer and mycofungicide formulation.

3. The nature and composition of PGPF

PGPF are common root-associated and soil-borne fungi from diverse genera. Fungi reported as PGPF include Ascomycetes, Basidiomycetes and Oomycetes [5]. Some strains of hypovirulent binucleate *Rhizoctonia* (HBNR) are known to be PGPF [13]. PGPF also include isolates of mycelial fungi that do not produce any spores, generally known as sterile black fungus (SBF), sterile dark fungus (SDF) and sterile red fungus (SRF) [14]. The non-sporulating PGPF are often difficult to identify and mostly lack formal taxonomic status. Among the PGPF *Aspergillus*, *Fusarium*, *Penicillium, Phoma* and *Trichoderma* have a wide distribution and are, by far, the most extensively reported (**Table 1**). Each of the genera has a variety of species. *Aspergillus, Fusarium, Penicillium* and *Phoma* were frequently found in the rhizosphere or in the roots of plants. Instead, *Trichoderma* were mostly isolated from soil. Among the rhizosphere population, PGPF have a high relative abundance. A total of 619 (44%) out of 1399 fungal isolates collected from rhizosphere of six different plants were PGPF, while frequency of occurrence of PGPF in zoysiagrass, wheat, corn and eggplant rhizosphere were 46, 47, 38 and 10%, respectively [4]. This indicates that abundance of PGPF varies largely according to the host rhizosphere. Similarly, the dominating fungal genus is not necessarily the dominating PGPF in the rhizosphere population. The order of the frequency of the main genera among 1399 fungal isolates was *Fusarium* > *Trichoderma* > sterile fungi > *Penicillium* > *Pythium* > *Rhizoctonia* > *Mucor,* while that of PGPF from each plant genus was: *Trichodema* (~82%) > *Pythium* (~75%) > *Penicillium* (~69%) > *Alternaria* (~63%) > *Fusarium* (~44%) > sterile fungi (40%) > *Mucor* (~38%) [4]. The important characteristics of these fungi are their high rhizosphere competence and ability to promote plant growth.

Initial search for identification of PGPF was concentrated to rhizosphere fungi. Recent studies have demonstrated the potential of phyllosphere fungi as PGPF. The phyllosphere, which consists of the above ground surfaces of plants, is one of the most prevalent microbial habitats on earth. Phyllosphere fungi can act as mutualists promoting plant growth and tolerance of environmental stressors [53]. A few of other fungi isolated from tree bark, decorticated wood and water damaged building functioned as PGPF [43, 49]. More interestingly, the fungal entomopathogens also show potential to be PGPF and promote plant growth [54]. PGPF seem to have a cosmopolitan occurrence.

PGPF	Original source of isolation	References
Alternaria sp.	*Zoysia tenuifolia*, *Rosa rugosa*, *Camellia japonica*, *Delonix regia*, *Dianthus caryophyllus*, *Rosa hybrid*	[4, 15]
Aspergillus sp., *As. fumigatus, As. niger, As. terreus, As. ustus, As. clavatus*	*Capsicum annuum*, *Glycine max, Cicer arietinum, Elymus mollis, Solanum tuberosum, Nymphoides peltata*	[16–21]
Aureobasidium pullulans	Dark chestnut soil	[22]
Chaetomium globosum	*Capsicum annuum*	[23]
Cladosporium sp., *Cladosporium sphaerospermum*	*Cucumis sativus, Glycine max*	[24, 25]
Colletotrichum sp.	*Rosa rugosa*, *Camellia japonica*, *Delonix regia*, *Dianthus caryophyllus*, *Rosa hybrid*	[15]
Exophiala sp.	*Cucumis sativus*	[26]
Fusarium sp., *F. equiseti*, *F. oxysporum, F. verticillioides*	*Cynodon dactylon*, *Lygeum spartum, Zoysia tenuifolia, Musa* sp. and other environment	[27–32]
Non-sporulating sterile fungi	*Zoysia tenuifolia*	[14]
Penicillium sp., *Pe. chrysogenum, Pe. citrinum, Pe. kloeckeri, Pe. menonorum, Pe. resedanum, Pe. simplicissimum, Pe. janthinellum, Pe. viridicatum*	Halophyte, *Ixeris repenes, Cicer arietinum, Elymus mollis, Capsicum annuum, Zoysia tenuifolia*	[9, 16, 22, 33–40]
Phoma sp., *Phoma herbarum, Phoma multirostrata*	*G. max, Rosa rugosa*, *Camellia japonica*, *Delonix regia*, *Dianthus caryophyllus*, *Rosa hybrid, Zoysia tenuifolia*	[4, 14, 15, 34, 41, 42]
Phomopsis sp., *Phomopsis liquidambari*	*Rosa rugosa*, *Camellia japonica*, *Delonix regia*, *Dianthus caryophyllus*, *Rosa hybrid, Bischofia polycarpa* bark	[15, 43]
Purpureocillium lilacinum	Soil	[44]
Rhizoctonia spp.	Orchid, *Lycopersicon lycopersicum*, and soil	[13, 45, 46]
Rhodotorula mucilaginosa	Soil	[22]
Talaromyces wortmannii	Soil	[40]
Trichoderma asperellum, *T. atroviride, T. hamatum, T. harzianum, T. longibrachiatum, T. pseudokoningii, T. viride, T. virens*	Soil, wood and damaged building	[34, 47–52]

Table 1.
Different fungi reported as plant growth promoting fungi (PGPF) with their original source of isolation.

4. Impact of PGPF on plant growth promotion

PGPF exhibit traits beneficial to plant and as such, their capacity to enhance plant growth and development is well founded. PGPF mediate both short- and long-term effects on germination and subsequent plant performance. Improvement in germination, seedling vigor, shoot growth, root growth, photosynthetic efficiency, flowering, and yield are the most common effects decreed by PGPF. A particular PGPF may condition plant growth by exerting all or one or more of these effects.

4.1 Impact of PGPF on seed germination and seedling vigor

Seed germination and germinant growth are critical developmental periods of the young plantlet until it begins producing its own food by photosynthesis. Treatment with PGPF, particularly of the genus *Aspergillus*, *Alternaria*, *Trichoderma*, *Penicillium, Fusarium, Sphaerodes* and *Phoma* has been reported to improve seed germination and seedling vigor in different agronomic and horticultural crops (**Table 2**). Scarified seeds inoculated with spores from *Aspergillus* and *Alternaria* had significant increases in germination of Utah milkvetch (*Astragalus utahensis*) *in vitro*, and in greenhouse and fall-seeded plots near Fountain Green and Nephi [55]. The *Aspergillus*-treated seeds performed out seeds inoculated with *Alternaria*. An increase of 30% in seedling emergence was observed in cucumber plant raised upon the treatment of *T. harzianum* [47]. Application of *T. harzianum* also significantly increased seed germination, emergence index, seedling vigor and successful transplantation percentage in muskmelon compared to the untreated controls [59]. Early seedling emergence and enhanced vigor were observed in bacterial wilt susceptible tomato cultivar treated with *T. harzianum*, *Phoma multirostrata*, and *Penicillium chrysogenum* compared to untreated controls [34]. The culture filtrate of *Penicillium* was as effective as the living inocula in improving seed germination of tomato [70]. Significantly, higher germination and vigor index were observed in Indian spinach, when seeds were sown in sterilized field soil amended with wheat grain inoculum of *Fusarium* spp. PPF1 [27]. *Sphaerodes mycoparasitica*, a biotrophic mycoparasite of *Fusarium* species, improved wheat seed germination and seedling growth *in vitro* compared to *T. harzianum,* while under phytotron conditions, both *S. mycoparasitica* and *T. harzianum* had positive impact on wheat seedlings growth in the presence of *F. graminearum* [12]. These results show the positive impact of PGPF on seed germination and seedlings growth of a wide arrays of hosts.

4.2 Impact of PGPF on shoot growth

The most common form of growth promotion by PGPF is the augmented shoot in colonized plants. Shoot growth promotion has been shown by a great diversity of PGPF across a large number of plant species. Isolates of *Aspergillus, Trichoderma, Penicillium*, and *Fusarium* were capable of enhancing the shoot growth in model plant Arabidopsis [9, 20, 28, 33, 48]. Different species of *Aspergillus* are known to support shoot growth in chickpea [16], Chinese cabbage [56], cucumber [17], soybean [18, 65] and wheat [76]. Species of nonpathogenic *Fusarium* were reported to stimulate shoot growth in Indian spinach [27] and banana [29]. Application of barley grain inoculum of *Penicillium viridicatum* GP15-1 to the potting medium resulted in 26–42% increase in stem length, 37–46% increase in shoot fresh weight and 100–176% increase in shoot dry weight of cucumber plants [35]. Similarly, inoculation of cucumber plants with *Pe. menonorum* KNU3 increased cucumber shoot dry biomass by as much as 52% [36]. Stimulated shoot growth by *Penicillium* spp. was also reported in tomato [69], Waito-c rice [37, 38], chili [23, 39] and sesame [74]. Application of *T. longipile* and *T. tomentosum* increased shoot dry weight of cabbage seedlings by 91–102% in glasshouse trials [57]. Likewise, cottonseeds pretreated with *T. viride* showed four-fold increases in shoot length elongation and an almost 40-fold increase in plant dry weight compared to the control [66]. Augmented shoot growth by *Trichoderma* has also been reported in chickpea [16], wheat [79], maize [78], cucumber [60] and other plant species (**Table 2**). Isolates of *Phoma* were found to be an efficient stimulator of plant shoot [15, 41, 62]. A few hypovirulent *Rhizoctonia* isolates were able to induce significantly higher fresh leaves and stems weights in tomato plants grown in greenhouse [13]. Enhancement of shoot growth was also observed

Test crop	PGPF strain	Improvement	References
Arabidopsis thaliana	*Trichoderma virens* Gv. 29-8	Biomass, lateral root development	[48]
	Penicillium janthinellum GP16-2	Shoot biomass, leaf number	[33]
	Pe. simplicissimum GP17-2	Shoot biomass, leaf number	[9]
	Fusarium oxysporum NRRL 38499, NRRL 26379 and NRRL 38335,	Shoot-root growth	[28]
	Aspergillus ustus	Shoot growth, lateral root, root hair numbers	[20]
Astragalus utahensis	*Aspergillus* spp., *Alternaria* spp.	Seed germination	[55]
Basella alba	*Fusarium* spp. PPF1	Germination, seedling vigor, shoot-root growth, leaf area, leaf chlorophyll content	[27]
Brassica campestris	*Talaromyces wortmannii* FS2	Shoot fresh weight	[40]
B. chinensis	*A. niger* 1B and 6A	Plant dry weight, N and P content	[56]
B. oleracea var. *capitata*	*T. longipile, T. tomentosum*	Shoot dry weight, leaf area	[57]
Capsicum annuum	*Pe. resedanum* LK6	Shoot length, biomass, chlorophyll content, photosynthesis	[39]
	Chaetomium globosum CAC-1G	Plant biomass, root-shoot growth	[23]
Cicer arietinum	*A. niger* BHUAS01, *Pe. citrinum* BHUPC01, *T. harzinum*	Plant growth	[16]
	T. harzianum T-75	Yield	[58]
Cucumis melo	*T. harzianum* Bi	Germination, seedling health, vigor	[59]
Cucumis sativus	*Pe. simplicissimum* GP17-2	Root-shoot growth	[4]
	Pe. viridicatum GP15-1	Root-shoot length, biomass	[35]
	T. harzianum GT3-2	Root-shoot growth	[60]
	F. equiseti GF19-1	Root-shoot growth	[61]
	Aspergillus spp. PPA1	Root-shoot length, biomass, leaf area, chlorophyll content	[17]
	Exophiala sp. *LHL08*	Plant growth under drought and salinity	[26]
	Phoma sp.	Root-shoot growth, yield in the field	[62]
	Phoma sp. GS8-2, GS8-3	Root-shoot growth	[63]
GiSeLa6® (*Prunus cerasus* × *P. canescens*)	*T. harzianum* T-22	Root growth, development	[64]
Glycine max	*A. fumigatus* HK-5-2	Shoot growth, biomass, leaf area, chlorophyll contents, photosynthetic rate	[65]
	A. fumigatus LH02	Shoot growth, biomass, leaf area, chlorophyll contents, photosynthetic rate	[18]
	*Phoma herbarum*TK-2-4	Plant length, biomass	[41]
Gossypium arboreum L	*T. viride*	Root-shoot length, plant dry weight	[66]

Test crop	PGPF strain	Improvement	References
Helianthus annuus	*Trichoderma* sp., *Aspergillus* sp., *Penicillium* sp., *Phoma* sp., *Fusarium* sp.	Seed germination, seedling vigor	[67]
Lactuca sativus	*F. oxysporum* MSA 35	Root-shoot growth, chlorophyll content	[68]
Lycopersicon lycopersicum	*T. harzianum* TriH_JSB27, *Phoma multirostrata* PhoM_JSB17, *T. harzianum* TriH_JSB36, *Pe. chrysogenum* PenC_JSB41	Seedling emergence, vigor	[34]
	T. harzianum T-22	Seed germination under stress	[69]
	Penicillium spp.	Seed germination, root-shoot growth	[70]
	F. equiseti GF19-1	Plant biomass, root-shoot growth	[71]
Musa sp.	*F. oxysporum* V5W2, Eny 7.11o, Emb 2.4o	Yield	[29]
Nicotiana tabacum	*Alternaria* sp., *Phomopsis* sp., *Cladosporium* sp., *Colletotrichum* sp., *Phoma* sp.	Root-shoot growth, chlorophylls, soluble sugars, plant biomass	[15]
Pinus sylvestris var. *mongolica*	*T. harzianum* E15, *T. virens* ZT05	Seedling biomass, root structure, soil nutrients, soil enzyme activity	[72]
Saccharum officinarum	*T. viride*	Yield	[73]
Sesamum indicum	*Penicillium* spp. NICS01, DFC01	Root-shoot growth, chlorophylls, proteins, amino acids, lignans	[74]
Solanum tuberosum	*A. ustus*	Root-shoot growth, lateral root, root hair numbers	[20]
Spinacia oleracea	*F. equiseti*	Plant biomass, root-shoot growth	[75]
Suaeda japonica	*Penicillium* sp. Sj-2-2	Plant length	[38]
	Cladosporium sp. MH-6	Shoot length	[24]
	Pe. citrinum IR-3-3	Root-shoot length	[37]
	Phoma herbarum TK-2-4	Plant length	[41]
Triticum aestivum	*T. harzianum*, *T. koningii*	Plant biomass, root-shoot growth.	[4]
	Sphaerodes mycoparasitica	Seed germination, seedling growth	[12]
	A. niger NCIM	Shoot and total plant length ratio	[76]
Vinca minor	*T. harzianum*	Flowering, plant height, weight	[77]
Zea mays	*T. harzianum* T22	Shoot growth, area and size of main and secondary roots	[78]

Table 2.
Effect of different plant growth promoting fungi (PGPF) on seed germination, plant growth and yield in various plants.

by *Talaromyces wortmannii* in cabbage [40], *Chaetomium globosum* in chili [23], *Colletotrichum* sp. in tobacco and *Exophiala* sp. in cucumber [26]. The results from these studies are consistent with numerous field and growth chamber experiments that have shown that PGPF inoculants can mediate shoot growth improvement.

4.3 Impact of PGPF on photosynthesis

The plant growth promotion in some plant-PGPF interaction is occasionally associated with improvement in state and function of the photosynthetic apparatus of plants. Treatment with *T. longipile* and *T. tomentosum* increased leaf area of cabbage by 58–71% in glasshouse trials [57]. Tomato plants grown with HBNR isolates had significantly higher leaf fresh weight than control plants in greenhouse [13]. Arabidopsis grown in soil amended with *Pe. simplicissimum* GP17-2 and *Pe. janthinellum* GP16-2 were more greener and had approximately 1 more leaflet per plant than control plants 4 weeks after treatment [9]. *Penicillium* spp. also enhanced leaf chlorophyll content in cucumber and chili [36, 39]. Soil amendment with *Aspergillus* spp. PPA1 and *Fusarium* spp. PPF1 significantly increased leaf area and leaf chlorophyll content in cucumber and Indian spinach, respectively [27]. Improvement in leaf number, leaf area and leaf chlorophyll levels would contribute to increases in photosynthesis rate and net accumulation of carbohydrate in plants.

4.4 Impact of PGPF on root growth and architecture

Roots are vital plant organs that remain below the surface of the soil. The root system is important for plant fitness because it facilitates the absorption of water and nutrients, provides anchorage of the plant body to the ground and contributes to overall growth of plants. Root functions as the major interface between the plant and the microbes in the soil environment. The bulk of previous studies have evidenced the immense ability of PGPF in enhancement of root growth in different plants (**Table 2**). Plants forming association with PGPF show faster and larger root growth resulting in a rapid increase in the root biomass [27, 35, 50, 57]. Moreover, root length, root surface area, root diameter and branch number are under direct influence of intimate interaction with PGPF. Application of *T. virens* ZT05 increased root length, root surface area, average root diameter, root tip number and root branch number of pines by 25.11, 98.19, 5.66, 45.89 and 74.42%, respectively [72]. *A. ustus* is known to cause alterations in the root system architecture by promoting the formation of secondary roots in Arabidopsis and potato [20]. In maize (*Zea mays*), *Trichoderma* inoculation enhanced root biomass production and increased root hair development [78]. The abundance in root hair formation significantly increases root surface area, suggesting that PGPF inoculants could enhance the potential for plant roots to acquire nutrients under nutrient-limited conditions.

4.5 Impact of PGPF on flowering

The application of PGPF may influence the number, size and timing of flower in flowering plants. *Tagetes* (marigolds) grown with companion of *Pe. simplicissimum* flowered earlier and had greater flower size and weight [80]. Steamed or raw soil infested with *T. harzianum* hastened flowering of periwinkle and increased the number of blooms per plant on chrysanthemums [77]. Under greenhouse conditions, *T. harzianum* TriH_JSB27 and *Pe. Chrysogenum* PenC_JSB41 accelerated the flowering time in tomato [34]. Similarly, root colonization by the nematophagous fungus *Pochonia chlamydosporia* hurried flowering in *Arabidopsis thaliana* [81]. Root colonization by *Piriformospora indica* also results in early flowering in *Coleus forskohlii*, bottle gourd and *Nicotiana tabacum* [82]. Flowering time has commercial significance for crops and ornamental plants by shortening crop duration and improving productivity. A short duration crop would have several advantages over a long duration crop, even with equal total yields such as require less water, expose less to stresses and increase the availability of the land for subsequent cropping. This indicates that PGPF improve the plasticity of complex plant traits.

4.6 Impact of PGPF on yield

PGPF show promising ability to promote growth through extensive improvements and betterment of fundamental processes operating in the plants, all of which directly and indirectly contributes to the crop yield increase. Inoculation of banana (cv. Giant Cavendish and Grand Nain) with *F. oxysporum* resulted in 20–36% yield increase in the field [29]. Soil treatment with *T. harzianum* alone or in combination with organic amendment and fungicide significantly improved seed yield in pea [83] and chickpea [58]. Similarly, soil treatment with *T. viride* produced significantly the highest number of fruits per plant, number of seeds per fruit, fruit weight and dry weight of 100 seeds as compared to untreated control [84]. The beneficial association of plants with nonpathogenic binucleate *Rhizoctonia* spp. resulted in increase in yield of carrot, lettuce, cucumber, cotton, radish, wheat, tomato, Chinese mustard and potato [13, 45, 46]. These results demonstrate that PGPF hold great promise in the improvement of agriculture yields.

5. Duration of sustained plant growth promotion effect by PGPF

The duration of biofunctional activities of PGPF in plants is a key factor for their effective application in the field. Naturally, a legitimate question may arise whether PGPF isolates that have shown promising effects on early growth stage of plants, could also affect the middle or late ontogenetic stages and ultimately contribute to yield increases at harvest. As for potato, an increase in leaf, shoot, and tuber weight was observed by a nonpathogenic isolate (No. 521, AG-4) of *Rh. solani* 63–70 days after planting, while it was not expressed in yield at harvest [85]. Conversely, increased growth responses of wheat plants treated with PGPF were observed during seedling (2 weeks after sowing), vegetative (4 weeks), preflowering (6 weeks), flowering (10 weeks) and seed maturation stages (14 weeks) [4]. The isolates of *Phoma* sp. (GS6-1, GS7-4) and non-sporulating fungus (GU23-3), increased plant height, ear-head length and weight, seed number and plant biomass at harvest [79]. Again, isolates of *Phoma* sp. and non-sporulating fungus significantly increased plant length, dry biomass, leaf number and fruit number of cucumber cv. Jibai until 10 weeks post planting in greenhouse trials [62]. These isolates were equally effective in promoting growth and increasing yield of cucumber at 6 and 10 weeks post planting in the field [62]. There are other PGPF, which as well have shown the ability to confer long-term growth benefits to different plants. Rice and pea plants inoculated with *Westerdykella aurantiaca* FNBR-3, *T. longibrachiatum* FNBR-6, *Lasiodiplodia* sp. FNBR-13 and *Rhizopus delemar* FNBR-19 showed a stimulatory increase of growth for 8 weeks in the greenhouse [86]. Similarly, a single inoculation with inoculum of *Penicillium* and *Pochonia* affected the whole life cycle of tomato and Arabidopsis, respectively, accelerating the growth rate, shortening their vegetative period and enhancing seed maturation [34, 81]. As such, majority of PGPF strains are able to induce sustained beneficial effects on plant growth. The basis of sustained effects of PGPF on plants is not fully understood. One possibility is that the fungus continues to colonize the root system and establishes a life-long colonization with crop roots. The ability of PGPF to confer sustained benefit to plant is of great agriculture importance in terms of improving crop yield.

6. Host specificity of the plant growth-promoting cooperation

Although plants harbor a diverse community of fungi, a preferential interaction exists between certain PGPF and a particular host. Once a particular host mutualizes this fungus, it undergoes host-specific adaptations. The outcome of such adaptations is a highly specialized and finely tuned mutualism, leading to improved responsiveness to each other needs. Evidences show that PGPF that induce growth in one plant species do not necessarily have the same effect in other species [5]. Some PGPF exert general growth promotion effects in several plant species, other fungi only do so in specific host plant. A field study showed that most of eight non-sporulating PGPF isolates enhanced the growth of one wheat variety, whereas a few isolates enhanced the growth of the other variety [87]. Moreover, at least four isolates increased yields of both varieties. Thus, the efficacy of the PGPF isolates depended upon the wheat variety in addition to their inherent growth promoting abilities. Similarly, many of the zoysiagrass PGPF isolates promoted growth of bentgrass [4], in contrast to a few isolates enhanced growth in soybean [88]. Similarly, nine isolates belonging to *Phoma* sp. and one non-sporulating fungus caused consistent plant length enhancement in cucumber cv. Shogoin fusiharii compared to nine isolates except the non-sporulating fungus in cv. Aodai Kyuri. Again, plant length enhancement in cv. Jibai was shown by eight *Phoma* sp. and one non-sporulating fungus compared to five *Phoma* sp. isolates in cv. Ociai fushinari [62]. Identically, *Pe. simplicissimum* GP17-2 and *F. equiseti* 19–1 demonstrated sufficient growth-promoting effects on different host plants [4, 9, 60], but did not have effect on *Lotus japonicas* [89]. The outcome of the plant-PGPF interaction, therefore, depends on the plant and PGPF species. It is likely that the specific interaction develops during long-term co-evolution, as it has been observed for compatible and incompatible interactions of pathogens with plants [90]. Moreover, certain components of root exudates may attract and interact microbe specifically and allow it colonize the roots.

7. Mechanisms of plant growth promotion

The course of plant growth promotion by PGPF is complex and often cannot be attributed to a single mechanism. Various mechanisms that are known to modulate plant growth and development can be either direct or indirect. Direct growth promotion occurs when substances produced by the fungi or nutrient available by them facilitate plant growth. On the other hand, the ability of fungi to suppress plant pathogens and to ameliorate stress are considered major indirect mechanisms of plant growth promotion by PGPF. A particular PGPF may affect growth and development of plants using one or more of these mechanisms (**Table 3**).

7.1 Phosphate solubilization

Phosphorus is the second most important and frequently limiting macronutrient for plant growth and productivity. It is an important component of the key macromolecules in living cells and thereby, required for wide array of functions necessary for the survival and growth of living organisms. Despite the abundance of phosphorus in agricultural soils, the majority occurs in an insoluble form. Phosphorus forms complex compounds by reacting with iron, aluminum or calcium depending on the soil types and becomes insoluble and unavailable to plants [102]. To circumvent this problem, phosphate-solubilizing PGPF can play an important role dissolving insoluble P into the soluble form and making it available for plants. PGPF produce

Mechanisms	Specific activities	PGPF strain	References
Phosphate solubilization	Solubilized P by acid phosphatase and alkaline phosphatase	*F. verticillioides* RK01, *Humicola* sp. KNU01	[30]
	Solubilized P from rock phosphate and Ca-P by organic acid	*A. niger* 1B and 6A	[56]
	Solubilize P from tricalcium phosphate (TCP)	*A. niger* BHUAS01, *Pe. citrinum* BHUPC01, *T. harzinum*	[16]
	Solubilized P by organic acid activities	*Pe. oxalicum* NJDL-03, *Aspergillus niger* NJDL-12	[91]
	Phytase-mediated improvement in phytate phosphorus	*A. niger* NCIM	[76]
	Increased HCO_3-extractable P (23% increase)	*Pe. bilaiae* RS7B-SD1	[92]
Mineralization of organic substrate	Increased production of NH4-N and NO_2-N in soil	*T. harzianum* GT2-1, *T. harzianum* GT3-1	[4]
	Increased availability of ammonium nitrogen from barley grain	*Phoma* sp.GS6-1, GS6-2, GS7-3, GS7-4, GS8-6, GS10-1, GS10-2, sterile fungus GU23-3	[87]
	Solubilize minerals such as MnO_2 and metallic zinc	*T. harzianum* Rifai 1295-22	[93]
	Increased availability of ammonium nitrogen from barley grain	*Phoma* sp. GS8-1, GS8-2, GS8-3, Sterile fungus GU21-1	[62]
	Increased concentration of Cu, P, Fe, Zn, Mn and Na in roots Increased concentration of Zn, P and Mn in shoot	*T. harzianum* strain T-203	[47]
	Increased soil organic carbon, N, P and K content	*T. viride*	[73]
	Increased availability of macro and micronutrients and organic carbon	*T. harzianum* strain Th 37	[94]
Phytohormone and enzyme production	Auxin-related compounds (indole-3-acetic acid, IAA)	*T. virensGv. 29-8*	[48]
	Gibberellins (GA1 and GA4) production	*A. fumigatus* HK-5-2	[65]
	GAs production	*Pe. resedanum* LK6	[39]
	GAs production	*Penicillium* sp. Sj-2-2	[38]
	GAs production	*Cladosporium* sp.MH-6	[24]
	GAs production	*Pe. citrinum* IR-3-3	[37]
	GAs and IAA production	*Chaetomium globosum*CAC-1G	[23]
	GAs production	*Exophiala* sp. *LHL08*	[26]
	GAs production	*Phoma herbarum* TK-2-4	[41]
	GAs production	*A. fumigatus* HK-5-2	[65]
	GAs production	*A. fumigatus* LH02	[18]
	IAA production	*T. harzianum* T-22	[64]
	Zeatin (Ze), IAA, 1-aminocyclopropane-1-carboxylic acid (ACC)	*T. harzianum*	[95]

Mechanisms	Specific activities	PGPF strain	References
Suppression of deleterious pathogens	Suppressed damping off caused by *Pythium irregular, Pythium* sp., *Pythium paroecandrum, Pythium aphanidermatum* and *Rhizoctonia solani* AG4	Sterile fungus GSP102, *T. harzianum* GT3-2, *F. equiseti* GF19-1, *Pe. simplicissimum* GP17-2	[4]
	Induced systemic resistance against *Colletotrichum graminicola*	*T. harzianum* T22	[78]
	Suppressed bacterial wilt disease caused by *Ralstonia solanacearum*	*T. harzianum* TriH_JSB27, *Phoma multirostrata* PhoM_JSB17, *T. harzianum* TriH_JSB36, *Pe. chrysogenum* PenC_JSB41	[34]
	Suppressed Fusarium wilt caused by *Fusarium oxysporum* f. sp. *ciceris*	*T. harzianum* T-75	[58]
	Suppressed *Fusarium graminearum*	*Sphaerodes mycoparasitica*	[12]
	Suppressed damping off caused by *Rhizoctonia solani* AG4	*Pe. viridicatum* GP15-1	[35]
	Suppressed nematodes *Pratylenchus goodeyi* and *Helicotylenchus multicinctus*	*F. oxysporum* V5W2, Eny 7.11o and Emb 2.4o	[29]
	Suppressed seedling mortality by *Rhizoctonia solani*	*T. harzianum* isolate T-3	[83]
Amelioration of abiotic stress	Increased tolerance to salt stress	*T. harzianum* T-22	[69]
	Mitigation of oxidative stress due to NaOCl and cold stress	*T. harzianum* Rifai strain 1295-22	[96]
	Enhanced maize seedling copper stress tolerance	*Chaetomium globosum*	[97]
	Minimized Cu-induced electrolytic leakage and lipid peroxidation	*Pe. funiculosum* LHL06	[98]
	Increased tolerance to drought stress	*T. atroviride ID20G*	[99]
Volatile organic compounds (VOCs)	Produced abundant classes of VOCs (sesquiterpenes and diterpenes)	*F. oxysporum* NRRL 26379, NRRL 38335	[28]
	Produced mainly terpenoid-like volatiles including β-caryophyllene	*Talaromyces wortmannii* FS2	[40]
	Produced 2-methyl-propanol and 3-methyl-butanol	*Phoma* sp. GS8-3	[100]
	Produced abundant amount of isobutyl alcohol, isopentyl alcohol, and 3-methylbutanal	*T. viride*	[101]

Table 3.
Different mechanisms of plant growth promotion used by various plant growth promoting fungi (PGPF).

phosphate-solubilizing enzymes such as phytases and phosphatases and organic acids, which liberate P from insoluble phosphates. The most efficient phytase and phosphatase producing PGPF belong to the genera *Aspergillus, Trichoderma,* and *Penicillium* [103]. The order in terms of phytate hydrolysis efficacy was *Aspergillus* > *Penicillium* > *Trichoderma* [104]. *Fusarium verticillioides* RK01 and *Humicola* sp. KNU01 solubilized phosphate by increasing activities of acid phosphatase and alkaline phosphatase, and promoted soybean growth significantly [30]. The phosphate solubilizing fungi possess greater phosphorus solubilization ability than bacteria, especially under acidic soil conditions [105]. The main reason is most fungi are eosinophilic, and have relatively higher growth in acidic environments than bacteria [106]. The acidity has significant influence on organic acid-mediated phosphate solubilizing activities of *Pe. oxalicum* NJDL-03 and *A. niger* NJDL-12 [91]. However, acidification is not always the major mechanism of P solubilization by *T. harzianum* Rifai 1295-22 (T-22), where pH of cultures never fell below 5.0 and no organic acids were detected [93]. Some of the reported PGPF such as *Aspergillus niger* has twin abilities of P mineralization and solubilization [104]. The fungus releases P both from organic and inorganic sources. These suggests that specific PGPF may have specific activity in solubilizing phosphate and making it available for crop growth.

7.2 Substrate degradation (mineralization)

Microorganisms primarily mediate soil nutrient pathways. Microbial mineralization of nutrients from organic matter is crucial for plant growth. Some PGPF promote plant growth, but do not produce plant hormones or solubilize fixed phosphate. Among *Pe. radicum*, *Pe. bilaiae* (strain RS7B-SD1) and *Penicillium* sp. strain KC6-W2, the strongest growth promotion in wheat, medic, and lentil was shown by *Penicillium* sp. KC6-W2, while the only significant P increase (~23% increase) was found in *Pe. bilaiae* RS7B-SD1-treated plants [92]. Similarly, seven *Trichoderma* isolates significantly improved the growth of bean seedlings; despite some of them do not possess any of the assessed growth-promoting traits such as soluble P, indole acetic acid (IAA) and siderophores [107]. These PGPF are believed to encourage plant growth by accelerating mineralization in the soil. Fungi have better substrate assimilation efficiency than any other microbes and are able to break down complex polyaromatic compounds such lignin and humic or phenolic acids [108]. A close relationship was found between the cellulose and starch degradation activity of PGPF for decomposing barley grain and their subsequent growth promotion effect in plants [109]. Application of *T. harzianum* strain Th 37 increased the availability of macro and micronutrients and organic carbonate in the ratoon initiation stage in sugarcane [94]. Colonization of *T. harzianum* in cucumber roots enhanced the availability and uptake of nutrients by the plants [47]. Cucumber plants grew better and produced more marketable fruits due to an increase in soil nutrients caused by PGPF, and accumulated more inorganic minerals like Ca, Mg, and K in aerial shoots [62]. PGPF are also directly involved the degradation of the nitrogenous organic materials through ammonization and nitrification. Formation of NH_4-N and NO_2-N in soil was accelerated during soil amendment with PGPF-infested barley grains [109]. More interestingly, the fungal entomopathogen *Metarhizium robertsii,* when established as a root endophyte, was shown to translocate nitrogen from a dead insect to a common bean plant host, suggesting this PGPF's potential to acquire mineral nutrients from organic matter and promote plant growth [54]. Nutrient release by mineralization could explain why PGPF other than mycorrhizae improve plant growth when added to soil.

7.3 Phytohormone production

Phytohormones are involved in many forms of plant-microbe interactions and also in the beneficial interactions of plants with PGPF. The commonly recognized classes of phytohormones produced by PGPF are the auxins (IAA) and gibberellins (GAs) (**Table 3**). IAA, the most studied auxin, regulates many aspects of plant growth, in particular, root morphology by inhibiting root elongation, increasing lateral root production, and inducing adventitious roots [48]. The *T. harzianum* T-22-mediated root biomass production and root hair development in maize is believed to operate through a classical IAA response pathway [78]. Similarly, a direct correlation exists between increased levels of fungal IAA and lateral root development in *Arabidopsis* seedlings inoculated with *T. virens* [48].

GAs are well known for their role in various developmental processes in plants, including stem elongation. Shoot elongation of waito-c rice seedlings by culture filtrates of *Pe. citrinum* IR-3-3 and *A. clavatus* Y2H0002 was attributed to the activity of physiologically active GAs existing in the culture filtrates [19, 37]. Biochemical analyses of *Penicillium* sp. LWL3 and *Pe. glomerata* LWL2 culture filtrates that enhanced the growth of Dongjiin beyo rice cultivar and in GA-deficient mutant *Waito-C* revealed the presence of IAA and various GAs [110]. Similarly, production of bioactive GAs correlated with enhanced growth of *Waito-C* under salinity by *Penicillium* sp. Sj-2-2 [38]. GA also played key roles during root colonization by *P. indica* in pea roots [111].

Another phytohormone through which PGPF mediate plant growth is cytokinin, especially the Zeatin. Zeatin production has been documented in *Piriformospora indica, T. harzianum* and *Phoma* sp., and the fungi that also produce other phytohormones [95, 112, 113]. *P. indica* produces low amounts of auxins, but high levels of cytokinins. *Trans*-Zeatin cytokinin biosynthesis was found crucial for *P. indica*-mediated growth stimulation in Arabidopsis [112]. This evidence suggests that PGPF often mediate the various growth and developmental processes in plants by influencing the balance of various plant hormones.

7.4 Microbial ACC deaminase

PGPF produces a crucial enzyme ACC (1-aminocyclopropane-1-carboxylic acid) deaminase. ACC deaminase cleaves the ethylene precursor, I-aminocyclopropane-1-carboxylic acid (ACC), into NH_3 (ammonia) and α-ketobutyrate [114]. The ACC deaminase regulates the plant growth by cleaving ACC produced by plants and thereby minimizing the ethylene level in the plant, which when present in high concentrations can lead to a reduced plant growth [115]. ACC deaminase is an inducible enzyme encoded by *acdS* genes of fungi and bacteria [116]. ACC deaminase appears to be central to the functional interactions of some plant-PGPF. *T. asperellum* T203 produced high levels of ACC deaminase and showed an average 3.5-fold induction of the *acds* gene [117]. When ACC deaminase expression is impaired in the fungus *T. asperellum* T203, the plant growth promotion abilities of this organism are also decreased [51]. The root colonizing bacteria *T. harzianum* T22 no longer promote canola root elongation after its *acdS* gene is knocked out [64]. Production of ACC deaminase was reported in some other fungi, which include *Issatchenkia occidentalis* [118], and *Penicillium citrinum* and a stramnopile, *Phytophthora sojae* [119, 120]. The ACC deaminase-producing microbes have competitive advantages in the rhizosphere over nonproducing microorganisms because the enzyme acts as a nitrogen source for them [116]. Moreover, bacteria and fungi that express ACC deaminase can lower the impact of a range of different stresses that affect plant growth and development [114]. These show that ACC deaminase is not only related to plant growth promotion abilities of the microbes, but also play additional roles in the rhizosphere.

7.5 Suppression of deleterious microorganisms by PGPF

The key indirect mechanism of PGPF-mediated plant growth promotion is through their activities as biocontrol agents. PGPF protect and empower plants to resist harmful pathogens and ensure their better growth. The mechanisms by which PGPF suppress growth or activity of invading pathogens in crop plants include antibiosis, competition for nutrient and space, mycoparasitism and induced systemic resistance (ISR) [121]. PGPF of diverse genera promoted growth of field-soil grown cucumber by counteracting damping off pathogen *Pythium* sp. through microbial antagonism [4]. Banana plants inoculated with PGPF *F. oxysporum* significantly suppressed nematode pathogens *Pratylenchus goodeyi* and *Helicotylenchus multicinctus* resulting in up to ~20 to 36% increase in banana yields [29]. The mycoparasite *Sphaerodes retispora* has been reported to improve the plant dry weight and to decrease plant mortality in the presence of *F. oxysporum* [122]. Similarly, under phytotron conditions, seed germination, root biomass, total biomass, root length, and total length of *F. graminearum*-infected wheat were noticeably increased with the treatments of *S. mycoparasitica* and *T. harzianum*, as compared to inoculation with *F. graminearum* alone. Both mycoparasites prevented colonization and reduction in root growth by the pathogen [12]. PGPF compete with the pathogen for colonization niche on roots [79]. Other mechanisms of disease suppression by PGPF are, therefore, likely to include competition with pathogens for infection sites on the root surface. Moreover, there is a long and growing list of PGPF such as *Trichoderma, Penicillium, Fusarium, Phoma*, and non-sporulating fungi, which can protect crop plants against pathogens by eliciting ISR [14, 31, 123, 124]. Although many fungal strains to act as PGPF and elicit ISR, it is not clear how far both mechanisms are connected. These microbes may use some of the same mechanisms to promote plant growth and control plant pathogens.

7.6 Rhizoremediation and stress control

The microbial association of plants has a major influence on plant adaptation to abiotic stresses such as salinity, drought, heavy metal toxicity, extreme temperatures and oxidative stress. Recent studies indicate that fitness benefits conferred by certain PGPF contribute plant adaption to stresses [125]. There are reports of enhanced plant growth because of the association of PGPF with plants, even when plants are under suboptimal conditions [126]. Root colonization by *T. atroviride* ID20G increased fresh and dry weight of maize roots under drought stress [99]. Supplementation of *T. harzianum* to NaCl treated mustard seedlings showed elevation by 13.8, 11.8, and 16.7% in shoot, root length and plant dry weight, respectively as compared to plants treated with NaCl (200 mM) alone [127]. The fungus *Pe. funiculosum* significantly increased the plant biomass, root physiology and nutrients uptake to soybean under copper stress [98]. These fungi have been known to produce plant growth regulators (like GAs and auxins) and extend plant tolerance to abiotic and biotic stresses [23, 125]. Recurrently, *T. harzianum* T22 has little effect upon seedling performance in tomato, however, under stress; treated seeds germinate consistently faster and more uniformly than untreated seeds [69]. A few other fungi like *Microsphaeropsis, Mucor, Phoma, Alternaria, Peyronellaea, Steganosporium*, and *Aspergillus* are known to grow well in polluted medium and protect plants from adverse effects of metal stress [128]. There are numerous similar examples of PGPF ameliorating abiotic stresses and promoting plant growth. Despite significant differences between different stresses, cellular responses to them share common features. Enhanced resistance of PGPF-treated plants to abiotic stresses is explained partly due to higher capacity to scavenge ROS and recycle oxidized ascorbate and glutathione [99, 127]. The increase in proline content is found to be very useful in providing tolerance to these plants under stress [129]. Both enzymatic (peroxidase, catalase, superoxide dismutase, ascorbate peroxidase, monodehydroascorbate reductase, dehydroascorbate reductase, glutathione reductase, glutathione S-sransferase and gaucol peroxidase), and non-enzymatic (ascorbic acid, reduced glutathione, oxidized glutathione) antioxidants are induced by PGPF further enhance the synthesis of these phytoconstituents and defend the plants from further damage [127].

7.7 Production of volatile organic compounds (VOCs)

Microorganisms produce various mixtures of gas-phase, carbon-based compounds called volatile organic compounds (VOCs) as part of their normal metabolism. The comparative analysis of experimental data has shown that volatile metabolites make a much greater contribution to the microbial interactions than non-volatile ones [130]. Recent studies reveal that VOC emission is indeed a common property of a wide variety of soil fungi, including PGPF. Some of these VOCs produced by PGPF exert stimulatory effects on plants. A PGPF, *Talaromyces wortmannii* emits a terpenoid-like volatile, β-caryophyllene, which significantly promoted plant growth and induced resistance in turnip [40]. The identified VOCs emitted by *Phoma* sp. GS8-3 belonged to C4-C8 hydrocarbons, where 2-methyl-propanol and 3-methyl-butanol formed the main components and promoted the growth of tobacco seedlings [100]. These two components were also extracted from PGPR [131]. On the other hand, 3-methyl-butanal has been reported from *T. viride* [101]. The other most abundant VOCs from *T. viride* were isobutyl alcohol, isopentyl alcohol, farnesene and geranylacetone. Arabidopsis cultured in petri plates in a shared atmosphere with *T. viride*, without direct physical contact was taller with more lateral roots, bigger with augmented total biomass (~45%) and earlier flowered with higher chlorophyll concentration (~58%) [101]. Moreover, volatile blends showed better growth promotion than individual compounds [132]. Volatile compounds produced by PGPF are also heavily involved in induce systemic resistance toward pathogens [100].

8. Pattern and process of root colonization by PGPF

Root colonization is considered as an important strategy of PGPF for plant growth promotion. Root colonization is the ability of a fungus to survive and proliferate along growing roots in the presence of the indigenous microflora over a considerable period [35]. The fungus that colonizes plant root effectively is more rhizosphere competent than others [107]. Rhizosphere competence is a necessary condition for a fungus to be an efficient PGPF. Re-isolation frequency of the fungus from the colonized roots is an indirect measure of its root colonizing ability and thereby, its rhizosphere competence. In such studies, *Pe. simplicissimum* GP17-2 and *Pe. viridicatum* GP15-1 were re-isolated from *Arabidopsis* Col-0 roots 3 weeks after planting at high frequencies which were found to be >90% (**Figure 2**). Similarly, the re-isolation frequency of *Pe. janthinellum* GP16-2 from the roots of Col-0 plants was recorded to be, on average, 85% [33]. *Aspergillus* spp. PPA1 was re-isolated from the roots of cucumber plants at a frequency of 95–100% 3 weeks after planting [17], indicating a rapid and efficient root colonization by the PGPF. However, a slow root colonization by PGPF was also reported, as it was the case with *Phoma* sp. GS8-2, which achieved maximum colonization on cucumber roots at 10 weeks [62]. The relative growth rate of the fungi and roots seems to determine the length of time required for maximum root colonization.

Some PGPF selectively colonize host roots and promote growth. Isolates of *Phoma* and sterile fungi showed poor ability to colonize the soybean roots and were unable to enhance the growth of soybean [79]. Similarly, *T. koningi* colonized roots and enhanced growth of *Lotus japonicas,* but *Pe. simplicissimum* and *F. equiseti* did not [89]. It was observed that *T. koningi* induced a transient and decreased level of defense gene expression in *L. japonicas* during its entry into the roots, while a

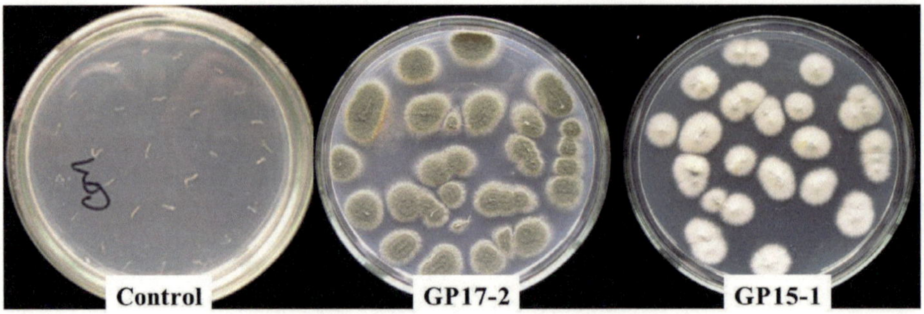

Figure 2.
Re-isolation of Penicillium simplicissimum *GP17-2 and* Penicillium viridicatum *GP15-1 at higher frequencies from colonized roots of* Arabidopsis thaliana *ecotype Col-0 3 weeks after sowing.*

stimulated expression of these genes was induced by *Pe. simplicissimum* and *F. equiseti* [89]. *T. koningi* resembles symbiotic fungi, while *Pe. simplicissimum* and *F. equiseti* act similar to fungal pathogens in activating host defense. This shows that legumes selectively avoid some PGPF and thus allow only specific PGPF to interfere.

There are also PGPF, in particular, the non-sporulating sterile fungi that lack root colonization ability, but they are able to promote growth and yield of plants [62, 133]. This indicates that root colonization is not an indispensable condition for growth promotion by all PGPF. Some chemical factor(s) produced by them might be responsible for growth promotion.

The colonization of the root system of by PGPF is not always homogenous; the density of PGPF varies in different parts of the root system. The colonization of roots by the majority of PGPF appears to be higher in the upper than in the middle and lower root parts of roots, [35, 133]. The lower part was always less colonized by PGPF, especially during first 2 weeks of colonization. This is probably due to the faster growth of the roots than of the hyphae. Moreover, the main zone of root exudation is located behind the apex [134]. However, some PGPF can keep up with root growth and colonize the entire root system [35]. Only fungi with large nutrient reserves can move to the root and along the root over larger distances [135].

Anatomical data show that PGPF may colonize root tissues internally and establish a mutualistic relationship with host. *F. equiseti* GF19-1 produced abundant hyphal growth on the root surface, formed appressoria-like structures and grew in the intercellular space, not inside the cell [31]. *T. harzianum* CECT 2413 exhibited profuse adhesion of hyphae to the tomato roots and colonized the epidermis and cortex. Intercellular hyphal growth and the formation of plant-induced papilla-like hyphal tips were also observed [136]. Hyphae of *T. koningi* penetrated the epidermis and entered the intercellular inner cortex tissues [89]. Sterile red fungus has been also demonstrated to invade the inner root regions that helped plants derive nutrients from the soil and protected roots from pathogens [137].

9. Formulation of PGPF

PGPF, especially *Trichoderma*, have many success stores as plant growth promoting agents and appear to have much potential as a commercial formulation. Different organic and inorganic carrier materials have been studied for effective delivery of bioinoculants. A talc-based formulation was developed for *T. harzianum* to supply concentrated conidial biomass of the fungus with high colony forming units (CFU) and long shelf life [138]. The concentrated formulation provided an extra advantage of smaller packaging for storage and transportation, and low product cost as compared to other carriers such as charcoal, vermiculite, sawdust and cow dung. Seed application of the formulation recorded significant increase in growth promotion in chickpea [138]. Corn and sugarcane bagasse were used as potential carriers for *Trichoderma* sp. SL2 inoculants. The corn formulation of SL2 significantly enhanced rice seedlings root length, wet weight and biomass compared to inoculum mixed with sugarcane bagasse and control [139]. A spray-dried flowable powder formulation was developed for biostimulant *Trichoderma* strains using a CO_2 generating dispersant system, based on polyacrylic acid, citric acid and sodium bicarbonate, polyvinyl alcohol as adhesives and lecithin as wetting agent [140]. Hydrolytic amino acids derived from pig corpses were used in the preparation of *T. harzianum* T-E5-containing bioorganic fertilizer. The resulting bioorganic fertilizer supported higher densities of *T. harzianum* T-E5 and substantially enhanced plant growth when applied as a soil amendment [141]. A composted cattle manure-based *Trichoderma* biofertilizer was developed and tested in the field. Plots fertilized with biofertilizer had the greatest aboveground biomass of any treatment and were significantly more productive than non-amended plots and plots fertilized with any rate of organic fertilizer [142]. Effective formulation of *P. indica* was prepared in talcum powder or vermiculite with 20% moisture. The talcum-based formulations performed significantly better as bioinoculant over vermiculite-based formulations in glasshouse experiments [143]. These show the feasibility of commercial level production and applicability of different PGPF formulations for plant growth promotion in the field.

10. Conclusions

Because of current concerns over the adverse effects of agrochemicals, there is a growing interest in improving our understanding of the role and application of beneficial microbes in agriculture. The plant-associated growth promoting fungi show excellent potential for wider use in sustainable agriculture as they improve plant growth and yield in an ecofriendly and cost-effective manner. However, the PGPF continue to be greatly underutilized, primarily due to some practical problems such as the inconsistency in field performance, which appears to be the greatest challenge in the development of microbial inoculants for plant growth until now and well into the future. If our understanding of complex rhizosphere environment, of the mechanisms of action of PGPF and of the practical aspects of mass production, inoculant formulation and delivery increase, more PGPF products will become available. Knowledge of multiple microbial interaction with different or complementary mode of actions is also of extreme value for development of bio-formulation.

Recent advances in biotechnological tools and reliable transformation system could be useful in engineering of the PGPF to confer improved benefits to the crop. Genetic transformation and overexpression of one or more of the plant growth promoting traits that act synergistically may lead to enhanced performance by the inoculant. Research may be required periodically in order to evaluate the genetic stability and ecological persistence of the genetically modified strain. Efforts should be strengthened to foster linkage between investigators and entrepreneurs in facilitating technology transfer, promotion and acceptance by end users.

Author details

Md. Motaher Hossain[1*] and Farjana Sultana[2]

1 Department of Plant Pathology, Bangabandhu Sheikh Mujibur Rahman Agricultural University, Gazipur, Bangladesh

2 College of Agricultural Sciences, International University of Business Agriculture and Technology, Dhaka, Bangladesh

*Address all correspondence to: hossainmm@bsmrau.edu.bd

References

[1] WPP. World Population Prospect: 2019. New York: Population Division, Department of Economic and Social Affairs, United Nations; 2019. p. 2

[2] Hunter MC, Smith RG, Schipanski ME, Atwood LW, Mortensen DA. Agriculture in 2050: Recalibrating targets for sustainable intensification. BioScience. 2017;**67**(4):386-391

[3] Vacheron J, Desbrosses G, Bouffaud ML, Touraine B, Moënne-Loccoz Y, Muller D, et al. Plant growth-promoting rhizobacteria and root system functioning. Frontiers in Plant Science. 2013;**4**:356

[4] Hyakumachi M. Plant-growth-promoting fungi from turf grass rhizosphere with potential for disease suppression. Soil Microorganisms. 1994;**44**:53-68

[5] Hossain MM, Sultana F, Islam S. Plant growth-promoting fungi (PGPF): Phytostimulation and induced systemic resistance. In: Singh D, Singh H, Prabha R, editors. Plant-Microbe Interactions in Agro-Ecological Perspectives, Volume 2: Microbial Interactions and Agro-Ecological Impacts. Singapore: Springer; 2017. pp. 135-191. DOI: 10.1007/978-981-10-6593-4

[6] Lyn ME, Burnett D, Garcia AR, Gray R. Interaction of water with three granular biopesticide formulations. Journal of Agricultural and Food Chemistry. 2010;**58**:1804-1814

[7] Aly AH, Debbab A, Proksch P. Fungal endophytes: Unique plant inhabitants with great promises. Applied Microbiology and Biotechnology. 2011;**90**:1829-1845

[8] Bent E. Induced systemic resistance mediated by plant growth-promoting rhizobacteria (PGPR) and fungi (PGPF). In: Tuzun S, Bent E, editors. Multigenic and Induced Systemic Resistance in Plants. New York: Springer; 2006. pp. 225-258

[9] Hossain MM, Sultana F, Kubota M, Koyama H, Hyakumachi M. The plant growth-promoting fungus *Penicillium simplicissimum* GP17-2 induces resistance in *Arabidopsis thaliana* by activation of multiple defense signals. Plant & Cell Physiology. 2007;**48**(12):1724-1736

[10] Hyakumachi M, Kubota M. Fungi as plant growth promoter and disease suppressor. In: Arora DK, editor. Mycology Series. Vol. 21. Fungal Biotechnology in Agricultural, Food, and Environmental Applications. New York: Marcel Dekker; 2004. pp. 101-110

[11] Kaewchai S, Soytong K, Hyde KD. Mycofungicides and fungal bio-fertilizers. Fungal Diversity. 2009;**38**: 25-50

[12] Vujanovic V, Goh YK. qPCR quantification of *Sphaerodes myco-parasitica* biotrophic mycoparasite interaction with *Fusarium graminearum*: *in vitro* and *in planta* assays. Archives of Microbiology. 2012;**194**(8):707-717

[13] Muslim A, Horinouchi H, Hyakumachi M. Biological control of *Fusarium* wilt of tomato with hypovirulent binucleate *Rhizoctonia* in greenhouse conditions. Mycoscience. 2003;**44**:77-84

[14] Sultana F, Hossain MM, Kubota M, Hyakumachi M. Elicitation of systemic resistance against the bacterial speck pathogen in *Arabidopsis thaliana* by culture filtrates of plant growth-promoting fungi. Canadian Journal of Plant Pathology. 2008;**30**(2):196-205

[15] Zhou Z, Zhang C, Zhou W, Li W, Chu L, Yan J, et al. Diversity and plant

growth-promoting ability of endophytic fungi from the five flower plant species collected from Yunnan, Southwest China. Journal of Plant Interactions. 2014;**9**(1):585-591

[16] Yadav J, Verma JP, Tiwari KN. Plant growth promoting activities of fungi and their effect on chickpea plant growth. Asian Journal of Biological Sciences. 2011;**4**:291-299

[17] Islam S, Akanda AM, Sultana F, Hossain MM. Chilli rhizosphere fungus *Aspergillus* spp. PPA1 promotes vegetative growth of cucumber (*Cucumis sativus*) plants upon root colonisation. Archives of Phytopathology and Plant Protection. 2014;**47**:1231-1238

[18] Khan AL, Hamayun M, Kim YH, Kang SM, Lee JH, Lee IJ. Gibberellins producing endophytic *Aspergillus fumigatus* sp. LH02 influenced endogenous phytohormonal levels, plant growth and isoflavone biosynthesis in soybean under salt stress. Process Biochemistry. 2011;**46**:440-447

[19] You YH, Kwak TW, Kang SM, Lee MC, Kim JG. *Aspergillus clavatus* Y2H0002 as a new endophytic fungal strain producing gibberellins isolated from *Nymphoides peltata* in fresh water. Mycobiology. 2015;**43**:87-91

[20] Salas-Marina MA, Silva-Flores MA, Cervantes-Badillo MG, Rosales-Saavedra MT, Islas-Osuna MA, Casas-Flores S. The plant growth-promoting fungus *Aspergillus ustus* promotes growth and induces resistance against different lifestyle pathogens in *Arabidopsis thaliana*. Journal of Microbiology and Biotechnology. 2011;**21**(7):686-696

[21] Waqas M, Khan AL, Hamayun M, Shahzad R, Kim YH, Choi KS, et al. Endophytic infection alleviates biotic stress in sunflower through regulation of defence hormones, antioxidants and functional amino acids. European Journal of Plant Pathology. 2015;**141**:803-824

[22] Ignatova LV, Brazhnikova YV, Berzhanova RZ, Mukasheva TD. Plant growth-promoting and antifungal activity of yeasts from dark chestnut soil. Microbiological Research. 2015;**175**:78-83

[23] Khan AL, Shinwari ZK, Kim YH, Waqas M, Hamayun M, Kamran M, et al. Role of endophyte *Chaetomium globosum* LK4 in growth of *Capsicum annum* by production of gibberellins and indole acetic acid. Pakistan Journal of Botany. 2012;**44**:1601-1607

[24] Hamayun M, Khan SA, Khan AL, Rehman G, Kim YH, Iqbal I, et al. Gibberellin production and plant growth promotion from pure cultures of *Cladosporium* sp. MH-6 isolated from cucumber (*Cucumis sativus* L.). Mycologia. 2010;**102**:989-995

[25] Hamayun M, Khan SA, Ahmad N, Tang D-S, Kang S-M, et al. *Cladosporium sphaerospermum* as a new plant growth-promoting endophyte from the roots of *Glycine max* (L.) Merr. World Journal of Microbiology and Biotechnology. 2009;**25**(4):627-632

[26] Khan AL, Hamayun M, Ahmad N, Waqas M, Kang SM, Kim YH, et al. *Exophiala* sp. LHL08 reprograms *Cucumis sativus* to higher growth under abiotic stresses. Physiologia Plantarum. 2011;**143**(4):329-343

[27] Islam S, Akanda AM, Prova A, Sultana F, Hossain MM. Growth promotion effect of *Fusarium* spp. PPF1 from Bermuda grass (*Cynodon dactylon*) rhizosphere on Indian spinach (*Basella alba*) seedlings are linked to root colonization. Archives of Phytopathology and Plant Protection. 2014;**47**:2319-2331

[28] Bitas V, McCartney N, Li N, Demers J, Kim JE, Kim HS, et al.

Fusarium oxysporum volatiles enhance plant growth via affecting auxin transport and signaling. Frontiers in Microbiology. 2015;**6**:1248

[29] Waweru B, Turoop L, Kahangi E, Coyne D, Dubois T. Non-pathogenic *Fusarium oxysporum* endophytes provide field control of nematodes, improving yield of banana (*Musa* sp.). Biological Control. 2014;**74**:82-88

[30] Radhakrishnan R, Khan AL, Kang SM, Lee IJ. A comparative study of phosphate solubilization and the host plant growth promotion ability of *Fusarium verticillioides* RK01 and *Humicola* sp. KNU01 under salt stress. Annales de Microbiologie. 2015;**65**(1):585-593

[31] Kojima H, Hossain MM, Kubota M, Hyakumachi M. Involvement of the salicylic acid signaling pathway in the systemic resistance induced in *Arabidopsis* by plant growth-promoting fungus *Fusarium equiseti* GF19-1. Journal of Oleo Science. 2013;**62**(6):415-426

[32] Maciá-Vicente JG, Jansson HB, Talbot NJ, Lopez-Llorca LV. Real-time PCR quantification and live-cell imaging of endophytic colonization of barley (*Hordeum vulgare*) roots by *Fusarium equiseti* and *Pochonia chlamydosporia*. New Phytologist. 2009;**182**:213-228

[33] Hossain MM, Sultana F, Kubota M, Hyakumachi M. Differential inducible defense mechanisms against bacterial speck pathogen in *Arabidopsis thaliana* by plant-growth-promoting-fungus *Penicillium* sp. GP16-2 and its cell free filtrate. Plant and Soil. 2008;**304**:227-239

[34] Jogaiah S, Abdelrahman M, Tran LSP, Shin-ichi I. Characterization of rhizosphere fungi that mediate resistance in tomato against bacterial wilt disease. Journal of Experimental Botany. 2013;**64**:3829-3842

[35] Hossain MM, Sultana F, Miyazawa M, Hyakumachi M. The plant growth promoting fungi *Penicillium* spp. GP15-1 enhances growth and confers protection against damping-off and anthracnose in the cucumber. Journal of Oleo Science. 2014;**63**(4):391-400

[36] Babu AG, Kim SW, Yadav DJ, Hyum U, Adhikari M, Lee YS. *Penicillium menonorum*: A novel fungus to promote growth and nutrient management in cucumber plants. Mycobiology. 2015;**43**:49-56

[37] Khan SA, Hamayun M, Yoon H, Kim HY, Suh SJ, Hwang SK, et al. Plant growth promotion and *Penicillium citrinum*. BMC Microbiology. 2008;**8**:231. DOI: 10.1186/1471-2180-8-231

[38] You YH, Yoon H, Kang SM, Shin JH, Choo YS, Lee IJ, et al. Fungal diversity and plant growth promotion of endophytic fungi from six halophytes in Suncheon Bay. Journal of Microbiology and Biotechnology. 2012;**22**:1549-1556

[39] Khan AL, Waqas M, Lee IJ. Resilience of *Penicillium resedanum* LK6 and exogenous gibberellin in improving *Capsicum annuum* growth under abiotic stresses. Journal of Plant Research. 2015;**128**(2):259-268

[40] Yamagiwa Y, Toyoda K, Inagaki Y, Ichinose Y, Hyakumachi M, Shiraishi T. *Talaromyces wortmannii* FS2 emits β-caryophyllene, which promotes plant growth and induces resistance. Journal of General Plant Pathology. 2011;**77**:336-341

[41] Hamayun M, Khan SA, Khan AL, Rehman G, Sohn EY, Shah AA, et al. *Phoma herbarum* as a new gibberellin-producing and plant growth-promoting fungus. Journal of Microbiology and Biotechnology. 2009;**19**:1244-1249. DOI: 10.4014/jmb.0901.030

[42] Hamayun M, Khan SA, Khan AL, Tang DS, Hussain J, Ahmad N, et al.

Growth promotion of cucumber by pure cultures of gibberellin-producing *Phoma* sp. GAH7. World Journal of Microbiology and Biotechnology. 2010;**26**:889-894

[43] Siddikee MA, Zereen MI, Li CF, Dai CC. Endophytic fungus *Phomopsis liquidambari* and different doses of N-fertilizer alter microbial community structure and function in rhizosphere of rice. Scientific Reports. 2016;**6**:32270

[44] Cavello IA, Crespo JM, García SS, Zapiola JM, Luna MF, Cavalitto SF. Plant growth promotion activity of keratinolytic fungi growing on a recalcitrant waste known as "Hair Waste". Biotechnology Research International. 2015;**2015**:952921. DOI: 10.1155/2015/952921

[45] Pascual CB, Raymundo AD, Hayakumachi M. Efficacy of hypovirulent binucleate *Rhizoctonia* sp. to control banded leaf and sheath blight in corn. Journal of General Plant Pathology. 2000;**66**:95-102

[46] Jiang J, Tam S, Toda T, Chen L. Controlling *Rhizoctonia* damping-off of Chinese mustard by using endomycorrhizal *Rhizoctonia* spp. isolated from orchid mycorrhizae. Plant Disease. 2015;**100**:85-91

[47] Yedidia I, Srivastva AK, Kapulnik Y, Chet I. Effect of *Trichoderma harzianum* on microelement concentrations and increased growth of cucumber plants. Plant and Soil. 2001;**235**:235-242

[48] Contreras-Cornejo HA, Macías-Rodríguez LI, Cortés-Penagos C, López-Bucio J. *Trichoderma virens*, a plant beneficial fungus, enhances biomass production and promotes lateral root growth through an auxin-dependent mechanism in *Arabidopsis*. Plant Physiology. 2009;**149**:1579-1592

[49] Lee S, Yap M, Behringer G, Hung R, Bennett JW. Volatile organic compounds emitted by *Trichoderma* species mediate plant growth. Fungal Biology and Biotechnology. 2016;**3**:7

[50] Zhang S, Gan Y, Xu B. Application of plant-growth-promoting fungi *Trichoderma longibrachiatum* t6 enhances tolerance of wheat to salt stress through improvement of antioxidative defense system and gene expression. Frontiers in Plant Science. 2016;**7**:1405

[51] Brotman Y, Landau U, Cuadros-Inostroza Á, Takayuki T, Fernie AR, et al. *Trichoderma*-plant root colonization: Escaping early plant defense responses and activation of the antioxidant machinery for saline stress tolerance. PLoS Pathogens. 2013;**9**(3):e1003221

[52] Studholme DJ, Harris B, Le Cocq K, Winsbury R, Perera V, Ryder L, et al. Investigating the beneficial traits of *Trichoderma hamatum* GD12 for sustainable agriculture—Insights from genomics. Frontiers in Plant Science. 2013;**4**:258

[53] Stone BWG, Weingarten EA, Jackson CR. The role of the phyllosphere microbiome in plant health and function. Annual Plant Reviews. 2018;**1**:1-24. DOI: 10.1002/9781119312994.apr0614

[54] Behie SW, Zelisko PM, Bidochka MJ. Endophytic insect-parasitic fungi translocate nitrogen directly from insects to plants. Science. 2012;**336**:1576-1577

[55] Eldredge SD, Geary B, Jensen SL. Seed isolates of *Alternaria* and *Aspergillus* fungi increase germination of *Astragalus utahensis*. Native Plants. 2016;**17**(2):89-94

[56] Chuang CC, Kuo YL, Chao CC, Chao WL. Solubilization of inorganic phosphates and plant growth promotion by *Aspergillus niger*. Biology and Fertility of Soils. 2007;**43**:575-584

[57] Rabeendran N, Moot DJ, Jones EE, et al. Inconsistent growth promotion of cabbage and lettuce from *Trichoderma* isolates. New Zealand Plant Protection. 2000;**53**:143-146

[58] Hossain MM, Hossain N, Sultana F, Islam SMN, Islam S, Bhuiyan MKA. Integrated management of *Fusarium* wilt of chickpea (*Cicer arietinum* L.) caused by *Fusarium oxysporum* f.sp. *ciceris* with microbial antagonist, botanical extract and fungicide. African Journal of Biotechnology. 2013;**12**(29):4699-4706

[59] Kaveh H, Vatandoost S, Aroiee H, Mazhabi M. Would *Trichoderma* affect seed germination and seedling quality of two muskmelon cultivars, Khatooni and Qasri and increase their transplanting success? Journal of Biological and Environmental Sciences. 2011;**5**:169-175

[60] Chandanie WA, Kubota M, Hyakumachi M. Interactions between the arbuscular mycorrhizal fungus *Glomus mosseae* and plant growth-promoting fungi and their significance for enhancing plant growth and suppressing damping-off of cucumber (*Cucumis sativus* L.). Applied Soil Ecology. 2009;**41**:336-341

[61] Saldajeno MGB, Hyakumachi M. The plant growth-promoting fungus *Fusarium equiseti* and the arbuscular mycorrhizal fungus *Glomus mosseae* stimulate plant growth and reduce severity of anthracnose and damping-off diseases in cucumber (*Cucumis sativus*) seedlings. The Annals of Applied Biology. 2011;**159**:28-40

[62] Shivanna MB, Meera MS, Kubota M, Hyakumachi M. Promotion of growth and yield in cucumber by zoysiagrass rhizosphere fungi. Microbes and Environments. 2005;**20**(1):34-40

[63] Chandanie WA, Kubota M, Hyakumachi M. Interaction between arbuscular mycorrhizal fungus *Glomus mosseae* and plant growth promoting fungus *Phoma* sp. on their root colonization and growth promotion of cucumber (*Cucumis sativus* L.). Mycoscience. 2005;**46**:201-204

[64] Sofo A, Tataranni G, Xiloyannis C, Dichio B, Scopa A. Direct effects of *Trichoderma harzianum* strain T-22 on micropropagated shoots of GiSeLa6® (*Prunus cerasus*×*Prunus canescens*) rootstock. Environmental and Experimental Botany. 2012;**76**:33-38

[65] Hamayun M, Khan SA, Khan MA, Khan AL, Kang SM, Kim SK, et al. Gibberellin production by pure cultures of a new strain of *Aspergillus fumigates*. World Journal of Microbiology and Biotechnology. 2009;**25**:1785-1792

[66] Shanmugaiah V, Mathivanan N, Varghese B. Purification, crystal structure and antimicrobial activity of phenazine-1-carboxamide produced by a growth-promoting biocontrol bacterium, *Pseudomonas aeruginosa* MML2212. Journal of Applied Microbiology. 2010;**108**:703-711. DOI: 10.1111/j.1365-2672.2009.04466.x

[67] Nagaraju A, Murali M, Sudisha J, Amruthesh KN, Murthy SM. Beneficial microbes promote plant growth and induce systemic resistance in sunflower against downy mildew disease caused by *Plasmopara halstedii*. Current Botany. 2012;**3**(5):12-18

[68] Minerdi D, Bossi S, Maffei ME, Gullino ML, Garibaldi A. *Fusarium oxysporum* and its bacterial consortium promote lettuce growth and expansin A5 gene expression through microbial volatile organic compound (MVOC) emission. FEMS Microbiology Ecology. 2011;**76**:342-351

[69] Mastouri F, Björkman T, Harman GE. Seed treatment with *Trichoderma harzianum* alleviates biotic, abiotic, and physiological stresses in germinating seeds and seedlings. Phytopathology. 2010;**100**:1213-1221

[70] Mushtaq S, Nasim G, Khokhar I, Mukhtar I. Effects of *Penicillium* extracts on germination vigour in subsequent seedling growth of tomato (*Solanum lycopersicum* L.). Archives of Phytopathology and Plant Protection. 2012;**45**(8):932-937

[71] Horinouchi H, Katsuyama N, Taguchi Y, Hyakumachi M. Control of *Fusarium* crown and root rot of tomato in a soil system by combination of a plant growth-promoting fungus, *Fusarium equiseti*, and biodegradable pots. Crop Protection. 2008;**27**:859-864

[72] Halifu S, Deng X, Song X, Song R. Effects of two *Trichoderma* strains on plant growth, rhizosphere soil nutrients, and fungal community of *Pinus sylvestris* var. *mongolica* annual seedlings. Forests. 2019;**10**(9):758

[73] Yadav RL, Shukla SK, Suman A, Singh PN. *Trichoderma* inoculation and trash management effects on soil microbial biomass, soil respiration, nutrient uptake and yield of ratoon sugarcane under subtropical conditions. Biology and Fertility of Soils. 2009;**45**:461-468

[74] Radhakrishnan R, Kang S, Baek I, Lee I. Characterization of plant growth-promoting traits of *Penicillium* species against the effects of high soil salinity and root disease. Journal of Plant Interactions. 2014;**9**(1):754-762

[75] Horinouchi H, Muslim A, Hyakumachi M. Biocontrol of *Fusarium* wilt of spinach by the plant growth promoting fungus *Fusarium equiseti* GF183. Journal of Plant Pathology. 2010;**92**(1):249-254

[76] Gujar PD, Bhavsar KP, Khire JM. Effect of phytase from *Aspergillus niger* on plant growth and mineral assimilation in wheat (*Triticum aestivum* Linn.) and its potential for use as a soil amendment. Journal of the Science of Food and Agriculture. 2013;**93**:2242-2247

[77] Chang YC, Baker R, Kleifeld O, Chet I. Increased growth of plants in the presence of the biological control agent *Trichoderma harzianum*. Plant Disease. 1986;**70**:145-148

[78] Harman GE, Petzoldt R, Comis A, Chen J. Interactions between *Trichoderma harzianum* strain T22 and maize inbred line Mo17 and effects of this interaction on diseases caused by *Pythium ultimum* and *Colletotrichum graminicola*. Phytopathology. 2004;**94**:147-153

[79] Shivanna MB, Meera MS, Hyakumachi M. Role of root colonization ability of plant growth promoting fungi in the suppression of take-all and common root rot of wheat. Crop Protection. 1996;**15**(6):497-504

[80] Hameed K, Couch HB. Effects of *Penicillium simplicissimum* on growth, chemical composition, and root exudation of axenically grown marigolds. Phytopathology. 1971;**62**:669

[81] Zavala-Gonzalez EA, Rodríguez-Cazorla E, Escudero N, Aranda-Martinez A, Martínez-Laborda A, Ramírez-Lepe M, et al. *Arabidopsis thaliana* root colonization by the nematophagous fungus *Pochonia chlamydosporia* is modulated by jasmonate signaling and leads to accelerated flowering and improved yield. The New Phytologist. 2016;**213**(1):351-364. DOI: 10.1111/nph.14106

[82] Varma A, Bakshi M, Lou B, et al. *Piriformospora indica*: A novel plant growth-promoting Mycorrhizal fungus. Agricultural Research. 2012;**1**:117-131

[83] Akhter W, Bhuiyan MKA, Sultana F, Hossain MM. Integrated effect of microbial antagonist, organic amendment and fungicide in controlling seedling mortality (*Rhizoctonia solani*) and improving yield in pea (*Pisum sativum* L.). Comptes Rendus Biologies. 2015;**338**:21-28

[84] Hammad R, Elbagory M. Using plant growth-promoting fungi (PGPF), as a biofertilizer and biocontrol agents against Tetranychus cucurbitacearum on Nubian watermelon (*Citrullus lanatus* L.). Journal of Advances in Microbiology. 2019;**16**(2):1-15

[85] Sneh B, Ichielevich-Auster M, Barash I, Koltin Y. Increased growth response induced by a nonpathogenic *Rhizoctonia solani*. Canadian Journal of Botany. 1986;**64**:2372-2378

[86] Srivastava PK, Shenoy BD, Gupta M, Vaish A, Mannan S, Singh N, et al. Stimulatory effects of arsenic-tolerant soil fungi on plant growth promotion and soil properties. Microbes and Environments. 2012;**27**(4):477-482

[87] Shivanna MB, Meera MS, Hyakumachi M. Sterile fungi from zoysiagrass rhizosphere as plant growth promoters in spring wheat. Canadian Journal of Microbiology. 1994;**40**:637-644

[88] Shivanna MB, Meera MS, Kageyama K, Hyakumachi M. Influence of zoysiagrass rhizosphere fungal isolates on growth and yield of soybean plants. Mycoscience. 1995;**36**:25-30

[89] Masunaka A, Hyakumachi M, Takenaka S. Plant growth-promoting fungus, *Trichoderma koningi* suppresses isoflavonoid phytoalexin vestitol production for colonization on/in the roots of *Lotus japonicus*. Microbes and Environments. 2011;**26**(2):128-134

[90] Desender S, Andrivon D, Val F. Activation of defence reactions in Solanaceae: Where is the specificity? Cellular Microbiology. 2007;**9**:21-30

[91] Li Z, Bai T, Dai L, Wang F, Tao J, Meng S, et al. A study of organic acid production in contrasts between two phosphate solubilizing fungi: *Penicillium oxalicum* and *Aspergillus niger*. Scientific Reports. 2016;**6**:25313

[92] Wakelin SA, Gupta VVSR, Harvey PR, Ryder MH. The effect of *Penicillium* fungi on plant growth and phosphorus mobilization in neutral to alkaline soils from southern Australia. Canadian Journal of Microbiology. 2007;**53**:106-115

[93] Altomare C, Norvell WA, Björkman T, Harman GE. Solubilization of phosphates and micronutrients by the plant-growth-promoting and biocontrol fungus *Trichoderma harzianum* Rifai 1295-22. Applied and Environmental Microbiology. 1999;**65**:2926-2933

[94] Singh V, Singh PN, Yadav RL, Awasthi SK, Joshi BB, Singh RK, et al. Increasing the efficacy of *Trichoderma harzianum* for nutrient uptake and control of red rot in sugarcane. Journal of Horticulture and Forestry. 2010;**2**(4):66-71

[95] Martïnez-Medina A, Roldán A, Albacete A, Pascual JA. The interaction with arbuscular mycorrhizal fungi or *Trichoderma harzianum* alters the shoot hormonal profile in melon plants. Phytochemistry. 2011;**72**:223-229

[96] Björkman T, Blanchard LM, Harman GE. Growth enhancement of shrunken-2 sweet corn with *Trichoderma harzianum* 1295-22: Effect of environmental stress. Journal of the American Society for Horticultural Science. 1998;**123**:35-40

[97] Alhamed MFA, Shebany YM. Endophytic *Chaetomium globosum* enhances maize seedling copper stress tolerance. Plant Biology. 2012;**14**:859-863

[98] Khan AL, Lee IJ. Endophytic *Penicillium funiculosum* LHL06 secretes gibberellin that reprograms *Glycine max* L. growth during copper stress. BMC Plant Biology. 2013;**13**:86

[99] Guler NS, Pehlivan N, Karaoglu SA, Guzel S, Bozdeveci A. *Trichoderma*

atroviride ID20G inoculation ameliorates drought stress-induced damages by improving antioxidant defence in maize seedlings. Acta Physiologiae Plantarum. 2016;**38**:132. DOI: 10.1007/s11738-016-2153-3

[100] Naznin HA, Kiyohara D, Kimura M, Miyazawa M, Shimizu M, Hyakumachi M. Systemic resistance induced by volatile organic compounds emitted by plant growth-promoting fungi in *Arabidopsis thaliana*. PLoS One. 2014;**9**(1):e86882

[101] Hung R, Lee S, Bennett JW. *Arabidopsis thaliana* as a model system for testing the effect of *Trichoderma* volatile organic compounds. Fungal Ecology. 2013;**6**:19-26

[102] Islam MT, Hossain MM. Plant probiotics in phosphorus nutrition in crops with special reference to rice. In: Maheshwari DK, editor. Bacteria in Agrobiology: Plant Probiotics. Berlin, Heidelberg: Springer; 2012. pp. 325-363

[103] Rajankar PN, Tambekar DH, Wate SR. Study of phosphate solubilization efficiencies of fungi and bacteria isolated from saline belt of Purna river basin. Research Journal of Agriculture and Biological Sciences. 2007;**3**(6):701-703

[104] Gaind S, Nain L. Soil-phosphorus mobilization potential of phytate mineralizing fungi. Journal of Plant Nutrition. 2015;**38**(14):2159-2175. DOI: 10.1080/01904167.2015.1014561

[105] Nahas E. Factors determining rock phosphate solubilization by microorganisms isolated from soil. World Journal of Microbiology and Biotechnology. 1996;**12**:567-572

[106] Rousk J, Brookes PC, Bååth E. Contrasting soil pH effects on fungal and bacterial growth suggests functional redundancy in carbon mineralisation. Applied and Environmental Microbiology. 2009;**75**:1589-1596

[107] Hoyos-Carvajal L, Orduz S, Bissett J. Growth stimulation in bean (*Phaseolus vulgaris* L.) by *Trichoderma*. Biological Control. 2009;**51**:409-416

[108] Ruess L, Ferris H. Decomposition pathways and successional changes. In: Proceeding of the Fourth International Congress of Nematology. Vol. 2. 2004. pp. 547-556

[109] Hyakumachi M. Microorganism Resources: Its Characteristics and Utilization. Tokyo: Soft Science Inc.; 2000. pp. 81-92

[110] Waqas M, Khan AL, Kamran M, Hamayun M, Kang SM, Kim YH, et al. Endophytic fungi produce gibberellins and indole acetic acid and promotes host-plant growth during stress. Molecules. 2012;**17**:10754-10773

[111] Foo E, Ross JJ, Jones WT, Reid JB. Plant hormones in arbuscular mycorrhizal symbioses: An emerging role for gibberellins. Annals of Botany. 2013;**111**:769-779

[112] Vadassery J, Ritter C, Venus Y, Camehl I, Varma A, Shahollari B, et al. The role of auxins and cytokinins in the mutualistic interaction between Arabidopsis and *Piriformospora indica*. Molecular Plant-Microbe Interactions. 2008;**21**:1371-1383

[113] Saxena S. Applied Microbiology. India: Springer Pvt. Ltd; 2015. p. 190

[114] Nascimento FX, Rossi MJ, Soares CRFS, McConkey BJ, Glick BR. New insights into 1-aminocyclopropane-1-carboxylate (ACC) deaminase phylogeny, evolution and ecological significance. PLoS One. 2014;**9**(6):e99168

[115] Glick BR, Cheng Z, Czarny J, Duan J. Promotion of plant growth by ACC deaminase-producing soil bacteria.

European Journal of Plant Pathology. 2007;**119**:329-339

[116] Glick BR. Bacteria with ACC deaminase can promote plant growth and help to feed the world. Microbiological Research. 2014;**169**:30-39

[117] Viterbo A, Landau U, Kim S, Chernin L, Chet I. Characterization of ACC deaminase from the biocontrol and plant growth-promoting agent *Trichoderma asperellum* T203. FEMS Microbiology Letters. 2010;**305**:42-48

[118] Palmer C, Golden K, Danniels L, Ahmad H. ACC deaminase from *Issatchenkia occidentalis*. Journal of Biological Sciences. 2007;7:188-193

[119] Jia YJ, Kakuta Y, Sugawara M, Igarashi T, Oki N, Kisaki M, et al. Synthesis and degradation of 1-aminocyclopropane-1-carboxylic acid by *Penicillium citrinum*. Bioscience, Biotechnology, and Biochemistry. 1999;**63**:542-549

[120] Singh N, Kashyap S. *In-silico* identification and characterization of 1-aminocyclopropane-1-carboxylate deaminase from *Phytophthora sojae*. Journal of Molecular Modeling. 2012;**18**:4101-4111

[121] Whipps JM. Microbial interactions and biocontrol in rhizosphere. Journal of Experimental Botany. 2001;**52**:487-511

[122] Harveson RM, Kimbrough JW, Hopkins DL. Novel use of a Pyrenomycetous mycoparasite for management of *Fusarium* wilt of watermelon. Plant Disease. 2002;**86**(9):1025-1030

[123] Koike N, Hyakumachi M, Kageyama K, Tsuyumu S, Doke N. Induction of systemic resistance in cucumber against several diseases by plant growth-promoting fungi: Lignification and superoxide generation. European Journal of Plant Pathology. 2001;**107**:523-533

[124] Hossain MM, Sultana F. Genetic variation for induced and basal resistance against leaf pathogen *Pseudomonas syringae* pv. *tomato* DC3000 among *Arabidopsis thaliana* accessions. Springerplus. 2015;**4**:296. DOI: 10.1186/s40064-015-1070-z

[125] Khan AL, Hamayun M, Kim YH, Kang SM, Lee IJ. Ameliorative symbiosis of endophyte (*Penicillium funiculosum* LHL06) under salt stress elevated plant growth of glycine max L. Plant Physiology and Biochemistry. 2011;**49**:852-862

[126] Bae H, Sicher RC, Kim MS, Kim SH, Strem MD, Melnick RL, et al. The beneficial endophyte *Trichoderma hamatum* isolate DIS 219b promotes growth and delays the onset of the drought response in *Theobroma cacao*. Journal of Experimental Botany. 2009;**60**:3279-3295

[127] Ahmad P, Hashem A, Abd-Allah EF, Alqarawi AA, John R, Egamberdieva D, et al. Role of *Trichoderma harzianum* in mitigating NaCl stress in Indian mustard (*Brassica juncea* L.) through antioxidative defense system. Frontiers in Plant Science. 2015;**6**:868

[128] Li B, Li Q, Xiong L, Kronzucker HJ, Krämer U, Shi W. *Arabidopsis* plastid AMOS1/EGY1 integrates abscisic acid signaling to regulate global gene expression response to ammonium stress. Plant Physiology. 2012;**160**:2040-2051

[129] Rasool S, Ahmad A, Siddiqi TO, Ahmad P. Changes in growth, lipid peroxidation and some key antioxidant enzymes in chickpea genotypes under salt stress. Acta Physiologiae Plantarum. 2013;**35**:1039-1050

[130] Kanchiswamy CN, Malnoy M, Maffei ME. Chemical diversity of

microbial volatiles and their potential for plant growth and productivity. Frontiers in Plant Science. 2015;**6**:151. DOI: 10.3389/fpls.2015.00151

[131] Farag MA, Ryu CM, Sumner LW, Pare PW. GC-MS SPME profiling of rhizobacterial volatiles reveals prospective inducers of growth promotion and induced systemic resistance in plants. Phytochemistry. 2006;**67**:2262-2268

[132] Naznin HA, Kimura M, Miyazawa M, Hyakumachi M. Analysis of volatile organic compounds emitted by plant growth-promoting fungus *Phoma* sp. GS8-3 for growth promotion effects on tobacco. Microbes and Environments. 2013;**28**(1):42-49

[133] Meera MS, Shivanna MB, Kageyama K, Hyakumachi M. Plant growth promoting fungi from zoysiagrass rhizosphere as potential inducers of systemic resistance in cucumbers. Phytopathology. 1994;**84**:1399-1406

[134] Rovira AD, Newman EI, Bowen HJ, Campbell R. Quantitative assessment of the rhizoplane microflora by direct microscopy. Soil Biology and Biochemistry. 1974;**6**:211-216

[135] Bowen GD. Microbial dynamics in the rhizosphere: Possible strategies in managing rhizosphere populations. In: Keister DL, Cregan PB, editors. The Rhizosphere and Plant Growth. Dordrecht: Kluwer; 1991. pp. 25-32

[136] Chacón MR, Rodríguez-Galán O, Benítez T, Sousa S, et al. Microscopic and transcriptome analyses of early colonization of tomato roots by *Trichoderma harzianum*. International Microbiology. 2007;**10**(1):19-27

[137] Dewan MM, Sivasithamparam K. Effect of colonization by a sterile fungus on viability of seed and growth and anatomy of roots. Mycological Research. 1990;**94**:553-557

[138] Singh PC, Nautiyal CS. A novel method to prepare concentrated conidial biomass formulation of *Trichoderma harzianum* for seed application. Journal of Applied Microbiology. 2012;**113**:1142-1450. DOI: 10.1111/j.1365-2672.2012.05426.x

[139] Doni F, Isahak A, Zain CRCM, Ariffin SM, Mohamad WNW, Yusoff WMW. Formulation of *Trichoderma* sp. SL2 inoculants using different carriers for soil treatment in rice seedling growth. Springerplus. 2014;**3**:532. DOI: 10.1186/2193-1801-3-532

[140] Oancea F, Raut J, Şesan TE, Cornea PC. Dry flowable formulation of biostimulants *Trichoderma* strains. Agriculture and Agricultural Science Procedia. 2016;**10**:494-502

[141] Zhang FG, Meng XH, Feng CL, Ran W, Yu GH, Zhang YJ, et al. Hydrolytic amino acids employed as a novel organic nitrogen source for the preparation of PGPF-containing bio-organic fertilizer for plant growth promotion and characterization of substance transformation during BOF production. PLOS One. 2016;**11**:e0149447

[142] Zhang FG, Huo YQ, Cobb AB, Luo GW, Zhou JQ, Yang GW, et al. *Trichoderma* biofertilizer links to altered soil chemistry, altered microbial communities, and improved grassland biomass. Frontiers in Microbiology. 2018;**9**:848

[143] Sarma MVRK, Kumar V, Saharan K, Srivastava R, Sharma AK, Prakash A, et al. Application of inorganic carrier-based formulations of fluorescent pseudomonads and Piriformospora indica on tomato plants and evaluation of their efficacy. Journal of Applied Microbiology. 2011;**111**:456-466

Section 4

Plant Protection

Chapter 6

Biological Efficacy of *Trichoderma* spp. and *Bacillus* spp. in the Management of Plant Diseases

Francisco Daniel Hernández-Castillo, Francisco Castillo-Reyes, Marco Antonio Tucuch-Pérez and Roberto Arredondo-Valdes

Abstract

This chapter will cover topics about the microbial antagonists *Trichoderma* spp. and *Bacillus* spp. from the perspective of use as potential biological control agents on plant diseases. Results obtained in the laboratory about from their isolation, microbial strain collections for both genera, taxonomic identification, antifungal activity in in vitro tests, obtained evaluation of the antifungal effect of secondary metabolites from microbial antagonists will be shown. Besides, results obtained from bioassays in the greenhouse and field are used as biopesticides in the control of diseases in fruit trees and vegetables and their effects on the promotion of plant growth and increased crop yield.

Keywords: inhibition, disease, plant pathogen, incidence, severity, antagonist microorganisms

1. Introduction

The agricultural production systems are generally based on technological dependence of high-yield varieties and the use of agrochemicals, causing an imbalance in different agroecosystems. The crops had turned to be susceptible to plague organisms to which before they were not and have proliferated because their natural enemies have eliminated the selection pressure has been favored by the monoculture or by the excessive use of plaguicides, what has conducted to the rupture of resistance of host and resistance toward the pesticides.

Those, as mentioned above, do rethink the actual technological management of the crops, researching less aggressive options with low or without environmental impact. This focus allows the searching of microbial alternatives as biological control agents of diseases, as a viable option to reduce their impact, enhancing the yield and quality of the agricultural products. The economic production losses caused by pests and diseases worldwide are estimated to be 36.5% on average, where 14.1% is caused by diseases, 10.2% by insects, and 12.2% by weeds, without considering the 6–12% of agricultural products postharvest losses. Although it estimated that in developing countries, these could reach up to 50% of production losses, considering only the disease, it estimates that annual losses worldwide can reach about 220 billion dollars. Overall, the diseases from plants can destroy crops before and after harvest or yield partial losses and cause loss of quality in the products harvested.

For example, the apple scab caused by *Venturia inaequalis* (Cook) Wint. (Anamorph: *Spilocaea did Fr.*) is the most important disease of this fruit at a worldwide level, which can cause significant economic losses until 100% of the production, affecting the commercial quality of fruits [1, 2]. Generally, its control is based on the use of agrochemicals. In vegetables, wilting of chili pepper and tomato crops is one of the main biological limitations in the production of these crops and can be caused by *Phytophthora capsici*, *Rhizoctonia solani*, and *Fusarium oxysporum* [3]; this disease is reported throughout Mexico, estimating losses of up to 80% due to root rot by invading the vascular system of plants. Likewise, chemical control is the most used method for disease management and is common to reduce the inoculum by disinfecting the soil with metam sodium, 2-thiocyanomethyl benzothiozole (TCMTB), metalaxyl, azoxystrobin, and propanocarp fungicide applications to control *P. capsici* [4]. *R. solani* and *Fusarium* spp. are controlled with tebuconazole, carbendazim, thiabendazole, and methyl thiophanate [5]. The use of this control method significantly increases the production costs and the negative impact it causes on the environment and to human health and induces resistance of the pathogens toward the active ingredients. An alternative is the use of biological control by microorganisms antagonistic to fungi and stramenopiles from the soil, which has little or no effect on the environment and human health.

2. Biopesticides market

The worldwide market of biopesticides was of 1213 million dollars in 2010 and 3222 million dollars in 2017; the annual rate increases to 15.8% since 2012 besides 2017. Within this market, bioinsecticides represented 46% in 2011, and biofungicides were of 600.5 million dollars, reaching 1447 million in 2017. The annual rate from 2012 to 2017 grows up at 16.1%. Given that there currently exists a market demand for free products of pesticide waste, huge agrochemical companies are in the market of bioproducts, acquiring biocontrol companies and developing new biotechnological products.

3. Isolation and identification of *Bacillus* spp. and *Trichoderma* spp.

Trichoderma and *Bacillus* are essential genera of antagonistic microorganisms for control of a large number of phytopathogens. *Trichoderma* is a cosmopolitan soil fungus, which is frequently on soil from the plant root system. This fungus is attractive for organic management of diseases because present different action modes against phytopathogens as competition for nutrients, mycoparasitism, and antibiosis by hydrolytic enzymes and metabolites also produce substances that promote plant growth [6, 7]. On the other hand, *Bacillus* spp. is a large and heterogeneous group of Gram-positive, rod-shaped, aerobic and facultative anaerobic, and endospore-forming bacteria; same as *Trichoderma*, *Bacillus* is an alternative of biological control of plant diseases due to its capability to inhibit phytopathogens and growth promotion in plants [8, 9].

Due to the abovementioned and because there is a large number of species from both microorganisms, their isolation and identification for their possible commercial use are necessary; some of the species of *Trichoderma* are *T. virens*, *T. harzianum*, and *T. viride* and of *Bacillus* spp. are reported as antagonists *B. amyloliquefaciens*, *B. licheniformis*, *B. subtilis*, and *B. pumilus* [10, 11]. Thus a correct identification of the species which needs work is necessary.

3.1 Isolation and identification of *Trichoderma* spp.

3.1.1 Isolation and morphological identification of Trichoderma spp.

The first step for correct identification of antagonist microorganisms depends on isolation. In the case of *Trichoderma* spp., they are present in a great variety of agricultural and natural soils. The soil sampling for its isolation is relatively simple; using a shovel at 10–20 cm depth, 500 g of soil is taken and deposited in plastic bags; after, the samples will be moved to a laboratory and placed on storage at 4°C until used. Purification of *Trichoderma* spp. it is essential on investigations and present many ways or techniques; nevertheless, the monosporic culture is suggested by Trichoderma on culture media as potato dextrose agar (PDA) or *Trichoderma* Specific Medium (TSM), and incubated at 28 ± 2°C for 96 h [11–13]. Once the monosporic culture is obtained, the identification of *Trichoderma* species can be realized using taxonomic keys through its morphological features or with molecular biology, extracting DNA and utilizing general or specific primers. In case of the use of taxonomic keys, structures as width and length of phialide, length and width of conidia, and presence of chlamydospores will be observed [7, 14].

3.1.2 Molecular identification

This kind of identification has gained acceptance because it presents more precision and reliability among several strains of *Trichoderma.* The phylogeny of this genus has been based in the sequence analysis of the internal transcribed spacers of ribosomal DNA using the universal primers ITS1 and ITS4 with a subsequent sequencing and analysis through databases [15] but also can be identified through specific primers which are a powerful tool that allows to identify a specific species of *Trichoderma* [16] (**Table 1**). In Mexico, diverse species of *Trichoderma* have been isolated and identified [6]; they identified *T. atroviride*, *T. asperellum*, *T. citrinoviride*, *T. ghanense*, *T. harzianum*, *T. inhamatum*, *T. longibrachiatum*, and *T. yunnanense* (**Figure 1**) from samples taken from several agricultural regions. In a similar research, Osorio et al. [23] identified the species as *T. asperellum*, *T. rossicum*, and *T. hamatum* from different localities of Mexican Northeast region.

Species specificity	Primer sequences (5′–3′)	Product size (bp)	References
T. harzianum	HAR-1.6F: GTACCTCGCGAATGCATCTA HAR-1.6R: GGCTATGACCATGATTACGC	1600	[17]
T. asperellum	T2AF: CTCTGCCGTTGACTGTGAACG T2AR: CGATAGTGGGGTTGCCGTCAA	507	[18]
T. virens	TvCTT56f: CTTGATGACAAGCCAAAAGG TvCTT56r: GAAGAGAGGACATAGGGTCTGG	289	[19]
T. atroviride	Q01_4F: GCACACCAACTGCTGGAGCTT Q01_4R: CACGCTGACAATGACCGACAC	1017	[20]
T. aggressivum	Th-F: CGGTGACATCTGAAAAGTCGTG Th-R: TGTCACCCGTTCGGATCATCCG	444	[21]
T. pleuroti	FPforw1: CACATTCAATTGTGCCCGACGA PSrev1: GCGACACAGAGCACGTTGAATC	218	[22]

Table 1.
Examples of species-specific primers for Trichoderma *spp.*

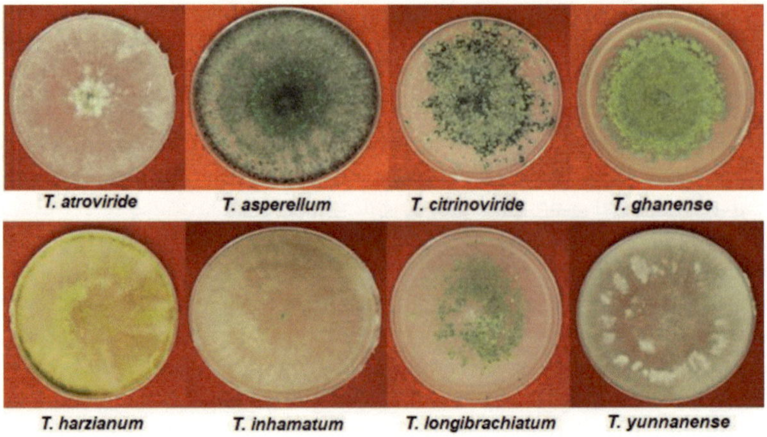

Figure 1.
Morphologic characteristics of different Trichoderma *spp. isolated from samples of different agricultural systems of Mexico.*

3.2 Isolation and identification of *Bacillus* spp.

Bacillus spp. is a genus present in the soil of a considerable amount of crops and naturally is on the rhizosphere; due to this, the traditional tools for determining the soil bacterial community and diversity are used [24]. The first step is to make the collecting of the rhizosphere soil, take 10 g of soil with a sterile spoon, and store the sample at 4°C. Heat or pasteurization treatments are the most commonly used techniques to select spores due to this, the sample is diluted in 90 mL of sterile normal saline and heated at 80°C for 10 min to eliminate vegetative cells; once heated, the sample is serially diluted (10^{-1}–10^{-4}) and placed on 1 mL nutrient agar (NA) medium with cycloheximide (100 mg mL^{-1}) to prevent fungal growth or carboxymethylcellulose (CMC) agar, and it is incubated at 37°C for 24 h [25, 26].

However, the treatment with heat can be different depending on the species because endospores of some strains are more resistant to heat than others [24]. Due to this, the drying treatment is considered more gentle; this method consists of placing the samples on a dryer at 70°C for 1 h [25]. The considerable variety of physiology of *Bacillus* spp. requires elaborate biochemical and morphological tests for species identification; as colony growth in artificial media, form cell unit, presence, number and orientation of flagella, Gram stain, spore form-position and specific environmental conditions of growth and finally the specific use of carbon sources gave its metabolic diversity [27].

3.2.1 Molecular identification of Bacillus spp.

Several molecular approaches are currently utilized for the identification of microorganisms; in this sense, the use of polymerase chain reaction (PCR) in combination with 16S rRNA is a tool frequently used for identification of *Bacillus* spp. from various environments including soil. Using the 16S rRNA sequence, five groups within biological control of root pathogens by plant growth-promoting *Bacillus* spp., the genus *Bacillus* spp., where group 1 comprises species *B. amyloliquefaciens*, *B. subtilis*, *B. pumilus*, and *B. licheniformis*, have been identified [24, 25].

Bacillus spp. can identify through specific primers (**Table 2**). Such as *Trichoderma* spp., the Mexican agricultural systems are an excellent source to obtain *Bacillus* spp., such as mentioned by Guillén-Cruz et al. [9] and Hernandez-Castillo

Species specificity	Primer sequences (5′–3′)	Product size (bp)	References
B. subtilis	p-gyrA-f: CAGTCAGGAAATGCGTACGTCCTT p-gyrA-r: CAAGGTAATGCTCCAGGCATTGCT	741	[28]
B. amyloliquefaciens	trpE(G) F: TTTGAATCCGAGCCCTTATG trpE(G) R: ACATACATTTCGGGGGATGA	78	[29]
B. pumilus	GC-U968(F): GCAACGCGAAGAACCTTAC L1401(R): GCGTGTGTACAAGACCC	490	[30]
B. licheniformis	BL8AF: TCACAACCCGTTGACGACAA BL8AR: CGTGTCCGAGTGTGCGTTATAT	247	[31]

Table 2.
Examples of species-specific primers for Bacillus *spp.*

et al. [32] whom identified several *Bacillus* spp. species as *B. amyloliquefaciens*, *B. pumilus*, *B. licheniformis*, and *B. subtilis* from samples coming from several regions of the center and northern of Mexico.

4. Antifungal activity in vitro of *Trichoderma* spp. and *Bacillus* spp.

The antifungal activity of *Trichoderma* species has been evaluated in in vitro studies against soilborne and foliar fungi, and there have been acceptable results. The antifungal activity can be determined such a direct manner as indirect manner. In the case of a direct manner, the most used technique is the dual culture where the inhibition percentage, Bell scale, and the days to contact are evaluated to determine the antagonistic activity of *Trichoderma* species. Dual culture consists of Petri dishes with PDA where a disk (5 mm in diameter) with mycelium of the plant pathogen is placed and, on the other side of the Petri dish equidistantly, a disk of mycelium of the same diameter of *Trichoderma* strains under study is placed. The plates inoculated are incubated at 27 ± 1°C until the growth of control treatment (with only plant pathogen disk) covered the Petri dish. The effect of *Trichoderma* strains on plant pathogens is determined by the percentage of mycelial growth inhibition. The days of contact between plant pathogen antagonistic and antagonistic ability of *Trichoderma* isolates according to the methodology proposed by Bell et al. [33] are also determined. Bell et al. [33] classified the antagonism produced by *Trichoderma* as follows: Class I, *Trichoderma* overgrows completely to pathogen and covers the whole surface of the medium; Class II, *Trichoderma* overgrows two-thirds of the surface of the medium; Class III, *Trichoderma* and pathogen colonized each half of the surface, and nobody seems to dominate the other; Class IV, the pathogen colonizes the two-third parts of the media surface and resists invasion by *Trichoderma*; and Class V, the plant pathogen overgrows completely to *Trichoderma* and covers an area total culture media [6]. In case of the desire to determine the antifungal activity of an indirect manner, the volatile compounds are an option; this method is realized as follows. In the center of a Petri dish having only PDA medium, a disk of 5 mm in diameter with active mycelia of the plant pathogen is placed, and the top of the dish is replaced with another Petri dish in which a disk with mycelia of *Trichoderma* strain is placed; in this case, the lid is pierced with a punch (10 mm in diameter), and the Petri dishes are joined and sealed with parafilm paper and incubated at 26 ± 1°C until each pathogen control covered the Petri dish. The effect of volatile compounds is measured considering the diameter of pathogen colonies and expressed as percentage of mycelial growth inhibition [6].

Several research were carried out to determine the antifungal activity of *Trichoderma* spp. strains, due to its potential as biocontroller of plant pathogens, as reported by Hernandez et al. [6] who evaluated several strains of *Trichoderma* spp. against *Sclerotinia sclerotiorum* and *Sclerotium cepivorum* through dual culture and observed rate of inhibition of 45–63.8% and 50.9–81.5 for *S. sclerotiorum* with *T. ghanense* and *T. longibrachiatum*, respectively, and 81.5 and 81.2% of *S. cepivorum* with *T. inhamatum* and *T. asperellum*, respectively. For the Bell scale and contact days for both phytopathogenic fungi, the mean was of 2 days and scale of I, II, and III with all the *Trichoderma* spp. strains. Some research about the inhibition of secondary metabolites, precisely the volatile compounds, present inhibition of *S. sclerotiorum* and *S. cepivorum* against *T. longibrachiatum* with 28.1 and 73.8%, respectively, followed by *T. harzianum* with 12.5 and 62.5%.

Some studies by Osorio et al. [23] mentioned a Trichoderma spp. as controller. They reported the overgrowth of *Trichoderma* spp. strains over *Phytophthora capsici*; in total 13 *Trichoderma* spp. strains showed level 1 according to the Bell scale. This effect can be attributed to enzyme production (β-1, 3-glucanase, chitinase, protease, and cellulose) by these *Trichoderma* spp. strains. As too volatile compounds produced by the *Trichoderma* spp. strains, they reported inhibition of *P. capsici* mycelial growth ranged between 4.3 and 48.8, the major effect observed with the *T. asperellum* and the least with *Trichoderma* sp. strain. Some studies mentioned that *Trichoderma* spp. produces volatile compounds, carbon dioxide, oxygen, ethylene, and 6-pentyl-α-pyrone (6PP) which cause adverse effect in the development of phytopathogenic fungus. Furthermore, *Trichoderma* spp. can inhibit the growth of foliar fungus as *Colletotrichum* spp., such as mentioned by Tucuch et al. [34] who determined the antifungal activity *Trichoderma* spp. strains and reported antagonism level 1 in Bell scale by *T. asperellum*, while for volatile compounds, the species *T. lignorum* affected the fungus inhibiting its developments in 24.02% (**Figure 2**).

Usually, *Trichoderma* spp. inhibit several phytopathogenic fungi due to their capacity to produce enzymes, volatile compounds and compete for nutrients against phytopathogens. Studies realized by Samaniego-Fernández et al. [35] and Kumar et al. [7] showed the antagonistic capacity of *Trichoderma* spp.; in the first case, the species *T. harzianum* and *T. viride* are controlled to *S. rolfsii* y *Fusarium* spp.; in the second study, several *Trichoderma* spp. showed mycelial growth inhibition of *S. rolfsii* more than 50%.

Likewise, then *Trichoderma* spp. strains, the antifungal activity of *Bacillus* spp., can be tested by dual culture; nevertheless, the method is different than with *Trichoderma* spp. PDA disk (5 mm) with active mycelium of the phytopathogen is placed in the center of a Petri dish with PDA; on the same plate, at a distance of 1.5 cm in the four cardinal points, a loopful of antagonistic bacterial isolates is placed. Plates inoculated with the pathogen culture serve as controls. In order to quantify the antagonistic potential of bacterial strains, the size of growth inhibition zones measured after 6 days of incubation at 25–28°C and the percent of radial growth inhibition (PICR) are calculated [36]. In this sense, study also showed the capacity of *Bacillus* spp. to inhibit the growth of phytopathogenic fungi. Thereby

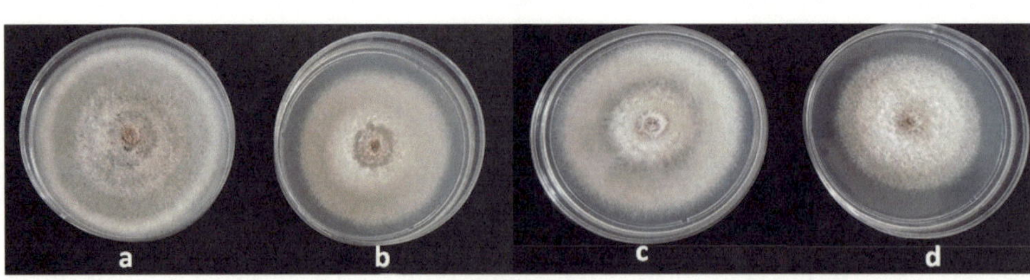

Figure 2.
Inhibition of Colletotrichum *spp. by volatile compounds produced by different* Trichoderma *spp. strains (a) control, (b)* T. asperellum, *(c)* T. yunnanense, *and (d)* T. lignorum.

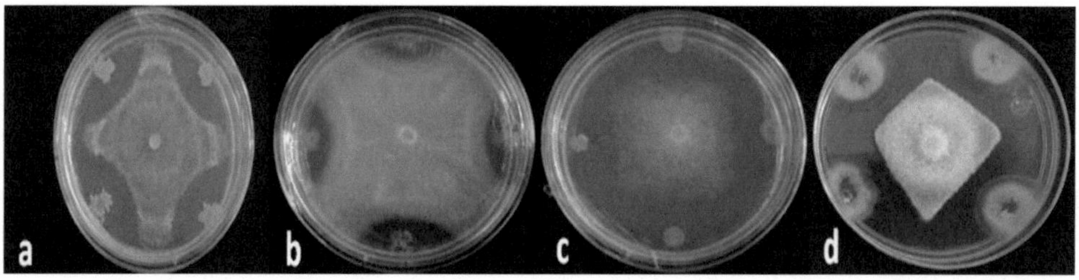

Figure 3.
Antagonistic effect of Bacillus *spp. strains against different phytopathogenic fungus (a)* Rhizoctonia solani*, (b)* Fusarium oxysporum*, (c)* Phytophthora capsici*, and (d)* Colletotrichum *spp.*

Jimenez et al. [36] reported the inhibition of *Venturia inaequalis* mycelia by *B. subtilis* and *B. licheniformis* ranged 33.4–41.3%, respectively, and Tucuch et al. [34] observed 50% of inhibition from *B. subtilis* against *Colletotrichum* spp. (**Figure 3**).

5. Obtaining secondary metabolites of *Trichoderma* spp. and *Bacillus* spp. and their antifungal activity in vitro (PDA methods and microplate dilution)

Generally, the production of the secondary metabolites from biological agents such as *Trichoderma* spp. and *Bacillus* spp. is carried out using liquid media by a fermentation process in a reactor, which can be of different types from a simple bottle to until an automated reactor, where the temperature and shaking rate are the key variables for the emission of secondary metabolites. The liquid medium is integrated by several components such as carbon sources and mineral salts, where the biological microorganism is inoculated [37]. The secondary metabolites obtained from the fermentation of *Bacillus* spp. and *Trichoderma* spp. are filtered with nitrocellulose membrane of 0.22 um; after the recovery of these metabolites, it is necessary to perform a screening to determine their ability to inhibit phytopathogenic microorganisms. There are many methods to determine the antifungal activity from secondary metabolites of antagonistic microorganisms, the most common is the poisoned medium, adding the substance to evaluate in the culture medium before solidification, which consists in adding 200 μl of the secondary metabolites on PDA medium in the center of Petri dish and 5 mm mycelium disk of the phytopathogen, then Petri dishes are incubated at 28 ± 1°C until the control treatment covers the petri dish, after that the percentage inhibition is determined [23].

However, our workgroup has standardized the method in microdilution on plate, which consists in an adaptation of the technique proposed by Masoko et al. [38]; polystyrene microplates of 96 wells are used; in all wells, 100 μl of liquid medium is placed; column 1 is the negative control, column 2 consists of the positive control, and column 3 is a control which consists of the fermentation medium. Starting in column 4, 100 μl of the secondary metabolites from strains is mixed in a pipette with 100 μl of the liquid medium, and then 100 μl of the mixture is transferred to the next column, discarding the last 100 μl from column 12, to get serial microdilutions to 50.00% of the secondary metabolites. Once the microdilutions are carried out, the growth developer 2,3,5-triphenyltetrazolium chloride is added in the whole plate; the concentration of the growth developer is the lowest reported in the literature, as an excess of this indicator can interfere with the growth of the pathogen or react with reagents from the medium; this indicator measures the respiratory activity associated with electron transport chains, and when reduced, it precipitates forming a complex, intense red color; its use is due to its high sensitivity to detect inhibition of microorganisms with deficient amounts of antimicrobial products; besides that, the red coloration is a visual indicator of the antimicrobial activity of the treatment. Finally, starting in column 2, 10 μl of a spore solution of the fungus at a concentration of 1×10^8 in all wells is added, keeping all the wells a volume of 150 μl in total; each microplate is considered a replicate. The microplates are incubated in agreement with the necessary conditions of the fungus on absorbance realized at 490 nm in a spectrophotometer. The secondary metabolites from the different strains placed in the rows A to F. To calculate the growth and inhibition percentage, the following formulas used: % Growth = (A – B/C)(100); where A is treatment absorbance, B is negative control absorbance, C is positive control absorbance, and % Inhibition = 100 – % Growth.

In general, the selection principle of strains is the determination of their antagonistic capacity; the method of microdilution in plate is to some extent interesting since it allows determined quickly and efficiently in time and costs its capacity of antifungal inhibition. Several studies demonstrate the effectiveness of secondary metabolites in the control of phytopathogenic fungi with PDA medium; Osorio et al. [23] mentioned that the inhibiting effect by *T. asperellum* and *T. hamatum* against *P. capsici* ranged to 15–20%; this inhibition attributed to the concentration of metabolites like glycotoxins, viridine, trichodermin, furanone, and 6-pentyl-α-pyrone (**Figure 4**).

Likewise, the PDA method or the microdilution in plate method can be an excellent technique to evaluate substance with antifungal activity as the secondary metabolites; in this sense, Jimenez et al. [36] observed that secondary metabolites obtained from *T. yunnanense* and T. *harzianum* at a concentration of 50 and 25% showed an inhibiting total effect of 100% of mycelial growth of *V. inaequalis*, while the metabolites obtained from *T. asperellum* and *T. lignorum* at a concentration of 50% showed an inhibiting effect from 90 to 84%, respectively (**Figure 5**). On the other hand, Jimenez et al. [36] reported that secondary metabolites obtained from *B. licheniformis* at a concentration of 50 and 25% showed an inhibiting effect in 100% against *V. inaequalis*, while the metabolites obtained from *B. subtilis* at a concentration of 50% showed an inhibiting effect near to 78% of the development of this pathogen (**Figure 5**).

In another study, Tucuch-Pérez et al. [39] reported six *Bacillus* spp. strains with antifungal activity against *F. oxysporum*; in this case, the species *B. licheniformis* and *B. subtilis* showed the highest inhibition percentages ranged from 80 to 100%, being the lowest inhibition percentage registered of 50% (**Figure 6**).

The results generated by the microplate dilution method are consistent with the results obtained from indirect test by confrontation or dual test, such as shown in the following results against fungi isolated from crops as melon, pepper, and others, which are produced in different zones from Mexico. In melon crop, the root and stem rot disease is a big problem; in this context, Espinoza-Ahumada et al. [40] studied the in vitro antagonist effects of *Trichoderma* spp. and found that *T. asperellum* have excellent biological activity against *Fusarium* strains, isolated from melon,

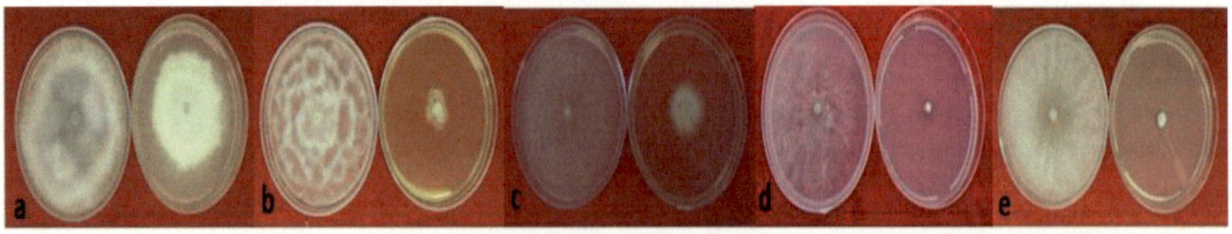

Figure 4.
Antifungal activity of secondary metabolites from different Trichoderma *spp. strains against phytopathogenic fungi; (a)* T. asperellum *vs.* Fusarium oxysporum, *(b)* T. yunannense *vs.* Phytophthora capsici, *(c)* T. longibrachiatum *vs.* Rhizoctonia solani, *(d)* T. asperellum *vs.* Sclerotinia sclerotiorum *and (e)* T. asperellum *vs.* Sclerotium cepivorum.

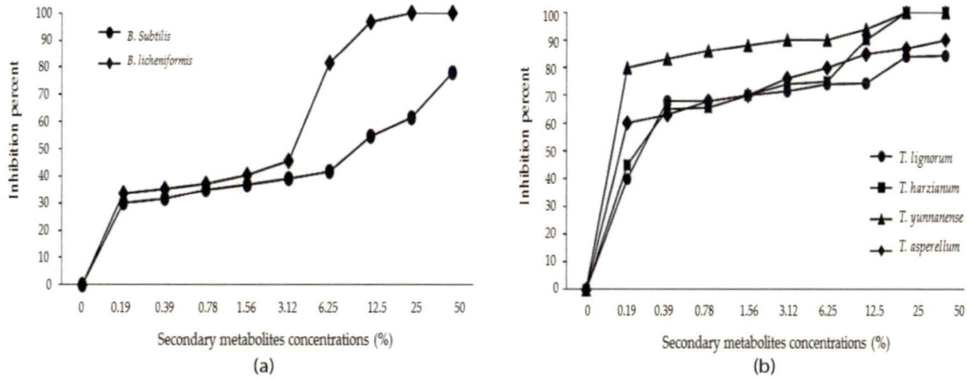

Figure 5.
Percentage of inhibition of secondary metabolites obtained from Bacillus *spp. (a) and* Trichoderma *spp. (b) against* Venturia inaequalis.

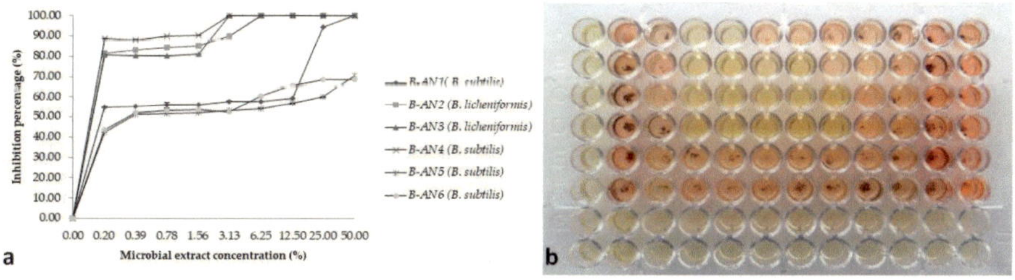

Figure 6.
(a) Percentage inhibition of microbial extracts from Bacillus *spp. metabolite dilutions against* F. oxysporum. *(b) Microplate with treatments to several concentrations, and the pathogen elapsed 48 h after incubation. Row A= B-AN1, B = B-AN2, C = B-AN3, D = B-AN4, E = B-AN5, F = B-AN6; column 1 = negative witness, column 2 = positive witness, 3 = growth medium of* Bacillus *spp., 5 = 50%, 5 = 25%, 6 = 12.50%, 7 = 6.25%, 8 = 3.13%, 9 = 1.56%, 10 = 0.78%, 11 = 0.39%, and 12 = 0.20%.*

shown in **Table 3**. In general, these authors report that the inhibition of *Fusarium* spp. is higher when *Trichoderma* spp. are used (62.4–54.8%), in contrast when *Bacillus* spp. (44.5–36.9%) is used (**Figure 7**).

In a work carried out by Francisco et al. [41], where the behavior of *Bacillus* spp. against *Fusarium* species was studied, it showed low inhibition values. However, they report that the species *B. pumilus* and *B. liquefaciens* can be used effectively against many *Fusarium* species. On the other hand, higher effectiveness of *Bacillus*

Antagonistic agent	***Fusarium* strain**					
	FRR-1	**FRG-2**	**FAF-3**	**FRE-4**	**FCA-5**	**FHA-6**
B. liquefaciens	$46.9^{A,ab}$	$38.4^{B,abc}$	$31.4^{BC,bc}$	$34.8^{CD,bc}$	$28.2^{CD,c}$	$41.6^{DE,ab}$
B. amyloliquefaciens	$51.4^{AB,ab}$	$38.9^{B,c}$	$37.8^{B,c}$	$42.9^{BC,bc}$	$41.2^{BC,bc}$	$55.2^{C,a}$
B. subtilis	$45.7^{B,a}$	$38.6^{B,a}$	$37.2^{B,a}$	$42.4^{B,ca}$	$34.6^{C,a}$	$46.4^{D,a}$
T. asperellum	$57.2^{A,b}$	$61.5^{A,ab}$	$64.3^{A,ab}$	$62.3^{A,ab}$	$54.9^{A,b}$	$74.1^{A,a}$
T. harzianum	$54.9^{AB,b}$	$55.4^{A,b}$	$56.1^{A,ab}$	$60.5^{A,ab}$	$57.9^{A,b}$	$64.7^{B,a}$
T. viride	$49.8^{A,ab}$	$58.3^{A,a}$	$54.9^{A,a}$	$51.1^{A,ab}$	$50.3^{BC,a}$	$64.7^{B,a}$

High letters indicate comparison between columns; low letters indicate comparison between rows. Percentages of inhibition with different letters are significantly different ($p \leq 0.5$).

Table 3.
Percentage of antagonism of different biological agents against Fusarium *spp. strains.*

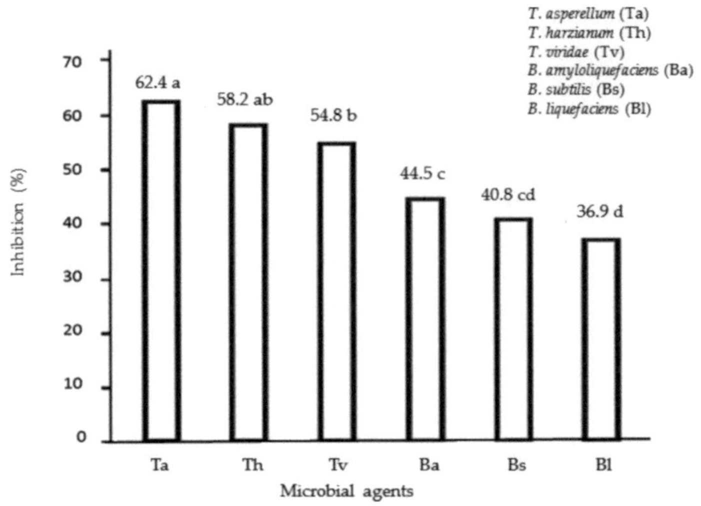

Figure 7.
Microbial agents antagonist to Fusarium oxysporum *(FAF-3, FRE-4, FHA-6) and* Fusarium solani *(FRR-1, FRG-2, FCA-5).*

spp. was observed when applied in the early stage of growth [42], showing that *B. cereus* was most effective against Fusarium dry rot when applied as young cultures (24 h), however *B. thuringiensis* strains was most effective when applied as older cultures (48–72 h). Nevertheless, different studies revealed that *B. pumilus* produced different antifungal compounds as "iturin" which inhibits the growth of *Aspergillus* sp. and their production of aflatoxins [30]. Osorio et al. [23] found an inhibition ranged between 4.3 and 48.8% of *P. capsici* mycelial growth induced by the volatile compounds produced by *Trichoderma* spp. strains. The Tukey test indicated that 21 *Trichoderma* spp. strains showed the highest percentage inhibition. *T. asperellum* (T25) strain present the best result for activity inhibition, strain (T9) being the one with the least inhibition activity. It observed that the 31 *Trichoderma* spp. strains were able to produce volatile compounds with inhibitory properties against *P. capsici*.

6. Antifungal activity bioassay under greenhouse conditions

From different research projects under greenhouse conditions, we have found satisfactory results both in the disease suppression and in the promotion of growth and quality in crops. Hernández-Castillo et al. [24], made an experiment under greenhouse conditions using silty clay soil from an experimental batch previously plot with chile crop and were symptoms of wilting incidence were express. The experiment included three bacterial strains of the genus *Bacillus* spp. (B1, B3, and B13), a chemical control (thiabendazole, T), and control (TA) without fungicide. Before the application of the suspension, an initial colony-forming units (CFU) count of the pathogen involved by the dilution method is performed. The application of spores of the bacterial strains is performed at the time of the transplant. The seedlings are immersed in a spore suspension at a concentration of 10^8 (CFU/mL for 15 min). Subsequently, at 20 and 40 days after transplantation, the same spore suspension was applied to the stem base. In the final evaluation of each treatment (10 adult plants), plant height, fresh fruit weight per cut, root length, dry root weight, and incidence and severity of wilt are measured. The determination of the severity of the disease was with the scale reported by Copes and Stevenson [43]. As a result of this work, a very low wilt incidence found for plants is inoculated with biological strains (B13 and B3) with an incidence of less than 10%, while values of 60 and 40% for TA and T, respectively, were observed (**Figure 8A**). Likewise, the wilting showed a reduction in severity in those treatments where three bacterial strains were applied (**Figure 8B**), in contrast to the control treatments where the severity of the damage was more considerable.

In **Figure 9**, we can see that the harmful microbiological population rate also reduced with the use of organisms considered as beneficial, according to the final count at the end of the experiment; that could be because antagonistic bacteria are capable of influencing biocontrol mechanisms against phytopathogenic fungi such as antibiosis, siderophores, competition for nutrients, and production of hydrolytic enzymes. Similarly, Ulacio et al. [44] evaluated organic matter and antagonistic microorganisms as management strategies against white rot in garlic cultivation. These authors reported that the fungus *Sclerotium cepivorum* is significantly reduced and there was a lower incidence of the disease in the treatments where the fungus *T. harzianum*, the bacteria *B. firmus*, and vermicompost were combined.

Some microorganisms posess the ability by several ways to reduce the incidence and severity of diseases in crops, and also can participate in the stimulation of plant growth, yield, and crop quality. **Figure 10A** and **B** shows the values related to the promotion of root length (A) and its weight (B), where this effect is clearly observed. In **Figure 10C** and **D**, it was observed that *Bacillus* spp. strains increase the height of the plant by 28% compared to treatment T, and 34.5% concerning the TA. These results coincide with previous work where the biological effectiveness

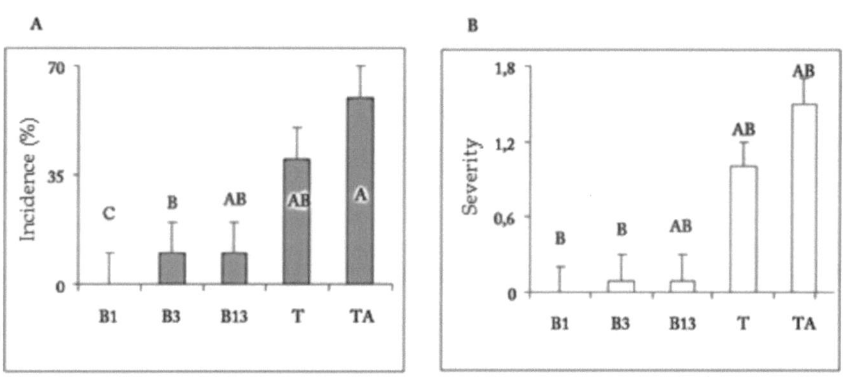

Figure 8.
Incidence (A) and severity (B) in plant traits with Bacillus *spp. strains (B1, B2, B3) in contrast with control (TA) and chemical control (T = thiabendazole).*

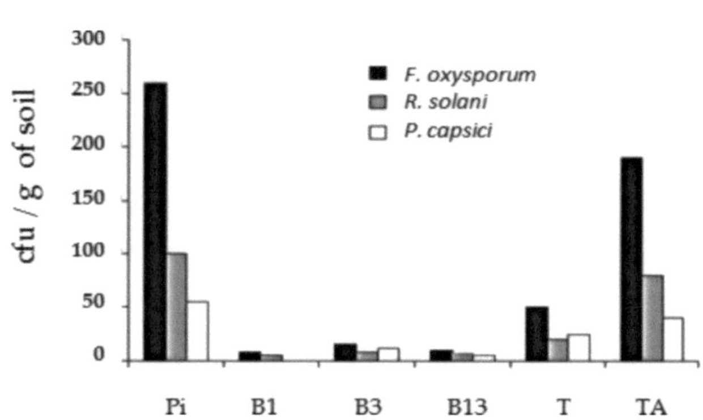

Figure 9.
Colony-forming units from initial (Pi) and final populations of phytopathogenic soil fungi after applying Bacillus *spp. strains (B1, B2, B3) against chemical (T = thiabendazole) and control (TA).*

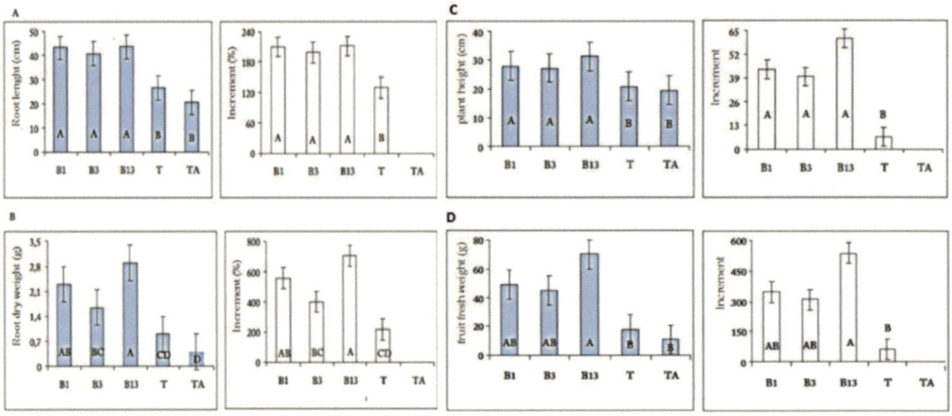

Figure 10.
Root length (A), dry rot weight (B), height (C), fresh fruit weight (D), and increments in chile pepper plant by effect Bacillus *strains (B1, B3, B13) against chemical (T = thiabendazole) and control (TA). Different letters with bars indicate significant differences among treatments ($p \leq 0.05$).*

of 57 strains of the genus *Bacillus* spp. isolated from the rhizosphere of commercial sowing chile plants in Northeast Mexico was analyzed, which showed an apparent antagonistic effect against *P. capsici*, *F. oxysporum*, and *R. solani* fungi. The plants inoculated with *Bacillus* spp. strains significantly increased height and dry weight in 191 and 60.2%, respectively [12]. The application of native *Bacillus* spp. strains shows a clear tendency to produce more biomass compared to chemical (T) and control (TA) treatments.

Likewise, del Ángel et al. [45] found a decrease in the incidence and severity of the disease caused by *Rhizoctonia solani* and *Fusarium oxysporum* with formulated endophytic bacteria, which induce a positive effect on the promotion of growth in the bean crop, increasing height and stem diameter in the treatments. Those formulated with bacteria in the absence of the phytopathogen stood out for their stimulating effect on the growth of the plants under study. This stimulating growth effect is observed in

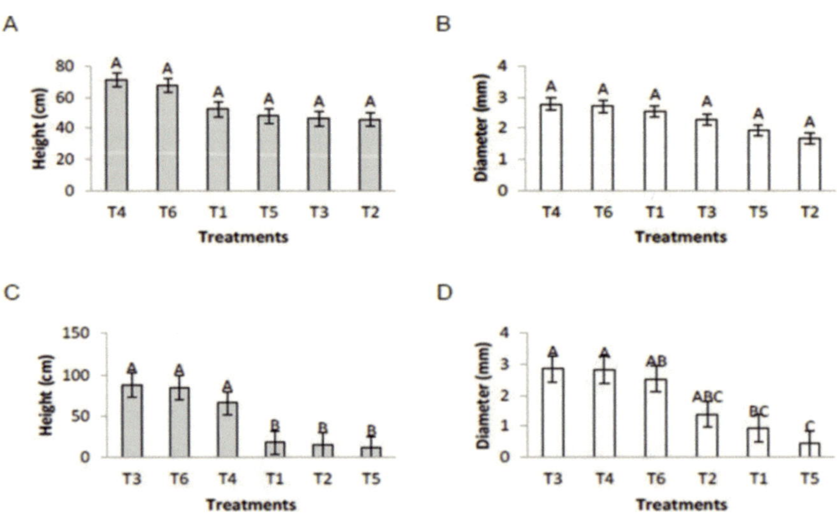

Figure 11.
Effect of endophytic bacteria on plant height and stem diameter in bean crop under greenhouse condition. Fusarium solani: *height (A), diameter (B), and* Rhizoctonia solani: *height (C), diameter (D). Means with the same letter are not significantly different according to the Tukey test ($p \leq 0.05$). Error bars are a standard error of the mean.*

plants treated with those formulated and inoculated at the same time with pathogens. It is essential to mention that the plants grew under no chemical treatment. Therefore, they did not receive fertilization by any chemical source (**Figure 11**).

7. Bioassays of antifungal activity under field conditions

7.1 Fruits results

Jimenez et al. [36] report results obtained on apple fruit and trees under the direct influence of the application of CFU from *Bacillus* spp., and *Trichoderma* spp., as control agents against the incidence and severity of *Venturia inaequalis* under field conditions in commercial apple cultivar. **Table 4** shows the incidence of fungus *Venturia inaequalis* in fruit, and this incidence varied from 5.6 to 6.25 when biological agents (*Trichoderma* spp. and *Bacillus* spp.) were used in maxima doses (2 L ha^{-1}) to 19.3% for the control, respectively, after 15 days of a first application. After 60 days from the start of the applications, the incidence is expressed in a range of 42.5–46.62% for *Bacillus* spp. and *Trichoderma* spp. at doses of 2 L ha^{-1} and for the control observed a 91.2%. The range of severity is observed between 1.8 and 2.6 of lesions per fruit by treatment *Trichoderma* spp. 2 L ha^{-1} and control, respectively, after 15 days of application initiation. After 60 days of treatment application appears first symptoms, so it was evaluated on a range of the number of lesions per fruit (severity) from 5.3 to 14.5 corresponding to *Bacillus* spp., 2 L ha^{-1}, and control, respectively (**Table 4**). The treatment with the best antagonism effect under field conditions was *Bacillus* spp., 2 L ha^{-1}, who expressed 42.5% by incidence and five lesions per fruit in contrast to the control, which showed 91.2% incidence and 14.5 lesions per fruit (**Figure 12**).

The field experiment is carried out to test biocontrol agents for control *V. inaequalis* in commercial apple cultivar; the statistical analysis showed highly significant differences between treatments ($p \leq 0.5$), the incidence in foliage treated with *Trichoderma* spp. 2 L ha^{-1} was lower in first evaluation (after 15 days of first application) and until harvest. This treatment expressed 10.6% incidence and two lesions per leaf, in contrast to the control which showed 31.8% and three lesions per leaf (**Table 5**). On the other hand, severity did not show significant differences among treatments.

7.2 Vegetable results

Espinoza-Ahumada et al. [40] aimed to find more environmentally friendly alternatives to the wilting of chile pepper; they evaluated the application of

Treatment	Incidence (%) in fruit		Severity (lesions) in fruit	
	15 days	60 days	15 days	60 days
Bacillus spp. 1 L ha^{-1}	18.12 ± 2.4a	51.87 ± 5.5b	1.82 ± 0.6ab	8.02 ± 0.7b
Bacillus spp. 2 L ha^{-1}	6.25 ± 5.2b	42.50 ± 6.5b	1.07 ± 0.8b	5.30 ± 0.5b
Trichoderma spp. 1 L ha^{-1}	15.00 ± 3.5a	55.00 ± 5.4b	1.77 ± 0.5ab	7.62 ± 0.2b
Trichoderma spp. 2 L ha^{-1}	5.62 ± 4.7b	45.62 ± 5.2b	1.00 ± 0.0b	6.32 ± 0.7b
Control	19.37 ± 4.7a	91.25 ± 4.3a	2.65 ± 0.5a	14.57 ± 0.3a

Treatments with the same letter are statistically equal to each other ($p < 0.05$).

Table 4.
The incidence in apple fruits by Venturia inaequalis.

Figure 12.
Expression of symptoms caused by Venturia inaequalis *in apple trees. (a) Without treatment, (b)* Bacillus *spp. effect, and (c)* Trichoderma *spp. effect.*

Treatments	Incidence (%) in leaves			
	15 days	30 days	45 days	60 days
Bacillus spp. 1 L ha^{-1}	8.12 ± 6.9ab	11.25 ± 1.2bc	13.12 ± 2.1bc	22.50 ± 4.8b
Bacillus spp. 2 L ha^{-1}	6.25 ± 2.1ab	11.25 ± 2.5bc	11.87 ± 2.4bc	17.50 ± 4.6b
Trichoderma spp. 1 L ha^{-1}	8.13 ± 3.8ab	13.12 ± 5.5bc	13.75 ± 4.7bc	20.62 ± 2.4b
Trichoderma spp. 2 L ha^{-1}	2.50 ± 1.7b	6.25 ± 1.4c	6.25 ± 1.4c	10.62 ± 1.3c
Control	18.12 ± 4.1a	21.87 ± 3.1a	23.75 ± 4.3a	31.87 ± 3.8a

Treatments with the same letter are statistically equal to each other ($p < 0.05$).

Table 5.
The incidence in apple leaves by Venturia inaequalis.

biological agents for this purpose under field conditions. For this, an experiment is established where different genotypes of chile pepper are evaluated (Serrano, HS-52, Coloso, HS-44, Centauro, Paraíso and Tampiqueño 74 cv.) generated by INIFAP-Mexico. In this experiment, the microbial agents *T. asperellum*, *T. harzianum*, *T. yunnanense* [23, 29], *B. amyloliquefaciens*, *B. licheniformis*, and *B. subtilis* [24] under a mixture of microbial propagative ferment (consortium ferment) are based on *Trichoderma* spp. and *Bacillus* spp. Treatments of bioassay by *Trichoderma* spp. were different: consortium treatment one consists of a *Trichoderma* spp. at $1{\times}10^8$ CFU; treatment two consists of ferment consortium; treatment three consists of a *B. consortium* at $1{\times}10^8$ CFU; treatment four consists of a chemical control by thiabendazole prepared at 60% W/V; and the treatment five consists of an absolute control. A dose of 1 L.ha^{-1} was applied for treatments one, two, and three, while the dose applied for thiabendazole was 0.5 kg.ha^{-1}. Field sowing is done with chile seedlings (10 cm), transplanted in 1.5 m double row beds. The application is made to drench with a manual sprinkler at 7, 28, and 49 days after the transplant (DDT). After 85, 105, 125, and 145 DDT, the yield per block (4.5 m^2) is determined and transformed to t ha. To determine yields and improvements of treatments, ten fruits were evaluated, where the weight (g) and size (mm) per fruit were determined. In the first and last harvest, the incidence assessed and transformed into a percentage. The severity is evaluated through the visual scale, where 0 = no visible symptoms; 1 = initial light chlorosis and presence of flowers and fruits; 2 = intermediate, partial wilt, severe chlorosis, and premature ripening of fruits; and 3 = advanced. For total wilt without recovery, the leaves and fruits remain stuck to the stem. The field results observed as the effects of biological agents are shown in **Table 6**. The disease incidence values between HS-52 and Coloso treatments were statistically different ($p \leq 0.05$); in the other varieties, there were no differences between treatments. The treatment based on *Trichoderma* is the biological one that suppresses in higher

Microbial agents	Serrano chile pepper varieties					
	HS-52	Coloso	HS-44	Centauro	Tampiqueño 74	Paraíso
Trichoderma spp.	10.67a	18.17ab	16.84a	19.17a	12.5a	10.00a
Consortium	26.67ab	15.5ab	10.50a	15.33a	19.83a	10.67a
Bacillus spp.	29.17ab	29.67b	20.07a	20.5a	21.83a	19.5a
Thiabendazole	21.00ab	6.83a	19.51a	24.17a	10.33a	16.17a
Control	31.83b	21.33ab	21.51a	23.33a	24.67a	22.33a

Mean values on the same column indicated by different letters are statistically different ($p < 0.05$) according to the LSD test.

Table 6.
Incidence of the disease (%) in serrano chile pepper varieties inoculated with microbial agents in the field.

percentage the incidence of wilting disease in chile pepper crops; in this case, the lowest incidence was in the HS-52 variety which showed a value of 10.67%, while that in the witness it was 31.87%, which represents a decrease of 71% concerning the latter.

Disease evaluation in the presence of treatments of consortium and *Trichoderma* demonstrates the lowest incidence percentage with values between 14.39 and 16.39%, while the control and *Bacillus* spp. were having high levels of the presence of symptoms (24.08 and 23.36%). In the case of severity, it also behaves differently between treatments. **Table 7** shows the values related to the severity of the disease

Treatment	Serrano chile pepper varieties					
	HS-52	Coloso	HS-44	Centauro	Tampiqueño 74	Paraíso
Trichoderma	11.3ab	18.43a	10.8a	6.83a	7.83ab	6.93a
Consortium	8.33a	16.28a	24.45a	12.76a	14.35ab	6.46a
Bacillus spp.	14.4ab	19.16a	15.09a	11.54a	17.6b	16.22ab
Thiabendazole	20.04b	14.07a	17.97a	15.74a	6.56a	18.24b
Control	19.45ab	24.35a	24.07a	13.65a	17.96b	18.43b

Mean values on same column indicated by different letters are statistically different ($p < 0.05$) according to LSD test.

Table 7.
Severity of the disease (%) in serrano pepper with respect to treatments.

Microbial agents	Total yield in chile pepper varieties (t ha^{-1})			
	HS-52	Centauro	Paraíso	HS-44
Trichoderma spp.	15.67a	13.22a	8.48b	7.55a
Consortium	10.37ab	11.52ab	10.59a	13.04a
Bacillus spp.	7.26b	8.18ab	5.41b	10.3a
Thiabendazole	10.02ab	8.69ab	7.44b	10.62a
Control	5.98b	5.15b	2.59b	6.94a

Mean values on the same column indicated by different letters are statistically different ($p < 0.05$) according to the LSD test.

Table 8.
Total yield, length, and weight of fruit of the serrano chile pepper crop obtained with the use of microbial agents.

Figure 13.
Expression of incidence of coffee rust. (a) Plants with treatment based on bio formulate based on Bacillus *spp., and (b) plants without treatment, where leaf defoliation is clearly expressed.*

as transformed percentages ($p \leq 0.05$). It can be seen that *Trichoderma* spp.-based treatments alone or in combination have lower severity values.

The effects on yield as the weight and size of the fruit showed by the use of microbial agents applied alone or in combination as shown in **Table 8**. When *Trichoderma* is used, the yield increased; for example, its increase in the production was 62% when used alone and up to 76% when used as a mixture in comparison with the control.

This behavior of positive effect has already evidenced with the use of different *Trichoderma* species on habanero pepper plants (*Capsicum chinense*) [46], lettuce (*Lactuca sativa*), and radish (*Raphanus sativus*) [47]. In the same context, Cubillos-Hinojosa et al. [48] tested *T. harzianum* in the passion fruit crops (*Passiflora edulis*) where they were able to determine an antagonist to *F. oxysporum* and *F. solani*, in addition to stimulating germination, increased biomass, and root length.

In other field tests with *Bacillus* spp. bioformulate prototypes, a reduction in incidence and severity of coffee rust (*Hemileia vastatrix*) was observed. It was observed that the control presented 38% of incidence; nevertheless, it showed defoliation compared with the prototype treatments, which present an incidence between 5 and 15%, while with the chemist, the incidence was 9% (**Figure 13**).

The positive interaction between *Trichoderma* spp. and the host plant is attributed to a complex chemical activity of volatile and diffusible secondary metabolites and release of phytohormones and antibiotics in the rhizosphere, which promote root development and increased nutrient absorption, which helps control phytopathogens and increase yield [49]; this effect explains the results produced in this research. Microbial extracts as biofertilizers can generate hormones that stimulate development and increase yield, which are verified with the application of exudates from consortium *Trichoderma* spp. and *Bacillus* spp., which showed an effect on disease control and crop development in the same or better percentage than when using microorganisms.

8. Resistance induction by *Trichoderma* spp. and *Bacillus* spp.

In addition to the above aspects, plants can develop an increase in resistance to pathogen infection by treatment with a wide variety of biotic and abiotic inducers. Among the biotic inducers, we have the same phytopathogens, the growth-promoting rhizobacteria, and the microbial agents of the species of the genera *Bacillus* spp., *Streptomyces*, *Pseudomonas*, *Burkholderia*, and *Agrobacterium* and nonpathogenic microorganisms such as *Trichoderma* species (antibiotics or siderophores that lead to induction of resistance).

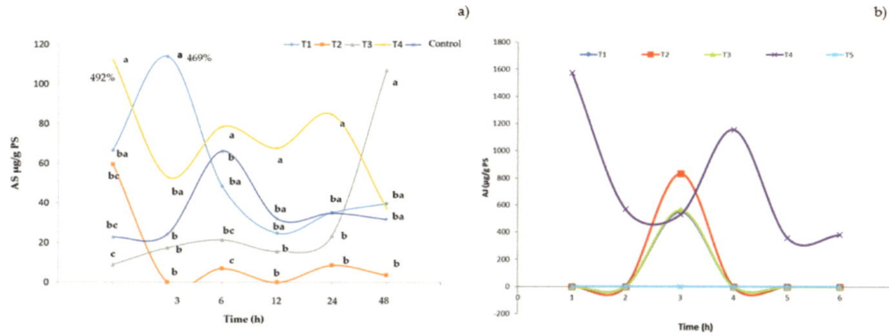

Figure 14.
(a) Salicylic acid production on potato leaves in a different time. T1 = Bacillus *spp. and* Pseudomonas fluorescens*, T2 = jasmonic ac. 1500 ppm, T3 = mezcla T1 0.5% + T2 0.1%, T4 = Milor®, and T5 = control (agua). Different letters indicate significant difference. (b) Jasmonic acid production on potato leaves in different time. T1 =* Bacillus *spp. and* Pseudomonas fluorescens*, T2 = jasmonic ac. 1500 ppm, T3 = mezcla T1 0.5% + T2 0.1%, T4 = Milor®, and T5 = control.*

Among the abiotic inducers are salicylic acid (SA), jasmonic acid (JAS), β-aminobutyric acid, ethylene, chitosan, potassium, sodium or magnesium phosphate, acibenzolar-S-methyl (ASM), menadione, sodium bisulfite, and phosphites. The application of these inducers causes specific biochemical changes that occur after their application such as expression of genes that code for PR proteins; the increase of certain defense-related enzymes such as polyphenol oxidase, lipoxygenase, peroxidase, superoxide dismutase, and phenylalanine ammonia-lyase (PAL); the accumulation of phytoalexins and phenolic compounds; and the reinforcement of the cell wall with lignin deposition.

In this regard, we have observed changes in the endogenous levels of salicylic acid and jasmonic acid in potato plants in response to foliar application of microbial consortiums based on *Bacillus* spp. and *Pseudomonas fluorescent*. The microbial consortium of *Bacillus* spp. significantly increased the production of SA 3 h after spraying raising to 114.02 µg/g DW. This is 496% more than the control (**Figure 14a**). Jasmonic acid is not detected in control plants but detected in plants treated with the microbial consortium. The level of jasmonic acid, 6 h later, reached a level of 550 µg/g DW (**Figure 14b**).

The resistance induction is associated with some defense gene expression as encoding pathogenicity-related proteins (PR), for example, phenylalanine ammonia-lyase, which is crucial in the synthesis of phytoalexins, because these constitute highly toxic compounds to the pathogen. On the other hand, PAL is part of the synthesis of salicylic acid and phenolic compounds that reduce the incidence of diseases in plants. It has also shown that *B. amylolicheniformis*, *B. subtilis*, *B. pumilus*, and *B. cereus* are capable of eliciting and activating the induced systemic resistance by increasing the levels of biochemical compounds related to resistance induction. Besides, it reported that some *Pseudomonas* species could induce systemic resistance in plants.

9. Conclusions

The results shown in this chapter allow to demonstrate the efficacy of *Bacillus* and *Trichoderma,* as agents of biological control of fungi and stramenopiles that are causatives of plant diseases; these beneficial microorganisms can be used under a sustainable agriculture program or under integrate management pest program in a conventional agriculture. The microbial agents also express other advantages due to their beneficial effects on the increase of the yields, growth, and development of plants, as well as the induction of systemic resistance in plants to phytopathogens. Currently our workgroup has any projects on the development of prototypes based on these microbial agents, alone or in consortium, as well as micro- and nanoencapsulated formulations.

Acknowledgements

The authors acknowledge the financial support for the development of research from the National Council of Science and Technology of Mexico (CONACYT), University Autonomy Agrarian Antonio Narro and Greencorp Biorganiks of Mexico, S.A. de C.V corporation.

Conflict of interest

The authors declare no conflict of interest.

Author details

Francisco Daniel Hernández-Castillo[1*], Francisco Castillo-Reyes[2], Marco Antonio Tucuch-Pérez[1] and Roberto Arredondo-Valdes[3]

1 Departamento de Parasitología, Universidad Autónoma Agraria Antonio Narro, Saltillo, Coahuila, Mexico

2 Instituto Nacional de Investigaciones Forestales, Agrícolas y Pecuarias, Saltillo, Coahuila, Mexico

3 Departamento de Nanobiociencias, Universidad Autónoma de Coahuila, Saltillo, Coahuila, Mexico

*Address all correspondence to: fdanielhc@hotmail.com

References

[1] Doolotkeldieva T, Bobusheva S. Scab disease caused by *Venturia inaequalis* on apple trees in Kyrgyzstan and biological agents to control this disease. Advances in Microbiology. 2017;7:450-466. DOI: 10.4236/aim.2017.76035

[2] Jha G, Thakur K, Thakur P. The Venturia apple pathosystem: Pathogenicity mechanisms and plant defense responses. BioMed Research International. 2010;**1**:1-10. DOI: 10.1155/2009/680160

[3] Albañil-Juárez JA, Mariscal-Amaro LA, Martínez-Martínez TO, López A, Luis J, Cisneros-López HC, et al. Estudio regional de fitopatógenos asociados a la secadera del chile en Guanajuato, México. Revista Mexicana de Ciencias Agrícolas. 2015;**6**:2191-2197. DOI: 10.29312/remexca.v0i11.797

[4] Perez L, Rodriguez ME, Rodriguez F, Roson C. Efficacy of acibenzolar-S-methyl, an inducer of systemic acquired resistance against tobacco blue mould caused by *Peronospora hyoscyami* f. sp. *tabacina*. Crop Protection. 2003;**22**(2):405-413. DOI: 10.1016/S0261-2194(02)00198-9

[5] Yossen VE, Conles MY. Eficacia de fungicidas in vitro para el control de *Fusarium oxysporum* y *F. proliferatum*, agentes causales de marchitamiento en el cultivo de orégano en la Argentina. Revista Industrial y Agrícola de Tucumán. 2014;**91**(1):19-25

[6] Hernandez C, Berlanga P, Gallegos M, Cepeda S, Rodriguez H, Aguilar G, et al. *In vitro* antagonist action of *Trichoderma* spp. strains against *Sclerotinia sclerotiorum* and *Sclerotium cepivorum*. American Journal of Agricultural and Biological Sciences. 2011;**6**(3):410-417. DOI: 10.3844/ajabssp.2011.410.417

[7] Kumar K, Amaresan N, Bhagat S, Madhuri K, Srivastava RC. Isolation and characterization of *Trichoderma* spp. for antagonistic activity against root rot and foliar pathogens. Indian Journal of Microbiology. 2012;**52**(2):137-144. DOI: 10.1007/s12088-011-0205-3

[8] Avşar C, Koyuncu H, Aras ES. Isolation and molecular characterization of *Bacillus* spp. isolated from soil for production of industrial enzymes. Biological and Chemical Research. 2017;**3**(9):72-86

[9] Guillén-Cruz R, Hernández-Castillo FD, Gallegos-Morales G, Rodríguez-Herrera R, Aguilar-González CN, Padrón-Corral E, et al. *Bacillus* spp. como biocontrol en un suelo infestado con *Fusarium* spp., *Rhizoctonia solani* Kühn y *Phytophthora capsici* Leonian y su Efecto en el Desarrollo y Rendimiento del Cultivo de Chile (*Capsicum annuum* L.). Revista Mexicana de Fitopatología. 2006;**24**(2):105-114. Available from: https://www.redalyc.org/pdf/612/61224204.pdf

[10] Hermosa MR, Grondona I, Iturriaga ET, Diaz-Minguez JM, Castro C, Monte E, et al. Molecular characterization and identification of biocontrol isolates of *Trichoderma* spp. Applied and Environmental Microbiology. 2000;**66**(5):1890-1898. DOI: 10.1128/AEM.66.5.1890-1898.2000

[11] Parvathi A, Krishna K, Jose J, Joseph N, Nair S. Biochemical and molecular characterization of *Bacillus* spp. *pumilus* isolated from coastal environment in Cochin, India. Brazilian Journal of Microbiology. 2009;**40**(2):269-275. DOI: 10.1590/S1517-83822009000200012

[12] Acurio-Vásconez RD, España-Imbaquingo CK. Aislamiento, caracterización y evaluación de *Trichoderma* spp. como promotor de crecimiento vegetal en pasturas de raygrass (*Lolium perenne*) y trébol

blanco (*Trifolium repens*) [thesis]. Universidad Politécnica Salsiana; 2017

[13] López CG, González PAG. Selección de cepas nativas de *Trichoderma* spp. con actividad antagónica sobre *Phytophthora capsici* Leonian y promotoras de crecimiento en el cultivo de chile (*Capsicum annuum* L.). Revista Mexicana de Fitopatología. 2004;**22**(1):117-124. Available from: https://www.redalyc.org/pdf/612/61222115.pdf

[14] Gams W, Bissett J. Morphology and Identification of Trichoderma. In: Kubicek C, Harman G, editors. Trichoderma and Gliocladium. 1st ed. Salisbury, UK: Christian P Gams; 1998. pp. 3-31

[15] Rivero MV, Maniscalco DP. Identificación de cepas del hongo *Trichoderma* spp. por métodos moleculares. Faraute. 2016;**8**(2):1-9

[16] Kredics L, Chen L, Kedves O, Büchner R, Hatvani L, Allaga H, et al. Molecular tools for monitoring *Trichoderma* spp. in agricultural environments. Frontiers in Microbiology. 2018;**9**:1-17. DOI: 10.3389/fmicb.2018.01599

[17] Meena SN. Development of species-specific SCAR markers for the identification of *Trichoderma* spp. species [thesis]. Dharwad: Department of Biotechnology, College of Agriculture, Dharwad University of Agricultural Sciences; 2009

[18] Devi TP, Prabhakaran N, Kamil D. Development of species specific markers for the identification of *Trichoderma* spp. *asperellum* and *Trichoderma* spp. *harzianum*. Vegetos—An International Journal of Plant Research. 2017;**30**:94-100

[19] Geistlinger J, Zwanzig J, Heckendorff S, Schellenberg I. SSR markers for *Trichoderma* spp. virens: Their evaluation and application to identify and quantify root-endophytic strains. Diversity. 2015;**7**(4):360-384. DOI: 10.3390/d7040360

[20] Skoneczny D, Oskiera M, Szczech M, Bartoszewski G. Genetic diversity of *Trichoderma* spp. *atroviride* strains collected in Poland and identification of loci useful in detection of within-species diversity. Folia Microbiologica. 2015;**60**:297-307. DOI: 10.1007/s12223-015-0385-z

[21] Chen X, Romaine CP, Ospina-Giraldo MD, Royse DJ. A polymerase chain reaction-based test for the identification of *Trichoderma* spp. *harzianum* biotypes 2 and 4, responsible for the worldwide green mold epidemic in cultivated *Agaricus bisporus*. Applied Microbiology and Biotechnology. 1999;**52**:246-250. DOI: 10.1007/s002530051516

[22] Kredics L, Kocsubé S, Nagy L, Komon-Zelazowska M, Manczinger L, Sajben E. Molecular identification of *Trichoderma* spp. species associated with *Pleurotus ostreatus* and natural substrates of the oyster mushroom. FEMS Microbioly Letter. 2009;**300**:58-67. DOI: 10.1111/j.1574-6968.2009.01765.x

[23] Osorio E, Hernández FD, Gallegos G, Rodríguez R, Castillo F. *In-vitro* behavior of *Trichoderma* spp. against *Phytophthora capsici* Leonian. African Journal of Agricultural Research. 2011;**6**(19):4594-4600. DOI: 10.5897/AJAR11.1094

[24] Hernández-Castillo FD, Lira-Saldivar RH, Gallegos-Morales G, Hernandez-Suarez M, Solis-Gaona S. Biocontrol of pepper wilt with three *Bacillus* species and its effect on growth and yield. Phyton, International Journal of Experimental Botany. 2014;**83**:49-55. Available from: http://ppct.caicyt.gov.ar/index.php/phyton/article/view/8684

[25] Foysal MJ, Lisa AK. Isolation and characterization of *Bacillus* sp. strain

BC01 from soil displaying potent antagonistic activity against plant and fish pathogenic fungi and bacteria. Journal of Genetic Engineering and Biotechnology. 2018;**16**(2):387-392. DOI: 10.1016/j.jgeb.2018.01.005

[26] Manzum AA, Al Mamun MA. Isolation of *Bacillus* spp. bacteria from soil for production of cellulase. Nepal Journal of Biotechnology. 2018;**6**(1): 57-61. DOI: 10.3126/njb.v6i1.22338

[27] Schaad NW, Jones JB, Chun W. Plant Pathogenic Bacteria. Estados Unidos: APS Press; 2001

[28] Chun J, Bae KS. Phylogenetic analysis of *Bacillus* spp. *subtilis* and related taxa based on partial gyrA gene sequences. Antonie Van Leeuwenhoek. 2000;**78**(2):123-127. DOI: 10.1023/A:102655583

[29] Johansson AH, Bejai S, Niazi A, Manzoor S, Bongcam-Rudloff E, Meijer J. Studies of plant colonisation by closely related *Bacillus* spp. *amyloliquefaciens* biocontrol agents using strain specific quantitative PCR assays. Antonie Van Leeuwenhoek. 2014;**106**(6):1247-1257. DOI: 10.1007/s10482-014-0295-0

[30] Huang X, Zhang N, Yong X, Yang X, Shen Q. Biocontrol of *Rhizoctonia solani* damping-off disease in cucumber with *Bacillus* spp. *pumilus* SQR-N43. Microbiological Research. 2012;**167**(3):135-143. DOI: 10.1016/j.micres.2011.06.002

[31] de Almeida MEM. *Bacillus* spp. *licheniformis* specific DNA markers for identification and culture-independent monitoring [thesis]. Universidad Do porto; 2014

[32] Hernandez-Castillo FD, De la Garza-Rodriguez R, Gallegos-Morales G, Padrón-Corral E, Sanchez-Arizpe A, Lira-Saldivar RH. Efectividad biológica de bacterias rizosféricas esporuladas sobre el complejo de hongos de la marchitez del chile. Phyton. 2005;**54**:171-180

[33] Bell D, Well H, Markham C. *In vitro* antagonism of *Trichoderma* species against six fungal plant pathogens. Phytopathology. 1982;**73**:379-382. DOI: 10.1094/Phyto-72-379

[34] Tucuch MA, Labastida A, Hernández FD, Roberto A. Determinación *in vitro* de la actividad antifúngica de productos biológicos y naturales contra *Colletotrichum* sp. (Glomerellales: Glomerellaceae) de café. In: Proceedings of the XLIIC Congreso Nacional de Control Biológico; November. Vol. 2019. México: Boca del rio, Veracruz; 2019. p. 139. DOI: 10.13140/RG.2.2.13709.08161

[35] Samaniego-Fernández LM, Harouna M, Corbea O, Rondón-Castillo AJ, Placeres-Espinosa I. Aislamiento, identificación y evaluación de cepas autóctonas de *Trichoderma* spp. spp. antagonistas de patógenos del suelo. Revista de Protección Vegetal. 2018;**33**(3):1-11

[36] Jimenez MDCM, Hernández FD, Alcalá EIL, Morales GG, Valdés RA, Reyes FC. Biological effectiveness of *Bacillus* spp. spp. and *Trichoderma* spp. spp. on apple scab (*venturia inaequalis*) in vitro and under field conditions. European Journal of Physical and Agricultural Sciences. 2018;**6**(2):7-17. Available from: http://www.idpublications.org/wp-content/uploads/2018/09/Full-Paper-BIOLOGICAL-EFFECTIVENESS-OF-BACILLUS-SPP.-AND-TRICHODERMA-SPP.-ON-APPLE-SCAB.pdf

[37] Ragazzo JA, Robles A, Lomelí L, Luna G, Calderón M. Selección de cepas de *Bacillus* spp. productoras de antibióticos aisladas de frutos tropicales. Revista Chapingo Serie Horticultura. 2011;**17**(1):5-11

[38] Masoko P, Picard J, Eloff JN. Antifungal activities of six south African Terminalia species (Combretaceae). Journal of Ethnopharmacology. 2005;**99**(2):301-308. DOI: 10.1016/j.jep.2005.01.061

[39] Tucuch-Pérez MA, Daniel F, Arredondo-Valdés R, Laredo-Alcalá EI. Antifungal activity of metabolites from *Bacillus* spp. against *Fusarium oxysporum* using micro dilution in plate method. European Journal of Biotechnology and Genetic Engineering. 2018;**5**(1):22-30

[40] Espinoza-Ahumada CA, Gallegos-Morales G, Ochoa-Fuentes YM, Hernández-Castillo FD, Méndez-Aguilar R, Rodríguez-Guerra R. Antagonistas microbianos para biocontrol de la marchitez y su efecto promotor en el rendimiento de chile serrano. Revista Mexicana de Ciencias Agrícolas. 2019;**23**:187-197. DOI: 10.29312/remexca.v0i23.2020

[41] Francisco FN, Gallegos-Morales G, Ochoa-Fuentes YM, Delgado-Ortiz JC, Hernández-Castillo FD. *In vitro* antagonism of *Bacillus* spp. strains against *Fusarium* species. Mycopathologia. 2016;**14**:15-19

[42] Sadfi N, Cherif M, Fliss I, Boudabbous A, Antoun H. Evaluation of bacterial isolates from salty soils and *Bacillus thuringiensis* strains for the biocontrol of *Fusarium* dry rot of potato tubers. Journal of Plant Pathology. 2001;**83**(2):101-117. Available from: https://www.jstor.org/stable/41998046

[43] Copes WE, Stevenson KL. A pictorial disease severity key and the relationship between severity and incidence for black root rot of pansy caused by *Thielaviopsis basicola*. Plant Disease. 2008;**92**(10):1394-1399. DOI: 10.1094/PDIS-92-10-1394

[44] Ulacio D, Zavaleta E, García R, Delgadillo F, Pedroza A, Martínez A. Materia orgánica y microorganismos antagonistas como estrategias de manejo de *Sclerotium cepivorum* Berk. y su impacto en el progreso de la pudrición blanca en ajo (*Allium sativum* L.). Revista Mexicana de Fitopatología. 2003;**21**(3):346-354

[45] del Ángel EC, Hernández FD, Fuentes YMO, Morales GG, Reyes FC, Tucuch FM. Endophytic bacteria controlling *Fusarium oxysporum* and *Rhizoctonia solani* in *Solanum tuberosum*. European Journal of Physical and Agricultural Sciences. 2017;**5**(1):29-39. Available from: https://pdfs.semanticscholar.org/8ce0/019bd26c7bc5177efc49bbf6aea8798fb5de.pdf

[46] Candelero DJ, Cristóbal AJ, Reyes RA, Gamboa AMM, Ruíz SE, Tun SJM. *Trichoderma* spp. fostering growth on *Capsicum chinense* Jacq. seedlings and antagonistic against *Meloidogyne incognita*. International Journal of Experimental Botany. 2016;**84**(1):113-119. Available from: http://ppct.caicyt.gov.ar/index.php/phyton/article/view/8136

[47] Ortuño N, Miranda C, Claros M. Selección de cepas de *Trichoderma* spp. generadoras de metabolitos secundarios de interés para su uso como promotor de crecimiento en plantas cultivadas. Journal of the Selva Andina Biosphere. 2013;**1**(1):16-32

[48] Cubillos-Hinojosa J, Valero N, Mejía L. *Trichoderma harzianum* como promotor del crecimiento vegetal del maracuyá (*Passiflora edulis* var. flavicarpa Degener). Agronomía Colombiana. 2009;**27**(1):81-86. Available from: https://www.redalyc.org/pdf/1803/180314730011.pdf

[49] López-Bucio J, Pelagio-Flores R, Herrera-Estrella A. *Trichoderma* as biostimulant: Exploiting the multilevel properties of a plant beneficial fungus. Scientia Horticulturae. 2015;**196**:109-123. DOI: 10.1016/j.scienta.2015.08.043

Section 5

Weed Management

Chapter 7

Nonchemical Weed Control in Winter Oilseed Rape Crop in the Organic Farming System

Aušra Marcinkevičienė, Marina Keidan, Rita Pupalienė, Rimantas Velička, Zita Kriaučiūnienė, Lina Marija Butkevičienė and Robertas Kosteckas

Abstract

A field experiment was conducted during the 2014–2017 period at Aleksandras Stulginskis University (now—Vytautas Magnus University Agriculture Academy) on a *Endocalcaric Endogleyic Luvisol* (*LV-can.gln*) according to the WRB 2014. The three nonchemical weed control methods were explored: (1) thermal (using wet water steam), (2) mechanical (interrow loosening), and (3) self-regulation (smothering). In the thermal and mechanical weed control treatments, winter oilseed rape was grown with an interrow spacing of 48.0 cm and in weed smothering (self-regulation) treatment with an interrow spacing of 12.0 cm. Winter oilseed rape was grown in the soil with a regular humus layer (23–25 cm) and with a thickened humus layer (45–50 cm). Annual weeds predominated in the winter oilseed rape crop. In the soil with both humus layers, regular and thickened, the most efficient weed control method was mechanical weed management both during the autumn (efficacy 26.7–75.1%) and spring (efficacy 37.1–76.7%) growing seasons. Thermal and mechanical weed control in combination with the bio-preparations in droughty years significantly reduced the number of weed seedlings. Dry matter mass of weeds most markedly decreased through the application of the mechanical weed management method.

Keywords: winter oilseed rape, weed control methods, organic farming system, bio-preparations, soil humus layer

1. Introduction

The development of organic farming was prompted by the environmental concerns, health-related issues, and the search for solutions to social problems.

In Lithuania, organic farms account for more than 5% of the area under cultivation and are in line with the EU average. According to the popularity of organic farming, Lithuania surpasses the neighboring Poland but is far behind the other Baltic States. Although organic farms represent a small proportion of the total number of farms, their number has increased rapidly over the past decade. In 2017, 2448 organic farms were certified in Lithuania; they cover about 244,000 hectares of the agricultural land [1]. Most of the organic production farms are 10–30 ha in size.

Oilseed rape is one of the world's most important oil crops [2, 3]. The cultivation of oilseed rape on organically managed farms was encouraged by a search for healthy, high-quality, and safe food. According to the data from the public organization "Ekoagros," in 2017, the total area devoted to oilseed rape production on organically managed farms in Lithuania amounted to 3962.2 ha, including 3250.98 ha of winter oilseed rape and 711.22 ha of spring oilseed rape. The main reasons why winter oilseed rape production area on organically managed farms is not increasing are the problems associated with plant nutrition [4, 5], weed, disease and pest control, and unstable plant overwinter survival, and all these factors result in low rapeseed yields [6, 7]. Many organic farms in Lithuania are located on infertile soils, and the erosion and productivity problems are relevant there. Organic crop production farms are prevalent in this region; therefore the problem of crop rotation, nutrient, and humus balance is highly relevant. In the organic production farms, the inclusion of oilseed rape in the crop rotation is very important because it is characterized by phytosanitary properties, is a good pre-crop for other crops, and improves soil properties [8].

In the organic production farms, in the absence of the possibility of controlling weeds with herbicides, a great deal of attention is paid to nonchemical methods of weed control—mechanical, thermal, and natural crop-weed competition/self-regulation. Weed control by using wet water steam has not been extensively studied in the world. More comprehensive studies on the thermal weed control by water steam have been carried out by Lithuanian scientists [9–13].

2. Material and methods

Field experiments were conducted in 2014–2017 at the Experimental Station (54°53′ N, 23°50′ E) of Aleksandras Stulginskis University on an *Endocalcaric Endogleyic Luvisol* (*LV-can.gln*) according to the WRB 2014. Agrochemical properties of the experimental soil with a regular humus layer (averaged data of 2014, 2015, and 2016) were as follows: pH, 7.30; humus, 1.79%; contents of available nutrients in the soil: P_2O_5, 199.0 mg kg^{-1}; K_2O, 97.7 mg kg^{-1}; total nitrogen, 0.079%. Agrochemical properties of the experimental soil with a thickened humus layer were as follows: pH, 7.20; humus, 2.19%; contents of available nutrients in the soil: P_2O_5, 277.7 mg kg^{-1}; K_2O, 123.0 mg kg^{-1}; total nitrogen, 0.115%.

Two-factor field experiments were established using a split-plot design. Winter oilseed rape was grown in the soil with a regular humus layer (23–25 cm) (experiment I) and in the soil with a thickened humus layer (45–50 cm) (in 1988, a thickened humus-rich layer was artificially formed using fertile soil delivered from elsewhere) (experiment II). The length of initial plots was 14 m, width 6 m, and the area 84 m^2. The length of the harvested plots was 10 m, width 2 m, and the area 20 m^2. The experiments included four replications. The winter oilseed rape crop was preceded by black fallow.

The study object was winter oilseed rape (*Brassica napus* L. spp. *oleifera biennis* Metzg.) agrocenosis.

Experimental treatments:

Factor A: nonchemical weed control methods:

1. Thermal (water steam)

2. Mechanical (interrow cultivation)

3. Self-regulation/smothering (natural weed-crop competition, sowing with narrow interrows)

Factor B: biological preparations (bio-preparations):

1. Without bio-preparations

2. With bio-preparations

A winter rape cultivar "Cult" (Sweden, SW Seed) was grown in the experiment. The crop was sown at a seed rate of 3 kg ha^{-1} with a Multidrill M300 sowing machine. In 2014, winter rape was sown on September 1, in 2015 on August 27, and in 2016 on August 29. In the thermal weed control treatment, the oilseed rape crop was grown with interrow spacing of 48 cm, and weeds were killed using a tractor-mounted wet water steam unit at a 3–4 leaf growth stage (BBCH 13–14) of winter rape. The thermal power of the device is 90 kW, with a capacity of 120 kg h^{-1} steam; the device is run on liquefied gas. The temperature of steam is 99°C, with thermal treatment time of 2 s (Sirvydas, Kerpauskas, 2012).

The principal scheme of the tractor-mounted thermal weed control unit using wet water steam is presented in **Figure 1**.

2.1 Description of operation of the mobile thermal weed control unit

Liquefied gas is fed through tube 6 into the combustion chamber 4 of demountable steam boiler 8. There the burning gas heats water present in the steam boiler 8. The wet water steam which has formed in steam boiler 8 gets into the steam separator, in which steam dampness is reduced. Then the wet water steam which has passed through the steam overheater 11 is fed through the tube 5 into steam diffusers 2, which spread/distribute steam in the environment of target weeds.

The height of the steam diffusers is adjusted by the height adjustment mechanism 3. To prevent the liquefied gas from cooling, the gas cylinder 9 is placed into the heating tank 10. Hot water from steam boiler 8 is fed into the gas cylinder's heating tank 10 through tube 7. The mobile thermal weed control unit is mounted on a tractor 12 with a mounting device 11.

In the mechanical weed control treatment, the interrows were loosened with an interrow cultivator (KOR-4.2-01, Ukraine) using two passes. In the weed control treatment involving weed smothering (self-regulation), the winter oilseed rape was

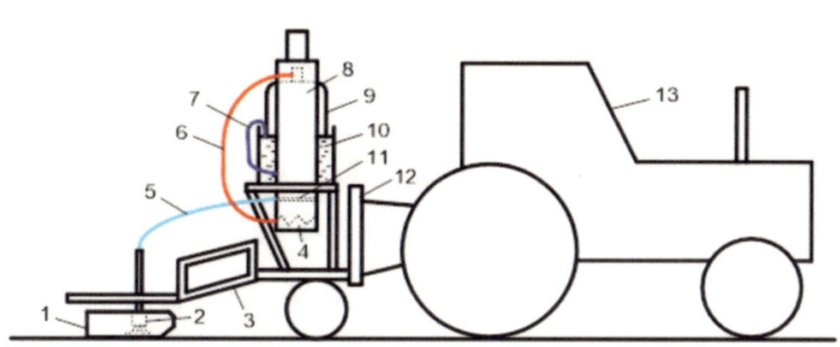

Figure 1.
The principal technological scheme of the tractor-mounted thermal weed control unit: (1) protectors of steam diffusers, (2) steam diffusers, (3) diffusers' height adjustment mechanism, (4) combustion chamber, (5) tube by which wet water steam is fed to steam diffusers and discharges, (6) tube by which gas is fed to combustion chamber, (7) tube by which hot water is fed to the heating tank of gas cylinder, (8) demountable steam boiler, (9) gas cylinder, (10) heating tank of gas cylinder, (11) steam overheater, (12) mounting device of thermal weed control unit, and (13) tractor.

grown with 12.0 cm interrow spacing. Winter rape was not fertilized, and no chemical plant protection products were applied.

In the treatments with the use of the bio-preparations, pre-sowing, the seeds of winter rape were coated with the bio-organic fertilizer Nagro (BioPlant) (0.5 l per ton of seeds and 10 l of water), and during the growing season, the winter rape crop was sprayed twice with the bio-preparations (in the autumn with Terra Sorb Foliar (BioIberica) (2 l ha^{-1}), in the spring with Terra Sorb Foliar (1 l ha^{-1}) and 0.3% Conflic (Atlantica Agricola)).

2.2 Assessment of weed incidence in the crop

The analysis of weed seedlings was carried out at winter rape 3–4 leaf growth stage (BBCH 13–14) in the autumn and after resumption of vegetation in the spring (BBCH 50) before the application of thermal and mechanical weed management methods. In each experimental plot, in four randomly selected 0.10 m^2 record plots, the number of weed seedlings and weed species composition were established. This analysis was done for the second time 5–7 days after application of the weed control methods in the marked record plots. The number of weed seedlings was recalculated as per m^2. The efficacy (E) of the weed management methods for the change in the number of weed seedlings was calculated according to the formula:

$$E = (S1 - S2)/S1 \times 100\,\%, \qquad (1)$$

where S1 is the number of weed seedlings per m^2 before application of the weed control methods and S2 is the number of weed seedlings per m^2 after application of the weed control methods.

At winter rape green silique stage (BBCH 79), the number of weeds and weed species composition were determined in each plot in four 0.25 m^2 record plots; he weeds were dried in a drying chamber at 60°C temperature and weighed [14].

2.3 Statistical analysis

The significance of the differences between the means was estimated using the *t* criterion; the interplay between the traits was determined by the correlation regression methods. The statistical analysis of the experimental data was performed using software STAT from the software package SELEKCIJA [15]. The experimental data that did not fit the normal distribution law, prior to the statistical evaluation, were transformed using the function $y = \ln x + 1$.

2.4 Meteorological conditions

In 2014 autumn was warm and long, so conditions for rape growing were favorable. In winter, meteorological conditions were favorable for rape over-wintering. In 2015, autumn was warm and humid. During the first decade of January 2016 of very cold weather and the absence of snow, over-wintering of rape was not successful. In 2016, the conditions for rape preparation for wintering and for over-wintering were favorable. In 2017, rape vegetation renewed on March 31. April was cold and humid with 35.3 mm more rainfall than usual. As a result, some winter rape has not over-wintered.

3. Experimental results

3.1 Weed incidence in the winter rape crop in the autumn and spring growing seasons

The following annual weed species predominated in the winter rape crop: *Chenopodium album* L., *Tripleurospermum perforatum* (Merat) M. Lainz, *Stellaria media* (L.) Vill., *Viola arvensis* Murray, *Veronica arvensis* L., *Sinapis arvensis* L., *Capsella bursa-pastoris* (L.) Medik, and *Poa annua* L.

In the soil with both regular and thickened humus layers, in the treatments where winter rape was grown with wide interrow spacings (48 cm), when the light and moisture conditions were favorable, the number of emerged weeds was higher than in the treatments with narrow interrows (12 cm), except for the spring growing season of 2017. The application of the bio-preparations in most cases reduced the number of weed seedlings in the winter rape crop both during the autumn and spring growing seasons.

Experiment I: in the soil with a regular humus layer. The most effective weed control method in organic winter rape crop was mechanical: efficiency 26.7–71.5% without biological preparations and 54.2–71.7% with biological preparations (**Figure 2**). The efficiency of the thermal weed control method was lower than the mechanical one. In Ref. [16], it was stated that the efficiency of mechanical and thermal weed control was 50–100%. Biological preparations enhanced the effectiveness of thermal and mechanical weed control techniques only in 2014. The effectiveness of the self-regulation method for the change of weed sprouts was negative throughout the study years.

Experiment II: in the soil with a thickened humus layer. The most effective weed control method in rapeseed crop was mechanical: efficiency was 39.8–75.1% without biological preparations and 53.0–68.9% with biological preparations (**Figure 3**). The efficiency of the thermal weed control method was lower than the mechanical one. Bio-preparations enhanced the effectiveness of mechanical weed control only in 2014. The method of self-regulation did not reduce the number of weed sprouts.

Figure 2.
The efficacy of the nonchemical weed control methods for the change in the number of weed seedlings in the winter oilseed rape crop, grown in the soil with a regular humus layer in the autumn (2014–2016). Note. T, thermal; M, mechanical; S, self-regulation; BP, bio-preparations.

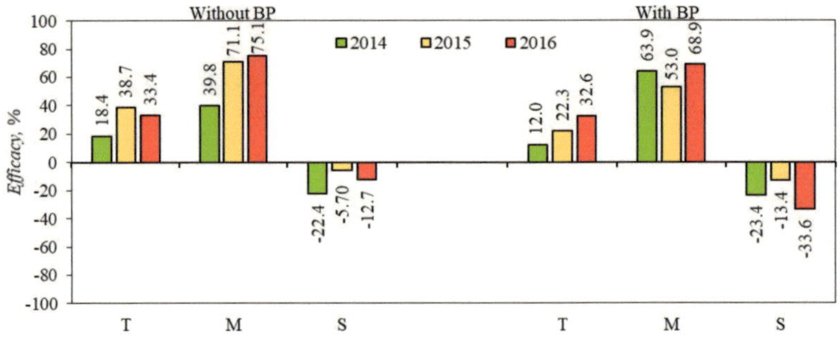

Figure 3.
The efficacy of the nonchemical weed control methods for the change in the number of weed seedlings in the winter oilseed rape crop, grown in the soil with a thickened humus layer in the autumn (2014–2016). Note. T, thermal; M, mechanical; S, self-regulation; BP, bio-preparations.

3.2 Weed incidence in the winter rape crop before harvesting

Experiment I: in the soil with a regular humus layer, the nonchemical weed management methods did not exert any pronounced effect on the weed number in the winter rape crop before harvesting throughout study years (**Figure 4**). Using bio-preparations in 2015 in experimental plots with mechanical weed control method, weed number was significantly 42.9% lower than in experimental plots with self-regulation. In 2016 and 2017, there were no significant differences in weed number between the different weed control methods.

The use of bio-preparations did not significantly affect the number of weeds in the oilseed rape crop during all study years compared to the treatment where they were not used.

There was no significant difference in weed dry matter mass between different weed control methods in plots with bio-preparations or without bio-preparations in 2015 (**Figure 5**). In 2016 and 2017, weed killing with interrow loosening and without biological preparations resulted in significantly lower weed dry mass than this in plots where thermal weed control and self-regulation methods were used, respectively, 2.0 and 2.5 times and 3.0 and 4.5 times. In 2017, the weed dry matter mass in plots with thermal weed control was significantly 32.3% lower than that of the self-regulation plots. In 2016 in plots with bio-preparations and mechanical

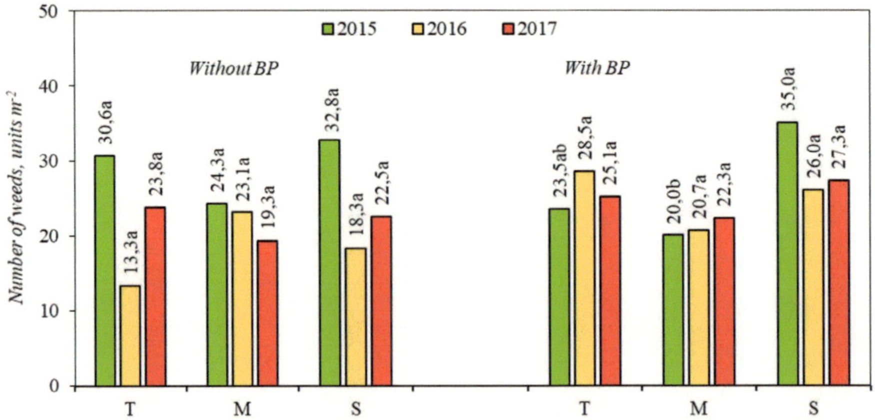

Figure 4.
The number of weeds in the winter oilseed crop, grown in the soil with a regular humus layer, before harvesting (2015–2017). Note. The differences between the averages of treatments of factor a, marked by not the same letter (a, b), are significant ($P < 0.05$). T, thermal; M, mechanical; S, self-regulation; BP, bio-preparations.

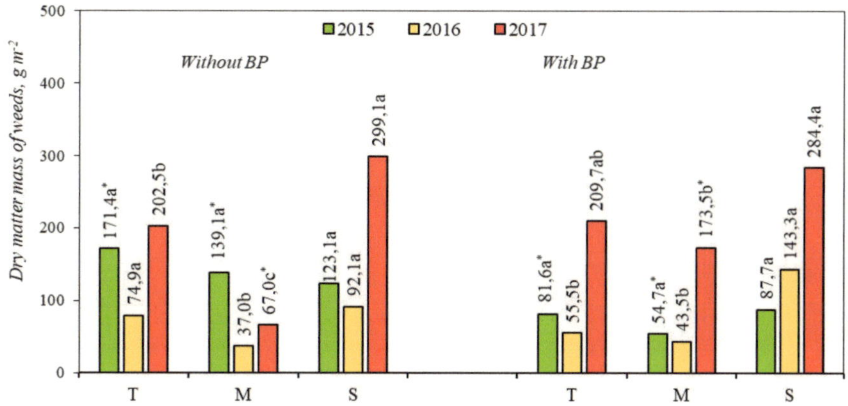

Figure 5.
The dry matter mass of weeds in the winter oilseed crop, grown in the soil with a regular humus layer, before harvesting (2015–2017). Note. The differences between the averages of treatments of factor a, marked by not the same letter (a, b, c), and between the averages of treatments of factor B, marked by an asterisk, are significant (P < 0.05). T, thermal; M, mechanical; S, self-regulation; BP, bio-preparations.

weed control method, the weed dry matter mass was significantly 2.6 and 3.3 times lower than that in the self-regulation weed control method plots. In 2017, mechanical weed control resulted in a significantly 39.0% lower dry mass of weeds than self-regulation method. It was found that thermal weed control resulted in a 44.0% reduction in dry mass of weeds using thermal weed control with water steam [16]. The similar results were obtained in the experiments in Lithuania—weed numbers were 3.2–4.4 times lower in plots with mechanical weed control method than this in plots with self-regulation weed control method and respectively weed dry mass by 2.2–3 times [17, 18].

With the use of biological preparations, the dry matter mass of weeds decreased significantly by 2.1 and 2.5 times in the fields of thermal and mechanical weed control methods only in the droughty 2015, as the more fertile rapeseed crop using biological preparations better suppressed weeds. K. Różyło and E. Pałys [19] found that as the assimilation leaf area of rape increased, the dry matter mass of weeds decreased. In 2017 with the use of biological agents, the dry weight of weeds increased significantly 2.6 times in the fields of mechanical weed control.

Experiment II: in 2015 in the soil with a thickened humus layer, interrow loosening without biological preparations significantly decreased weed number by 38.8 and 36.8% compared to thermal weed control and self-regulation (**Figure 6**). In 2016 and 2017, there were no significant differences in the number of weeds 0in the plots with various weed control methods without biological preparations. Using bio-preparations in 2015 in plots with mechanical weed control method, and in 2017 in plots with thermal weed control method, the weed numbers were significantly lower than in the plots of self-regulation weed control method by 42.9 and 34.7%, respectively. In 2016, there was no significant difference in the number of weeds in the plots with different weed control methods and with biological preparations.

Without biological preparations in 2015 and 2016, there were no significant differences in weed dry mass between the different weed control methods (**Figure 7**). Different weed control techniques used in plots with biological preparations in 2015 had no significant effect on the dry matter mass of the weeds. In 2016 in plots with mechanical weed control and with biological preparations, weed dry matter mass was significantly 4.1 and 5.1 times lower than that in plots with thermal weed control and self-regulation. In 2017 the dry matter mass of weeds was 6.7 and 5.7

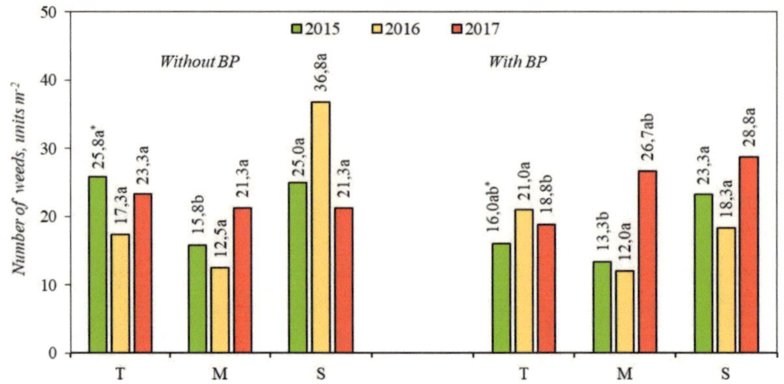

Figure 6.
The number of weeds in the winter oilseed crop, grown in the soil with a thickened humus layer, before harvesting (2015–2017). Note. The differences between the averages of treatments of factor a, marked by not the same letter (a, b), and between the averages of treatments of factor B, marked by an asterisk, are significant (P < 0.05). T, thermal; M, mechanical; S, self-regulation; BP, bio-preparations.

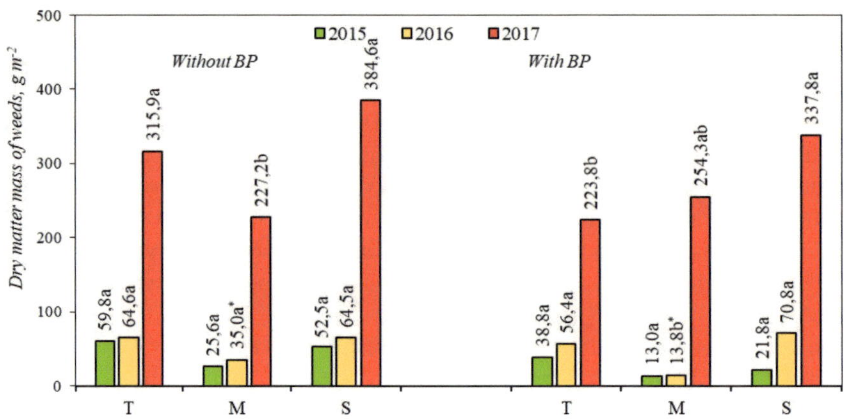

Figure 7.
The dry matter mass of weeds in the winter oilseed crop, grown in the soil with a thickened humus layer, before harvesting (2015–2017). Note. The differences between the averages of treatments of factor a, marked by not the same letter (a, b), and between the averages of treatments of factor B, marked by an asterisk, are significant (P < 0.05). T, thermal; M, mechanical; S, self-regulation; BP, bio-preparations.

times higher, respectively, without the use of biological preparations and 11.1 and 5.8 times higher with the use of biological preparations than in 2015 and 2016. Apparently, the higher weed dry matter content was caused by the lower rape crop and the humid and cold weather during rape vegetation. The mechanical weed control without biological preparations were found to have a significantly lower mass of weed dry mass than the thermal weed control and self-regulation, by 28.1 and 40.9% lower. The use of biological preparations and thermal weeds to control using wet water vapor resulted in a significantly by 33.7% lower dry mass of weeds than that in the self-regulation plots.

The use of biological agents significantly reduced weed dry matter in mechanical weed control fields only by 2016.

In 2017, negative, very strong, significant correlations were determined between the winter rape plant population density and weed dry matter mass: in the soil with a regular humus layer $r = -0.95$, $P < 0.01$; in the soil with a thickened humus layer $r = -0.91$, $P < 0.05$. [20]. R. Kosteckas (2011) [20] also found that the dry matter mass of weeds correlates with the density of rape crop.

4. Conclusions

Annual weeds predominated in the winter oilseed rape crop: *Chenopodium album* L., *Tripleurospermum perforatum* (Merat) M. Lainz, *Stellaria media* (L.) Vill., *Viola arvensis* Murray, *Veronica arvensis* L., *Sinapis arvensis* L., *Capsella bursa-pastoris* (L.) Medik., and *Poa annua* L.

In the soil with both humus layers, regular and thickened, the most efficient weed control method was mechanical weed management both during the autumn (efficacy 26.7–75.1%) and spring (efficacy 37.1–76.7%) growing seasons.

Thermal and mechanical weed control in combination with the bio-preparations in droughty years significantly reduced the number of weed seedlings. Dry matter mass of weeds most markedly decreased through the application of the mechanical weed management method.

Author details

Aušra Marcinkevičienė[1], Marina Keidan[2], Rita Pupalienė[1]*, Rimantas Velička[1], Zita Kriaučiūnienė[1], Lina Marija Butkevičienė[1] and Robertas Kosteckas[1]

1 Vytautas Magnus University, Kaunas, Lithuania

2 Public Institution "Ekoagros"

*Address all correspondence to: rita.pupaliene@vdu.lt

References

[1] Melnikienė R, editor. Agriculture and Food Sector in Lithuania. Vol. 2019. Vilnius: Lietuvos Agrarinės Ekonomikos Institutas/Lithuanian Institute of Agrarian Ecomnomics; 2018. p. 216

[2] Crnobarac J, Marinković B, Jeromela-Marjanović A, Balalić I, Jaćimović G, Latković D. The effect of variety and sowing date on oilseed rape yield and quality. Agriculture and Food. 2015;**3**:241-245

[3] Vincze E. The effect of sowing date and plant density on yield elements of different winter oilseed rape (Brassica napus var. napus f. biennis L.) genotypes. Columella. 2017;**4**(1):21-25. DOI: 10.18380/SZIE.COLUM.2017.4.1.suppl

[4] Alaru M, Talgre L, Eremeev V, Tein B, Luik A, Nemvalts A, et al. Crop yields and supply of nitrogen compared in conventional and organic farming systems. Agricultural and Food Science. 2014;**23**:317-326. DOI: https://doi.org/10.23986/afsci.46422

[5] Engstörm L, Stenberg M, Wallenhammar AC, Stahl P, Gruvaeus I. Organic winter oilseed rape response to N fertilisation and preceding agroecosystem. Field Crops Research. 2014;**67**:94-101. DOI: 10.1016/j.fcr.2014.07.01

[6] Valantin-Morison M, Meynard JM. Diagnosis of limiting factors of organic oilseed rape yield. A survey of farmers' fields. Agronomy for Sustainable Development. 2008;**28**(4):527-539. DOI: 10.1051/agro:2008026

[7] Zihlmann U, Scherrer C, Krebs H, Oberholzer HR, Vögeli GA, Nemecek T, et al. Integrierter und biologischer Anbau im Vergleich. ART-Bericht. 2010;**722**:1-16

[8] Velička R, Rimkevičienė M, Marcinkevičienė A. Changes in the properties of a loamy gleyic cambisol as related to the saturation of crop rotations with rape. Eurasian Soil Science. 2006;**39**(9):1002-2293

[9] Sirvydas A, Kerpauskas P, Čekanauskas S, Čingienė R. Vandens garo sklaida profiliuotame dirvos paviršiuje termiškai naikinant piktžoles. Vandens ūkio inžinerija. LŽŪU mokslo darbai (in Lithuanian). 2006;**30**(50):28-35

[10] Čekanauskas S. Aukštatemperatūrės aplinkos veiksniai termiškai naikinant piktžoles/High Temperature Environment Factors in Weed Thermal Control [thesis]. Akademija: Lithuanian University of Agriculture; Čekanauskas; 2007

[11] Staniulienė R. Aukštatem-peratūrinės aplinkos poveikis sunkiai termiškai sunaikinamoms piktžolėms [thesis]. Akademija: Lithuanian University of Agriculture; 2010

[12] Sirvydas K. Terminis piktžolių Naikinimas/Thermal Weed Control. Monograph. Akademija: Lithuanian University of Agriculture; 2012. p. 327

[13] Mockevičienė R. Necheminių piktžolių kontrolės priemonių ir biologinių prepratų įtaka vasarinių rapsų Agrocenozei / the Influence of Non-chemical Weed Control and Biopreparations on the Agrocenosis of Spring Oilseed Rape [Thesis]. Akademija: Aleksandras Stulginskis University; 2017

[14] Stancevičius A. Piktžolių apskaita ir laukų piktžolėtumo kartografavimas/ Weed accounting and field weed mapping. Akademija; 1979

[15] Raudonius S. Application of statistics in plant and crop research: Important issues. Žemdirbystė = Agriculture. 2017;**104**(4):377-382

[16] Raffaelli M, Fontanelli M, Frasconi C, Sorelli F, Ginanni M, Peruzzi A. Physical weed control in processing tomatoes in Central Italy. Renewable Agriculture and Food Systems. 2011;**26**(2):95-103. DOI: 10.1017/S1742170510000578

[17] Kerpauskas P, Sirvydas AP, Lazauskas P, Vasinauskienė R, Tamošiūnas A. Possibilities of weed control by water steam. Agronomy Research. 2006;**4**:221-225

[18] Velička R, Mockevičienė R, Marcinkevičienė A, Pupalienė R, Butkevičienė LM, Kriaučiūnienė Z, et al. Necheminių piktžolių kontrolės būdų efektyvumo palyginimas vasarinių rapsų pasėlyje ekologinės žemdirbystės sąlygomis/ the comparison of non-chemical weed control methods efficiency in spring oilseed rape crop under the conditions of organic farming system. Žemės ūkio mokslai. Agricultural Sciences. 2015;**22**(44):189-197. DOI: 10.6001/zemesukiomokslai.v22i4.321

[19] Różyło K, Pałys E. Impact of plant density on the canopy area index and weed infestation depending on different cultivars of winter oilseed rape (*Brassica napus* L., ssp. oleifera Metzg.). Annales Universitatis Mariae Curie-Skłodowska Lublin–Polonia. 2014;**LXIX**(2):44-55

[20] Kosteckas R. Skirtingu Intensyvumu tręštų vasarinių rapsų (*Brassica napus* L.) Biopotencialo Formavimosi dėsningumai įvairaus Tankumo pasėliuose/the Regularities of Biopotential Formation of Spring Oilseed Rape (*Brassica napus* L.) Fertilized at Different Intensities in Different Crop Stand Densities [Thesis]. Akademija: Lithuanian University of Agriculture; 2011

Section 6

Socio-Economic Aspects

Chapter 8

Socio-Economic Dimensions of Adoption of Conservation Practices: What Is Needed to Be Done?

Nisa Sansel Tandogan and Haluk Gedikoglu

Abstract

Promoting sustainable agricultural production requires farmers to adopt new technologies such as organic farming to increase the agricultural productivity, while conserving the environment. Adoption and diffusion of new technologies need a long process, as experienced in the past. There are social and economic factors, identified in the literature, and those could cause delays in farmers' use of new technologies. Hence, technology adoption and diffusion are important policy issues in agriculture. For that reason, this paper provides a literature review including factors influencing the adoption and diffusion of technology in agriculture and aims to contribute to the future studies and policies, especially focusing on the social capital or the social aspects, which are proven not to be analyzed by the previous studies comprehensively. The results show that interaction with neighbors and relatives, and membership in a group or organization, which represent the social aspects, has a positive influence on adoption and diffusion of new technologies. Hence, policy-makers should incorporate the social aspects when designing the policies, such as cost sharing programmes, to promote adoption and diffusion of new technologies.

Keywords: technology adoption, diffusion of innovations, conservation practices, sustainable agriculture, social capital

1. Introduction

Promoting sustainable agricultural production requires farmers to adopt new technologies to increase the agricultural productivity, while conserving the environment. Since the seminal study by Griliches, adoption of new technologies has been widely analyzed in the agricultural economics literature [1]. Technology adoption in agriculture is also analyzed in the rural sociology literature. The studies in rural sociology mostly focused on diffusion of new technologies in a region, whereas studies in agricultural economics focused on adoption of new technologies by an individual farmer. Initially, the focus of agricultural technology adoption was to increase the productivity of the farmers or the profitability, especially during the Green Revolution. Hence, profitability of the technology was found to be an important factor for adoption [1].

 IntechOpen

Previous studies incorporated factors related to technology, farm and farmer characteristics into the analysis to explain why a technology is adopted or not. Hence, starting with profitability of the new technology, many factors were identified in the literature influencing the farmers' use. Since the 90's environmental concerns becoming important, and recently the global warming, the focus of technology adoption has shifted mostly from productivity-increasing technologies to sustainable agricultural technologies and the conservation practices. Especially for the developed countries, the focus is more on conservation practices and technologies such as precision agriculture that would both increase profitability and conserve the environment. The 2030 Agenda for Sustainable Development has mainly focused on the global aims, which include demanding unprecedented actions and efforts across multiple interconnected social, economic and environmental issues. In this sense, science, technology and innovation have a big share to realize these aims. The contributions of technology with the innovations to the economies provide the opportunities to improve living standards through rise in productivity, fall in costs and prices, and contributions to

Factor category	+	—	D	A	N	Total	References
Age	3	2	1	0	0	6	[22, 24–27, 49]
Gender	3	0	0	0	0	3	[24, 26, 28]
Income	2	0	1	0	0	3	[17, 26, 29]
Off-farm income	4	0	1	0	0	5	[30–34]
Ownership and wealth	6	0	1	0	0	7	[10, 18, 22, 24, 26, 36]
Education	14	1	0	0	0	15	[7, 11, 18, 23, 24, 27, 37, 38, 42, 43, 50, 68]
Learning	6	0	2	1	0	9	[18, 28, 44–49, 69]
Information	9	0	0	0	0	9	[7, 15, 17, 25, 26, 38, 41, 50]
Social capital	3	0	0	3	0	6	[57, 59, 60, 61, 63, 64]
Norms	1	2	0	0	0	3	[57, 59, 66]
Neighborhoods	4	0	0	2	0	6	[22, 24, 25, 44, 46, 48]
Relatives	4	1	0	0	0	5	[25, 26, 28, 36, 61]
Contacts with extension agents	4	0	0	0	0	4	[27, 43, 60, 68]
Membership in a group and organization	3	0	0	1	0	4	[25, 27, 36, 61]
Experience	4	0	0	0	0	4	[41, 46, 50, 69]
Farm size	5	0	1	0	0	6	[11, 15, 27, 29, 37, 41]
Farm location	3	2	0	0	0	5	[6, 22, 25, 36, 60]
Characteristics of soil and land	3	0	0	2	0	5	[10, 36, 37, 59, 60]
Prices of inputs	3	0	0	0	0	3	[10, 19, 70]
Risk	3	1	1	1	0	6	[7, 11, 29, 45, 50, 61]
Credit accessibility	5	1	1	1	0	8	[17, 19, 26, 27, 29, 59, 60, 61]

+: *affect positively.*
-: *affect negatively.*
D: significance of the effect changes depending on situation.
A: the sign of the effect changes based on situation.
N: no information.

Table 1.
Factors influencing technology adoption in agriculture.

the real wages [2]. It also plays a crucial role in the field of agricultural development [3, 4]. Technology decreases the risk of diseases and pest and increases the productivity and developments in agriculture by providing more information about the crop and soil structure for the farmers [5–9]. Additionally, new technologies are shown as a solution to the impeding conditions in agriculture, such as water scarcity, drainage and pollution [10–13]. Not only for the crops and soils but also for communication and information, technology contributes to the agriculture by providing several communication infrastructures and digital portals, reducing human intervention and eliminating technology breaks [14]. All of the effects mentioned above are important factors for sustainable agriculture. However, in this respect, there are two important issues: the adoption and the diffusion of technology. In the field of agriculture, with the well-known study of Feder and Slade, there have been many studies about agricultural innovations and their adoption [15]. As it is understood from the definition of technology for adaptation by UNFCCC (2005), which is stated as 'the application of technology in order to reduce the vulnerability, or enhance the resilience, of a natural or human system to the impacts of climate change', technology adoption influencing agricultural productivity significantly is an important issue in the economic sense [16]. In parallel with the adaptation, the diffusion of technology in agricultural field has been approached in many studies [17–19] beginning with Griliches [1]. Some tools have also been used to predict the speed and level of adoption [20]. As a reason, technology should have been used and diffused to benefit from its advantages; however, to do that, profitability needs to be understood by adopters [21]. When analyzing these studies, it is seen that there are many factors affecting technology adoption and diffusion (see **Table 1**). The main object of this paper is to review the factors influencing the adoption and diffusion of technology in agriculture comprehensively and contribute to the future studies.

2. Methodology

While this review handles the factors influencing the adoption and diffusion of technology in agriculture, it specifically focuses on the social capital including norms, neighborhoods, relatives, contacts with extension agents and membership in a group or an organization. Accordingly, the literature from 1974 to 2018 was analyzed by taking the social capital as a base. Our paper consists of 78 reviews, which cover 56 articles in journals, 10 book chapters, 4 selected papers for presentation, 3 working papers, 3 reports, 1 conference and 1 discussion paper. Rather than examining a specific time period or a specific region, 44 empirical studies, which cover various time periods and places, were analyzed to find the factors influencing the adoption and diffusion of technology. The revealing 21 factors were collected under the three main titles: the characteristics of farmers, the characteristics of farms and external incentives or disincentives. The effects of factors analyzed were tabulated depending on their positivity, negativity and variability to ease the future studies.

3. Factors influencing technology adoption in agriculture

3.1 Characteristics of farmers

3.1.1 Age

In the decision-making process, age is another influencing factor because it affects attitudes and perspectives on the new technology. Many studies about the effect of age on technology adoption indicate different results. The analysis of Case shows that if the age of the additional male household member is between 0 and 15, farmers more adopt the new technology but if it is above the age of 55, the adoption rate decreases [22]. Hua et al. support that idea with their study by showing that while farmers under the age of 50 are more likely to adopt one type of new technology, farmers over 60 have lesser tendency to the adoption [23]. However, for another kind of new technology, the same study also shows that there is no significant difference between farmers under the age of 40 and over 59. Hence its effect can be attributed to the type of technology. From a different perspective, Weir and Knight found that farmers tend to be more influenced by someone in the same age group; however, there is a significant difference in this point [24]. If the farmers are educated, they have been more affected by older people rather than being influenced by people in the same age group. In this sense, younger farmers are more likely to be educated because they have more role models than older farmers. Contrary to most studies, some studies assert that older farmers have more tendency to adopt new technology even earlier than younger farmers due to their extensive experience in farm, and increase in the age shows positive effect on the adoption [24–26]. The study of Abdulai et al. differentiates the effect of age on adoption by determining criteria rather than differentiating it as older and younger [27]. According to their studies, although an increase in age at the younger ages affects adoption positively, increase at the older ages affects adoption negatively.

3.1.2 Gender

Differences stemming from gender have been observed in many subjects, the adoption of new technology in the agricultural field also shows different results depending on the gender. The study of Weir and Knight demonstrates that people are influenced by the people of the same gender while trying the new inputs [24]. Although the percentage being influenced is different among males and females, it is clear that they have a tendency to copy what people of the same gender do. The findings in their study show that female-headed households and male-headed households are influenced by the households of the same gender by 53% and 94% respectively. The adoption rate of female and male households differs from each other. Bandiera and Rasul studied on the social networks and technology adoption in Mozambique, which is mentioned as sunflower adoption in the study, and concluded that female-headed households have more tendency to adopt [28]. Other findings by Deressa et al. support the effect of gender differences on the decisions of the farmer by showing more adoption of male-headed households to climate change such as conserving soil, changing crop varieties and planting trees [26].

3.1.3 Income

The rises and falls in income have an effect on the decisions of adoption because farmers may change their preferences according to their new income level. Using new agricultural products causes changes in income, and income is doubled when modern crop yields per acre are used rather than traditional technology [17]. However, the use of new agricultural products or technology also depends on income because it needs financial facilities. In regard to this issue, the study of Feder and O'Mara indicates that the larger landowners benefit from the innovations more than smaller farmers because of their different income levels [17]. While experiencing the innovation is too risky for smaller farmers and hence they abstain from the adoption, when there is a fall or stability in the relative risk aversion with income, larger farmers allocate more land for new yields. They also suggest that income distribution that occurred with the different income levels is improved as the smaller farms switch to new technology. Correspondingly, the diffusion process is completed when the incomes of all farmers increase and thus it is possible to worsen income distribution in the initial stages until the smaller farmers participate in the adopters' ranks. From a different perspective, Feder shows that the larger farmers allocate relatively less land for the risky activity and associate income with risk aversion [29]. He asserts that, in the case of binding credit constraint, increasing farm size leads to decrease in the allocated land for the modern crop in the situation of increasing relative risk aversion with income. Then, the rise of relative risk aversion with income decreases the expected income per acre and this causes the improvement of income distribution via the introduction of modern crop. On this issue, Deressa et al. analyze the influence of farm income and non-farm income on adoption [26]. It is shown that when farming is the main source for the income, farmers have the tendency to make an investment on productivity, and farm income affects conserving soil and using different crop varieties. On the other hand, although non-farm income has a negative effect on these modern agricultural methods even if it is not significant, it has a positive effect on the possibility of planting trees and using irrigation as an adoption option. Both farm income and non-farm income also increase the possibility of changing planting dates.

3.1.4 Off-farm income

Off-farm income, as in income provided by farm, is also a significant factor for the adoption of technology because it contributes to the economic performance of the farm household [30]. It helps farmers to increase capital availability and financial resources to invest in new inputs, practices or technology [31, 32]. Hence, various studies indicate that off-farm income has a positive impact on the adoption of new technology [30, 31, 33]. However, this impact has shown differences depending on the technologies. Gedikoglu et al. assert that although there is a positive relationship between the adoption of capital-intensive technology and operator's off-farm employment, it cannot be supported for the relatively labour-intensive practice [32]. Fernandez-Cornejo et al. confirm that farms with labour-intensive enterprises less prefer off-farm work [30]. Moreover, the decisions taken by the operators to work off-farm have an influence on the decisions of their spouses and that decision for off-farm work has also a positive relationship with adoption. This relationship is not one-sided, there is a correlation because adoption also enables the off-farm work [34]. The study of Fernandez-Cornejo et al. indicates that the more off-farm income increases, the more the probability of adoption of technologies increases for better time management [30]. For lower off-farm income, it is seen that fall in the off-farm income directs farmers to adopt yield monitors. From another perspective, this study emphasizes that the increase in off-farm work activities leads to a decrease in the farm-level efficiency because of allocating less time for farm management. This situation affects the adoption of management-intensive technologies negatively. In this sense, the study by Goodwin and Bruer provides an explanation that because crop producers make seasonal production, they are more advantageous than livestock product producers who have to work year-round [35]. Although there are some factors influencing off-farm employment stated in the literature such as farm size and wages, the positive effect of off-farm work on the adoption of technologies is apparent [30, 32, 35].

3.1.5 Ownership and wealth

Incomes and livestock owned represent wealth in agriculture [26]. Therefore, most of the studies taking livestock and machinery as a measure for wealth show that as the household wealth increases, the farmers have more tendency to experience new agricultural products and adopt them [22, 36]. However, this relationship is not stable, it changes as less wealthy households adopt in time [18]. The study of Weir and Knight on adoption and diffusion of agricultural innovations draws a conclusion that a great majority of households are influenced more by richer households in the process of adoption decision and this factor is valid also for the adoption decisions of educated households [24]. In parallel with the effect of wealth, ownership has an impact on adoption. While the study of Deressa et al. shows the positive effect of livestock ownership on adoption methods [26], Kassie et al. emphasize its positive and significant effects on the adoption of improved seeds [36]. Land tenure is also evaluated as a factor within the scope of ownership. Kassie et al. who examined this issue in detail indicate that land tenure has an influence on adoption in terms of security because while better tenure security increases the probability on farmers' investments, worse security decreases the adoption of some agronomic practices on rented plots [36]. Additionally, being a tenant is also effective in the decision process because while tenants focus on short-term soil fertility by overusing chemical fertilizers, owners take a long-term decision about soil fertility on their plots, thus its effect in the short term is stated as ambiguous.

3.1.6 Education

As in most of other fields, education has an impact on technology adoption in agriculture. There have been many studies that show the positive effects and contributions of the education on technology adoption from different perspectives. First of all, education increases the decision-making efficiency of adoption [37] and improves the systematic and creative thinking skill for making innovative decisions [38, 39]. It helps the understanding of the effects and the results of the technology adoption [38]. Cotlear indicates that all three types of education, which are formal, nonformal and informal education, play a great role in the diffusion of innovations through the rise in farm productivity, explanation of the information and shaping of behaviors, beliefs and habits [40]. From a different perspective, education has a significant effect on the initial adoption of innovations [24] because it leads to decrease in adoption costs and uncertainty and hence the timing of adoption becomes shorter and the probability of early adoption increases [41]. Rise in productivity also has been seen as being related with high education because of its contribution on the information acquisition and the accessibility of improved technologies [42]. The study of Huffman indicates that benefiting from the available information totally depends on the education level of decision-makers [43]. He also states that education improves the skill to obtain and process information and accelerates the changes, and the production increases at the end of the process.

3.1.7 Learning

Learning is another element influencing technology adoption. Social learning has an important role in the farmers' decisions and knowledge diffusion, and the investments in learning about technology are related with the technology adoption [44]. Besley and Case explain the cause of the slow technology diffusion as the lack of learning its profitability [18]. In line with this idea, learning is explained as the fall in the likelihood of allocative error by obtaining more information about the likelihood of output [45]. There have been many studies about the understanding of how learning comes true. A study conducted by Foster and Rosenzweig examines both learning by doing and learning from others [46]. Their study emphasizes that if the learning realizes with both neighbors' and own experiences, profitability occurs more rapidly. In this sense, while Krishnan and Patnam support the power of social learning in adoption, they claim that the effect of learning from farmers' own and farmers' neighbors' experiences on adoption is greater than the effect of learning from extension services, in their study [47]. Conley and Udry also assert that neighbors influence the behaviors of individuals in the lack of learning, and individuals may behave according to their preferences because of being subjected to an unknown thing [49]. Munshi states that when the technology performance depends on the latent characteristics of neighbors, social learning will be weaker [48]. As a reason, farmers take into account not only their own direct observations of realizations but also those learned by neighbors [49]. Even, Besley and Case provide the reason of early adoption as being forward looking by learning more about the new technology [18].

3.1.8 Information

There is an extensive literature about the effect of information on the adoption in agriculture because it is an important stage for the adoption and diffusion of technology. It has been considered that the resource allocation skills and efficiency of adoption decisions can be increased via information gathering [7]. However, this information-gathering process can be affected by the adopters' specific attitudes [50] and hence the differences in the interpretations of information lead to different adoption decisions [38]. In this sense, especially, uncertainty and the lack of information are the subjects, which are mostly touched on. Imperfect information and uncertainty have an influence on the adoption of the decision and information diffusion affects the adoption positively by reducing the uncertainty [50]. Fall in uncertainty also decreases the cost over time [17] and hence the farmers who have more information tend to adopt earlier more than other farmers who do not [15, 41] because they consider other options less valuable to wait [7]. However, information access is different for each farmer and adoption changes accordingly. Feder and Slade indicate that the resources used for obtaining information by larger farmers are more than those used by smaller farmers and this leads larger farmers to have more knowledge and adopt earlier [15]. From a different perspective, Wozniak states that information increases innovative ability and while having more information makes farmers innovators, having less information makes them operator [38]. As a source of information, agricultural extension services and private agricultural supply firms have an important role in the agricultural sector [41] but information obtained from agricultural extension services has been seen as more valuable than information obtained from private agricultural supply firms because while the private firms can provide service for profit of the firm, agricultural extension services exist for giving technical information [38]. He added that the diffusion of the information has influenced production and welfare positively.

3.1.9 Social capital

The concept of social capital, which was first mentioned by Hanifan, has been defined in many different ways since there is no consensus about the definition [51–54]. One of the most important differences differentiating this term with other kinds of capital has been seen as its existence in social relationship [55]. In this sense, one of the most accepted definitions about the social capital was made by

Smithson as 'social capital is a person's or group's sympathy toward another person or group that may produce a potential benefit, advantage, and preferential treatment for another person or group of persons beyond that expected in an exchange relationship' [56]. Thus, its measures have been generally taken as networks, trust and norms [57, 58]. Many studies about the social capital have shown that it also affects technology adoption in different ways [57, 59]. According to these studies, firstly, social capital enables farmers financially in the lack of credit accessibility [60, 61]. Secondly, social networks which are a part of social capital decrease the transaction costs [62] because the major driver of that costs is the lack of information and contract enforcement assistance [63]. In the same way, it fills the information gap leading to market inefficiency [64]. Social capital also contributes to the well-being of an agent by influencing that person's relationship with others [65]. All these effects cause the rise in production and adoption. On the other side, some studies about the influence of social capital on adoption indicate that social capital may also influence adoption negatively because it depends on technology [60]. However, the studies about its positive sides on the adoption are quite a few than its negative sides. By benefitting from the previous studies, the instruments of social capital, in this paper, have been taken as norms and networks including neighborhoods, relatives, contact with extension agents and membership in a group or organization.

3.1.9.1 Norms

As it is shown many times in the literature, individuals are affected by the opinions and decisions of others. Social norm, in this sense, is another factor that influences the likelihood of adoption. Läpple and Kelley examined this issue and suggested that the decision of adoption is made by not only farmers but also others [66]. It is found that belief-based subjective norms are a prominent motivation for farmers to convert their intentions in their study. Hunecke et al. studied on the topic of understanding the role of social capital in adoption decisions and concluded that norms affect the adoption of scheduling significantly and negatively [57].

3.1.9.2 Neighborhoods

One way to obtain information is from neighbors, and the effects of neighbors on adoption should not be overlooked to obtain unbiased and valid results because it influences both decisions and actions of farmers [22]. It is seen that if there is no opportunity for social information, individuals have to experience on their own but individual information does not compensate for the information gained from neighbors [48]. While more educated household heads or novice farmers have more tendency to learn using new inputs or obtain information from their neighbors [24, 44], farmers, who have the lower ability for information obtained from neighbors, have less reaction and slower adoption speed [44]. Also, if there is not a feasible environment, the person under the social pressure may prefer to act like a neighbor to improve the productivity [44, 46]. Neighbors contribute to farmers in terms of teaching input use and they take an active role in providing information with agricultural extension services [24, 46]. Not only for using input but also for changes in the use of inputs, neighbors are an important information source. The study of Conley and Udry (2010) emphasizes the strong relationship between changes in use of inputs and neighbors and indicates that if an information neighbor makes higher profits than expected by using more inputs, the farmers have a tendency to increase their input uses [44]. In this sense, the experience of the neighbors is one of the main points. According to the findings of Munshi, if the experiences of neighbors cannot be observed well by the individuals, weak social learning and slow diffusion rate will occur [48]. Similarly, with this study, Foster and Rosenzweig indicate that profitability stemming from new technology increases with the individuals' own experiences and those of their neighbors; even an increase in the experience of neighbors approximately doubles the profitability according to the same increase in individuals' own experience [46]. The study also shows that decisions taken by neighbors play an important role in the decisions of farmers, future decisions taken by farmers are influenced by the past decisions of neighbors, and this shapes the planning decisions of farmers. Even this influence may outweigh the influence of extension services; Krishnan and Patnam assert that the influence of adoption by neighbors is approximately three times higher than the influence of extension agents [47]. Being approved by the society is another reason encouraging the farmer to adopt new agricultural technology [25]. In this sense, access to neighbors is an important factor in the adoption of new agricultural technology. Findings indicate that farmers who have access to neighbors have more tendency to adopt the new technology and adopt more quickly with information obtained [48, 59]. Wollni and Andersson state that if the neighbors are adopters, farmers become more disposed to adopt [25]. However, Munshi supports that even if an individual observes the decisions of neighbors well, that person may not obtain the same result with theirs because of the effect of different characteristics on performance [48]. On the other hand, although there are many benefits of the connection with neighbors, some farmers prefer to abstain from adoption because of the idea that the adoption also provides benefit to their neighbors' plots [25]. Nowak, whereas, defends the opinion that collective work is important to solve local resource management problems [67].

3.1.9.3 Relatives

When it is considered that the adoption decisions of farmers are affected by external factors, the effect of family on the decisions taken is inevitable. The findings of Bandiera and Rasul indicate that there is a strong correlation between adoption decisions and family [28]. As the number of relatives increases, the adoption possibility of households increases because of the opportunity for experiencing new technologies with lower risk; however, at the same time, great numbers of relatives lead to decrease in the work efficiency [36]. The factor that increases the probability of adoption can be shown as the labour supply provided by family and lower opportunity costs [25]. Deressa et al. confirm this positive relationship although the coefficients are not significant in their study and show that having more relatives in a local place affects adoption positively [26]. The study of Bandiera and Rasul examines the effect in terms of being the adopters of family members and concludes that increasing the number of adopters in the family increases the adoption probability of farmers [28]. However, this situation is found as valid just up to 10, then the marginal effect of the network is negative, so the relationship is shown in an inverse-U shape. Contrary to these studies, Wossen et al. show that compulsory sharing and strong loyalty among kin members have a negative effect on the adoption of farmers because they cause the free-riding behaviors [61].

3.1.9.4 Contacts with extension agents

As in mentioned in the effect of gathering information on adoption, the sources providing information also have an influence on the adoption; one of them is agricultural extension services. The study of Huffman who addresses the extension activity as an indicator of information availability indicates that there is a positive and significant effect of the availability of information on the information gathering and processing it if adjustment is needed [43]. Also, agricultural extension services can substitute education in allocative efficiency. The information provided by the extension services helps the farmers in terms of understanding the process and using new technology; hence farmers who are in contact with agricultural extension services adopt the new technologies more [27]. The study of Husen et al. on this issue proves that when it is compared to the farmers without contact with agricultural extension services, the adoption of farmers having contacts with extension services increased by 28.85% for productivity-enhancing technologies [60]. From a different perspective, Huffman examines the role of human capital in the farm and off-farm work decisions in his study by including the effect of agricultural extension as an input [68]. His findings support that there is a positive relationship between the off-farm labour supply and the agricultural extension input because the extension services increase the productivity of farm and shorten the time for gathering information. This situation leads to shift in the demand of farmers for farm work and increases the farmers' days of off-farm work.

3.1.9.5 Membership in a group or an organization

Being a member of an organization or an institution is important in terms of accessing information and knowledge about new practices and technologies [25]. As shown in the literature, the better and easier access to information mostly has a positive effect on the adoption; hence the membership in an organization and institution also plays a role in the adoption. The study of Wollni and Anderson proves that membership in a farmer group, which provides assistance and information for farmers, increases the adoption of organic agriculture by 26% [25]. Wossen et al. examined this issue in detail, indicating that being a member of an association is not only important for gathering information but also for providing financial resources and having labour-exchange options [61]. As a reason, while membership in informal credit and saving associations helps farmers by solving liquidity problems, a member in labour-sharing arrangements may benefit from the opportunities of labour resources provided by those arrangements. These opportunities relax farmers and facilitate the adoption. The findings of their study show that these associations increase the adoption; when a farmer becomes a member of a local saving credit association, the likelihood of adoption of land management practices increases by 19.4%. Corresponding to the result of this study, Kassie et al. and Abdulai et al. confirm the positive relationship between being a member in an institution or association and adoption of new agricultural technologies or practices [27, 36]. On the other hand, the study of Wossen et al. emphasizes that the effect of this relationship depends on the type of institution or arrangement because it is found that being a member in funeral insurance arrangements or having a great number of relatives has a negative effect on the adoption of improved land management practices [61]. For the reason, funeral insurance arrangements direct farmers to make their social commitments and make money for funeral expenses, leading to abstaining from the agricultural innovation. The kinship also decreases the expectations from the adoption and so the likelihood of adoption because of the low incentives for collective sharing of benefits from adoption.

3.1.10 Experience

Although there are many beliefs and opinions about the new technology, these change with experience in time [18]. Especially in the adoption process, prior adoption experience has a significantly positive effect on adoption intensity [50].

The study of Wozniak shows that while uncertainty and the fixed costs of adoption obstruct early adoption, gaining experience increases the probability of being the early adopter of farmers because the more farmers gain experience, the more they cope with the difficulties of the adoption process [41]. Experience also affects the productivity positively [69]. Foster and Rosenzweig, who studied on the topic of learning by doing and learning from others, conclude that the experience of farmers and their neighbors allows them to take better decisions about the use of new technologies, and these decisions affect the profitability of adoption [46]. It is shown that experienced neighbors make more profit for farmers than inexperienced neighbors, and moreover, an increase in average experience of a farmer's neighbor affects the profitability of a farmer by approximately twice more than own experience. Conflictingly, although Wozniak supports the positive relationship between experience and the adoption time, he asserts that the relationship between experience and adoption shows differences because experience depreciates in a technological environment in time and hence the result may be biased in case of no any measurement for specific experience of adoption behavior [41].

3.2 Characteristics of farm

3.2.1 Farm size

The likelihood of adoption of new technologies has been also associated to farm size, in the literature, because the probability of adoption has changed significantly depending on farm size [37, 41]. One of the reasons for this difference has been shown as information accumulation. Feder and Slade emphasize that there is a positive relationship between the accumulation of information and farm size. The more farm size becomes large, the faster critical level of information is reached because larger farmers allocate more resources to get information [15]. All of these lead to the earlier adoption. Supportively, Wozniak and Rahm and Huffman confirm that larger producers have a relatively greater incentive to obtain information about innovations; hence they spend more time and more money for better quality information [37, 41]. This more allocation for information makes large-scale producers early adopters by allowing them to adjust the inputs according to the innovations. When profitability and costs of the inputs are taken into consideration, Torkamani and Shajari also support that the larger farms adopt new technologies more rapidly than smaller farms to prevent water cost and derive more profit [11]. In contrast with these studies, the study of Feder suggests that larger farmers tend to allocate relatively less land to the modern crop so as to not endanger their wealth because of the risk factor [29]. Besides all these, Abdulai et al. emphasize that farm size is not the only factor influencing adoption, cropping patterns and physical characteristics also have an effect on the adoption because while farm size has a positive and significant effect on one product, it may not have any significant effect on another product such as onion [27]. However, the effect of farm size on the selection of product for growing is also shown in their study in that while larger farmers tend to grow some kind of agricultural products such as onions and cabbage, smaller farmers tend to grow other products such as lettuce.

3.2.2 Farm location

When obtaining information about the new technology, the location of farm is important in terms of ease of accessibility and availability of information. While opportunities that stem from the farm location make adoption easier and shorten the adoption time, negative things caused by farm location make adoption difficult and extend the time. The study of Khanna, which is one of the studies on this issue, handles the farm location as a main factor influencing adoption of soil testing [6]. It is shown that the more proximity of farmers to professional services increases, the more likelihood of soil testing adoption increases because most farmers trust the services provided by professional dealers. By confirming this relationship between farm location and adoption, the study of Kassie et al. indicates that the location has an effect on the investment decisions of farmers and hence being far from a village or a household that has more opportunities in terms of input and output decreases the likelihood of adoption of sustainable agricultural projects [36]. Case and Husen et al., who handle the subject in terms of accessibility to agricultural extensions, also emphasize that while proximity to agricultural centre affects adoption positively, parcel distance and being distant from their farm affect their adoption and farm management negatively [22, 60]. Due to these positive effects on adoption, it is shown that proximity to main market center provides better access to organic market outlets and hence their adoption becomes easier [25]. Eventually, when information is considered as positively related with the adoption, it is clearly seen that the accessibility to information, which means the closeness of farm location to the information sources, increases the likelihood of adoption.

3.2.3 Characteristics of soil and land

Many research studies conducted on technology adoption in agriculture prove that soil characteristics have a certain effect on the yield [37]. The study by Rahm and Huffman indicates that reduced tillage practices, which are used to measure adoption, influence the yield positively on the soils having poor characteristics [37]. Even, they expect to be dependent of per acre profitability on soil characteristics [10] because technologies decrease problems, which can stem from climatic conditions and natural events, by providing necessary conditions as required by the soil [37]. Larger and unfavorable fields for agriculture have more tendency to be equipped with modern technologies [10]. The probability of adoption changes depending on the soil characteristics and hence adoption shows the differences among farms [37]. Rahm and Huffman and Isham indicate that the probability of adoption under better soil conditions is higher [37, 59]. The benefits of reducing unfavorable conditions and improving soil characteristics provided by adoption affect the crop production value, input expenditures, productivity and sustainability significantly [36]. However, in this sense, the type of technology also causes differences in adoption. For instance, the study of Husen et al., examining the adoption of soil and water conservation practices (SWCs) and productivity-enhancing technologies (PETs), indicates that although land slope affects the adoption of SWC positively, it has a negative effect on the adoption of PET [60]. The adoption differences in terms of soil fertility, parcel distance and agricultural extension had also been observed in this study.

3.3 External incentives or disincentives

3.3.1 Prices of inputs

In the process of diffusion and adoption, the prices of inputs have an effect on the decisions of farmers. Dinar and Yaron suggest that past and future price expectations for inputs and outputs have importance in investment decisions for new equipment, which is mentioned in their study as irrigation equipment [70]. Moreover, the rise in these prices affects the use of modern technologies positively. In contrast to this study, the findings of Dinar et al. indicate the possibility that although the price of cotton increases and thus the area allocated for cotton increases, the amount of farms equipped with modern technologies decreases in their study [10]. With these two opposite results, the study of Abdulai and Huffman concludes that the effect of price on the adoption and diffusion depends on the phases [19]. While the expected price of the new technology affects the diffusion process positively and significantly in the early adoption phase, it is shown that the price of new technology in the second phase does not have a significant effect on the adoption. The positive effect of the expected price can be shown by the result of its diminishing time and accelerating adoption effect, in their study.

3.3.2 Risk

Risk is one of the most commonly addressed issues about technology adoption in agriculture, in the literature. The study by Koundouri et al. indicates that risk has an important effect on the adoption decision process for a new technology [7]. They assert that farmers tend to invest in and adopt new technology more to avoid the production risk they encounter with the risk of extreme outcomes. In their study, it is shown that farmers who face adverse climatic conditions adopt new technologies to decrease the risk level. Uncertainty about the profitability, which is a risk factor, also increases the probability of farmers' adoption because the adoption of new technology decreases the production risk, risk premium and relative risk premium [7, 11]. If the producer is decisive in adopting, the adoption degree is also affected by the risk factors [50]. For instance, the larger farmers allocate relatively less land and hence smaller proportion of their incomes to risky activities, which means the higher input of fertilizer per acre [29]. In this sense, Hiebert suggests that while risk-preferring farmers tend to use more land and fertilizer for production than the risk-neutral farmers, risk-neutral farmers tend to use more inputs than risk-averting farmers [45]. Correspondingly, Wossen et al. assert that, in the adoption decision process, risk-averse households trust their social capitals in terms of adoption decision more than risk-loving households [61]. Networks and traditional sharing norms such as the social capital affect the risk-mitigating measures negatively [71]. On the other hand, Wossen et al. suggest that the relationship between social capital and risk aversion changes among households [61]. While some results show a significant and negative effect on risk aversion, some show a positive effect in their study.

3.3.3 Credit accessibility

No doubt that having a financial potential is necessary while experiencing an innovation. Financial assets also have an influence on adoption, and hence credit constraint leads to different adoption rates [17]. In the literature, there are many studies that show the importance of credit accessibility in the adoption. Access to credit facilitates investment because it provides the support for liquidity requirements [61]. Through this support, farmers may buy the inputs such as fertilizer and benefit from the facilities more easily; it also influences the change in planting dates and using irrigation systems positively [26]. By depending on its contributions to the agriculture, most studies indicate that credit availability encourage people to adopt, and increase the adoption by accelerating [19, 27, 59]. However, on the other hand, Husen et al. state that the effect of credit availability on adoption depends on the technology because they show that while credit access has a positive effect on the adoption of one of the technologies, it has a negative effect on the adoption of other technology in their study [60]. This negativity has been explained in that credit access may direct people to non-agricultural sectors for the investment. From a different perspective, Abdulai and Huffman emphasize that if credit constraint can be substituted with another financial source such as household savings, the constraint does not influence the adoption [19]. Correspondingly, in the literature, traditional community networks, friends and relatives have been also considered as a financial source in case of lack of credit [29, 61]. Husen et al. confirm the positive financial effect of social capital on adoption by supporting this idea [60]. Relaxing effect of credit on the liquidity can be provided to a member in a credit or saving organization, these kinds of organizations relax the farmers in terms of cash constraints [61]. Abdulai et al. show that credit access has an effect not only on the investment but also on the crop choices because the farmers suffering from credit access have a liquidity problem and this directly affects the crop choices [27]. When considered from this aspect, smaller farmers who have limited credit have less advantage than larger farmers having better credit opportunities in terms of adoption [17]. Moreover, the study of Abdulai and Huffman indicates that, in the situation of credit constraint, if farmers just consider the current generation, higher adoption is expected in the middle-aged farmers; but if there is no credit constraint and farmers behave by considering future generations, higher adoption is expected in younger farmers [19].

4. Policy implications and future directions

As shown in the table, when the literature about the adoption and diffusion of technology in the agriculture is reviewed, it is clearly seen that while some factors have an exactly positive effect on adoption, some have negative effects and some are changeable depending on the situation. This review reveals the significant factors and makes the policies that may be implemented to expedite the technology adoption and diffusion more explicit.

When the knowledge of people is taken into consideration, the factors including education, learning and information play a big role in adoption. In this sense, governments should provide a good and an extensive extension service not only in central locations but also in remote locations because as it can be seen in the literature, being closer to opportunities increases the likelihood of adoption and diffusion. Removing the disadvantages of farmers who live in remote areas and providing accessibility to information will contribute to the adoption.

Some organizations and activities that strengthen social relations and facilitate communication among people also need to be set up regularly by governments because social capital is another significant part of the adoption and diffusion of technology. Encouraging farmers to be a member in a group or an organization develops social capital and this helps to obtain information and learning by others. Besides, when the effect of neighbors and relatives is considered, some key persons who can be trained and affect more people can be chosen by the government. These key persons who communicate with their immediate circles such as relatives and neighbors can produce a 'butterfly effect' and increase the adoption and diffusion of technology. In this sense, finding the key person on the farmers is important. For that reason, governments need to prepare specific and different programmes for households having different social identities. These aim-targeted plans will affect the tendency of farmers for adoption positively.

On the other hand, financial aid should be provided for farmers to ease technology use because the literature shows the dominant positive effect of ownership and wealth on adoption and diffusion of technology. Liquidity problem of farmers can be solved with special funds, credit facilities with low interest rate and subsidies provided for new investment and new crops. These opportunities both relax farmers and increase the tendency of farmers to adopt the new technology. Undoubtedly, all of these are related to good and reliable governance; thus, if it exists, the adoption and diffusion of technology can be expedited.

Although education and learning are very big parts in the adoption and diffusion of technology in agriculture, it is necessary to do more detailed research on the dilemmas that 'Although communication and network are very effective on the adoption and diffusion, does learning by others without official agents cause mislearning and misuse of technology and hence to detract farmers from technology adoption or not?'

Future research may also analyze the effects of communication among relatives or neighbors of farmers on adoption because although they have mostly positive effects on adoption, there are also negative effects because of the idea of refusing to provide benefit for others. In brief, more precise results can be obtained about the effect of social capital on adoption and diffusion of technology by examining social relations in detail.

5. Conclusion

In these times when the competition among firms, sectors and countries is very strong, the importance of technology as a formula of long-run economic growth is great [72]. Technology contributes to the sectors in terms of real wages, increasing productivity and decreasing costs and prices [2]. These positive effects make it a big player in the agricultural sector for development [3]. Thus, the adoption and diffusion of technology in agriculture are very important to maintain continuity in production, increase production and generate more income.

In this paper, the factors influencing the adoption and diffusion of technology in agriculture had been analyzed and the results and policy that can be implemented were presented. The literature review includes 44 analyses and 21 factors on the adoption and diffusion of technology in agriculture. Analyses had been handled on a large scale rather than examining a specific time period or a specific region. For the section of characteristics of farmers, the results show that there is a predominantly positive relationship between income, off-farm income, ownership and wealth, education, learning, information, neighborhoods, relatives, contacts with extension agents, membership in a group or organization and experience with the adoption and diffusion of technology. General inference by looking at the effects of age, gender and norms cannot be made because although they have an effect on the adoption and diffusion of technology, the positivity and negativity vary by situation. For the section of the characteristics of farms, it is seen that both factors which cover the farm size and the closeness to the farm have positive effects on adoption. The following section, on external incentives and disincentives, indicates that prices of inputs and credit accessibility show a predominantly positive effect on the adoption and diffusion of technology in agriculture whereas the effect of risks changes. Although the factors examined reveal the main elements for the adoption and diffusion of technology in agriculture, future research may show more precise results for the uncertain factors depending on the situation.

Author details

Nisa Sansel Tandogan* and Haluk Gedikoglu
Department of Economics, Faculty of Social Sciences and Humanities, Konya Food and Agriculture University, Konya, Turkey

*Address all correspondence to: sansel.tandogan@gidatarim.edu.tr

References

[1] Griliches Z. Hybrid corn: An exploration in the economics of technological change. Econometrica. 1957;**25**:501-522. DOI: 10.2307/1905380

[2] UNCTAD. Technology and Innovation Report 2018: Harnessing Frontier Technologies for Sustainable Development. New York, Geneva: United Nations Publication; 2018

[3] Hayami Y, Ruttan VW. Agricultural Development: An International Perspective. Baltimore MD: John Hopkins University Press; 1971. p. 367

[4] Kulshreshta SN. Irrigation and prairie agricultural development. In: Schmitz A, editor. Free Trade and Agricultural Diversification: A Study of Canada and the United States. Boulder, CO: Westview Press; 1989. pp. 222-248

[5] Feder G, Just RE, Zilberman D. Adoption of agricultural innovations in developing countries, a survey. Economic Development and Cultural Change. 1985;**33**:255-298. DOI: 10.1086/451461

[6] Khanna M. Sequential adoption of site-specific technologies and its implications for nitrogen productivity: A double selectivity model. American Journal of Agricultural Economics. 2001;**83**:35-51. DOI: 10.1111/0002-9092.00135

[7] Koundouri P, Nauges C, Tzouvelekas V. Technology adoption under production uncertainty: Theory and application to irrigation technology. American Journal of Agricultural Economics. 2006;**88**(3):657-670. DOI: 10.1111/j.1467-8276.2006.00886.x

[8] OECD. Adoption of technologies for sustainable farming systems. In: Wageningen Workshop Proceedings. Paris: OECD; 2001

[9] Rehman A, Jingdong L, Khatoon R, Hussain I. Modern agricultural technology adoption its importance role and usage for the improvement of agriculture. American-Eurasian Journal of Agricultural & Environmental Sciences. 2016;**16**(2):284-288. DOI: 10.5829/idosi.aejaes.2016.16.2.12840

[10] Dinar A, Campbell MB, Zilberman D. Adoption of improved irrigation and drainage reduction technologies under limiting environmental conditions. Environmental and Resource Economics. 1992;**2**:373-398. DOI: 10.1007/BF00304968

[11] Torkamani J, Shajari S. Adoption of new irrigation technology under production risk. Water Resources Management. 2008;**22**:229-237. DOI: 10.1007/s11269-007-9154-3

[12] Wilson PN. First-order economizing: Irrigation technology adoption and the farm. Agrekon. 2001;**40**(2):231-248. DOI: 10.1080/03031853.2001.9524947

[13] Zilberman D, Khanna M, Lipper L. Economics of new technologies for sustainable agriculture. Australian Journal of Agricultural and Resource Economics. 1997;**41**(1):63-80. DOI: 10.1111/1467-8489.00004

[14] Schiefer G. New technologies and their impact on agriculture, environment and the food industry. In: European Federation for IT in Agriculture (EFITA) Conference. Hungary; 5-9 July 2003. pp. 3-11

[15] Feder G, Slade R. The acquisition of Information and the adoption of new technology. American Journal of Agricultural Economics. 1984;**66**(3):312-320. DOI: 10.2307/1240798

[16] UNFCC. Report on the Seminar on the Development and Transfer of Technologies for Adaptation to Climate Change. Tobago; 2005, 14-16 June 2005

[17] Feder G, O'Mara GT. Farm size and the diffusion of green revolution technology. Economic Development and Cultural Change. 1981;**30**:59-76. DOI: 10.1086/452539

[18] Besley T, Case A. Diffusion as a learning process: Evidence from HYV cotton. In: Research Program in Development Studies RPDS Discussion Paper No. 174. Center of International Studies, Woodrow Wilson School of Public and International Affairs; 1994

[19] Abdulai A, Huffman W. The diffusion of new agricultural technologies: The case of crossbred-cow technology in Tanzania. American Journal of Agricultural Economics. 2005;**87**:645-659. DOI: 10.1111/j.1467-8276.2005.00753.x

[20] Kuehne G, Llewellyn R, Pannell DJ, Wilkinson R, Dolling P, Ouzman J, et al. Predicting farmer uptake of new agricultural practices: A tool for research, extension and policy. Agricultural Systems. 2017;**156**:115-125. DOI: 10.1016/j.agsy.2017.06.007

[21] Lindner RK. Adoption and diffusion of technology: An overview. In: Technological Change in Postharvest Handling and Transportation of Grains in the Humid Tropics: ACIAR Proceeding Series No. 19. Bangkok: Australian Centre for International Agricultural Research; 1987. pp. 144-151

[22] Case A. Neighbourhood influence and technological change. Regional Science and Urban Economics. 1992;**22**:491-508. DOI: 10.1016/0166-0462(92)90041-X

[23] Hua W, Zulauf C, Sohngen B. To adopt or not to adopt: Conservation decisions and participation in watershed groups. In: American Agricultural Economics Association Annual Meeting. Denver, CO; 1-4 August 2004. DOI: 10.22004/ag.econ.20076

[24] Weir S, Knight J. Adoption and Diffusion of Agricultural Innovations in Ethiopia: The Role of Education. CSAE Working Paper Series 5. University of Oxford: Centre for the Study of African Economies; 2000

[25] Wollni M, Andersson C. Spatial patterns of organic agriculture adoption: Evidence from Honduras. Ecological Economics. 2014;**97**:120-128. DOI: 10.1016/j.ecolecon.2013.11.010

[26] Deressa TT, Hassan RM, Ringler C, Alemu T, Yesuf M. Determinants of farmers' choice of adaptation methods to climate change in the Nile Basin of Ethiopia. Global Environmental Change. 2009;**19**(2):248-255. DOI: 10.1016/j.gloenvcha.2009.01.002

[27] Abdulai A, Owusu V, Bakang JE. Adoption of safer irrigation technologies and cropping patterns: Evidence from Southern Ghana. Ecological Economics. 2011;**70**(7):1415-1423. DOI: 10.1016/j.ecolecon.2011.03.004

[28] Bandiera O, Rasul I. Social networks and technology adoption in Northern Mozambique. The Economic Journal. 2006;**116**(514):869-902. DOI: 10.1111/j.1468-0297.2006.01115.x

[29] Feder G. Farm size, risk aversion and the adoption of new technology under uncertainty. Oxford Economic Papers. 1980;**32**(2):263-283. DOI: 10.1093/oxfordjournals.oep.a041479

[30] Fernandez-Cornejo J, Mishra A, Nehring R, Hendricks C, Southern M, Gregory A. Off-Farm Income, Technology Adoption, and Farm Economic Performance. USDA Economic Research Service, Economic Research Report Number 36. Washington DC: United States

Department of Agriculture; 2007. DOI: 10.22004/ag.econ.7234

[31] Diiro GM. Impact of Off-Farm Income on Agricultural Technology Adoption Intensity and Productivity. IFPRI Working Paper 11. International Food Policy Research Institute; 2013

[32] Gedikoglu H, McCann L, Artz G. Off-farm employment effects on adoption of nutrient management practices. Agricultural and Resource Economics Review. 2011;**40**(2):293-306. DOI: 10.1017/S1068280500008078

[33] Gedikoglu H, Parcell JL. Impact of earned and unearned off-farm income on adoption of new technologies. In: Selected Paper for Presentation at the Agricultural & Applied Economics Association's AAEA & CAEA Joint Annual Meeting. Washington. DC; 4-6 August 2013. DOI: 10.22004/ag.econ.149702

[34] Fernandez-Cornejo J, Hendricks C. Off-farm work and the adoption of herbicide-tolerant soybeans. In: Selected Paper for Presentation at the Southern Agricultural Association Annual Meeting. Mobile. Alabama; 1-5 February 2003. DOI: 10.22004/ag.econ.35133

[35] Goodwin BK, Bruer SM. An empirical analysis of farm structure and off-farm work decisions. In: Selected Paper for Presentation at the American Agricultural Economics Association Annual Meeting. Montreal. Canada; 27-30 July 2003. DOI: 10.22004/ag.econ.22164

[36] Kassie M, Jaleta M, Shiferaw B, Mmbando F, Mekuria M. Adoption of interrelated sustainable agricultural practices in smallholder systems: Evidence from rural Tanzania. Technological Forecasting and Social Change. 2013;**80**(3):525-540. DOI: 10.1016/j.techfore.2012.08.007

[37] Rahm MR, Huffman WE. The adoption of reduced tillage: The role of human capital and other variables. American Journal of Agricultural Economics. 1984;**66**(4):405-413. DOI: 10.2307/1240918

[38] Wozniak GD. The adoption of interrelated innovations: A human capital approach. The Review of Economics and Statistics. 1984;**66**(1):70-79. DOI: 10.2307/1924697

[39] Nelson RR, Phelps ES. Investment in humans, technological diffusion, and economic growth. The American Economic Review. 1966;**56**:69-75

[40] Cotlear D. The effects of education on farm productivity. In: Griffin K, Knight K, editors. Human Development and the International Development Strategy for the 1990s. 1st ed. UK: Palgrave Macmillan; 1990

[41] Wozniak GD. Human capital, information, and the early adoption of new technology. The Journal of Human Resources. 1987;**22**(1):101-112. DOI: 10.2307/145869

[42] Norris PE, Batie S. Virginia farmers' soil conservation decisions: An application of Tobit analysis. Southern Journal of Agricultural Economics. 1987;**19**(1):79-90. DOI: 10.1017/S0081305200017404

[43] Huffman WE. Decision making: The role of education. American Journal of Agricultural Economics. 1974;**56**(1):85-97. DOI: 10.2307/1239349

[44] Conley TG, Udry CR. Learning about a new technology: Pineapple in Ghana. The American Economic Review. 2010;**100**(1):35-69. DOI: 10.1257/aer.100.1.35

[45] Hiebert LD. Risk, learning, and the adoption of fertilizer responsive seed varieties. American Journal of Agricultural Economics. 1974;**56**(4):764-768. DOI: 10.2307/1239305

[46] Foster AD, Rosenzweig MR. Learning by doing and learning from others: Human capital and technical change in agriculture. Journal of Political Economy. 1995;**103**(6):1176-1209. DOI: 10.1086/601447

[47] Krishnan P, Patnam M. Neighbours and extension agents in Ethiopia: Who matters more for technology diffusion? American Journal of Agricultural Economics. 2013;**96**(1):308-327. DOI: 10.1093/ajae/aat017

[48] Munshi K. Social learning in a heterogeneous population: Technology diffusion in the Indian green revolution. Journal of Development Economics. 2004;**73**(1):185-213. DOI: 10.1016/j.jdeveco.2003.03.003

[49] Conley T, Udry U. Social learning through networks: The adoption of new agricultural technologies in Ghana. American Journal of Agricultural Economics. 2001;**83**(3):668-673. DOI: 10.1111/0002-9092.00188

[50] Saha A, Love HA, Schwart R. Adoption of emerging technologies under output uncertainty. American Journal of Agricultural Economics. 1994;**76**(4):836-846. DOI: 10.2307/1243745

[51] Hanifan LJ. The rural school community center. Annals of the American Academy of Political and Social Science. 1916;**67**:130-138. DOI: 10.1177/000271621606700118

[52] Coleman J. Foundations of Social Theory. Cambridge: Belknap Press/Harvard University Press; 1990. p. 993. DOI: 10.1007/978-3-658-08184-3_19

[53] Burt G. Structural Holes: The Social Structure of Competition. 1st ed. Cambridge: Harvard University Press; 1992. p. 311

[54] Portes A. Economic sociology and the sociology of immigration: A conceptual overview. In: Portes A, editor. The Economic Sociology of Immigration: Essays on Networks, Ethnicity, and Entrepreneurship. New York: Sage; 1995. pp. 1-41

[55] Robison LJ, Schmid AA, Siles ME. Is social capital really capital? Review of Social Economy. 2002;**60**(1):1-21. DOI: 10.1080/00346760110127074

[56] Smithson CW. Capital, a factor of production. In: Greenwald D, editor. Encyclopedia of Economics. New York: McGraw-Hill; 1982. pp. 111-112

[57] Hunecke C, Engler A, Jara-Rojas R, Poortvliet PM. Understanding the role of social capital in adoption decisions: An application to irrigation technology. Agricultural Systems. 2017;**153**:221-231. DOI: 10.1016/j.agsy.2017.02.002

[58] Grootaert C. Quantitative analysis of social capital data. In: Grootaert C, van Bastalaer T, editors. Understanding and Measuring Social Capital: A Multidisciplinary Tool for Practitioners. Washington D.C.: The World Bank; 2002. p. 41-84. DOI: 10.1596/0-8213-5068-4

[59] Isham JT. The effect of social capital on fertilizer adoption: Evidence from rural Tanzania. Journal of African Economies. 2002;**11**:39-60. DOI: 10.1093/jae/11.1.39

[60] Husen NA, Loos TK, Siddig HA. Social capital and agricultural technology adoption among Ethiopian farmers. American Journal of Rural Development. 2017;**5**:65-72. DOI: 10.12691/ajrd-5-3-2

[61] Wossen T, Berger T, Falco SD. Social capital, risk preference and adoption of improved farm land management practices in Ethiopia. Agricultural Economics. 2015;**46**(1):81-97. DOI: 10.1111/agec.12142

[62] Henning CHCA, Henningsen G, Henningsen A. Networks and

transaction costs. American Journal of Agricultural Economics. 2012;**94**(2):377-385. DOI: 10.1093/ajae/aar099

[63] Henningsen G, Henningsen A, Henning CHCA. Transaction costs and social networks in productivity measurement. Empirical Economics. 2015;**48**(1):493-515. DOI: 10.1007/s00181-014-0882-y

[64] Abdulai A, Monnin P, Gerber J. Joint estimation of information acquisition and adoption of new technologies under uncertainty. Journal of International Development. 2008;**20**:437-451. DOI: 10.1002/jid.1422

[65] Robison LJ, Shupp RS, Jin S, Siles ME, Ferrarini TH. The relative importance of selfishness and social capital motives. Journal of Socio-Economics. 2012;**41**(1):118-127. DOI: 10.1016/j.socec.2011.10.008

[66] Läpple D, Kelley H. Understanding the uptake of organic farming: Accounting for heterogeneities among Irish farmers. Ecological Economics. 2013;**88**:11-19. DOI: 10.1016/j.ecolecon.2012.12.025

[67] Nowak P. The subversive conservationist. Journal of Soil and Water Conservation. 2009;**64**(4):113A-115A. DOI: 10.2489/jswc.64.4.113A

[68] Huffman WE. Farm and off-farm work decisions: The role of human capital. The Review of Economics and Statistics. 1980;**62**:14-23. DOI: 10.2307/1924268

[69] Grossman SJ, Kihlstrom RE, Mirman LJ. A Bayesian approach to the production of information and learning by doing. The Review of Economic Studies. 1977;**44**(3):533-547. DOI: 10.2307/2296906

[70] Dinar A, Yaron D. Adoption and abandonment of irrigation technologies. Agricultural Economics. 1992;**6**(4):315-332. DOI: 10.1016/0169-5150(92)90008-M

[71] Di Falco S, Bulte E. The impact of kinship networks on the adoption of risk-mitigating strategies in Ethiopia. World Development. 2013;**43**(3):100-110. DOI: 10.1016/j.worlddev.2012.10.011

[72] Solow RM. A contribution to the theory of economic growth. Quarterly Journal of Economics. 1956;**70**(1):65-94. DOI: 10.2307/1884513

Supporting open minds since 2005

Internet of Things (IoT) for Automated and Smart Applications
http://dx.doi.org/10.5772/intechopen.77404
Edited by Yasser Ismail

Contributors
Menachem Domb, Yasser Ismail, Mohammed Ahmed Dauwed, Ahmed Meri, Avirup Dasgupta, Asif Gill, Farookh Hussain, Thierry Edoh, Jules Degila, Yoshifumi Nishida, Jesús Jaime Moreno Escobar

First published in London, United Kingdom, 2019 by IntechOpen
IntechOpen is the global imprint of INTECHOPEN LIMITED, registered in England and Wales, registration number: 11086078, 7th floor, 10 Lower Thames Street, London, EC3R 6AF, United Kingdom
Printed in Croatia

British Library Cataloguing-in-Publication Data
A catalogue record for this book is available from the British Library

Additional hard and PDF copies can be obtained from orders@intechopen.com

Internet of Things (IoT) for Automated and Smart Applications
Edited by Yasser Ismail
p. cm.
Print ISBN 978-1-78984-095-7
Online ISBN 978-1-78984-096-4
eBook (PDF) ISBN 978-1-78923-817-4

For EU product safety concerns:
IN TECH d.o.o., Prolaz Marije Krucifikse Kozulić 3, 51000 Rijeka, Croatia, info@intechopen.com or visit our website at intechopen.com.

Meet the editor

Dr. Yasser Ismail received his M.S. and Ph.D. degrees in Computer Engineering from the University of Louisiana at Lafayette - USA in 2007 and 2010. Dr. Ismail is currently working as an assistant professor in the Electrical Engineering Department, Southern University and A&M College-Baton Rouge, Louisiana, USA. His area of expertise is digital video processing algorithms/ architectures levels, Internet of Things (IoT), VLSI and FPGA design (low-power and high-speed performance embedded systems), automotive transportation, robotics, RFID, and wireless and digital communication systems. He has published two books, three book chapters, and more than 35 articles in related journals and conferences. Dr. Ismail served as a PI and Co-PI for several funded grants from international fund agencies.

Contents

Preface

Internet of Things (IoT) is a recent technology paradigm that creates a global network of machines and devices that are capable of communicating with each other. Security cameras, sensors, vehicles, buildings, and software are examples of devices that can exchange data bewteen each other. IoT is recognized as one of the most important areas of future technologies and is gaining vast recognition in a wide range of applications and fields related to smart homes and cities, military, education, hospitals, homeland security systems, transportation and autonomous connected cars, agriculture, intelligent shopping systems, and other modern technologies. This book explores the most important IoT automated and smart applications to help the reader understand the principle of using IoT in such applications.

A smart home system is one of the applications discussed in this book. A smart home system uses the IoT infrastructure to remotely control all home devices. Smart home systems have achieved great popularity in the last decades as they increase the comfort and quality of life. Most smart home systems are controlled by smartphones and microcontrollers. Remote healthcare is also discussed as it has become a vital service with the growing rate of senior citizens. Remote healthcare offers promising technology for health monitoring, rehabilitation, and assisted living for elderly and medically challenged patients. Securing the transmitted data between devices is a very important issue that will be discussed in the book. Securing the transmitted data between devices that use the IoT infrastructure requires an optimized Internet of Things Network (IoTN) topology that allows fast and secure data transmission between devices. The book will also discuss the use of Artificial Intelligence (AI) to facilitate and secure connecting devices through the IoT infrastructure. In the centenarian era, it has become even more imperative to address the physical and cognitive changes faced by children, the elderly, and disabled persons.

The editor of this book acknowledges the support of the IntechOpen team, the authors of the book chapters and the Electrical Engineering Department at Southern University and A&M College-Baton Rouge - USA for their support to finalize this book.

Dr. Yasser Ismail
Assistant Professor,
Electrical Engineering Department,
Southern University and A&M College,
Baton Rouge, LA, USA

Section 1

Internet of Things (IoT) Importance and Its Applications

Chapter 1

Introductory Chapter: Internet of Things (IoT) Importance and Its Applications

Yasser Ismail

1. What is the Internet of Things?

Internet of Things is a recent technology that creates a global network of machines and devices that are capable of communicating and exchanging data with each other through the Internet. There is a difference between the Internet of Things and the Internet. Internet of Things can create information about the connected objects, analyze it, and make decisions; in other words, one can tell that the Internet of Things is smarter than the Internet. Security cameras, sensors, vehicles, buildings, and software are examples of things that can exchange data among each other.

2. Internet of Things challenges

As the number of real-time applications (devices) that need smart connections among each other is increased, the Internet of Things challenges will also increase. Such challenges are as follows.

2.1 Smart connectivity

Sensors and devices that are connected and communicated together through the Internet of Things infrastructure may need to update their trends or feature to be suited to the changes of surrounding environments. The Internet of Things is a smart infrastructure that can process the collected data and make the required decisions to improve itself and to change the trends or features of the connected devices to accommodate the surrounding environments' changes. Internet of Things is a smart technology that helps all connected devices to update themselves according to changes in the surrounding environment and to be able to be adopted and work in any other strange environment with high accuracy. As a result, smart connected system can be produced if smart infrastructure is well designed to treat the collected data from devices correctly, to make the needed decisions.

2.2 Keeping high privacy and security of all connected devices

The main idea of using the Internet of Things is to have a smart system and to connect billions of devices over the whole world. The expected devices to be connected together are expected to be 50 billion through the Internet of Things devices by 2020. **Figure 1** shows the growth of the world population and the connected devices by 2020. Connecting such a huge number of devices required high-security levels to prevent scams and to allow high level of data protection. As a result, achieving high level of security is a big challenge to get the needed trust from both industries and individuals to share their data utilizing the Internet of Things.

2.3 Treating big data

The most important challenge of using the Internet of Things is the tremendous growth of the data transmitted between the connected devices. As seen in **Figure 2**, The basic three sources of data are (1) the database used in the business process; (2) the human daily activities such as email, Facebook, and weblogs; and (3) the connection of physical devices such as cameras and microphones. It is worth mentioning that a full 90% of all the data in the world has been generated over the last 2 years. This makes it more challenging for the Internet of Things infrastructure designers to deal with such growth of the generated data.

2.4 Reducing the overall data latency among machine-to-machine interactions

While connecting many devices through the Internet of Things infrastructure, the shared data among them will also tremendously increase. This will cause some delay or latency of data delivery among the connected devices. This opens a new challenge to be addressed by the Internet of Things to reduce such latency to be sure a robust Internet of Things infrastructure will be obtained.

2.5 Reducing bandwidth and power consumption

Both bandwidth and power consumption of the numerous number of devices that are connecting, communicating, and sharing data among each other through the Internet of Things are tremendously increased. This is why when designing an Internet of Things infrastructure, both bandwidth and power consumption challenges should be considered. The main trend nowadays is to reduce the size of the connected devices,

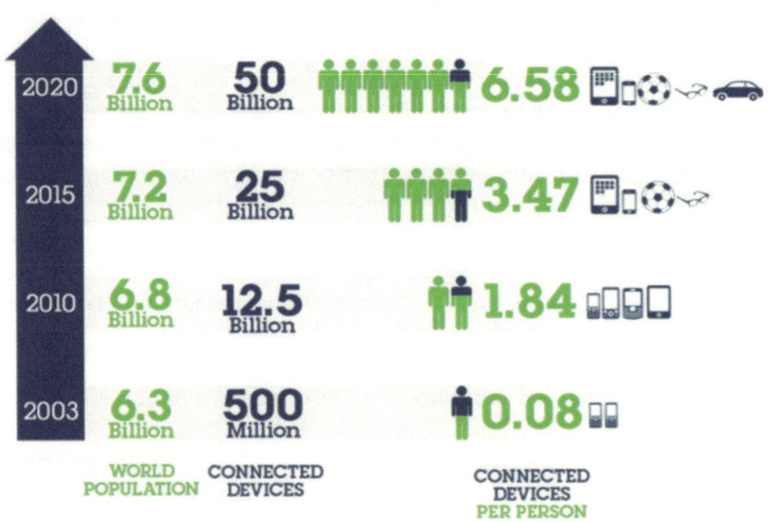

Figure 1.
The overall world population and the connected devices by 2020.

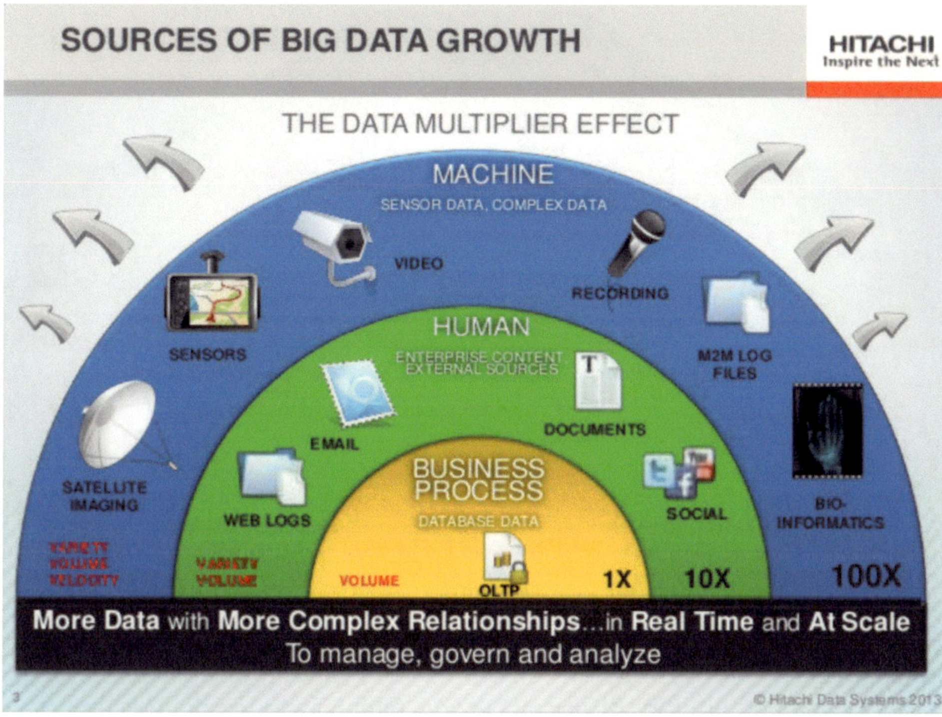

Figure 2.
Different sources of data growth (https://www.slideshare.net/bjorna/big-data-in-oil-and-gas).

and as a result, the power consumption will decrease. The transmitted data rate is still an issue to be solved due to the huge shared data among devices.

2.6 Complexity

Sharing data and connecting devices together through the Internet of Things can be implemented through several levels and layers of software/hardware and some standard protocols. With the tremendous increase in the shared data and connected devices, the used software/hardware and standard protocols will be more complicated. As a result, there is a challenge to reduce the complexity of the Internet of Things technology as the number of connected devices increases.

3. Internet of Things applications

The Internet of Things is recognized as one of the most important areas of future technologies and is gaining vast recognition in a wide range of applications and fields related to smart cities, military, education, hospitals, homeland security systems, transportation and autonomous connected cars, agriculture, intelligent shopping systems, and other modern technologies. The smart home is one of the main applications that use the Internet of Things infrastructure to connect several sensors. The sensors can sense and collect surrounding information that is used to fully control different home systems such as lighting and security as seen in **Figure 3**.

There are many other applications that use the Internet of Things infrastructures such as smart bridges and smart tunnels. Temperature and vibration sensors, as well as video surveillance cameras, can be fixed on a bridge to detect any abnormal

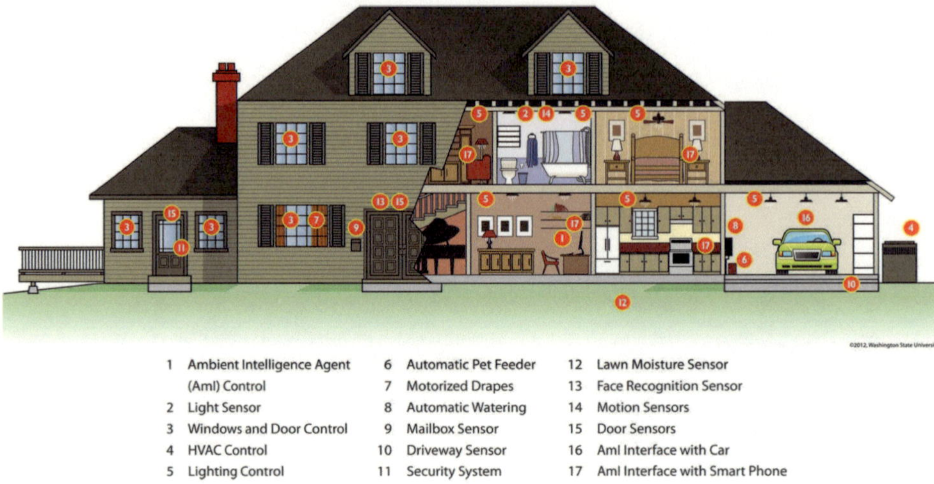

Figure 3.
SMART home sensors communication through the internet of things. (http://www.nibib.nih.gov/sites/default/files/SMART-HOUSE_2_DCook.jpg).

activity and send warnings via SMS. Also video processing analysis can be performed to control the traffic density on a bridge. The smart tunnel can use several sensors to monitor humidity, displacement, and temperature to call for appropriate maintenance if a problem is detected. All of these applications are using sensors to detect and collect data that are used to give a proper decision that maintains a high level of security of the installations.

Author details

Yasser Ismail
Electrical Engineering Department, Southern University and A&M College, Baton Rouge, Louisiana, USA

*Address all correspondence to: yasser_ismail@subr.edu

Section 2

Smart Home Applications

Chapter 2

Privacy of IoT-Enabled Smart Home Systems

Avirup Dasgupta, Asif Qumer Gill and Farookh Hussain

Abstract

Digital ecosystems are going through a period of change due to the advancement in technologies such as Internet of Things (IoT) as well as proliferation of less expensive hardware sensors. Through this chapter, we present current emerging trends in IoT in different industry sectors as well as discuss the key privacy challenges impeding the growth of IoT to reach its potential in the smart home context. The majority of the existing literature on IoT smart home platforms focuses on functionalities provided by smarter connected devices; however, it does not address the concerns from a consumer's viewpoint. Thus, the key questions are: What are the privacy concerns related to IoT, particularly from a "smart home device" consumer viewpoint? What are the existing remedial approaches for privacy management? This chapter proposes a framework to assist smart home user and IoT device manufacturer to make informed privacy management decisions. The findings of this research intend to help practitioners and researchers interested in the privacy of IoT-enabled smart systems.

Keywords: IoT, smart sensors, data governance, privacy, framework

1. Introduction

In last few years, we have observed a growing interest in IoT applications, which are being developed for the industries and ecosystems such as healthcare, smart home, manufacturing and agriculture ecosystems [1]. Presently, it is anticipated that there are about 16 billion IoT units installed worldwide generating vast amount of data. According to forecast reports from Frost and Sullivan, the number of interconnected objects is expected to increase above 60 billion by 2024 [2]. Aggregated data collected from different sensors are being used by organizations increasingly to gain data-driven business insights.

The growth of IoT has been possible due to the advancement of technologies like cheaper hardware sensors, ipv6, wireless coverage, smartphones and processing power of CPU [3]. While the use of IoT worldwide has been high, the maturity level of the solutions using this technology is varied. In this chapter, we highlight the various components making up IoT, evolution of IoT and the concerns related to privacy. We particularly focus on the IoT uses in the smart home context.

IoT ecosystem stands on the building blocks of multiple underlying technologies such as sensing (sensors and actuators), connectivity (mobile), analytics and computing. A typical IoT ecosystem involves the following stages [4].

- Things are fitted with electronics, software, actuators and sensors. They can be battery operated, electricity powered or use RFID transponders. Things collect raw data from the environments. Each thing has a unique identifiable address and of varying computational capability and complexity.
- Data collected from things are processed by applications.
- Using various connectivity technologies such as Wi-Fi, Zigbee, NFC, Bluetooth, cellular (2G/3G/4G/5G) and low-powered WAN, data are transmitted.
- Applications collect data in real time from different things to store, process and analyze in computing platforms.
- Insights are derived from the collated data using robust analytics enabling informed business decisions to be taken involving process and people.

The term "Internet of Things" was officially introduced in 1998–1999 by Kevin Ashton of automatic identification center (Auto-Id) at Massachusetts Institute of technology (MIT). Kevin suggested that Internet-connected RFID technologies can be used in supply chains to keep track of items without human involvement [5]. The philosophy of IoT further gained momentum in 2005, thanks to the formal acceptance of IoT in a world summit on information society (WSIS) in Tunisia [6]. However, the concept of IoT applications can be traced back to 1982 when one of the first attempts of an IoT application was developed at Carnegie Mellon University. The Internet-connected coke machine was able to report the drinks contained and whether the drinks were cold [7] (**Table 1**).

Year	Discovery
1747	Electricity (lightning)
1819	Practical electromagnetism
1831	Faraday: Electromagnetic induction
1873	Maxwell: Theory of electromagnetism
1887	Hertz: Radio waves
1895	Marconi: Radio telegraph
1907	First public use of radio
1911	First mobile transmitter (Zeppelin)
1915	First wireless voice transmission
1927	First car radio
1928	First TV broadcast
1933	First mobile phone (Germany, in-car)
1950s	UNIVAC(UNIVersal Automatic Computer) Ia mainframe
1958	First hand-held mobile phone
1961	Cloud computing precursor (John McCarthy)
1969	Internet precursor (ARPANET)
1973	1G cellular mobile (NTT, Japan)
1981	First wireless IoT connection(Coke machine, GSM)
1982	International Internet
1988/89	World Wide Web
1990	2G cellular mobile (GSM)

Year	Discovery
1991	Bluetooth
1994	Wi-Fi (CSIRO, IEEE)
1997	3G cellular mobile (UMTS)
1998	4G cellular mobile (LTE)
1999	IoT term coined
2005	United Nations mention IoT
2008	5G cellular mobile
2008	Cloud computing term coined
2012	Cisco introduces fog computing
2020	Industry expects 20 billion IoT devices worldwide

Table 1.
IoT evolution (adapted from [3, 8, 9]).

Industry	Use case
Smart City	Smartbin offers Smart Waste Monitoring through Smart Sensors & Route Optimization Technologies [10].
Transport	Spanish train operator RENFE uses Siemens' high-speed train and monitors train developing abnormal patterns and sends them back for inspection to prevent failure on the track [11].
Agriculture	Semios uses sensors and machine vision technology to track pest populations in orchards, vineyards, and other agricultural settings [12]
Financial Sector	Progressive Insurance uses Snapshot to determine Insurance premium for car drivers [13].
Healthcare	Abilify MyCite (aripiprazole tablets with sensor) has an ingestible sensor embedded in the pill that records that the medication was taken [14].
Government	US municipality has implemented smart meter monitoring for the entire town's residential and commercial water meters. The project involved placing water meter sensors on 66,000 devices that used to be manually read and recorded [15].
Utility	US oil and gas company is optimizing oilfield production with the Internet of Things. In this IoT example, the company is using sensors to measure oil extraction rates, temperatures, well pressure and more for 21,000 wells [15].
Environment	Autonomous sailboats and watercraft are already patrolling the oceans carrying sophisticated sensor instruments, collecting data on changes in Arctic ice [16].

Table 2.
IoT applications.

IoT has produced a number of sophisticated solutions that are growing in popularity among businesses. Many sectors have already graduated to this technology, and are putting IoT to use for digitizing their daily activities. The prominent adapters of IoT are Smart City, Retail, and Manufacturing. Some of the most notable applications rolled out in the marketplace are given in **Table 2**.

Although, there is a growing interest in IoT applications in different industry sectors, challenges in adoption exist. The key questions are: What are the privacy concerns related to IoT, particularly from a "smart home device" consumer viewpoint? What are the existing remedial approaches for privacy management? This chapter aims to address the above-mentioned questions. The remainder of this chapter is organized as follows. Section 2 presents various privacy concerns of IoT before proposing a novel framework to address IoT concerns from a consumer's perspective in Section 3. This is followed by an initial validation of the framework in Section 4 before we draw conclusions in Section 5.

2. Review of privacy literature with specific IoT focus

Privacy is defined by Clarke as the attention that individuals have in sustaining a personal space, free from interference by other people and organizations [17]. An intrinsic part of privacy issue is the exposure of sensitive data such as Personal Identifiable Information (PII) to non-intended recipients. Personal Identifiable Information (PII) comprises of details such as title, first name, last name, date of birth, address, and phone number, constituting some of the sensitive personal information (SPI). In addition, financial and health details and the geophysical location of an IoT user are also considered as sensitive information.

Internet of Things devices may collect data including sensitive data and store the data for further use for commercial purposes. It comprises of several stakeholders such as customer whose PII is collected; manufacturers who develop the sensors and other networking components and third parties who create IoT mobile apps or use the data for commercial advantage. According to McKinsey global report from 2015 [18], consumers are cautious about embracing IoT-based systems, particularly due to lack of privacy and the data at risk. OECD reported [19] that privacy incidents are growing in both number and sophistication. Similar concerns are expressed by several academic articles which suggest lack of privacy including unauthorized surveillance or eavesdropping [20] as a major concern for individuals.

Some researchers or practitioners confuse privacy with security. While security deals with the management of controlling who can access information, privacy is predominantly focused on granular control of what data can be collected, who can access what, when they can access specific data, and how long the data should be retained.

Protecting user's privacy comprises of technical, human and legal aspects. Other relevant aspects can also be considered.

2.1 Potential scenarios of privacy violation in smart home

Smart home segment comprises of connected appliances like TV set, thermostat, refrigerator, oven, home security, self-guided vacuum cleaners, cleaning and maintenance devices. Additionally, cameras, motion sensors and light sensors also collect data. Most of these data contain private and/or sensitive information such as locations, addresses, pictures and network access information. The data can be accessible to device manufacturer, mobile application owner, third-party vendors or public depending on use cases. There are several scenarios involving data collection such as:

- Movement of individuals (unauthorized surveillance) using motion sensors, camera and GPS tracker.

- Monitoring of actions of customers.

- Sharing of health data publicly from wearable devices or implantable devices such as Abilify MyCite, and Bluetooth-enabled oximeter [18].

- Sharing of data (e.g., financial, health, PII, Payment Card Information and geophysical data) to third party without explicit consent [21].

- Search query of user shows his preference traits (**Figure 1**).

There are very few contributions that address privacy in the context of smart home [22]. While several studies conducted surveys and interviews with IoT end user consumers to investigate the factors affecting privacy including data processing and information risk [23], none proposed a feasible solution to fix them.

Figure 1.
Key data exchanges in smart home.

2.2 Legal aspects of privacy in IoT era

Government organizations are taking significant interest in IoT security, privacy and interoperability from legal aspects. This is in alignment with the studies which advocated further collaboration and dialogs between the regulators and manufacturers of IoT devices to develop appropriate methods to tackle the relevant problems [24]. From regulatory perspective, some of the most important legislative requirements are HIPAA for healthcare, MA risk for supply chain management, California's Senate Bill 327, IoT Cybersecurity Improvement Act of 2017 and General Data Protection Regulation (GDPR).

Data privacy requirements are complex and differ by jurisdictions in regard to the definition of data and the relevant laws/regulations. In Europe, GDPR was introduced on May 25, 2018. GDPR is a new regulation approved by EU parliament, Council and European Commission. It aims to safeguard the personal data rights of EU citizens and residents in this era of new technological advancements. As per GDPR, organizations are required to

- Get explicit and affirmative consent before processing personal data. This includes financial, economic and health data and online information.
- Notify within 72 hours to the regulator and individual about any data breaches.
- Facilitate customers and employees' right to the removal of data from their system.
- Give the right of portability, and increased right of access and right to be forgotten by customer.
- Maintain records of processing activities.

GDPR non-compliance instances may incur penalties up to 2–4% of global revenue or 20 million Euros [25] to organizations based on the infringement. GDPR applies to any company, irrespective of their geographic location, that offers goods and services to European citizens and handles their data including IoT ecosystem-generated data.

In USA, California Senate Bill 327 [26] was introduced recently which allows the State of California ability to bring enforcement complaints against those companies that do not build adequate security safeguards into their Internet-connected IoT devices [27]. It provides the state the right to hold IoT device makers more accountable for consumer's data security. IoT Cybersecurity Improvement Act of 2017 [28] requires: (i) that IoT devices are patchable, (ii) that devices do not contain known vulnerabilities, (iii) that devices rely on standard protocols, (iv) that devices do not contain hard-coded password and (v) technical aspects of privacy in IoT era.

At present, different privacy-enhancing technologies (PETs) exist to protect privacy. Prevention, by means of access restrictions, is an effective way to safeguard customer privacy. In [29], the authors put forward a concept of using access control list (ACL) and data classification model, to classify data according to its sensitivity and assign tag value to each category. In [30], the authors presented the idea of using Certification Authority (CA)-based encryption to confirm the authenticity of sensor. Some authors argue that it adds overheads and hence it cannot be used as a viable solution. Instead, they proposed incorporating a chaos-based cryptographic scheme and Message Authentication Codes (MAC) for data transmission. In a recent research, authors from IT service firm Tata Consultancy Services recommended that the IoT stakeholders can adopt Preventive Privacy (3P) Framework [31] in order to build trust and confidence among end users.

Privacy by Design (PbD) is defined as another popular approach that enables privacy to be "built in" to the design of the information systems and business processes, ensuring that privacy is considered before, and throughout, the development and implementation of all initiatives that involve personal information [32]. Dr. Ann Cavoukian first proposed it in Canada in the 1990s. PbD is one of the highly recommended approaches to protect individual's privacy [31, 33] concerns in IoT. Unfortunately, even though the USA Federal Trade commission (FTC) and the European Commission accepted PbD to be effective [34], not all manufacturers consider PbD when developing IoT devices and applications.

2.3 People aspects of privacy in IoT era

According to a survey conducted by Cisco in 2017, "human factors" such as organization, culture and leadership contributed to the success of IoT implementations 75% of the time—which was higher than technical aspects [35]. A number of stakeholders are involved in IoT digital ecosystem such as the end users, product suppliers, Internet service providers, cloud storage functionalities and retailers. As mentioned earlier, a significant aspect of the value of IoT for consumers refers to: aggregating data collected from many source systems, generating new knowledge and making fact-based choices. The utilization of data to add value is best explained by the well-known DIKW hierarchy from Ackoff [36]. DIKW is a four-layer hierarchy comprising of data, information, knowledge and wisdom where each layer adds certain characteristics over and above the previous one. **Table 3** shows DIKW in an IoT context.

Hierarchy level	Description
Data	Most basic level of facts. Collected from things and used for storage and processing.
Information	Computing platform adds context to data (who, what, where, when) ingested.
Knowledge	This layer answers the question on how data are used. Analytics is applied in computing platform.
Wisdom	Evaluated understanding of when and why data are used

Table 3.
DIKW in an IoT context.

3. Consumer-centric approach

While IoT organizations are aware of the need for adopting PET and incorporating PbD, there is little guidance available on how to do so. Though there are PbD-driven frameworks available [34], no concrete solutions to establish auditing mechanism or control method systems have been developed (**Table 4**).

The lifecycle of an IoT service or product is shown in **Figure 2**.

Figure 3 summarizes: what can be done, at the minimal level, by consumer to safeguard his/her privacy. This provides the basis for the further development of

	Pre-purchase	Setup/post purchase	Decommission
Awareness omni channels	**Research + solution purchase**	**Use + feedback**	
• Web • Social • Mobile • In-store • Media • Advertising • Direct Marketing	• Products which provide audit mechanism while dealing with PII [20] • Products which notify user to provide dynamic consent for data use [37] • Products which stop working properly when consent is not given by user [38] • Firmware upgrade and patchability of IoT devices [24] are available. • Products transparent on how disclosed data are used by the developer of the IoT system or application [39] • Established reputed product with no or negligible data breach history	• Setting up, configuring and registering to IoT services • Signing Consents authorizing data to be collected and used by IoT service provider. • Update Firmware and mobile applications	• Remove authoriza-tion of IoT vendor to use your data • Deregister and destroy data.

Table 4.
Consumer's perspective of IoT product lifespan.

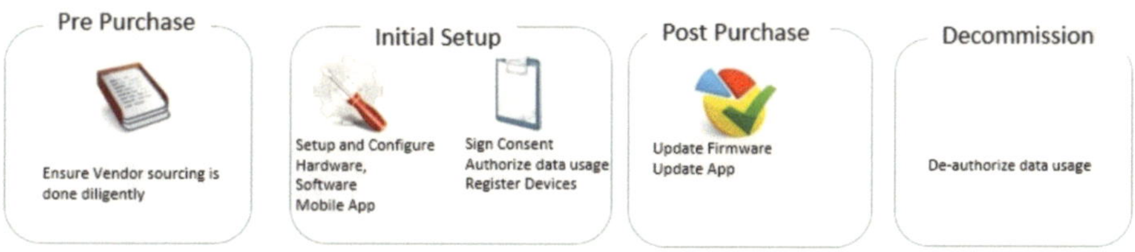

Figure 2.
IoT product lifecycle.

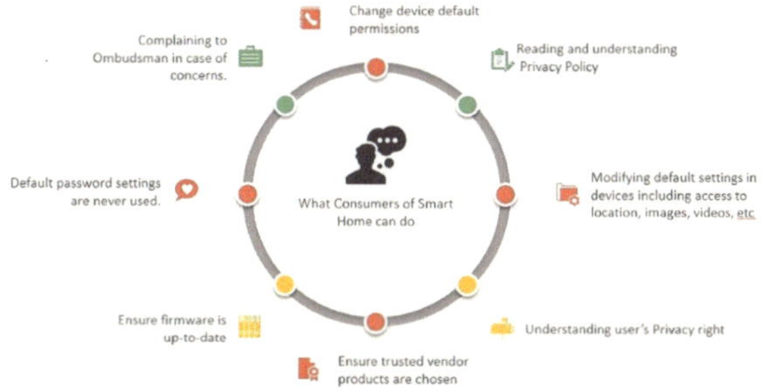

Figure 3.
Mitigation options for consumers (based on [31, 46, 47]).

Consumer questions	Risk factors
Who has access to the data? Will third parties have access to the data? What information can be inferred from the data?	Unacceptable usage of data without consent such as spamming
Will my data be shared outside my country?	Data sovereignty constraints
How will I be informed in case my data are compromised?	Data detection and notification of breach
Are there any known breaches or vulnerabilities about this device?	Outdated Firmware
What happens when I stop using the product or service?	De-authorization
Can my location be tracked from data?	Unveiling of physical address
What can I do if my PII is compromised?	Password renewal
How can I rectify my data?	Outdated information
Can I get a copy of my data or access my data?	Portability
Can I ask you to remove my digital footprint captured by the IoT service?	Unaligned data erasure

Table 5.
Key questions in IoT for consumers.

an IoT privacy tool or framework, which can address the concerns of the consumer [31, 40–45] compiled in **Table 5**.

4. Proposed framework

As mentioned in earlier section, the existing frameworks are relevant primarily for thing manufacturer and do not involve end thing consumers. Through this chapter, we seek to provide answers to the questions mentioned in **Table 5** by leveraging a four-phased data governance-driven 4I framework (Identify, Insulate, Inspect and Improve). The Identify phase of the 4I framework (**Figure 4**) comprises of seven key dimensions such as risk, compliance, policy, process, people, data asset and technology (**Table 6**).

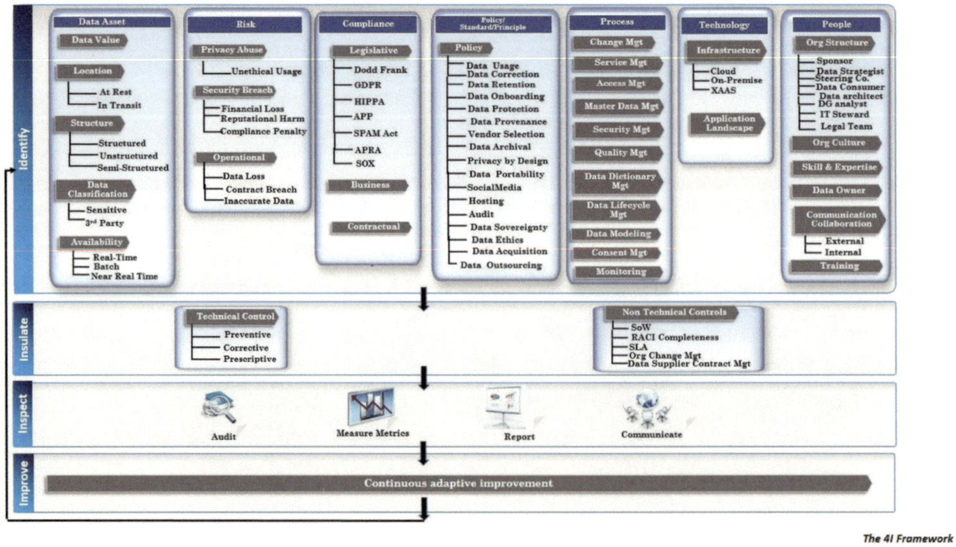

Figure 4.
The 4I framework.

Dimension	Description
Risk	Risk dimension comprises of the factors that influence both the IoT end user and thing manufacturer. It includes attributes such as lack of consent data breach, legal penalties, service level agreement violation, and lack of upgradability, interoperability and security [20, 41, 48, 49].
Compliance	Includes legal requirements (e.g., user consent), controls and baselines to be operationally compliant. There are a number of regulations such as SOX, GDPR, SPAM Act, Australian APP Privacy law, HIPPA and COPPA which are relevant for IoT [50].
Policy, standards and principles	This dimension spans the lifecycle from inception to deletion of data including items such as data sharing, acceptable use of data, data classification and storing rules. A well-defined and enforced governance providing the structure that works for the benefit of everyone concerned by ensuring that the IoT stakeholders adhere to accepted ethical standards [44, 51].
Data asset	Describes the benefit of the data and the salient features of the data [52, 53].
Process	Defines how various interfaces and functionalities are to deliver a functioning and solution [54].
People	The different stakeholders and their accountability in the IoT ecosystem such as consumer, ombudsman, policy maker, IoT thing manufacturer, IoT cloud provider, Internet service provider and the IoT service operators. People dimension also includes leadership and organization structures [55, 56].
Technology	This dimension includes hardware infrastructure, platforms and software agents that notify potential compliance violation through monitoring and workflows [34, 53].

Table 6.
Key dimensions of the Identify phase of the 4I framework.

Identify stage or phase refers to the key risks, requirements and context. Insulate stage refers to the precautionary measures taken to prevent lapses using technologies and non-technical risk remediation techniques. Inspect stage contains the essential toolkits such as maturity models, audit mechanisms, software agents required to continuously monitor, report and assess the IoT Data Governance Maturity from risk and value perspectives. The final stage focuses on continuous improvement.

4.1 The 4I framework applied to privacy context

To illustrate how the proposed 4I framework will work in an IoT-enabled home, a use case involving smart refrigerator is discussed in this section. Currently, when consumers buy an IoT device directly from vendors or service providers, they may have very little understanding when agreeing to the privacy policy (PP) and terms and conditions (T&C) before they start using the product or services or application. However, there are several risks associated with the data collected to render the services.

For example, the smart refrigerator can track our food preferences, search and order food from online stores [31]. Various traits of the fridge owners' eating behaviors can be inferred based on the search queries. If these data are sent to third-party business, they can use the information for the purpose of undesirable targeted advertisements. This can lead to the potential breach of privacy violating regulatory laws if explicit consent was not obtained from the consumer (**Figure 5**).

The **Identify** phase of the 4I framework discerns the potential risks associated with the consumer's data shared among the data processors in data supply chain. For example, it reviews the laws such as GDPR to understand the data protection rights of a smart home user [57] and ascertains the risk related to privacy and security breach. Policies related to data retention, service level agreement with vendors and data management are implemented in the **Insulation** phase of the framework. For instance, an agent called checkmyprivacyrules (CPMRs) can be installed at user's home router to ensure privacy policy and laws like GDPR are not violated based on a search query (**Figure 6**).

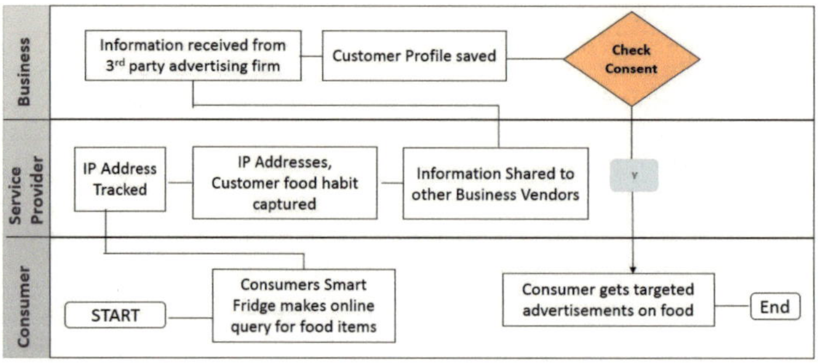

Figure 5.
Business process in smart home refrigerator.

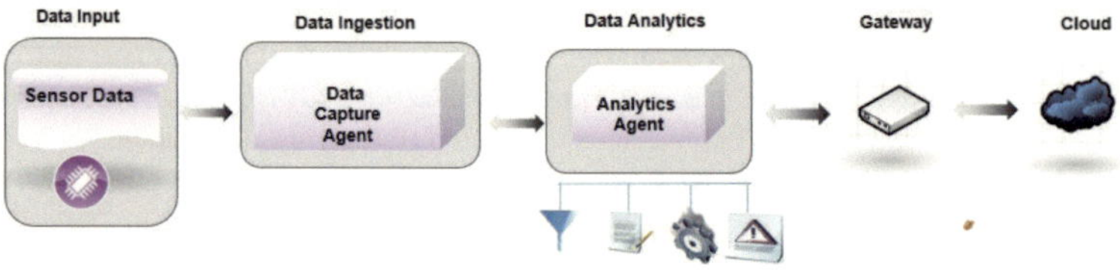

Figure 6.
Smart home 4I (filters, policies, rules, permissions) (adopted from [58]).

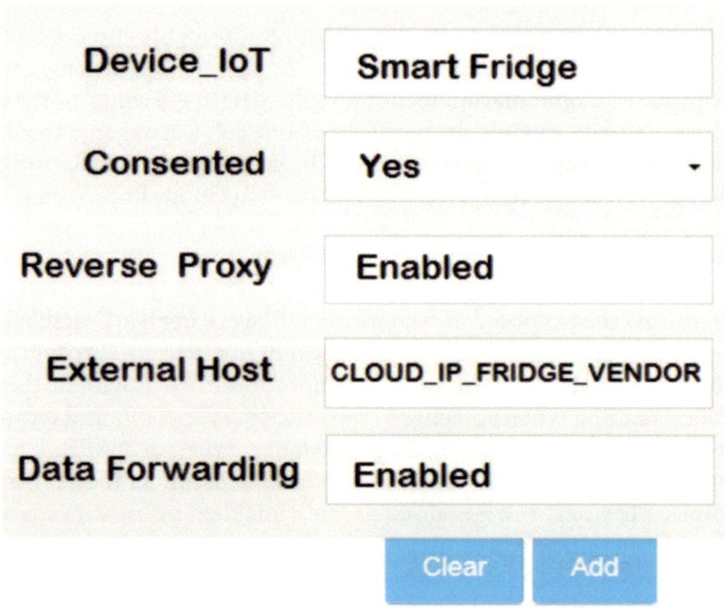

Figure 7.
Web configuration to add privacy rule for smart fridge.

Figure 7 shows a screen where smart fridge user can setup who can access the data. With the above settings, smart fridge can send data to cloud if

a. Device has latest firmware updates. This can be verified from Firmware update version captured periodically from Vendor Website by the agent installed in the router

b. Intended address to push data in the packet states matches external host IP address

c. Consent is set to "Yes"

d. Reverse Proxy is enabled. This will ensure even if the ISP or business gets IP address, it will not be accurate.

Listing 1 shows the Pseudo code of the agent.

CheckMyPrivacyRules (Di)
*Di - > device_*
Rij - > rule j for device Di
Begin
For each Di in domain D
For each rule Rij
If substring(Pi)- = Rij.
Transmit Data;
Else
Send SMS/email to user
Stop polling Di
Endif
Endif
	Di is the set of all smart IoT devices in a "Smart Home" and Di ⊂ D
	Rij is the ruleset j applied to Device Di before it leaves home network
	Pi is the packet send by Di to the router.

Listing 1.
Pseudocode for CPM.

The Inspection phase comprises of performing audit reviews periodically to assure the compliance of the process, systems and data flow. The **Inspect** phase can comprise of automated data quality checks and data access log monitoring. In the **Improve** phase, continuous improvement is done to ensure the continuous adaptation in response to changing data privacy requirements and landscape. For example, improving the agent to ensure software is not only patched to current version, but also data are secured using tokenization techniques [59] can be an outcome of this final phase of the 4I framework.

5. Conclusion

IoT's business growth potential is undeniable. Advancement in IoT has opened up new prospects for growth in the diversified areas such as health, energy, transport and smart home. In this chapter, we provided an overview of the IoT technology and real-life examples of usage of this technology. Next, we discussed the privacy problems in IoT from a consumer's perspective. A review of the related work was presented along with research gaps. Next, we proposed and provided an overview of a data governance-driven 4I framework. Finally, we provided the pseudocode and demonstrated the applicability of the 4I framework to address the privacy concerns in a smart home refrigerator context. This involved the policies, rules and configurations using time-tested data governance principles. In future, we intend to further test and improve the 4I framework in the overall context of data governance in digital ecosystem.

Acknowledgements

This research is supported by an Australian Government Research Training Program Scholarship.

Conflict of interest

There is no conflict of interest.

Author details

Avirup Dasgupta*, Asif Qumer Gill and Farookh Hussain
University of Technology Sydney, Ultimo, NSW, Australia

*Address all correspondence to: avirup.dasgupta@student.uts.edu.au

References

[1] Bader A, Ghazzai H, Kadri A, Alouini MS. Front-end intelligence for large-scale application-oriented internet-of-things. IEEE Access. 2016;**4**:3257-3272

[2] Global Internet of Things (IoT) Device Market, Forecast to 2024; 2018

[3] Rose K, Eldridge S, Chapin L. The Internet of Things: An Overview. The Internet Society (ISOC); 2015. pp. 1-50

[4] Jalali S, Bhatnagar I, editors. Leveraging Internet of Things Technologies and Equipment Data for an Integrated Approach to Service Planning and Execution. 2015 IEEE Region 10 Symposium; 2015 13-15 May 2015

[5] Ma H-D. Internet of things: Objectives and scientific challenges. Journal of Computer Science and Technology. 2011;**26**(6):919-924

[6] Dong L, Mingyue R, Guoying M. Application of internet of things technology on predictive maintenance system of coal equipment. Procedia Engineering. 2017;**174**:885-889

[7] Department CMUCS. The "Only" Coke Machine on the Internet [Internet]. Available from: https://www.cs.cmu.edu/~coke/history_long.txt [Accessed: 2018-11-29]

[8] Minerva R, Chebudie AB, Rotondi D. Towards a definition of the Internet of Things (IoT) 2015 [Internet]. Available from: http://iot.ieee.org/definition.html [Accessed: 2018-11-29]

[9] Gill AQ. Adaptive Cloud Enterprise Architecture. Hackensack, New Jersey: World Scientific; 2015

[10] Sharma N, Singha N, Dutta T. Smart bin implementation for smart cities. International Journal of Scientific and Engineering Research. 2015;**6**(9):787-791

[11] Tracy P. Case study: Siemens reduces train failures with Teradata Aster 2016 [Internet]. Available from: http://www.rcrwireless.com/20160912/big-data-analytics/siemens-train-teradata-tag31-tag99 [Accessed: 2018-11-29]

[12] Kshetri N. The economics of the internet of things in the global south. Third World Quarterly. 2017;**38**(2):311-339

[13] Handel P, Skog I, Wahlstrom J, Bonawiede F, Welch R, Ohlsson J, et al. Insurance telematics: Opportunities and challenges with the smartphone solution. IEEE Intelligent Transportation Systems Magazine. 2014;**6**(4):57-70

[14] FDA approves pill with sensor that digitally tracks if patients have ingested their medication [press release]; 2018

[15] Alison Bolen SIE. 3 Internet of Things examples from 3 industries [Internet]. Available from: https://www.sas.com/en_us/insights/articles/big-data/3-internet-of-things-examples.html [Accessed: 2018-11-29]

[16] Hughes RB. The autonomous vehicle revolution and the global commons. SAIS Review of International Affairs. 2016;**36**(2):41-56

[17] Clarke R. Internet privacy concerns confirm the case for intervention. Communications of the ACM. 1999;**42**(2):60-67

[18] Manyika J, Chui M, Bughin J, Dobbs R, Bisson P, Marrs A. Disruptive Technologies: Advances that Will Transform Life, Business,and the Global Economy. San Francisco, CA: McKinsey Global Institute; 2013

[19] OECD. Managing Digital Security and Privacy Risk; 2016

[20] Hernández-Serrano J, Muñoz JL, Bröring A, Esparza O, Mikkelsen L, Schwarzott W, et al., editors. On the Road to Secure and Privacy-Preserving IoT Ecosystems. Cham: Springer International Publishing; 2017

[21] Mullen M. The Internet and Public Policy: Privacy and Consumer Protection; 2018

[22] Kraemer MJ, Flechais I, editors. Researching privacy in smart homes: A roadmap of future directions and research methods. In: Living in the Internet of Things: Cybersecurity of the IoT - 2018. 2018 28-29 March 2018

[23] Almusaylim ZA, Zaman N. A review on smart home present state and challenges: Linked to context-awareness internet of things (IoT). Wireless Networks. 2018

[24] Sivaraman V, Gharakheili HH, Fernandes C, Clark N, Karliychuk T. Smart IoT devices in the home: Security and privacy implications. IEEE Technology and Society Magazine. 2018;**37**(2):71-79

[25] Commission E. European Commission. Eu Data Protection Reform What Benefits for Businesses in Europe; 2016

[26] Jackson S. SB-327 Information privacy: connected devices; 2018

[27] California Bill Mandates Privacy By Design For IoT Devices. ICT Monitor Worldwide. 2017 04/28/2017 Apr 28

[28] Senators Mark Warner CG, Ron Wyden, and Steve Daines. Cybersecurity Improvement Act 2017 [Internet]. Available from: https://www.warner.senate.gov/public/_cache/files/8/6/861d66b8-93bf-4c93-84d0-6bea67235047/8061BCEEBF4300EC702B4E894247D0E0.iot-cybesecurity-improvement-act---fact-sheet.pdf [Accessed: 2018-11-29]

[29] Muzzammil H, Neha K. An improvised framework for privacy preservation in IoT. International Journal of Information Security and Privacy (IJISP). 2018;**12**(2):46-63

[30] Jie Y, Pei JY, Jun L, Yun G, Wei X, editors. Smart home system based on IOT technologies. In: 2013 International Conference on Computational and Information Sciences. 2013 21-23 June 2013

[31] Chaudhuri A, Cavoukian A. The proactive and preventive privacy (3P) framework for IoT privacy by design. EDPACS. 2018;**57**(1):1-16

[32] Commissioner for Privacy and Data Protection V, Australia. Guidelines for Sharing Personal Information; 2016

[33] Caron X, Bosua R, Maynard SB, Ahmad A. The internet of things (IoT) and its impact on individual privacy: An Australian perspective. Computer Law and Security Review. 2016;**32**(1):4-15

[34] Porambage P, Ylianttila M, Schmitt C, Kumar P, Gurtov A, Vasilakos AV. The quest for privacy in the internet of things. IEEE Cloud Computing. 2016;**3**(2):36-45

[35] Cisco Survey Reveals Close to Three-Fourths of IoT Projects Are Failing [press release]; 2017

[36] Jifa G, Lingling Z. Data, DIKW, Big Data and Data Science. Procedia Computer Science. 2014;**31**:814-821

[37] Infosys. IoT Connected World; 2018

[38] Cushing T. LG Will Take The 'Smart' Out Of Your Smart TV If You Don't Agree To Share Your Viewing And Search Data With Third Parties Techdirt; 2014 [Internet]. Available from: https://www.techdirt.com/articles/20140511/17430627199/lg-will-take-smart-out-your-smart-tv-if-you-dont-agree-to-share-your-viewing-search-data-with-third-parties.shtml [Accessed: 2018-11-29]

[39] Baldini G, Botterman M, Neisse R, Tallacchini M. Ethical Design in the Internet of things. Science and Engineering Ethics. 2018;**24**(3): 905-925

[40] Michael S. Smith. Protecting Privacy in an IoT-Connected World. Tech Trend; 2015

[41] Wachter S. Normative challenges of identification in the internet of things: Privacy, profiling, discrimination, and the GDPR. Computer Law and Security Review. 2018

[42] Al-Ruithe M, Mthunzi S, Benkhelifa E, editors. Data governance for security in IoT & cloud converged environments. In: 2016 IEEE/ACS 13th International Conference of Computer Systems and Applications (AICCSA). 2016 Nov. 29 2016-Dec. 2 2016

[43] Sicari S, Rizzardi A, Grieco L, Coen-Porisini A. Security, privacy and trust in Internet of Things: The Road Ahead; 2015

[44] Chauhan S, Agarwal N, Kar AK. Addressing big data challenges in smart cities: A systematic literature review. Info. 2016;**18**(4):73-90

[45] Zarkout B. Making the Case for Governance IoT Data; 2017

[46] Barcena MB, Wueest C. Insecurity in the Internet of Things. Security Response, Symantec; 2015

[47] Miettinen M, Marchal S, Hafeez I, Asokan N, Sadeghi A, Tarkoma S, editors. IoT SENTINEL: Automated device-type identification for security enforcement in IoT. In: 2017 IEEE 37th International Conference on Distributed Computing Systems (ICDCS). 2017 5-8 June 2017

[48] Mansfield-Devine S. Data protection: Prepare now or risk disaster. Computer Fraud & Security. 2016;**2016**(12):5-12

[49] Maple C. Security and privacy in the internet of things. Journal of Cyber Policy. 2017;**2**(2):155-184

[50] van den Broek T, van Veenstra AF. Governance of big data collaborations: How to balance regulatory compliance and disruptive innovation. Technological Forecasting and Social Change. 2018;**129**:330-338

[51] Weber R. Governance of the internet of things—From infancy to first attempts of implementation? Laws. 2016;**5**(3):28

[52] Liu C, Yang C, Zhang X, Chen J. External integrity verification for outsourced big data in cloud and IoT: A big picture. Future Generation Computer Systems. 2015;**49**:58-67

[53] Karkouch A, Mousannif H, Al Moatassime H, Noel T. A model-driven framework for data quality management in the internet of things. Journal of Ambient Intelligence and Humanized Computing. 2018;**9**(4):977-998

[54] Singh VK, Guo J, editors. Improving service levels using internet of things infrastructure in data Centers. In: 2016 IEEE International Conference on Smart Computing (SMARTCOMP). 2016 18-20 May 2016

[55] Gantait A, Patra J, Mukherjee A. Defining your IoT governance practices: IBM; 2018 [Internet]. Available from: https://www.ibm.com/developerworks/library/iot-governance-01/index.html [Accessed: 2018-11-29]

[56] Al-Ruithe M, Benkhelifa E, Hameed K. Data governance taxonomy: Cloud versus non-cloud. Sustainability. 2018;**10**(1):95

[57] Sharma S, Chen K, Sheth A. Toward practical privacy-preserving analytics for IoT and cloud-based healthcare systems. IEEE Internet Computing. 2018;**22**(2):42-51

[58] Adler S. The IBM Data Governance Council Maturity Model: Building a Roadmap for Effective Data Governance. Somers, NY, USA: IBM Corporation; 2007

[59] Tokenization P. Securing Sensitive Data for PCI. HIPAA and Other Data Security Initiatives. 2011. p. 13

Chapter 3

Smart Home Systems Based on Internet of Things

Menachem Domb

Abstract

Smart home systems achieved great popularity in the last decades as they increase the comfort and quality of life. Most smart home systems are controlled by smartphones and microcontrollers. A smartphone application is used to control and monitor home functions using wireless communication techniques. We explore the concept of smart home with the integration of IoT services and cloud computing to it, by embedding intelligence into sensors and actuators, networking of smart things using the corresponding technology, facilitating interactions with smart things using cloud computing for easy access in different locations, increasing computation power, storage space and improving data exchange efficiency. In this chapter we present a composition of three components to build a robust approach of an advanced smart home concept and implementation.

Keywords: smart home, IoT, cloud computing, event processing, home appliances, rule-based event processing

1. Introduction

Classic smart home, internet of things, cloud computing and rule-based event processing, are the building blocks of our proposed advanced smart home integrated compound. Each component contributes its core attributes and technologies to the proposed composition. IoT contributes the internet connection and remote management of mobile appliances, incorporated with a variety of sensors. Sensors may be attached to home related appliances, such as air-conditioning, lights and other environmental devices. And so, it embeds computer intelligence into home devices to provide ways to measure home conditions and monitor home appliances' functionality. Cloud computing provides scalable computing power, storage space and applications, for developing, maintaining, running home services, and accessing home devices anywhere at anytime. The rule-based event processing system provides the control and orchestration of the entire advanced smart home composition.

Combining technologies in order to generate a best of breed product, already appear in recent literature in various ways. Christos Stergioua et al. [1] merge cloud computing and IoT to show how the cloud computing technology improves the functionality of the IoT. Majid Al-Kuwari [2] focus on embedded IoT for using analyzed data to remotely execute commands of home appliances in a smart home. Trisha Datta et al. [3] propose a privacy-preserving library to embed traffic shaping in home appliances. Jian Mao et al. [4] enhance machine learning algorithms to play a role in the security in a smart home ecosystem. Faisal Saeed et al. [5] propose using sensors to sense and provide in real-time, fire detection with high accuracy.

In this chapter we explain the integration of classic smart home, IoT and cloud computing. Starting by analyzing the basics of smart home, IoT, cloud computing and event processing systems. We discuss their complementarity and synergy, detailing what is currently driving to their integration. We also discuss what is already available in terms of platforms, and projects implementing the smart home, cloud and IoT paradigm. From the connectivity perspective, the added IoT appliances and the cloud, are connected to the internet and in this context also to the home local area network. These connections complement the overall setup to a complete unified and interconnected composition with extended processing power, powerful 3rd party tools, comprehensive applications and an extensive storage space.

In the rest of this chapter we elaborate on each of the four components. In Section 1, we describe the classic smart home, in Section 2, we introduce the internet of things [IoT], in Section 3, we outline cloud computing and in Section 4, we present the event processing module. In Section 5, we describe the composition of an advanced smart home, incorporating these four components. In Section 6, we provide some practical information and relevant selection considerations, for building a practical advanced smart home implementation. In Section 7, we describe our experiment introducing three examples presenting the essence of our integrated proposal. Finally, we identify open issues and future directions in the future of advanced smart home components and applications.

2. Classic smart home overview

Smart home is the residential extension of building automation and involves the control and automation of all its embedded technology. It defines a residence that has appliances, lighting, heating, air conditioning, TVs, computers, entertainment systems, big home appliances such as washers/dryers and refrigerators/freezers, security and camera systems capable of communicating with each other and being controlled remotely by a time schedule, phone, mobile or internet. These systems consist of switches and sensors connected to a central hub controlled by the home resident using wall-mounted terminal or mobile unit connected to internet cloud services.

Smart home provides, security, energy efficiency, low operating costs and convenience. Installation of smart products provide convenience and savings of time, money and energy. Such systems are adaptive and adjustable to meet the ongoing changing needs of the home residents. In most cases its infrastructure is flexible enough to integrate with a wide range of devices from different providers and standards.

The basic architecture enables measuring home conditions, process instrumented data, utilizing microcontroller-enabled sensors for measuring home conditions and actuators for monitoring home embedded devices.

The popularity and penetration of the smart home concept is growing in a good pace, as it became part of the modernization and reduction of cost trends. This is achieved by embedding the capability to maintain a centralized event log, execute machine learning processes to provide main cost elements, saving recommendations and other useful reports.

2.1 Smart home services

2.1.1 Measuring home conditions

A typical smart home is equipped with a set of sensors for measuring home conditions, such as: temperature, humidity, light and proximity. Each sensor is dedicated to capture one or more measurement. Temperature and humidity may be measured by one sensor, other sensors calculate the light ratio for a given area and the distance from it to each object exposed to it. All sensors allow storing the data and visualizing it so that the user can view it anywhere and anytime. To do so, it includes a signal processer, a communication interface and a host on a cloud infrastructure.

2.1.2 Managing home appliances

Creates the cloud service for managing home appliances which will be hosted on a cloud infrastructure. The managing service allows the user, controlling the outputs of smart actuators associated with home appliances, such as such as lamps and fans. Smart actuators are devices, such as valves and switches, which perform actions such as turning things on or off or adjusting an operational system. Actuators provides a variety of functionalities, such as on/off valve service, positioning to percentage open, modulating to control changes on flow conditions, emergency shutdown (ESD). To activate an actuator, a digital write command is issued to the actuator.

2.1.3 Controlling home access

Home access technologies are commonly used for public access doors. A common system uses a database with the identification attributes of authorized people. When a person is approaching the access control system, the person's identification attributes are collected instantly and compared to the database. If it matches the database data, the access is allowed, otherwise, the access is denied. For a wide distributed institute, we may employ cloud services for centrally collecting persons' data and processing it. Some use magnetic or proximity identification cards, other use face recognition systems, finger print and RFID.

In an example implementation, an RFID card and an RFID reader have been used. Every authorized person has an RFID card. The person scanned the card via RFID reader located near the door. The scanned ID has been sent via the internet to the cloud system. The system posted the ID to the controlling service which compares the scanned ID against the authorized IDs in the database.

2.2 The main components

To enable all of the above described activities and data management, the system is composed of the following components, as described in **Figure 1**.

a. Sensors to collect internal and external home data and measure home conditions. These sensors are connected to the home itself and to the attached-to-home devices. These sensors are not internet of things sensors, which are attached to home appliances. The sensors' data is collected and continually transferred via the local network, to the smart home server.

b. Processors for performing local and integrated actions. It may also be connected to the cloud for applications requiring extended resources. The sensors' data is then processed by the local server processes.

c. A collection of software components wrapped as APIs, allowing external applications execute it, given it follows the pre-defined parameters format. Such an API can process sensors data or manage necessary actions.

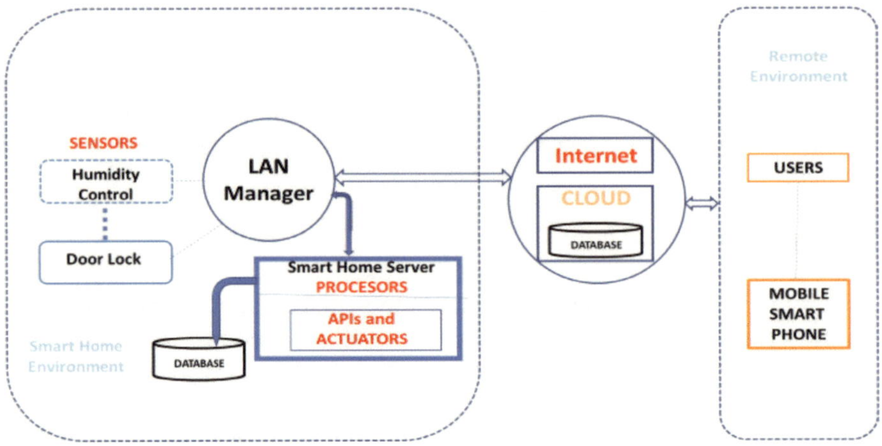

Figure 1.
Smart home paradigm with optional cloud connectivity.

d. Actuators to provision and execute commands in the server or other control devices. It translates the required activity to the command syntax; the device can execute. During processing the received sensors' data, the task checks if any rule became true. In such case the system may launch a command to the proper device processor.

e. Database to store the processed data collected from the sensors [and cloud services]. It will also be used for data analysis, data presentation and visualization. The processed data is saved in the attached database for future use.

3. Internet of things [IoT] overview

The internet of things (IoT) paradigm refers to devices connected to the internet. Devices are objects such as sensors and actuators, equipped with a telecommunication interface, a processing unit, limited storage and software applications. It enables the integration of objects into the internet, establishing the interaction between people and devices among devices. The key technology of IoT includes radio frequency identification (RFID), sensor technology and intelligence technology. RFID is the foundation and networking core of the construction of IoT. Its processing and communication capabilities along with unique algorithms allows the integration of a variety of elements to operate as an integrated unit but at the same time allow easy addition and removal of components with minimum impact, making IoT robust but flexible to absorb changes in the environment and user preferences. To minimize bandwidth usage, it is using JSON, a lightweight version of XML, for inter components and external messaging.

4. Cloud computing and its contribution to IoT and smart home

Cloud computing is a shared pool of computing resources ready to provide a variety of computing services in different levels, from basic infrastructure to most sophisticated application services, easily allocated and released with minimal efforts or service provider interaction [6, 7]. In practice, it manages computing, storage, and communication resources that are shared by multiple users in a virtualized and isolated environment. **Figure 2** depicts the overall cloud paradigm.

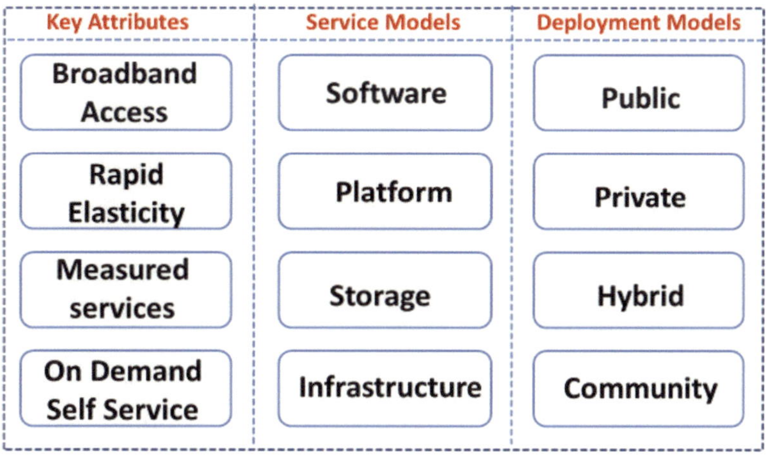

Figure 2.
Cloud computing paradigm.

IoT and smart home can benefit from the wide resources and functionalities of cloud to compensate its limitation in storage, processing, communication, support in pick demand, backup and recovery. For example, cloud can support IoT service management and fulfillment and execute complementary applications using the data produced by it. Smart home can be condensed and focus just on the basic and critical functions and so minimize the local home resources and rely on the cloud capabilities and resources. Smart home and IoT will focus on data collection, basic processing, and transmission to the cloud for further processing. To cope with security challenges, cloud may be private for highly secured data and public for the rest.

IoT, smart home and cloud computing are not just a merge of technologies. But rather, a balance between local and central computing along with optimization of resources consumption. A computing task can be either executed on the IoT and smart home devices or outsourced to the cloud. Where to compute depends on the overhead tradeoffs, data availability, data dependency, amount of data transportation, communications dependency and security considerations. On the one hand, the triple computing model involving the cloud, IoT and smart home, should minimize the entire system cost, usually with more focus on reducing resource consumptions at home. On the other hand, an IoT and smart home computing service model, should improve IoT users to fulfill their demand when using cloud applications and address complex problems arising from the new IoT, smart home and cloud service model.

Some examples of healthcare services provided by cloud and IoT integration: properly managing information, sharing electronic healthcare records enable high-quality medical services, managing healthcare sensor data, makes mobile devices suited for health data delivery, security, privacy, and reliability, by enhancing medical data security and service availability and redundancy and assisted-living services in real-time, and cloud execution of multimedia-based health services.

5. Centralized event processing, a rule-based system

Smart home and IoT are rich with sensors, which generate massive data flows in the form of messages or events. Processing this data is above the capacity of a human being's capabilities [8–10]. Hence, event processing systems have been developed and used to respond faster to classified events. In this section, we focus on rule management systems which can sense and evaluate events to respond to changes in values or interrupts. The user can define event-triggered rule and to control the proper delivery of services. A rule is composed of event conditions, event pattern and correlation-related information which can be combined for modeling complex situations. It was implemented in a typical smart home and proved its suitability for a service-oriented system.

The system can process large amounts of events, execute functions to monitor, navigate and optimize processes in real-time. It discovers and analyzes anomalies or exceptions and creates reactive/proactive responses, such as warnings and preventing damage actions. Situations are modeled by a user-friendly modeling interface for event-triggered rules. When required, it breaks them down into simple, understandable elements. The proposed model can be seamlessly integrated into the distributed and service-oriented event processing platform.

The evaluation process is triggered by events delivering the most recent state and information from the relevant environment. The outcome is a decision graph representing the rule. It can break down complex situations to simple conditions, and combine them with each other, composing complex conditions. The output is a response event raised when a rule fires. The fired events may be used as input for other rules for further evaluation. Event patterns are discovered when multiple events occur and match a pre-defined pattern. Due to the graphical model and modular approach for constructing rules, rules can be easily adapted to domain changes. New event conditions or event patterns can be added or removed from the rule model. Rules are executed by event services, which supply the rule engine with events and process the evaluation result. To ensure the availability of suitable processing resources, the system can run in a distributed mode, on multiple machines and facilitate the integration with external systems, as well. The definition of relationships and dependencies among events that are relevant for the rule processing, are performed using sequence sets, generated by the rule engine. The rule engine constructs sequences of events relevant to a specific rule condition to allow associating events by their context data. Rules automatically perform actions in response when stated conditions hold. Actions generate response events, which trigger response activities. Event patterns can match temporal event sequences, allowing the description of home situations where the occurrences of events are relevant. For example, when the door is kept open too long.

The following challenges are known with this model: structure for the processed events and data, configuration of services and adapters for processing steps, including their input and output parameters, interfaces to external systems for sensing data and for responding by executing transactions, structure for the processed events and data, data transformations, data analysis and persistence. It allows to model which events should be processed by the rule service and how the response events should be forwarded to other event services. The process is simple: data is collected and received from adapters which forward events to event services that consume them. Initially the events are enriched to prepare the event data for the rule processing. For example, the response events are sent to a service for sending notifications to a call agent, or to services which transmit event delay notifications and event updates back to the event management system.

5.1 Event processing languages

Event processing is concerned with real-time capturing and managing predefined events. It starts from managing the receptors of events right from the event occurrence, even identification, data collection, process association and activation of the response action. To allow rapid and flexible event handling, an event processing language is used, which allows fast configuration of the resources required to handle the expected sequence of activities per event type. It is composed of two modules, ESP and CEP. ESP efficiently handles the event, analyzes it and selects the appropriate occurrence. CEP handles aggregated events. Event languages describe complex event-types applied over the event log.

5.2 Rediscovering workflow from events

In some cases, rules relate to discrepancies in a sequence of events in a workflow. In such cases, it is mandatory to precisely understand the workflow and its associated events. To overcome this, we propose a reverse engineering process to automatically rediscover the workflows from the events log collected over time, assuming these events are ordered, and each event refers to one task being executed for a single case. The rediscovering process can be used to validate workflow sequences by measuring the discrepancies between prescriptive models and actual process executions. The rediscovery process consists of the following three steps: (1) construction of the dependency/frequency table. (2) Induction of dependency/frequency graphs. (3) Generating WF-nets from D/F-graphs.

6. Advanced smart home

In this section, we focus on the integration of smart home, IoT and cloud computing to define a new computing paradigm. We can find in the literature section [11–14] surveys and research work on smart home, IoT and cloud computing separately, emphasizing their unique properties, features, technologies, and drawbacks. However, our approach is the opposite. We are looking at the synergy among these three concepts and searching for ways to integrate them into a new comprehensive paradigm, utilizing its common underlying concepts as well as its unique attributes, to allow the execution of new processes, which could not be processed otherwise.

Figure 3 depicts the advanced smart-home main components and their interconnectivity. On the left block, the smart home environment, we can see the typical

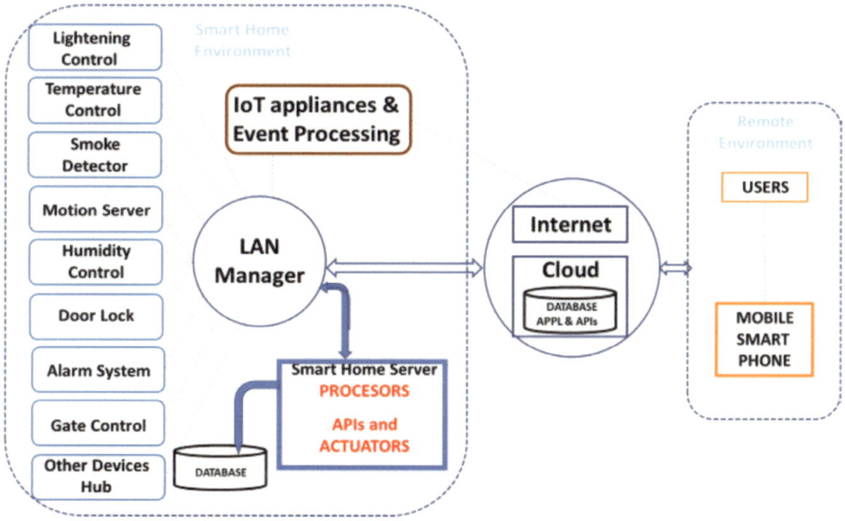

Figure 3.
Advanced smart home—integrating smart home, IoT and cloud computing.

devices connected to a local area network [LAN]. This enables the communication among the devices and outside of it. Connected to the LAN is a server and its database. The server controls the devices, logs its activities, provides reports, answers queries and executes the appropriate commands. For more comprehensive or common tasks, the smart home server, transfers data to the cloud and remotely activate tasks in it using APIs, application programming interface processes. In addition, IoT home appliances are connected to the internet and to the LAN, and so expands smart home to include IoT. The connection to the internet allows the end user, resident, to communicate with the smart home to get current information and remotely activate tasks.

To demonstrate the benefits of the advanced smart home, we use RSA, a robust asymmetric cryptography algorithm, which generates a public and private key and encrypts/decrypts messages. Using the public key, everyone can encrypt a message, but only these who hold the private key can decrypt the sent message. Generating the keys and encrypting/decrypting messages, involves extensive calculations, which require considerable memory space and processing power. Therefore, it is usually processed on powerful computers built to cope with the required resources. However, due to its limited resources, running RSA in an IoT device is almost impossible, and so, it opens a security gap in the Internet, where attackers may easily utilize. To cope with it, we combine the power of the local smart home processors to compute some RSA calculations and forward more complicated computing tasks to be processed in the cloud. The results will then be transferred back to the IoT sensor to be compiled and assembled together, to generate the RSA encryption/decryption code, and so close the mentioned IoT security gap. This example demonstrates the data flow among the advanced smart home components. Where, each component performs its own stack of operations to generate its unique output. However, in case of complicated and long tasks it will split the task to sub tasks to be executed by more powerful components. Referring to the RSA example, the IoT device initiates the need to generate an encryption key and so, sends a request message to the RSA application, running in the smart home computer. The smart home computer then asks the "prime numbers generation" application running on cloud, to provide p and q prime numbers. Once p and q are accepted, the encryption code is generated. In a later stage, an IoT device issues a request to the smart home computer to encrypt a message, using the recent generated RSA encryption key. The encrypted message is then transferred back to the IoT device for further execution. A similar scenario may be in the opposite direction, when an IoT device gets a message it may request the smart home to decrypt it.

To summarize, the RSA scenarios depict the utilization of the strength of the cloud computing power, the smart home secured computing capabilities and at the end the limited power of the IoT device. It proves that without this automatic cooperation, RSA would not be able to be executed at the IoT level.

A more practical example is where several detached appliances, such as an oven, a slow cooker and a pan on the gas stove top, are active in fulfilling the resident request. The resident is getting an urgent phone call and leaves home immediately, without shutting off the active appliances. In case the relevant IoTs have been tuned to automatically shut down based on a predefined rule, it will be taken care at the IoT level. Otherwise, the smart home realizes the resident has left home [the home door has been opened and then locked, the garage has been opened, the resident's car left, the main gate was opened and then closed, no one was at home] and will shut down all active devices classified as risk in case of absence. It will send an appropriate message to the mailing list defined for such an occasion.

7. Practical aspects and implementation considerations for IoT and smart home

Smart home has three components: hardware, software and communication protocols. It has a wide variety of applications for the digital consumer. Some of the areas of home automation led IoT enabled connectivity, such as: lighting control, gardening, safety and security, air quality, water-quality monitoring, voice assistants, switches, locks, energy and water meters.

Advanced smart home components include: IoT sensors, gateways, protocols, firmware, cloud computing, databases, middleware and gateways. IoT cloud can be divided into a platform-as-a-service (PaaS) and infrastructure-as-a-service (IaaS). **Figure 4** demonstrates the main components of the proposed advanced smart home and the connection and data flow among its components.

The smart home application updates the home database in the cloud to allow remote people access it and get the latest status of the home. A typical IoT platform contains: device security and authentication, message brokers and message queuing, device administration, protocols, data collection, visualization, analysis capabilities, integration with other web services, scalability, APIs for real-time information flow and open source libraries. IoT sensors for home automation are known by their sensing capabilities, such as: temperature, lux, water level, air composition, surveillance video cameras, voice/sound, pressure, humidity, accelerometers, infrared, vibrations and ultrasonic. Some of the most commonly used smart home sensors are temperature sensors, most are digital sensors, but some are analog and can be extremely accurate. Lux sensors measure the luminosity. Water level ultrasonic sensors.

Float level sensors offer a more precise measurement capability to IoT developers. Air composition sensors are used by developers to measure specific components in the air: CO monitoring, hydrogen gas levels measuring, nitrogen oxide measure, hazardous gas levels. Most of them have a heating time, which means that it requires a certain time before presenting accurate values. It relies on detecting gas

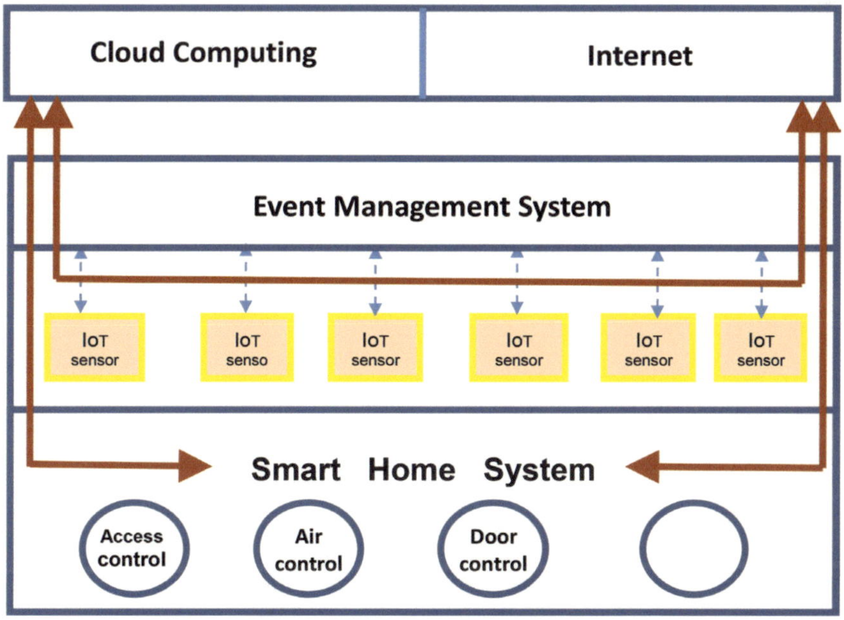

Figure 4.
Advanced smart home composition.

components on a surface only after the surface is heated enough, values start to show up. Video cameras for surveillance and analytics. A range of cameras, with a high-speed connection. Using Raspberry Pi processor is recommended as its camera module is very efficient due to its flex connector, connected directly to the board.

Sound detectors are widely used for monitoring purposes, detecting sounds and acting accordingly. Some can even detect ultra-low levels of noise, and fine tune among various noise levels.

Humidity sensors sense the humidity levels in the air for smart homes. Its accuracy and precision depend on the sensor design and placement. Certain sensors like the DHT22, built for rapid prototyping, will always perform poorly when compared to high-quality sensors like HIH6100. For open spaces, the distribution around the sensor is expected to be uniform requiring fewer corrective actions for the right calibration.

Smart home communication protocols: bluetooth, Wi-Fi, or GSM. Bluetooth smart or low energy wireless protocols with mesh capabilities and data encryption algorithms. Zigbee is mesh networked, low power radio frequency-based protocol for IoT. X10 protocol that utilizes powerline wiring for signaling and control. Insteon, wireless and wireline communication. Z-wave specializes in secured home automation. UPB, uses existing power lines. Thread, a royalty-free protocol for smart home automation. ANT, an ultra-low-power protocol for building low-powered sensors with a mesh distribution capability. The preferred protocols are bluetooth low energy, Z-wave, Zigbee, and thread. Considerations for incorporating a gateway may include: cloud connectivity, supported protocols, customization complexity and prototyping support. Home control is composed of the following: state machine, event bus, service log and timer.

Modularity: enables the bundle concept, runtime dynamics, software components can be managed at runtime, service orientation, manage dependencies among bundles, life cycle layer: controls the life cycle of the bundles, service layers: defines a dynamic model of communication between various modules, actual services: this is the application layer. Security layer: optional, leverages Java 2 security architecture and manages permissions from different modules.

OpenHAB is a framework, combining home automation and IoT gateway for smart homes. Its features: rules engine, logging mechanism and UI abstraction. Automation rules that focus on time, mood, or ambiance, easy configuration, common supported hardware:

Domoticz architecture: very few people know about the architecture of Domoticz, making it extremely difficult to build applications on it without taking unnecessary risks in building the product itself. For example, the entire design of general architecture feels a little weird when you look at the concept of a sensor to control to an actuator. Building advanced applications with Domoticz can be done using OO based languages.

Deployment of blockchain into home networks can easily be done with Raspberry Pi. A blockchain secured layer between devices and gateways can be implemented without a massive revamp of the existing code base. Blockchain is a technology that will play a role in the future to reassure them with revolutionary and new business models like dynamic renting for Airbnb.

8. Smart home and IoT examples

We can find in the literature and practical reports, many implementations of various integrations among part of the main three building blocks, smart home, IoT and cloud computing. For example, refer to [12–14]. In this section we outline three implementations, which clearly demonstrate the need and the benefits of interconnecting or integrating all three components, as illustrated in **Figure 5**. Each component is numbered, 1–6. In the left side, we describe for each implementation, the sequence of messages/commands among components, from left to right and from bottom up. Take for example the third implementation, a control task constantly runing at the home server (2) discovers the fact that all residents left home and automatically, initiates actuators to shut down all IoT appliances (3), then it issues messages to the relevant users/residents, updating them about the situation and the applied actions it took (6).

The use of (i) in the implementations explanation, corresponds to the circled numbers in **Figure 5**.

8.1 Discovery of water leaks and its prevention

First step is deploying water sensors under every reasonable potential leak source and an automated master water valve sensor for the whole house, which now means the house is considered as an IoT.

In case the water sensor detects a leak of water (3), it sends an event to the hub (2), which triggers the "turn valve off" application. The home control application then sends a "turn off" command to all IoT (3) appliances defined as sensitive to water stopping and then sends the "turn off" command to the main water valve (1). An update message is sent via the messaging system to these appearing in the notification list (6). This setup helps defending against scenarios where the source of the water is from the house plumbing. The underlying configuration assumes an integration via messages and commands between the smart home and the IoT control system. It demonstrates the dependency and the resulting benefits of combining smart home and IoT.

8.2 Smoke detectors

Most houses already have the typical collection of smoke detectors (1), but there is no bridge to send data from the sensor to a smart home hub. Connecting these sensors to a smart home app (2), enables a comprehensive smoke detection system. It is further expanded to notify the elevator sensor to block the use of it due to fire condition (1), and so, it is even further expanded to any IoT sensor (3), who may be activated due to the detected smoke alert.

In [5] they designed a wireless sensor network for early detection of house fires. They simulated a fire in a smart home using the fire dynamics simulator and a language program. The simulation results showed that the system detects fire early.

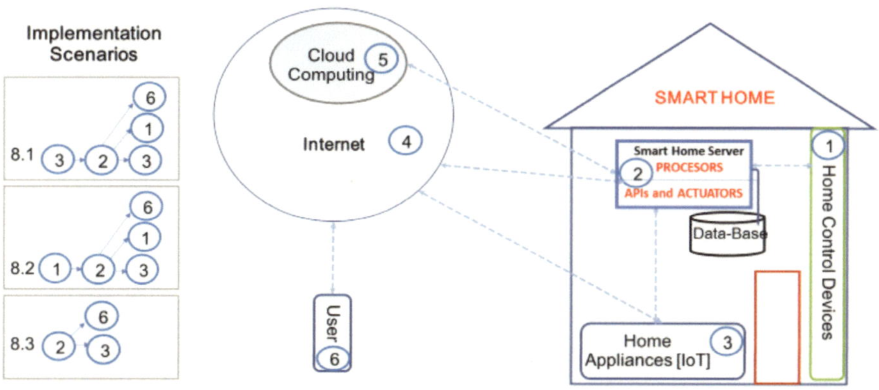

Figure 5.
Advanced smart home implementations chart.

8.3 Incident management to control home appliances

Consider the scenario where you leave home while some of the appliances are still on. In case your absence is long enough, some of the appliances may over heat and are about to blowout. To avoid such situations, we connect all IoT appliances' sensors to the home application (2), so that when all leave home it will automatically adjust all the appliances' sensors accordingly (3), to avoid damages. Note that the indication of an empty home is generated by the Smart Home application, while the "on" indication of the appliance, is generated by IoT. Hence, this scenario is possible due to the integration between smart home and IoT systems.

9. Conclusions and summary

In this chapter we described the integration of three loosely coupled components, smart home, Iot, and cloud computing. To orchestrate and timely manage the vast data flow in an efficient and balanced way, utilizing the strengths of each component we propose a centralized real time event processing application.

We describe the advantages and benefits of each standalone component and its possible complements, which may be achieved by integrating it with the other components providing new benefits raised from the whole compound system. Since these components are still at its development stage, the integration among them may change and provide a robust paradigm that generates a new generation of infrastructure and applications.

As we follow-up on the progress of each component and its corresponding impact on the integrated compound, we will constantly consider additional components to be added, resulting with new service models and applications.

Author details

Menachem Domb
Computer Science Department, Ashkelon Academic College, Ashkelon, Israel

*Address all correspondence to: dombmnc@edu.aac.ac.il

References

[1] Stergioua C, Psannis KE, Kimb B-G, Gupta B. Secure Integration of IoT and Cloud Computing. Elsevier, Future Generation Computer Systems, Vol. 78. Part 3. January 2018. pp. 964-975

[2] Al-Kuwari M, Ramadan A, Ismael Y, Al-Sughair L, Gastli A, Benammar M. Smart-Home Automation Using IoT-Based Sensing and Monitoring Platform, IEEE. 2018. Available from: ieeexplore.ieee.org

[3] Datta T, Apthorpe N, Feamster N. Developer-friendly library for smart home IoT privacy preserving traffic obfuscation, IoT S&P 18. In: Proceedings of the 2018 Workshop on IoT Security and Privacy. ACM; 2018. pp. 43-48

[4] Mao J, Lin Q , Bian J. Application of Learning Algorithms in Smart Home IoT System Security. American Institute of Mathematical Sciences; 2018. DOI: 10.3934/mfc.2018004

[5] Saeed F, Paul A, Rehman A, Hong WH, Seo H. IoT-based intelligent modeling of smart home environment for fire prevention and safety. Journal of Sensor and Actuator Networks. 2018;**7**(1):11. DOI: 10.3390/jsan7010011

[6] Botta A, de Donato W, Persico V, Pescapé A. Integration of cloud computing and internet of things: A survey. Future Generation Computer Systems. 2016;**56**:684-700

[7] Soliman M, Abiodun T, Hamouda T, Zhou J, Lung C-H. Smart home: Integrating internet of things with web services and cloud computing. In: International Conference on Cloud Computing Technology and Science; IEEE. 2013

[8] Paschke A, Kozlenkov A. Rule-Based Event Processing and Reaction Rules. London: Betfair Ltd; 2009. DOI: 10.1007/978-3-642-04985-9_8

[9] Khan NS, Ghani S, Haider S. Real-time analysis of a sensor's data for automated decision making in an IoT-based smart home. Sensors. 2018;**18**:1711. DOI: 10.3390/s18061711

[10] Malik R, Parameswaran N, Ghose U. Rule based event management systems. In: Proceedings of the 25th International Florida Artificial Intelligence Research Society Conference. Association for the Advancement of Artificial Intelligence; 2012

[11] Vinodhan D, Vinnarasi A. IOT based smart home. International Journal of Engineering and Innovative Technology (IJEIT). 2016;**5**(10):35-38

[12] Jian MS, Wu JY, Chen JY, Li YJ, Wang YC, Xu HY. IOT Base Smart Home Appliances by Using Cloud Intelligent Tetris Switch; 19-22 February 2017; ICACT, ISBN 978-89-968650-9-4, 2017

[13] Risteska Stojkoska BL, Trivodaliev KV. A review of internet of things for smart home: Challenges and solutions. Journal of Cleaner Production. Part 3. 2017 January 1;**140**(3):1454-1464

[14] Lia B, Yub J. Research and application on the smart home based on component technologies and internet of things. Elsevier, Procedia Engineering; Vol. 15. 2011. pp. 2087-2092. 2011:1877-7058. DOI: 10.1016/j.proeng.2011.08.390

Section 3

Healthcare Applications

Chapter 4

IOT Service Utilisation in Healthcare

Mohammed Dauwed and Ahmed Meri

Abstract

Utilising the new trend technologies in healthcare sector could offer alternative ways in managing the patients' health records and also improve the healthcare quality. As such, this chapter provides an overview of utilising the Internet of Things (IoT) technology in healthcare sector as an emerging research and practical trend nowadays. The main benefits and advantages have been discussed in this chapter. On the other hand, it has been found that most of the hospitals in different countries are still facing many issues regarding their health information exchange. Recently, various studies in the area of healthcare information system mentioned that the fragmentations of the health information are one of the most important challenges with the distribution of patient information records. Therefore, in this chapter, we gave an in detail overview regarding the current issues facing the health sector in line with the IoT technologies. Additionally, a full description of advantages and disadvantages has been highlighted for using IoT in healthcare that can be considered as solutions for the mentioned issues.

Keywords: IoT, Internet of Things, healthcare, e-Health, health information exchange

1. Introduction

The Internet of Things (IoT) is a new technology that aims to connect the world via smart devices or objects with capabilities of collecting and sharing various types of information at any location, time, media and environments. By assigning a unique identification to each object in the network, IoT allows its users to live smart, safe lives. In healthcare systems, IoT is mainly used to gain quick access to health information. IoT can be defined as an interconnected network that links a large number of devices to one another for purposes of making large-scale information accessible to all. This technology can be seen as a grid of computers that deliver software and data via the Internet. As illustrated in **Figure 1**, Cisco defines IoT as a revolution of the 'Internet of Everything' that involves people, processes, data and things [1].

Many health organisations need to exchange data with one another to address their problems and to improve their performance [1]. Health-related data are especially important for these organisations to provide their patients with better healthcare services. The exchange of health information among these organisations has been termed 'health information exchange (HIE)', which has become a pervasive global phenomenon [2, 3]. Although not a novel concept in the health industry, HIE needs to reinvent itself every 2.5 years to adapt to the current technological advancements and the changes in the environment [4]. According to the 'Evolution of State Health Information Exchange in the U.S. (2006)', HIE offers

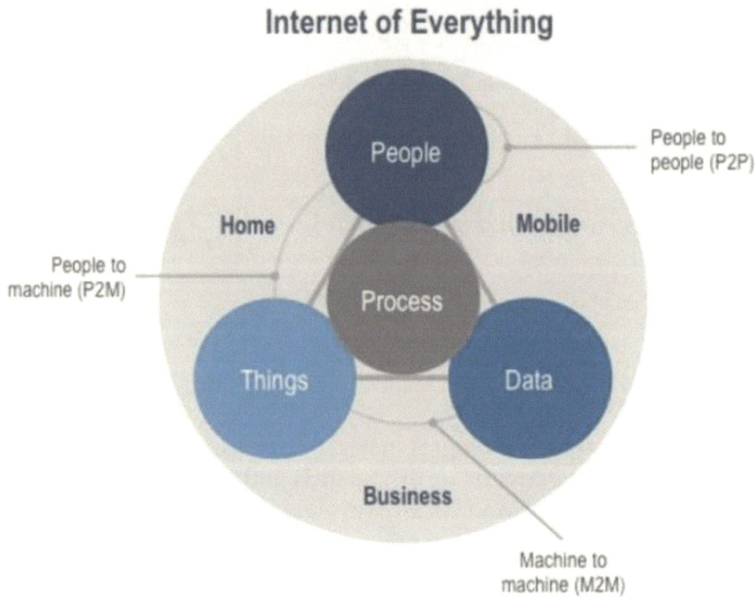

Figure 1.
IoT revolution [1].

many significant contributions to the designing of different projects, such as financing, identifying patterns of success, ensuring programmatic sustainability and highlighting challenges, trends and best practices [5]. HIE also provides many opportunities to improve the quality and reduce the cost of healthcare, improve the workflows of clinical organisations and facilitate the administration of data within the healthcare system [5]. However, HIE also poses one of the most complex problems in electronic health record (EHR) management [6]. Therefore, dissemination and communication are essential attributes of health information systems [7].

The medical records of each patient are stored in physical and electronic databases. However, when patients decide to move to new healthcare providers, the latter have no tools or directories that they can use to check where the medical records of these patients are stored. Such inaccessibility of medical records can lead to unnecessary procedures, duplicate tests and many other problems, such as adverse drug interaction. According to Tharmalingam et al. [8], Canada faces many difficulties related to HIE, including complex systems, lack of knowledge as to the location of patients' medical records, lack of access to information and lack of data standards that allow the exchange of clinical information. Some non-technological barriers also exist, including care burden, issues related to patient consent, differences in business models, limited understanding of procedures and loss of competitive advantage [2, 9].

Virtually storing patient data and making them ubiquitously accessible to all healthcare personnel is the first step in HIE [10]. Recent years have witnessed an increasing interest in the application of sensing technologies and widely available smart devices for monitoring personal health, fitness and activity. Continuously recording key physiological parameters via sensors can provide healthcare practitioners with the necessary data to produce rich longitudinal records [11]. Meanwhile, data from physical examinations provide doctors with comprehensive information that allows them to measure the physiological and metabolic states of their patients. Accessing a large number of observation data via health information systems can also help doctors improve their prognosis for their patients and recommend effective treatment, intervention and lifestyle choices to improve their health quality [12].

With the massive advancements in communication and computer technologies, organisations must urgently apply and utilise these technologies to compete effectively and survive in the market. IoT cannot improve the performance of hospitals if such technology is not being utilised to measure the success of a system [13]. A vast and multi-layered infrastructure of ubiquitous computing technologies and applications is also emerging. Mobile phones, laptops, Wi-Fi, Bluetooth, personal digital assistants and various forms of sensing devices based on digital and radio frequency identification (RFID) technologies have also penetrated the healthcare industry. IoT establishes connections among different entities, including humans (e.g. patients and medical staff), medical devices, intelligent wheelchairs, wireless sensors and mobile robots. People in the healthcare industry also rely on this technology to provide high-quality and affordable healthcare services, minimise medical errors, guarantee the safety of their patients and optimise their healthcare processes [14].

However, despite the wide availability of smart devices and novel communication technologies, healthcare professionals and patients are still generally unwilling to exchange health information while a large number of hospitals are yet to implement advanced technologies to promote their HIE capability [15–17]. IoT provides new opportunities for healthcare professionals to deliver health information to hard-to-reach populations. Utilising such technology often requires an organisation to spend a considerable amount of resources at different stages [18]. Unfortunately, most health organisations in developing countries only have few resources to spare for using new technologies, including IoT [19]. Many other issues also prevent these hospitals from receiving financial incentives that will enable them to adopt new technologies for facilitating HIE.

In sum, using IoT is in great demand in the healthcare sector. To effectively utilise IoT, hospitals must possess the necessary resources to produce the maximum value possible and to prevent failure [20]. Therefore, this chapter focuses on those problems being faced by the healthcare industry in its implementation of advanced technologies. Over the past 5 years, many health information systems have faced several concerns with regard to medical records. Most of these systems have focused on accelerating their provision of services to patients and improving the performance of hospitals by reconstructing their current workflows.

2. Internet of Things (IoT) in healthcare

The rapid proliferation of smart devices offers unprecedented opportunities for patients and health care professionals to exchange health information electronically [16]. The IoT is one of the smart technologies to integrate the smart devices on network. On the other hand, IoT is a global information infrastructure that enables advanced services by interconnecting devices based on existing and evolving interoperable information and communication technologies [21]. Thus, it is a collection of several opportunities that have wellness providing for the hospitals such as optimising the resources through automated workflows as well as process excellence. For instance, a majority of hospitals use IoT services for asset management and controlling humidity and temperature within operating rooms [22]. The collection of health data has multiple benefits to interdisciplinary healthcare collaboration, while most of the research focuses on the personal fitness plan and has a lack of compatibility and extensibility among a large number of devices and their business models. Compatibility involves in information exchanging, communication and events processing. There is a strong need for an efficient interface mechanism to simplify the management and interconnection of things. However,

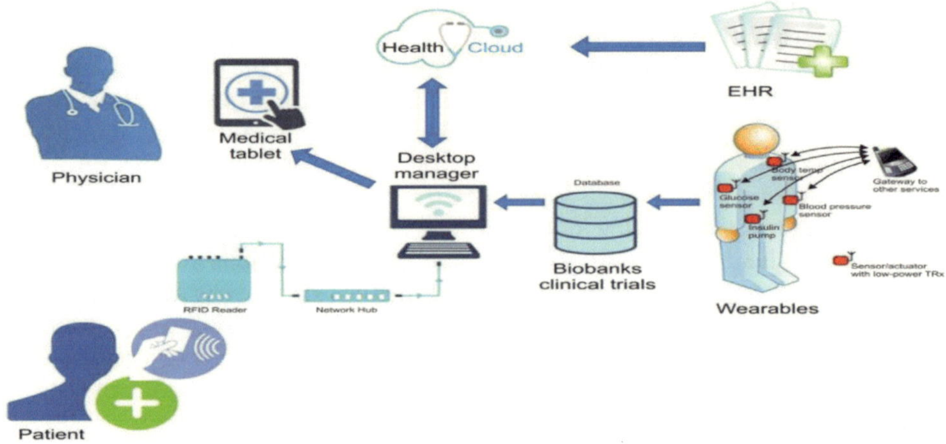

Figure 2.
IoT-hospital scenario [66].

the compatibility issue among the heterogeneous devices should be taken into consideration and addressed for the interactions among things [23].

Figure 2 illustrates how this revolution in the medical will look in a typical IoT hospital, in practice. The patient will have an ID card, which, when scanned, links to a secure cloud which stores their electronic health record vitals and lab results and medical and prescription histories.

The IoT has the potential to several benefits for health applications such as remote health monitoring, fitness programs, chronic diseases, children care and elderly care. Furthermore, it allows sharing and controlling the information between human to human or human-object or between objects using the Internet via ubiquitous sensors [24]. Therefore, various medical devices, sensors, and diagnostic and imaging devices can be viewed as smart devices or objects constituting a core part of the IoT [12]. The IoT-based e-Health monitoring method will help in reducing the number of visits to a doctor, and even the doctor can monitor his or her patient from anywhere. As this is a technology not so feasible now, but in coming years, this technology will meet the physical world definitely. The e-Health solutions provided through IoT devices are more accurate and accountable in the emerging IoT business landscape, which offers and provides various opportunities and challenges to an industry [25].

The IoT technology is still understudy to utilise it in the health sector in different regions in order to combine the information with control and monitoring such as China, US, Canada, etc. As a historical background, the Internet of Things was discovered by Kevin Ashton in 1998 to facilitate information exchange over the wide-world where every physical object connected through the Internet with a unique identification and can be monitored everywhere. One of the facilities of IoT for information systems is that it can provide services anywhere, anytime, and on any media [24]. In healthcare, the Internet of Things enables the potential benefits to achieve a high rate of exchange of massive information among organisations and organisation itself.

Some advantages of using Internet treatments included self-paced, interactive, of tailored service, multimedia format, greater accuracy reporting symptoms, timely information, accessibility, low cost, standardisation and increased user and supplier control of the intervention. Sensor technology and automated data collection enable passive monitoring of psychological states that can alert patients and healthcare providers to acute and chronic stress states [26]. These sensors

		Explanation	Sources
Advantage	Monitoring	Remote patient monitoring continues to grow and help physicians diagnose and treat illnesses and diseases with obtaining reliable information with a negligible error rate.	[27, 28]
	Sensing	IoT with intelligent medical sensors will enhance the quality of life significantly and prevent the occurrence of health problems.	[22]
	Low-cost solutions	Reduce unnecessary visits by doctors, and readmissions come from patients with chronic diseases and reduce testing cost.	[29]
	Ubiquitous access	Allow and increase the accessibility from anywhere, any time and any media allowing flexibility and mobility to the users. Enable real-time access services to the healthcare provider to access patient information and help them to make better decisions.	[30]
	Better quality of healthcare management	Increase the care quality and control by enhancing the management of drugs, reduce the medical error, enhance the patient experience, improve the disease management and improve outcome of treatment.	[31, 32]
	Unified information	Automated data collection enabled from health information resources such as monitoring, first aid, tracking, analysis, diagnosis, alarm-triggering, locating and collaboration with medical healthcare under unified communication platform and exchanged the health record.	[27]
	Time	This facilitates the interaction among the parts of an enterprise and allows for reducing the time necessary to adapt itself to the changes imposed by the market evolution.	[33]
Disadvantage	Complexity	The IoT is a diverse and complex network. There is a need of multiple services to grow device counts, massive increases of Internet bandwidth with a need to drive requirements for lower latency, greater determinism and processing closer to the edge of the network. Thus, any failure or bugs in the software or hardware will have serious consequences. Even power failure can cause a lot of trouble.	[34, 35]
	Compatibility	Although different manufacturers will be interconnected, the problem issue of compatibility when manufacturers do not agree to a common standard will make the people buy appliances from a certain manufacturer, leading to its monopoly in the market.	[23]
	Security and privacy	A location tracking and collect inappropriately information for any person considering as a challenge in the using of IoT services in the healthcare system. The patient concern of attacks his personal identity and privacy maybe arise. Therefore, bring big data from millions of things in a healthcare system can cause many security challenges.	[36, 37]
	Massive health data	In IoT, devices assemble and communicate information directly with each other via Internet and the cloud manages to collect record and analyse data blocks. But the 'things or devices' which are producing a massive amount of data are blowing out day-to-day, which needs to be treated and managed.	[38, 39]

Table 1.
The IoT advantages and disadvantages.

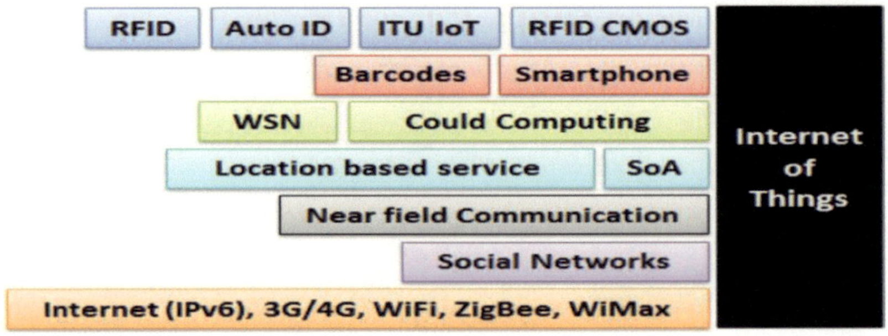

Figure 3.
Technologies associated with IoT [67].

can be used in monitoring patients, tracking daily activities, and caring for the chronic disease people or patients who have special states [27]. This information offers treatment that is evidence based from the information obtained from sensors and monitoring activities. All the applications of this technology culminated in increased comfort, convenience, and better management, thereby improving the quality of life. **Table 1** shows the multiple advantages and disadvantages of Internet of Things-based healthcare monitoring and management of health system.

Many open challenges need to be addressed by new research and investigation, mostly due to the complex deployment characteristics of such systems and the stringent requirements imposed by various services wishing to make use of such complex systems. Thus, it becomes critically important to study how the current approaches to standardisation in this area can be improved and at the same time better understand the opportunities for the research community to contribute to the IoT field [36]. In addition, many other technologies and devices such as barcodes, smart phones, social networks, and cloud computing are being used to form an extensive network for supporting IoT [12, 23] (as shown in **Figure 3**).

3. HIoT applications and device features

The healthcare applications and system have adopted several types of innovation technologies/devices in order to enhance the performance of healthcare services delivered. Most of these systems and applications are contributing to use IoT or smart technology devices to perform better advantage in healthcare services. These IoT applications and healthcare devices are called HIoT. The healthcare device implements dedicated sensor, and holds high collecting precision advantage, while it is also having a number of disadvantages such as insufficient portability, high cost and usability. This type of device possesses the following features:

- Wearability: most of the HIoT applications offer sensing on the human body so they collect data exactly and take vital signs of the human body as collecting targets. Thus, most of the existing medical health devices make the wearability as the basic requirement of collection of human body vital signs. On this vein, the users feel more comfortable and can be enhanced and the accuracy of the collected health data can be guaranteed through the collecting procedure. The layout of common human body sensors is shown in **Figure 4**.

- Long working time: the ways of dedicated health collecting data are several for instance universal mobile devices, wearable devices, pedometer, etc. The

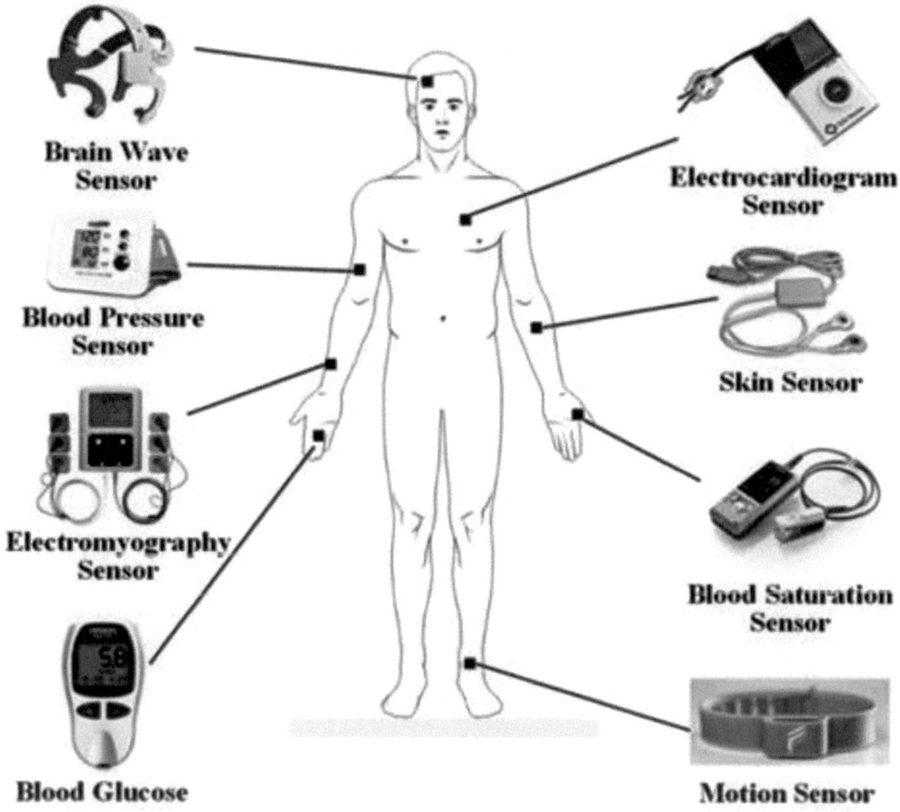

Figure 4.
Layout of common human body sensors.

purpose of these devices is to collect data from the human body for a relatively long time period that requires high power and capability.

- Constancy or stability: HIoT has high ability to collect data very normally even though the users are under strenuous exercise or in an extreme environment.

- Low participation degree of users: the functionality of HIoT applications and devices are relatively independent, as well as most HIoT devices do not require the intervention of users during the collecting data procedure. In addition, the users need to start up the power source only, and the HIoT device will start collecting data.

- Possessing data interim storage mechanism: the dimensions and weight of HIoT maybe limited strictly in order to meet the wearable feature. Thus, most HIoT devices do not integrate the data transmission module, but can select the data storage module with relatively small dimensions and adopt the data interim storage mechanism in order to store the collected data in advance, and then transmit the data through other network access devices accurately.

4. IoT scenario in the healthcare industry

Using IoT can improve and modify the delivered healthcare services in the following aspects:

1. Relying on sensing-based screening and assessment technologies in home and community environments can reduce the physical pressure on the environment of hospitals and turn this information into an electronic flow of information.

2. Changing the medication process from a reactive model to a proactive and preventative model can significantly minimise the hospital admission expenses for acute events.

3. Improving the personalisation of healthcare processes allows individuals to monitor and identify their risk factors, seek preventative intervention and treatment and live independently. In this way, personalising healthcare processes has a significant positive impact on the psychological and physiological states of patients.

4. Improving the management of clinical workloads can allow healthcare systems to effectively prioritise those patients who have the highest need for medical services.

5. Supporting self-care diagnostic processes for monitoring vital signs and other various measurements can produce data that are shared with physicians either personally or by phone in order for them to make effective diagnoses. These diagnoses can sometimes be automated for simple illnesses, such as flu.

Point-of-care tests can be optimised by reducing the time of diagnosis, which in turn can be achieved by reducing the requirements for sending samples to be tested. For example, automatic testing by using blood pressure cuffs and digital thermometers can help physicians review the history of their patients while performing the necessary measurements. Among its practical advantages, IoT can encourage the development of smart systems that support and improve biomedical and healthcare processes. Monitoring the physiological parameters of patients in real time can also facilitate the early detection of clinical deterioration, automatic people identification and tracking by using biomedical devices in smart hospitals and monitoring drug-patient associations [40].

Figure 5 illustrates the IoT scenario in smart hospitals. A patient with an emergency case is given a wearable device that detects the nearest ED that offers the required services. Upon being notified of an emergency case, the ED dispatches an ambulance to the location of the patient and delivers the necessary care services.

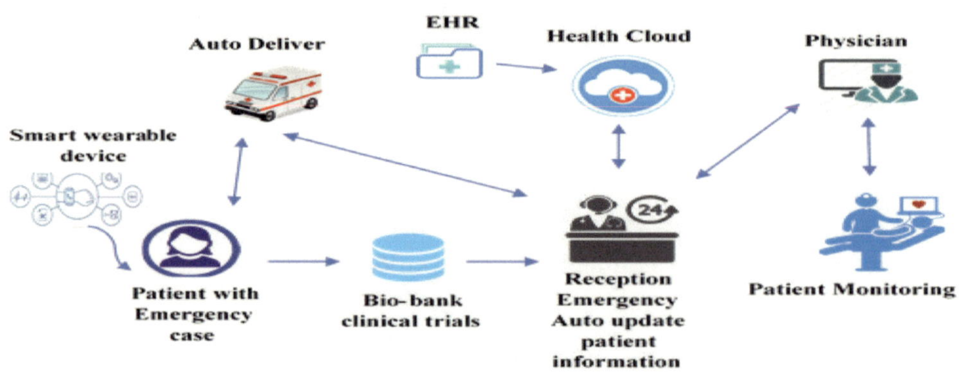

Figure 5.
IoT in the healthcare scenario.

Upon its arrival, the ambulance links the bio-bank of patient information to a secure cloud that stores the EHRs, laboratory test results and medical and prescription histories of the patient. This process can help health practitioners understand the status of their patients quickly, easily and effectively.

IoT in HIE systems is mostly designed to store, enter, receive and exchange health information. This system increases the number of devices and enhances the mobility of information to support health professionals in their consultations. Despite the benefits of using IoT in hospitals, several challenges related to availability, reliability, mobility, performance, management scalability, interoperability, security and privacy must be considered during its application [41].

5. Healthcare system challenges

Collecting and exchanging health information have become challenging due to the increasing population and demands for health services. These challenges can hinder the successful adoption of HIE. The following issues and challenges related to HIE adoption have been identified from the literature:

- Unified patients' data: this challenge refers to the combination of patient's data that are obtained from EHR systems that are being operated by healthcare providers (e.g. aged care providers, hospitals and healthcare specialists) for the purpose of sharing information. The unification of patient data provides excellent opportunities in continuing care, improving care quality and analysing and monitoring care service delivery and patient health outcomes.

- Teamwork of care: teamwork refers to collaboration among healthcare practitioners with the shared aim of exchanging information [42]. The communication deficiency among groups of healthcare professionals, departments or clinics has been identified as the main driver of critical safety incidents in tertiary care clinics. However, with the growing complexity of healthcare provision, the availability of patient information has been considered highly significant in the healthcare industry. Therefore, teamwork places less effort in promoting the availability of information. A survey of primary care doctors from 10 countries identified the overall communication, coordination of healthcare and teamwork as common challenges in HIE adoption. The lack of integration among primary care, specialty care and hospitals can also put patients at risk and lead to duplicative care, particularly for those patients suffering from complex chronic illnesses [43]. The full potential of teamwork is seldom realised due to training problems and the lack of trust in the reliability of healthcare services. Physicians are also often blamed for the errors that may occur during the provision of these services.

- Security and privacy: due to security and privacy concerns [44–46], many physicians and healthcare providers prefer to store patient records on computers or local systems that are not connected to the Internet [47]. Despite the benefits of large-scale HIE, a comparative study of the medical record exchange practices in Australia, Canada Germany, Netherlands, New Zealand, the UK and the US [48] revealed that Germany lacks a single approach for HIE and that healthcare software companies have achieved minimal success in their development of infrastructures where physicians can exchange clinical data due to security concerns. Similar to other countries, the substantial privacy and security concerns in the UK and the Netherlands have driven the resistance of healthcare professionals to HIE despite the benefits of this practice.

- Address shortage: another important issue that hinders the adoption of HIE is the storage of health information in a single pool. Cloud computing or other related technologies may be used as storage to allow healthcare practitioners to access and utilise health information at any time and place. Storing information online emerges as the most popular choice even though most users have expressed their concerns about storing their personal information on the Internet. In addition, the collected data must be managed and comply with standard formats and protocols in order for them to be retrieved and used by other healthcare providers. However, a common standard protocol for these data is yet to be devised [45]. Furthermore, patients should be allowed to access to their own data and be given the right to dispose of these data freely and ensure that their information is kept secure.

- Patient consent: the success of HIE also depends on public support, the willingness of patients to share their health information and their consent to have their health information shared with other parties via HIE [49, 50]. A study that examined the attitudes of patients towards giving consent revealed that the majority (91%) of the participating adult patients expect to be asked for their consent before their identifiable records are accessed and used for health provision, research or planning while only 9.2% of these respondents do not expect to be asked for their consent [51].

- Compatibility: compatibility refers to the degree to which the potential adopters perceive innovation as consistent with their values, previous experiences and needs. Therefore, based on physicians' expectations, the HIE system should be compatible with their work style and needs to motivate them to adopt such technology. This issue has a significant effect on the usage of innovation to promote HIE among hospitals [52].

- Hospital workflow: Healthcare professionals need to transform the HIE system to satisfy their demand for a faster access to patient information, which in turn can reduce their workflow. Issues related to workflow are important barriers that prevent the implementation of technologies in some health practices. Physicians in practices without EMR are generally reluctant to use computers to write prescriptions because these technologies are unavailable in many examination rooms. Therefore, HIE must promote consistency in workflows by facilitating staff training to improve their efficiency and by providing clinical information with minimal effort at any time or location [53, 54].

- System capacity: in order to facilitate HIE, the systems being used in hospitals should be effective and sustainable. Zhang et al. [15] attributed the limitation of system capacity to the following causes:

    - The failure to implement tecihnological advancements in most hospitals and the need to upgrade the HIE system to improve its capability.

    - The delayed development of a standard-compliant HIE system in many hospitals.

    - The overlapping functions among the needs of several regions, which reduce the need for information exchange.

However, exchanging patient records, including summaries and test results, among healthcare practitioners is not yet considered a norm in many countries. In the US, New Zealand and Canada, the current capacity of healthcare practitioners to share health information only ranges between 14 and 55% [43, 55]. With the technological advancements in networking, EHRs can be accessed by using various devices and stored in remote data centres.

6. Models/frameworks for IoT use in healthcare

To further understand the current utilisation of IoT in the healthcare sector, the related models/frameworks are reviewed as follows:

6.1 Tyagi et al.

Given the increasing demand of health organisations for access to patient records around the world, Tyagi et al. developed a cloud IoT-based healthcare framework and proposed *Platform as a Service* (PaaS) and *Infrastructure as a Service* (IaaS), which help patients find the best care at the optimal cost by allowing them to securely store and share their health information to healthcare organisations [56]. Patients can perform self-assessment to monitor their conditions and find hospitals that provide the healthcare services they need the most. However, the benefits of the cloud-IoT-based healthcare framework are offset by issues related to trust, privacy and security, all of which must be addressed before healthcare providers decide to adopt this framework. Moreover, the security requirements for the implementation of this model are yet to be fulfilled and its results need to be tested [56]. **Figures 6** and 7 summarise this framework.

6.2 Manate et al.

Collecting data from things, devices and multiple sources presents a significant problem. Patients can be classified into those patients who are having elective treatment and those emergency patients who require immediate treatment [57]. Those elective patients who do not require emergency treatment may experience health deterioration and eventually require emergency treatment or tests. A hospital setting is characterised by dynamic uncertainty and a frequent need to dynamically change the treatment pathway. Manate et al. proposed the intelligent context-aware

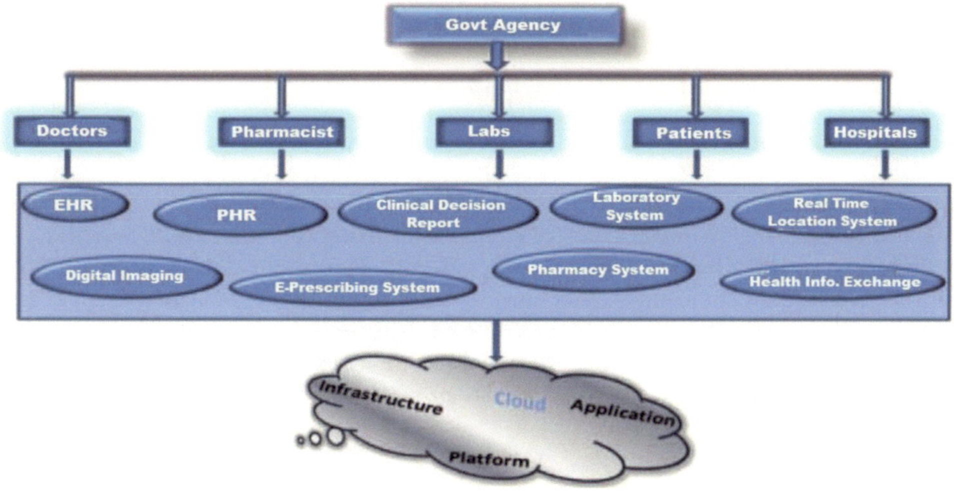

Figure 6.
Cloud-IoT-based healthcare framework [56].

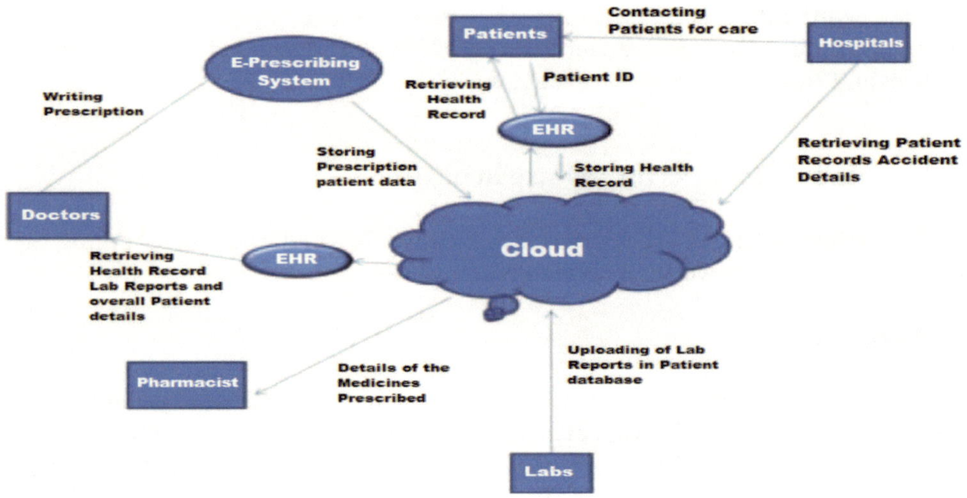

Figure 7.
Actors in the cloud-IoT-based healthcare framework [56].

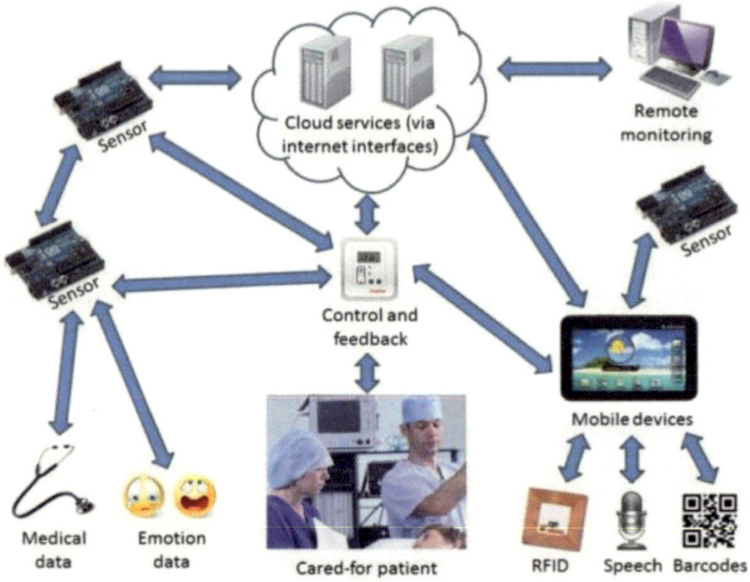

Figure 8.
Model of a typical ICADS [31].

decision support (ICADS) system, which provides an effective basis for rescheduling and prioritising essential services while maximising the effectiveness of the staff in knowing the health status of their patients, planning emergency treatment requirements and providing quality care. Even though this system can produce exciting benefits for the stakeholders of the healthcare industry, several complexities and challenges in hospital settings need to be addressed before implementing ICADS [31]. **Figure 8** summarises this system.

6.3 Datta et al.

Many mobile health applications are still operating offline and are yet to be integrated into the semantic Web technologies for e-Health services [58].

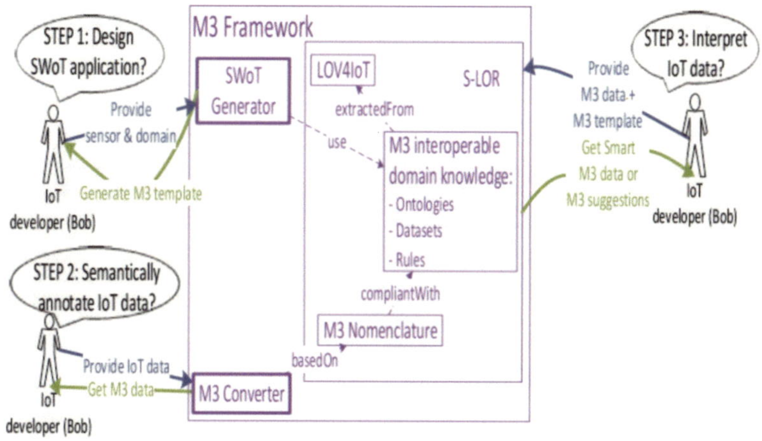

Figure 9.
Operational flow of the M3 framework [58].

Moreover, a unified rationale for developing healthcare development applications and middleware solutions is lacking. Therefore, users must build generic IoT applications to combine several domains. Datta et al. proposed the machine-to-machine (M3) framework, which enables the provision of smart, connected and personalised healthcare and wellness services to people living in smart homes [59]. This framework involves the use of wearable devices that collect patient data, which are then transmitted to smartphones that act as intermediate gateways. These data are then transmitted to remote cloud Web interfaces to maintain end-to-end security. The cloud computing platform is mainly targeted to manage patient data. However, this method does not allow patients to receive a high-level abstraction of the data collected by wearable devices [58]. **Figure 9** summarises this framework.

6.4 Prayoga and Abraham

Prayoga and Abraham iteratively tested, applied, refined and validated the behavioural intention in technology acceptance model (TAM) as one of the most prominent models used in Greater Jakarta to identify those variables that could predict the intention of individuals to utilise IoT health devices and integrate them into a theoretical model [60]. They analysed technology acceptance from the perspective of TAM and used perceived usefulness as the main predictor of behavioural intention. They also proposed a theoretical model to outline some important predictors of the behavioural intention of individuals to use IoT health devices. They performed a questionnaire survey among 186 college students from different faculties to test the hypothesised relationships between factors. As shown in the survey results, 91% of the respondents agreed that health trackers can help them achieve their personal health goals, 89% believed that these devices can change their health patterns and 90% thought that these devices will revolutionise healthcare systems. Although 87% of these respondents had searched for health-related information online while 35% had heard about such technology, only 13% of them had actually used health trackers [60]. **Figure 10** summarises the IoT behavioural intention model.

6.5 Roy et al.

Roy et al. proposed a model that facilitates the adoption of IoT-based innovations in urban poor communities [21]. This model identifies five sources of innovation,

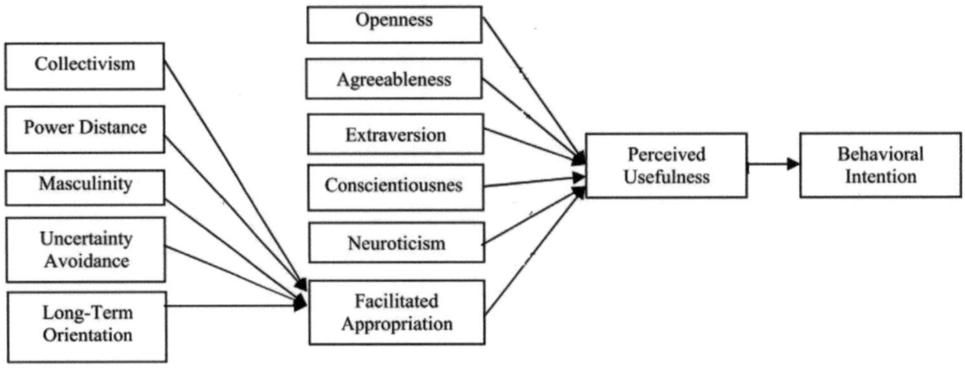

Figure 10.
IoT behavioural intention model [60].

namely, nutrition, healthcare, employment, education and finances. They also argued that IoT can positively affect the urban poor by providing them access to various types of services, including healthcare, education and food security. Their study was conducted in four stages, including a literature review, a survey of the target users, interviews with experts and a usability test of a prototype technology system. They assumed that the implemented system needs to provide quality service to its users and that users should experience tangible benefits and receive some training. These factors can help service providers deliver excellent services to their consumers and subsequently drive a higher consumer satisfaction [21]. This model is summarised in **Figure 11**.

6.6 Jagatheesan et al.

Jagatheesan et al. argued that multiple sensors with various applications from each manufacturer are easily configurable yet are generally not preferred by their users [61]. Therefore, they proposed the multiple producer multiple consumer (MPMC) network that aggregates human interfaces to allow users to control any part of the data distribution framework. This framework includes a scenario where IoT-based multiple sensors are used as producers of data and multiple IoT services are used as consumers of these data. Their findings highlighted how the experiences and perspectives of users affect the data framework design in MPMC environments by using the drop data framework infrastructure. However, this network does not serve the needs of IoT users, and service providers are unable to choose among multiple options and the security or actual data transfer protocols are usually lacking [61]. The MPMC framework is illustrated in **Figure 12**.

6.7 Bui et al.

The researchers investigated a case of a diabetic patient in an emergency situation [29]. They proposed the IoT communication framework as the main enabler of distributed worldwide healthcare applications. The main actors in this model include the monitored patients, physicians and distributed information databases. Their findings contribute to the actual implementation of a comprehensive healthcare system within IoT. They also highlighted the importance of using different devices, networks and processes in analysing diabetes progression. However, this framework is not yet completely available, the components presented in the use case are at different stages of realisation and the proposed framework does not integrate runtime sensing information into healthcare records [29]. This model is summarised in **Figures 13** and **14**.

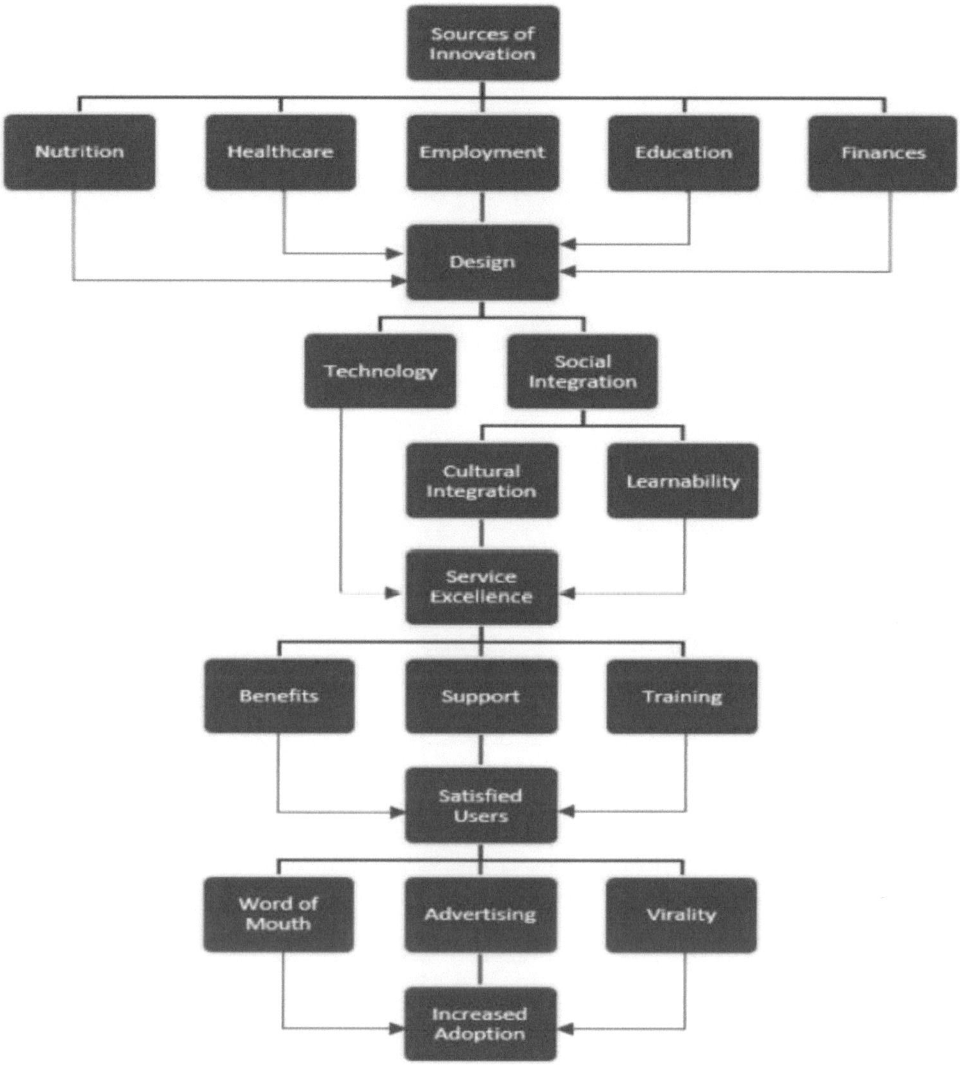

Figure 11.
Model of IoT-based innovations for the urban poor [21].

6.8 Manashty et al.

Manashty et al. aimed to fill the gap between symptoms and diagnosis trend data in order to predict health anomalies accurately and quickly [62]. Not one of the existing systems can act as a bridge between different systems to facilitate knowledge transfer and to enhance their detection and prediction capabilities. These systems are also unable to use the data and knowledge provided by similar systems due to the complexity involved in the data sharing process. Storing information also presents a challenge due to the high volume of data generated by each sensor. Therefore, Manashty et al. proposed the healthcare event aggregation lab (HEAL) model, a platform that provides services to developers and leverages the previously processed data and the corresponding detected symptoms. The proposed architecture is cloud-based and provides services for input sensors, IoT devices and context providers. The HEAL platform is an integrated system for high-level behaviour monitoring that supports many users and systems in their long-term analysis, thereby bridging the gap among many systems. However, Manashty et al. did not

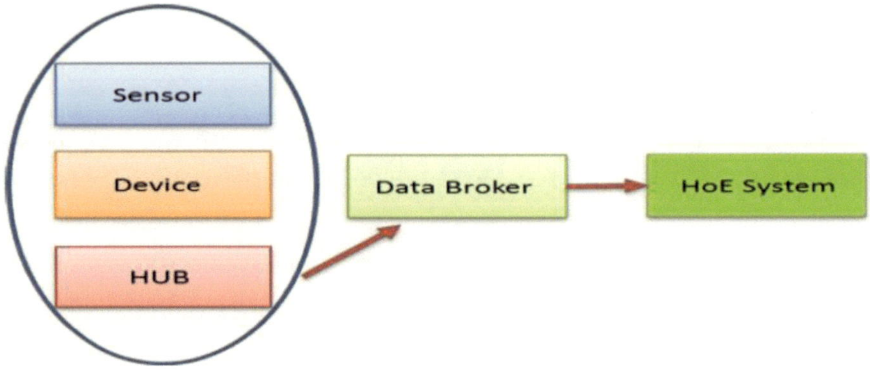

Figure 12.
MPMC framework [61].

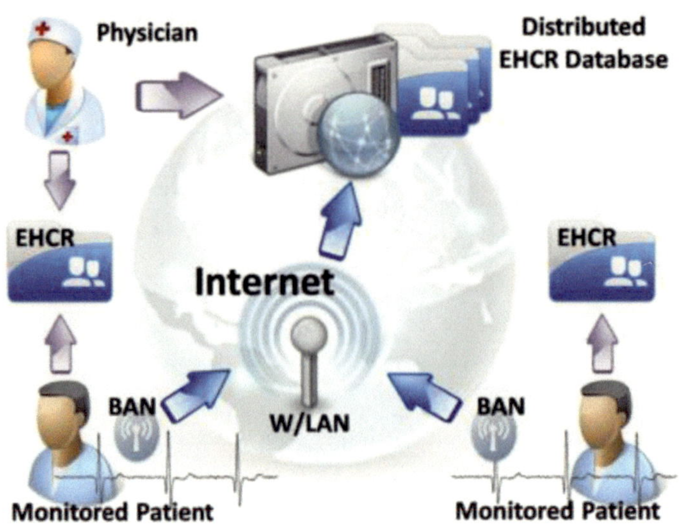

Figure 13.
IoT e-Health system model [29].

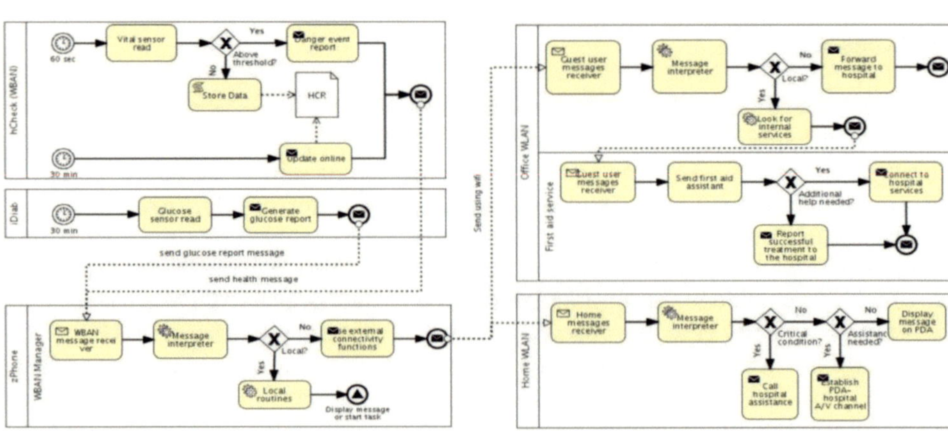

Figure 14.
IoT e-Health process model [29].

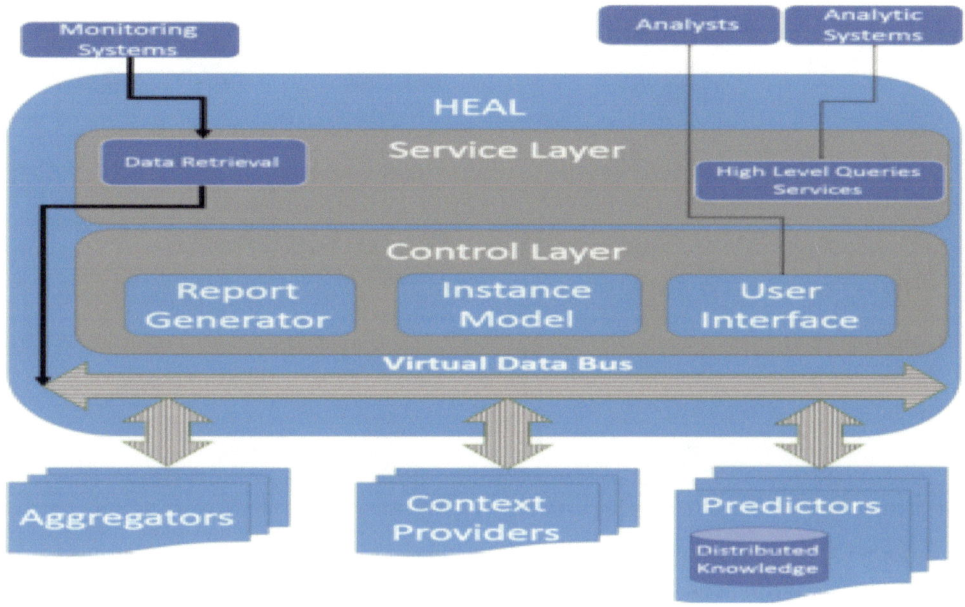

Figure 15.
Cloud-based HEAL platform model [62].

perform multiple case studies to evaluate the performance of the proposed system in complex heterogeneous scenarios with knowledge sharing [62]. This model is summarised in **Figure 15**.

6.9 Sheriff et al.

Sheriff et al. proposed a reference framework for healthcare informatics and analytics by integrating IoT, complex event processing (CEP) and big data analytics [63]. This framework can serve as a reference in implementing a holistic healthcare informatics and analytics ecosystem. Integrating IoT, CEP and big data analytics technologies can solve specific problems. Specifically, CEP can support the real-time and near-real-time analytical processing of patient events from different sources by using big data and ubiquitous communication via IoT. In the future, Sheriff et al. are planning to use this framework as a foundation for developing a healthcare application system that can address the informatics and analytic needs of healthcare and other dependent industries. However, they did not test the performance of this framework [63]. This framework is illustrated in **Figure 16**.

6.10 Pir et al.

Pir et al. developed the HMIS framework with context awareness for developing the management systems of smart hospitals based on IoT [64]. They introduced context awareness as a middleware of the IoT architecture to overcome the problems in large data management. This framework consists of three layers, including a physical layer, network layer and application layer. The physical layer, also known as the perception layer, collects data and communicates them to the network layer. The network layer then processes and transmits these data to the application layer. Context awareness, which is located above the network layer as middleware, analyses the data and transfers only the required data to the application layer. Afterwards, the application layer defines the context of the data based on the

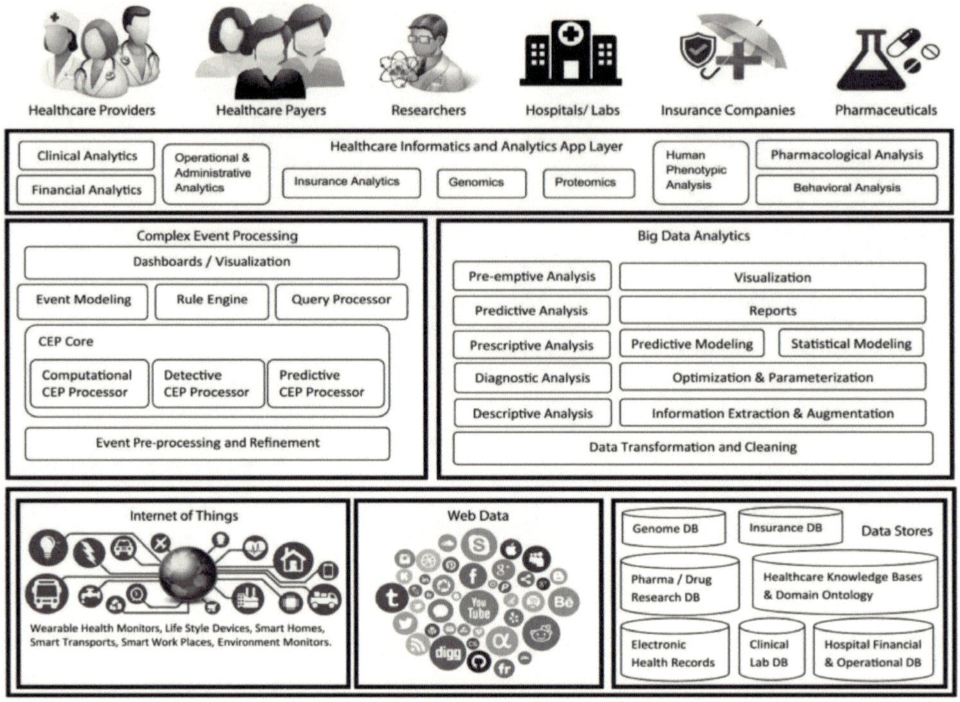

Figure 16.
Reference framework [63].

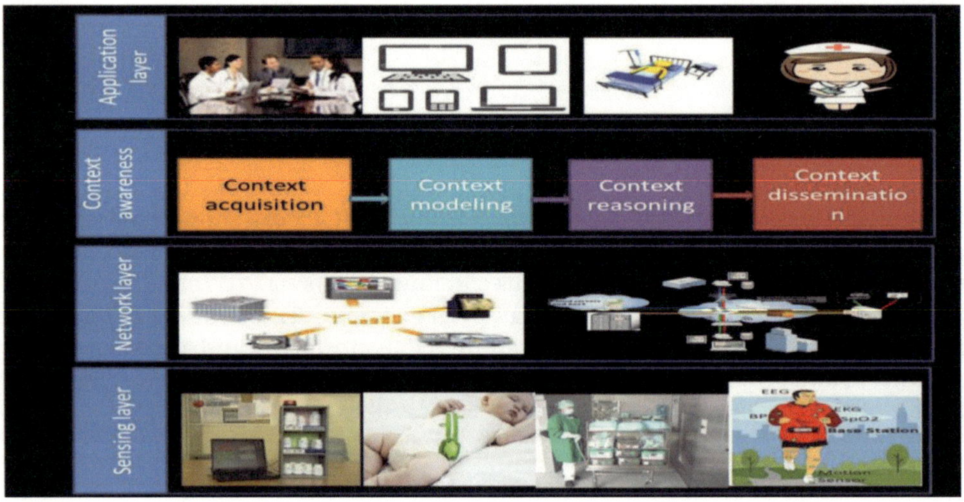

Figure 17.
HMIS framework [64].

problems faced by users when interacting with the system. However, Pir et al. did not test the applicability of this framework for users from a specific hospital [64]. Their proposed HMIS framework is presented in **Figure 17**.

6.11 Chatterjee and Armentano

Chatterjee and Armentano identified several issues, such as the availability of a live data connection and the security structure of a system, which prompted them to

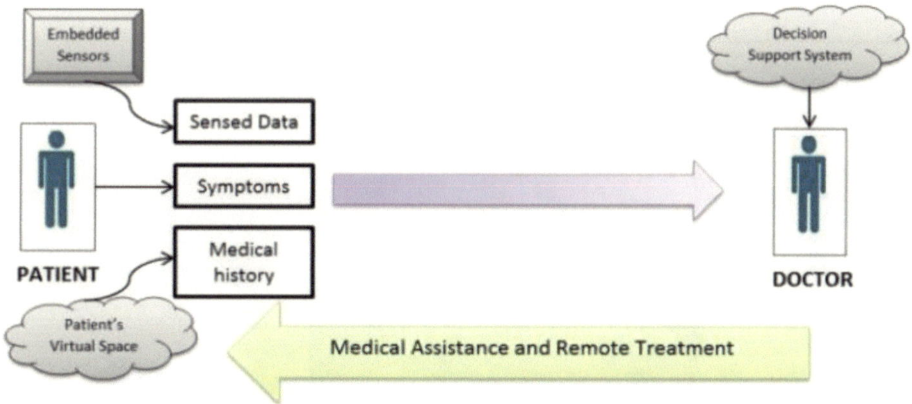

Figure 18.
Schematic diagram of the IoT-based remote treatment model [10].

develop a system for a smart medical environment that provides ubiquitous services [10]. Specifically, they proposed a model with an inclusive approach for applying IoT in a smart medical environment that provides ubiquitous services. This model virtually stores patient data and makes them ubiquitously accessible to the concerned healthcare personnel in order to be shared. Another important aspect of using these data lies in the design of an intelligent clinical decision support system that can help doctors when delivering treatment. However, Chatterjee and Armentano failed to address the requirements for adopting IoT and only focused on the inclusion of technologies in the healthcare sector, thereby limiting the generalisability of the factors that they proposed for different types of hospitals in various countries [10]. The schematic diagram of their IoT-based remote treatment model is summarised in **Figure 18**.

6.12 Gupta et al.

Gupta et al. examined the design and implementation of an IoT-based health monitoring system for emergency medical services [65]. This system demonstrates the flexible collection, integration and interoperation of IoT data that can provide support to emergency medical services. Their proposed model allows users to improve health-related risks and reduce healthcare costs by collecting, recording, analysing and sharing large amounts of data in real time. This system uses smart sensors that collect and send raw data to a database server where they are further analysed and statistically maintained to be used by medical experts. The results are deployed and tested on a patient whose personal details are inputted into a Web portal. This patient is then connected to a health monitoring system that includes a heart rate sensor and a temperature sensor. However, Gupta et al. did not consider in their work some factors in the organisational and system domain as identified in the literature review. They also did not consider the actual examination of healthcare professionals [65]. The proposed health monitoring system is illustrated in **Figure 19**.

The aforementioned models/framework for IoT use in healthcare can be classified based on the technological, system and individual aspects as summarised in **Table 2**.

In sum, most studies on IoT use in healthcare have some limitations related to their context of use, antecedents of implementation and need of use. Moreover, these studies have only focused on specific domains to achieve certain needs for using IoT in the healthcare context. Their models/frameworks are only designed for certain circumstances and environments related to the context and needs for which they are developed. Meanwhile, very few researchers have examined the actual

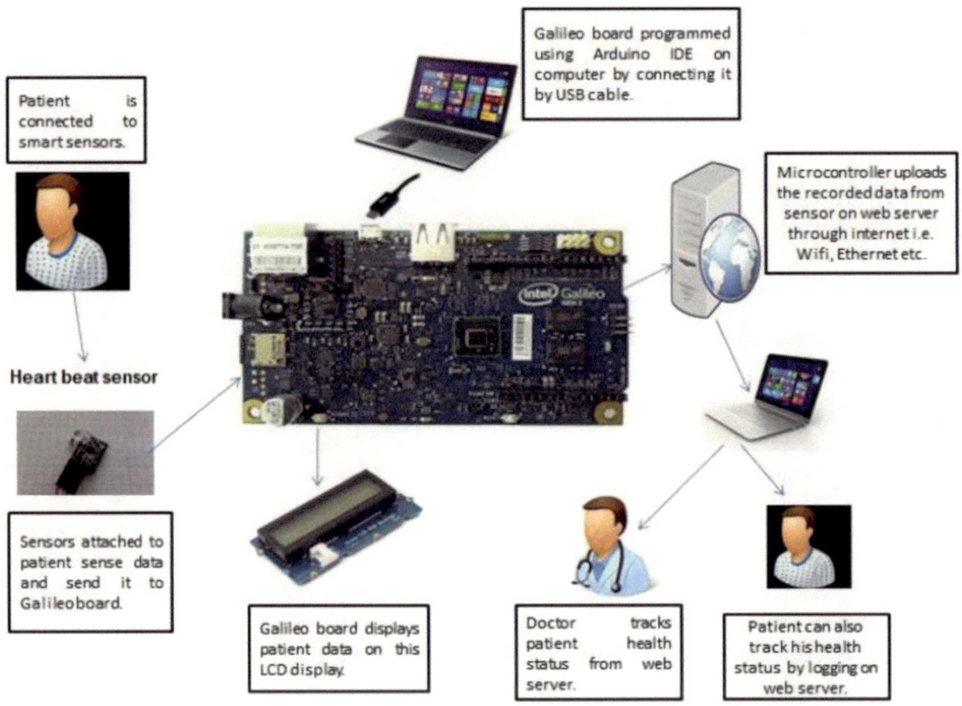

Figure 19.
IoT-based smart healthcare kit model [65].

Source	Technological	System	Individual	Context
[56]	x			Healthcare-based cloud computing network
[31]	x			Intelligent support system in hospitals
[58]	x	x		Healthcare in smart homes
[60]		x	x	Behaviour of using IoT health devices
[21]	x			Use of IoT in urban poor communities
[61]	x			Controlling via a human interface system
[24]	x			Monitoring via smartphones
[62]		x		Knowledge-based healthcare system
[29]		x		Healthcare application process
[63]	x	x		Healthcare informatics and analytics
[64]	x	x		Context awareness
[10]		x		Remote treatment
[65]		x		Healthcare monitoring

Table 2.
Models/frameworks for IoT use in healthcare.

implementation of IoT in hospitals. Therefore, further study must be conducted to generalise the application of these frameworks for hospitals. The literature review shows that the direct effect of technological and system-related factors on the utilisation behaviour of individuals has received no to limited input from previous research. The limitations of the aforementioned frameworks/models are summarised in **Table 3**.

No.	Source	Limitation
1	[29]	Requires the availability of several elements, including interoperability, reliability, privacy, authentication and integrity for exchanging EHRs across the network.
2	[31]	The system should identify the patients' conditions and notify the responsible staff who then review if the case of a patient needs to be treated as an emergency case depending on the information collected by sensors. Hospitals are facing several challenges in their implementation of IoT that should be acknowledged when designing an ICADS system.
3	[58]	This model focuses on smart home healthcare and the data collected from individuals must be managed and stored by decision makers in hospitals.
4	[61]	This study did not examine the requirements of IoT users and some issues related to data security transfer protocols.
5	[63]	This framework only focuses on the health information scenario and ignores those critical issues and challenges that may be faced by healthcare professionals.
6	[56]	Despite offering the benefits of trust and privacy to healthcare providers, several issues related to security remain unaddressed. This model needs to improve its security and test its results.
7	[60]	Model behavioural intention has been tested, applied, refined and validated many times in TAM to identify those variables that can predict the intention of individuals to use IoT health devices and integrate them into a theoretical model.
8	[21]	This model focuses on the use of IoT in urban poor communities, which is not considered part of a healthcare context.
9	[24]	This model focuses on the collection and uploading of health data by using smartphones as part of personal monitoring. The full utilisation of IoT has not been taken in consideration in this model.
10	[62]	Multiple case studies are not performed to assess the performance of the actual system in complex heterogeneous scenarios with knowledge sharing.
11	[64]	These results may satisfy certain hospitals in which no testing is performed in order to address the issues that they are facing.
12	[10]	This model only focuses on the inclusion of technologies in the health sector. Moreover, no experimental study has been performed, thereby limiting the generalisability of the proposed factors for different types of hospitals in various countries.
13	[65]	Those factors identified in the previous literature have not been considered and no actual examination of healthcare professionals has been performed.

Table 3.
Limitations of models/frameworks for IoT use in healthcare.

7. Conclusion

IoT use has become an urgent need for public hospitals and their technical and management activities. A successful IoT use is influenced by how well this technology fulfils the expectations of its users. The implementers of this technology must identify the implementation requirements from the management's perspective and align the implementation with the goals of hospitals in order to ensure a successful implementation and utilisation. **Table 3** shows that most studies on IoT use in healthcare have some limitations related to their identified factors as well as their context and purpose of use. These factors are also limited to certain developed and developing countries. In addition, the actual use of IoT in HIE has never been reviewed in the literature.

Specifically, some models and frameworks have been designed only for specific contexts, circumstances and environments. Meanwhile, other scholars have merely proposed models/frameworks without any post examination or evaluation, thereby making these models/frameworks unsuitable for examining IoT use in HIE for different reasons. These studies also do not focus on the HIE context and ignore the organisational, technological and individual aspects. Some of the proposed models have merely focused on security and privacy concerns and ignored all the other aspects related to organisational and technological issues. Very few studies have examined e-Health and m-Health architectures that use smartphone sensors and wearable devices to sense and transmit important patient data.

As a summary, this chapter shows that a model/framework specifically for IoT use in HIE is yet to be developed and that only few studies have examined the use of IoT in this type of exchange. However, most of the extant studies have identified HIE as a huge challenge for most countries and that the HIE among healthcare providers is very limited at present.

This study was motivated by the gaps in the literature and several issues related to HIE, including the limited capabilities of clinical centres and the perceived need for early detection. Another concern related to the interoperability of various smart electronic devices has also been raised. The findings presented in this chapter offer a foundation for future work on this topic. Proposing a process or framework may also be considered in future research from the perspectives of healthcare providers and management to offer solutions for the development of successful IoT services in the health sector.

This finding offers a foundation for further researchers in several ways. The success factors and proposed IoT implementation process identified and revealed in this study may be considered in future research from perspectives of healthcare providers and management, and thus offer a solution to develop successful IoT services in the health sector.

Author details

Mohammed Dauwed[1]* and Ahmed Meri[2]

1 Faculty of Information Science and Technology, Universiti Kebangsaan Malaysia, Selangor, Malaysia

2 Department of Medical Instrumentation Techniques Engineering, Al-Hussain University College, Karbala, Iraq

*Address all correspondence to: altaae@siswa.ukm.edu.my

References

[1] Mitchell S, Villa N, Stewart-Weeks M, Lange A. The Internet of Everything for Cities. Connect. People, Process. Data, Things to Improv. 'Livability'of Cities Communities; Cisco. 2013. pp. 1-21

[2] Downing NL et al. Health information exchange policies of 11 diverse health systems and the associated impact on volume of exchange. Journal of the American Medical Informatics Association. 2016;**34**(13):150-160

[3] Vest J.R. Health information exchange: national and international approaches. In Health Information Technology in the International Context. Vol. 12. Statesboro, GA, USA: Emerald Group Publishing Ltd; 2012:3-24

[4] Pagliari C. Health information exchange as a complex and adaptive construct: Scoping review. Journal of Innovation in Health Informatics. 2016;**23**(4): 633-683

[5] Rosenfeld S, Koss S, Caruth K. Evolution of state health information exchange/A study of vision, strategy, and progress. In: AHRQ Heal. IT Publ. No. 06. 2006

[6] Miller AR, Tucker C. Health information exchange, system size and information silos. Journal of Health Economics. 2014;**33**(1):28-42

[7] WHO. Health Information. World Health; 2008

[8] Tharmalingam S, Hagens S, Zelmer J. The value of connected health information: Perceptions of electronic health record users in Canada. BMC Medical Informatics and Decision Making. 2016;**16**(1):93

[9] Ghane K. Healthcare information exchange system based on a hybrid central/federated model. In: 36th Annual International Conference of the IEEE Engineering in Medicine and Biology Society (EMBC). 2014. pp. 1362-1365

[10] Chatterjee P, Armentano RL. Internet of things for a smart and ubiquitous eHealth system. In: 2015 International Conference on Computational Intelligence and Communication Networks (CICN). 2016. pp. 903-907

[11] Hassanalieragh M et al. Health monitoring and management using internet-of-things (IoT) sensing with cloud-based processing: Opportunities and challenges. In: Proc. of 2015 IEEE Int. Conf. Serv. Comput. SCC. 2015. pp. 285-292

[12] Riazul Islam SM, Kwak D, Humaun Kabir M, Hossain M, Kwak K-S. The internet of things for health care: A comprehensive survey. IEEE Access. 2015;**3**:678-708

[13] Davis FD, Bagozzi RP, Warshaw PR. User acceptance of computer technology: A comparison of two theoretical models. Management Science. 1989;**35**(8):982-1003

[14] Turcu CE, Turcu CO. Internet of things as key enabler for sustainable healthcare delivery. Procedia-Social and Behavioral Sciences. 2013;**73**:251-256

[15] Zhang H, Han BT, Tang Z. Constructing a nationwide interoperable health information system in China: The case study of Sichuan Province. Health Policy and Technology. 2017;**6**(142151):142-151

[16] Serrano KJ et al. Willingness to exchange health information via mobile devices: Findings from a population-based survey. Annals of Family Medicine. 2016;**14**(1):34-40

[17] Vest JR, Zhao H, Jasperson J, Gamm LD, Ohsfeldt RL. Factors motivating and affecting health information exchange usage. Journal of the American Medical Informatics Association. 2011;**18**(2):143-149

[18] Koru G, Alhuwail D, Topaz M, Norcio AF, Mills ME. Investigating the challenges and opportunities in home care to facilitate effective information technology adoption. Journal of the American Medical Directors Association. 2016;**17**(1):53-58

[19] Ghani MKA, Jaber MM, Suryana N. Barriers faces telemedicine implementation in the developing countries: Toward building Iraqi telemedicine framework. ARPN Journal of Engineering and Applied Sciences. 2015;**10**(4):1562-1567

[20] Kadhum AM, Hasan MK. Assessing the determinants of cloud computing services for utilizing health information systems: A case study. International Journal on Advanced Science, Engineering and Information Technology. 2017;**7**(2):503-510

[21] Roy A, Zalzala AMS, Kumar A. Disruption of things: A model to facilitate adoption of IoT-based innovations by the urban poor. Procedia Engineering. 2016;**159**:199-209

[22] Sallabi F, Shuaib K. Internet of things network management system architecture for smart healthcare. In: 2016 Sixth International Conference on Digital Information and Communication Technology and its Applications (DICTAP). 2016. pp. 165-170

[23] Li S, Da Xu L, Zhao S. The internet of things: A survey. Information Systems Frontiers. 2015;**17**(2):243-259

[24] Ullah K, Shah MA, Zhang S. Effective ways to use internet of things in the field of medical and smart health care. In: 2016 International Conference on Intelligent Systems Engineering (ICISE). Islamabad, Pakistanpp: IEEE; 2016:372-379

[25] Jog Y, Sharma A, Mhatre K, Abhishek A. Business approach for IoT based health solutions in India with respect to osterwalder framework. International Journal of Bio-Science and Bio-Technology. 2015;**7**(6):173-188

[26] Lim AK, Thuemmler C. Opportunities and challenges of internet-based health interventions in the future internet. In: 2015 12th International Conference on IEEE Information Technology-New Generations (ITNG). 2015. pp. 567-573

[27] Keh HC et al. Integrating unified communications and internet of M-health things with micro wireless physiological sensors. Journal of Applied Science and Engineering. 2014;**17**(3):319-328

[28] Rghioui A, L'Aarje A, Elouaai F, Bouhorma M. The internet of things for healthcare monitoring: Security review and proposed solution. In: Information Science and Technology (CIST). 2014. pp. 384-389

[29] Bui N, Zorzi M. Health care applications: A solution based on the internet of things. In: Int. Symp. Appl. Sci. Biomed. Commun. Technol. 2011. pp. 0-4

[30] Rasid MFA et al. Embedded gateway services for internet of things applications in ubiquitous healthcare. In: International Conference on Information and Communication Technology (ICoICT); Figure. 2014. pp. 145-148

[31] Manate B, Munteanu VI, Fortis TF, Moore PT. An intelligent context-aware decision-support system oriented towards healthcare support. In: Complex, Intelligent and Software

Intensive Systems (CISIS): 2014 Eighth International Conference on IEEE. 2014. pp. 386-391

[32] Zdravkoví M, Noran O, Trajanoví M. On pervasive healthcare information systems in the internet of things. In: 25th Australasian Conference on Information Systems (ACIS2014); Auckland, NZ. 2014

[33] Atzori L, Iera A, Morabito G. The internet of things: A survey. Computer Networks. 2010;**54**(15):2787-2805

[34] Felipe Fernandez GCP. Opportunities and challenges of the internet of things for healthcare. In: 2014 EAI 4th International Conference on Wireless Mobile Communication and Healthcare (Mobihealth); IEEE. 2014

[35] Rathore MM, Ahmad A, Paul A, Wan J, Zhang D. Real-time medical emergency response system: Exploiting IoT and big data for public health. Journal of Medical Systems. 2016;**40**(12):283

[36] Santos A, Macedo J, Costa A, Nicolau MJ. Internet of things and smart objects for M-health monitoring and control. Procedia Technology. 2014;**16**:1351-1360

[37] YIN Y. The internet of things in healthcare: An overview. Journal of Industrial Information Integration. 2016;**1**:3-13

[38] Nandyala CS, Kim HK. From cloud to fog and IoT-based real-time U-healthcare monitoring for smart homes and hospitals. International Journal of Smart Home. 2016;**10**(2):187-196

[39] Blake MB. An Internet of Things for Healthcare; 2015

[40] Catarinucci L et al. An IoT-aware architecture for smart healthcare systems. IEEE Internet of Things Journal. 2015;**2**(6):515-526

[41] Al-Fuqaha A, Guizani M, Mohammadi M, Aledhari M, Ayyash M. Internet of things: A survey on enabling technologies, protocols, and applications. IEEE Communication Surveys and Tutorials. 2015;**17**(4):2347-2376

[42] Park H et al. Can a health information exchange save healthcare costs? Evidence from a pilot program in South Korea. International Journal of Medical Informatics. 2015;**84**(9):658-666

[43] Song Z, et al. A Survey of Primary Care Doctors In Ten Countries Shows Progress in Use of Health Information Technology, Less in Other Areas; 2012

[44] Latif A, Othman M, Suliman A, Daher A. Current status, challenges and needs for pilgrim health record management sharing network, the case of Malaysia. International Archives of Medicine. 2016;**9**(1):1-10

[45] Sujansky W, Kunz D. A standard-based model for the sharing of patient-generated health information with electronic health records. Personal and Ubiquitous Computing. 2015;**19**(1):9-25

[46] Wang J-Y, Ho H-Y, Chen J-D, Chai S, Tai C-J, Chen Y-F. Attitudes toward inter-hospital electronic patient record exchange: Discrepancies among physicians, medical record staff, and patients. BMC Health Services Research. 2015;**15**(666):264

[47] Hollis KF. To share or not to share: Ethical acquisition and use of medical data. AMIA Joint Summits on Translational Science Proceedings. 2016;**2016**:420-427

[48] Jha AK, Doolan D, Grandt D, Scott T, Bates DW. The use of health information technology

in seven nations. International Journal of Medical Informatics. 2008;**77**(12):848-854

[49] Everson J et al. Health information exchange associated with improved emergency department care through faster accessing of patient information from outside organizations. Journal of the American Medical Informatics Association. 2016;**169**(10):1023-1028

[50] Abomhara M, Køien GM. Towards an access control model for collaborative healthcare systems. In: Proc. 9th Int. Jt. Conf. Biomed. Eng. Syst. Technol. (BIOSTEC 2016). Vol. 5. 2016. pp. 213-222

[51] Riordan F, Papoutsi C, Reed JE, Marston C, Bell D, Majeed A. Patient and public attitudes towards informed consent models and levels of awareness of electronic health records in the UK. International Journal of Medical Informatics. 2015;**84**(4):237-247

[52] Hsieh P. Physicians' acceptance of electronic medical records exchange: An extension of the decomposed TPB model with institutional trust and perceived risk. International Journal of Medical Informatics. 2015;**84**(1):1-14

[53] Mac McCullough J et al. Electronic health information exchange in underserved settings: Examining initiatives in small physician practices & community health centers. BMC Health Services Research. 2014;**14**:415

[54] Swain MJ, Kharrazi H. Feasibility of 30-day hospital readmission prediction modeling based on health information exchange data. International Journal of Medical Informatics. 2015;**84**(12):1048-1056

[55] Nguyen L, Bellucci E, Nguyen LT. Electronic health records implementation: An evaluation of information system impact and contingency factors. International Journal of Medical Informatics. 2014;**83**(11):779-796

[56] Tyagi S, Agarwal A, Maheshwari P. A conceptual framework for IoT-based healthcare system using cloud computing. In: 2016 6th International Conference—Cloud System and Big Data Engineering (Confluence). 2016. pp. 503-507

[57] Manate B, Fortis T-F, Negru V. Optimizing cloud resources allocation for an internet of things architecture. Scalable Computing: Practice and Experience. 2014;**15**(4):345-355

[58] Datta SK, Bonnet C, Gyrard A, Ferreira da Costa RP, Boudaoud K. Applying internet of things for personalized healthcare in smart homes. In: 2015 24th Wireless and Optical Communication Conference (WOCC). 2015. pp. 164-169

[59] Gyrard A, Datta SK, Bonnet C, Boudaoud K. Standardizing generic cross-domain applications in internet of things. In: 2014 IEEE Globecom Work (GC Wkshps). 2014. pp. 589-594

[60] Prayoga T, Abraham J. Behavioral intention to use IoT health device: The role of perceived usefulness, facilitated appropriation, big five personality traits, and cultural value orientations. International Journal of Electrical and Computer Engineering. 2016;**6**(4):1751-1765

[61] Jagatheesan A, Maragathavel S, Sivapurapu M, Lee J. Drops: A multi-producer and multi-consumer data sharing framework with human experience. In: 2015 12th Annual IEEE Consumer Communications and Networking Conference (CCNC). 2015. pp. 601-602

[62] Manashty A, Light J, Yadav U. Healthcare event aggregation lab (HEAL), a knowledge sharing platform for anomaly detection and prediction.

In: 2015 IEEE 17th International Conference on e-Health Networking, Applications and Services (Healthcom). 2016. pp. 648-652

[63] Sheriff CI, Naqishbandi T, Geetha A. Healthcare informatics and analytics framework. In: 2015 International Conference on Computer Communication and Informatics (ICCCI). 2015. pp. 1-6

[64] Pir A, Akram MU, Khan MA. Internet of things based context awareness architectural framework for HMIS. In: 2015 17th International Conference on E-Health Networking, Application & Services (HealthCom). 2016. pp. 55-60

[65] Gupta P, Agrawal D, Chhabra J, Dhir PK. IoT based smart healthcare kit. In: International Conference on Computational Techniques in Information and Communication Technologies (ICCTICT). 2016. pp. 237-242

[66] Dimitrov DV. Medical internet of things and big data in healthcare. Healthcare Informatics Research. 2016;**22**(3):156-163

[67] Da Xu L, He W, Li S. Internet of things in industries: A survey. IEEE Transactions on Industrial Informatics. 2014;**10**(4):2233-2243

Chapter 5

IoT-Enabled Health Monitoring and Assistive Systems for in Place Aging Dementia Patient and Elderly

Thierry Edoh and Jules Degila

Abstract

Hospitals and nursing care homes are facing severe challenges such as lack of skilled workforces and cost explosion, among others. Especially, the western healthcare systems are headed over a cliff. German nursing care houses, hospitals, and government are working hard on solutions to overcome the crucial workforce's crises. Therefore, they are planning to hire nursing-workforces from abroad. They would also like to motivate families with monetary incentives, such as tax-reduction, if they can care for their elderly and/or dementia family members at home. However, caring for a sick person at home is challenging for a working family member especially in the case the patient requests around-the-clock nursing care since elderly with cognitive impairments, people with physical disabilities, and people with dementia need (medical) assistance around the clock. The present study reveals an increasing tendency for family members to care for their sick family member at home despite the challenges and issues faced like the lack of time to intensively care for the sick person due to today's family lifestyle, challenging employment market, financial constraints, etc. The study reported in this chapter aims at providing family members willing to care for their sick member at home as long as possible with smart home automation enabled Patient@Home solution to effectively and efficiently care for their sick member despite working full time. People living in nursing care homes can also take benefit of the proposed system. Evidence shows the potential of the proposed system to effectively and efficiently assist in caring for people requesting nursing care around the clock. Though the health-related quality of life is positively impacted, patient's satisfaction is increased (better quality of experience).

Keywords: remote care, dementia, Parkinson, elderly patient, Internet of things, smart, home automation, age-related impairments, cognitive impairments, health-related quality of life, patient's satisfaction and quality of experience

1. Introduction

Nursing care is facing (is going through) an unprecedented crisis in terms of lack of skilled (health) care workforces. Like all high-income countries (HIC), Germany is going through the said crisis, which is continuously accentuating year after year.

The workforce's needs for long-term care in German nursing houses have been estimated in [1] for the period from 2009 to 2030, expecting an increase:

1. from 94.000 to 331.000 professional nursing and

2. from 157.000 to 298.000 care staff.

Elke Peters et al. have estimated in [2] the number of people living in Germany requesting nursing care to 3 million and to 5 million by 2050. The authors present a recent assessment of the nursing care services at nursing homes and at patient's home and point out the needs for patients to live at home despite the benefit of all care services.

In November 2016, in Mondorf-les-Bains (Germany), a workshop [3] on nursing care had taken place. The topic of the workshop was: Nursing Care at the (German) border Regions? (Ger. Pflege an der Grenze?). The workshop's main objectives were to strengthen the social aspect of nursing care and to more consider the nursing care to be taken place at the place of residence of the patient because of the demographic change and economic as well as employment market policy change. The said workshop pointed out that the share of family nursing care (also called care at home) is very small in comparison with ambulant nursing care. This means, family members do not care for their member requesting for nursing care. One of the main reasons leading to this situation is that the person requesting nursing care at home is living alone. Furthermore, direct family members are requested to participate in care costs. In order to participate, they must work to gain the necessary financial means to face the costs. This situation drives sometimes the family members to employ care staff without or with beginner's care skills to care their parents at home or they send their parents aboard to East-European countries since the nursing care costs are cheaper there though caring at home for a person is not as easy as one can think. Prof. Dr. Eckart Hammer points out in [4] that many dementia patients are subjected to violence by family members who are caring for them at home. By analyzing this book section, one can understand why the German government put effort to solve the care workforce issues faced in order to admit enough nursing care requesting people to the care or nursing homes. Thus, family members who are not able to care for their patients can send them to a nursing/care home. They obviously also want to help the family to decently and lovingly care for their patients and protect the patient against as well as prevent violence. Violence can result from stress faced by the caring person. And the causes of stress are multiple. Violence also occurs in nursing care houses.

People with advanced dementia have complex needs [5]. Schmidt et al. have investigated the needs in a recent study. The study shows the evidence that people with advanced dementia are requesting monitoring round the clock even for a simple activity like "food intake." At nursing care house, monitoring is guaranteed. But what happens if those people are living at their regular residence? This research question is justified by the results carried out by [6], which point out the causes of nursing care workforce shortage and provide recommendations to overcome the issues faced. In [7], the author recommends a series of solutions to fix the workforce shortage. One of these solutions is to use telemedicine to overcome the shortage of issues faced. He writes

Other solutions proposed to reduce the effects of shortages include the use of telemedicine to reach far-away neurologists (though this is unlikely to reduce workloads), the development of artificial intelligence to help in making diagnoses, and expanding neurological care to include non-neurologist physicians and advanced practitioners (specially trained nurses and physicians' assistants)…

The recommended solutions are intended for neurology, though some of them can suite other medical fields.

Using telemedicine to overcome workforce shortage implies to keep a patient at his residence or at the care unit with only primary care services. A further research question rising here is what is the quality of life (QoL) of patients treated at home? Is it worth treating dementia patients at home instead of a nursing care home? Rebecca Palm et al. investigated in [8] the environment as a factor impacting the health-related QoL regarding nursing care for dementia patients. The study reveals that the structural and organizational characteristics of care units may impact the QoL though the study does clearly prove through empirical evidence that the care unit's structure and organization influence the QoL. However, to our best knowledge, no study has investigated the impact of homecare on the health-related QoL based on the QoL measurement metrics pointed out in [8] such as *temperature, noise, lightning, familiarity, adequate space, and opportunities to participate in domestic activities*; it could be subsequently deduced that if the patient's residence place also provides the same environmental criteria as temperature, familiarity, sufficient food, and water, etc., the patient treated at home will undergo the same health-related QoL. In [9], the authors investigated the impact small-scaled nursing care homes have on health outcome-related QoL. They found out that moving from large-scaled to a small-scaled nursing house can improve the aspect of the QoL by reducing the anxiety. This study allows us to conclude that a patient treated at home in his family circle and habituated residence place has less anxiety and better QoL.

It is obvious that patients requesting nursing care can receive nursing care at their residence places with a better health-related QoL. The factors impacting the QoL are well known though caring for dementia, Parkinson's disease, and elderly patients suffering from possible cognitive impairments is a challenging task. The research question raised here is how to assess the factors impacting the QoL for better health outcome?

The literature review on technologies in nursing care or commonly in healthcare reveals that nursing care at home for dementia and elderly patients can take benefit of the technology (cf. section methodology/literature review).

1.1 Study objectives

Homecare is increasingly getting attention among the population for multiple reasons such as the nursing care crisis. This research mainly aims at proposing smart home automation enabled personalized homecare solution for a better quality of life (QoL) for the patient and for assisting the patient's family members to cost-effectively and efficiently care for their patients at home without any impairment of QoL. Furthermore, this study pursues the objectives to assess the impacts of being assisted by home automation system on the QoL of all involving family members.

1.2 Study contribution

This study contributes to the *multidimensionality of the concept of the smart* home where many dimensions of home automation have been considered. The study focuses on many aspects of home automation such as energy saving [10, 11], temperature management, and regulation, security, and safety by managing the entrance, control doors, and windows.

Additionally, the study creates an environment for well-being for people limited in the movement.

1.3 Structure of the chapter

The remainder of the chapter presents in Section 2 some backgrounds and definitions. The research methodology, consisting of a literature review, research data, and system design, is presented in Section 3. Research findings and discussion are presented in Section 4. Section 5 handles a daily personal assistance system, which is designed and implemented to assist patients receiving nursing care at home and who is most of the time alone, and Section 6 concludes the study.

2. Backgrounds and definitions

2.1 Nursing houses and nursing care

According to NIH-UK (National Institute of Health United Kingdom), a nursing home provides hospital-like care services to people (outpatients, elderly, palliative, etc.) that cannot stay in the hospital for any aftercare or for elderly care.

A nursing home is a place for people who don't need to be in a hospital but can't be cared for at home. Most nursing homes have nursing aides and skilled nurses on hand 24 hours a day. (NIH-UK)

It is worth noting the main risk factors of being admitted to nursing and/or care homes (both are similar but are different regarding the qualification of the care-staff—*see care homes vs nursing homes*).

- *Age*: elderly people have more chance of being admitted to a nursing/care home.

- *Low income*: people with low income are vulnerable and have not enough possibilities to hire private care workforce to care for them at home.

Precisely for these reasons, they have a higher chance of being admitted to a nursing home.

- *Poor family support*: especially in cases where the older adult lacks a spouse or children.

- *Low social activity*: isolated people because of cognitive or age-related impairment.

- *Functional or mental difficulties*.

Regarding the risk factors of being admitted to nursing or care homes, it is obvious that a group of people can be excluded from being admitted since they would not meet the conditions.

2.1.1 Nursing homes versus care homes

According to [12], nursing homes have been recommended to employ higher skilled nurse staffing in their homes, with 24-hour registered nursing care.

As the Balcombe Care Homes defines on its website:[1]

---

[1] https://balcombecarehomes.co.uk/about-us/

A nursing home will provide all the day-to-day care that you would expect from any care home, but the care is supervised by registered nurses who are on duty all day and all night.

while

Care homes are staffed 24 hours a day and a proportion of the staff will be qualified care assistants with NVQs (National Vocational Qualifications) at Level 2 or 3.

2.1.2 What is nursing care?

Segen's Medical Dictionary defines nursing care as

A nonspecific term in medicine; among medically qualified doctors in the UK, nursing care generally refers to procedures or medications which are solely or primarily aimed at providing comfort to a patient or alleviating that person's pain, symptoms or distress, and includes the offer of oral nutrition and hydration

Based on the Segen's Medical Dictionary definition of nursing care, nursing care can be assimilated to palliative as well as elderly care. Most elderly people are requesting nursing care due to health conditions such as cognitive impairments that include dementia, Parkinson, blindness, etc. [13, 14]. Though their chance of being admitted to a nursing home is low, modern technology, as well as methodology such as remote care, can assist to provide them with the needed nursing care at their residence place. The question raised is how will this work?

2.2 Elderly and age-related impairments

The demographic structure of the developed countries (DC) or high-income countries (HIC) contains a large number of older (from 85+ years) and elderly (from 60+ years) people than young people (up to 59 years) and a very small number of teenagers (up to 15 years) in their population. The population of older adults is fastly growing in HIC [15], whereas the population in developing or low- and middle-income countries (LMIC) is remaining younger, although the number of young people is decreasing (see the example of Uganda—**Figure 1**). The median age in LMIC is around 15 years (see **Figure 1**), while the median age in the European Union (EU) is predicted to pass from 36.5 years in 1995 to 47.6 years in 2060 with an increasing tendency [16]. Thus, EU countries are facing an increasingly elderly population with all related needs like nursing and care homes, accommodated elderly healthcare services, etc.

The term "Elderly people" is defined as adults aged 60+ years, while people aged 65+ years are considered as an elder. Orimo, Hajime et al. had reviewed the definition of the term "elderly" in [17] and found out a correlation between elderly and the request or need of medium to severe nursing care.

According to the conventional definition presented by the authors in contrary to the definition above, the elderly is from 65+ years.

Conventionally, "elderly" has been defined as a chronological age of 65 years old or older, while those from 65 through 74 years old are referred to as "early elderly" and those over 75 years old as "late elderly." [17].

Though the World Health Organization (WHO) considers people aged 60+ years as elderly.

At the moment, there is no United Nations standard numerical criterion, but the UN agreed cutoff is 60+ years to refer to the older population [18].

And arguments follow such as why no one can exactly determine the age at which one has to be considered as elderly.

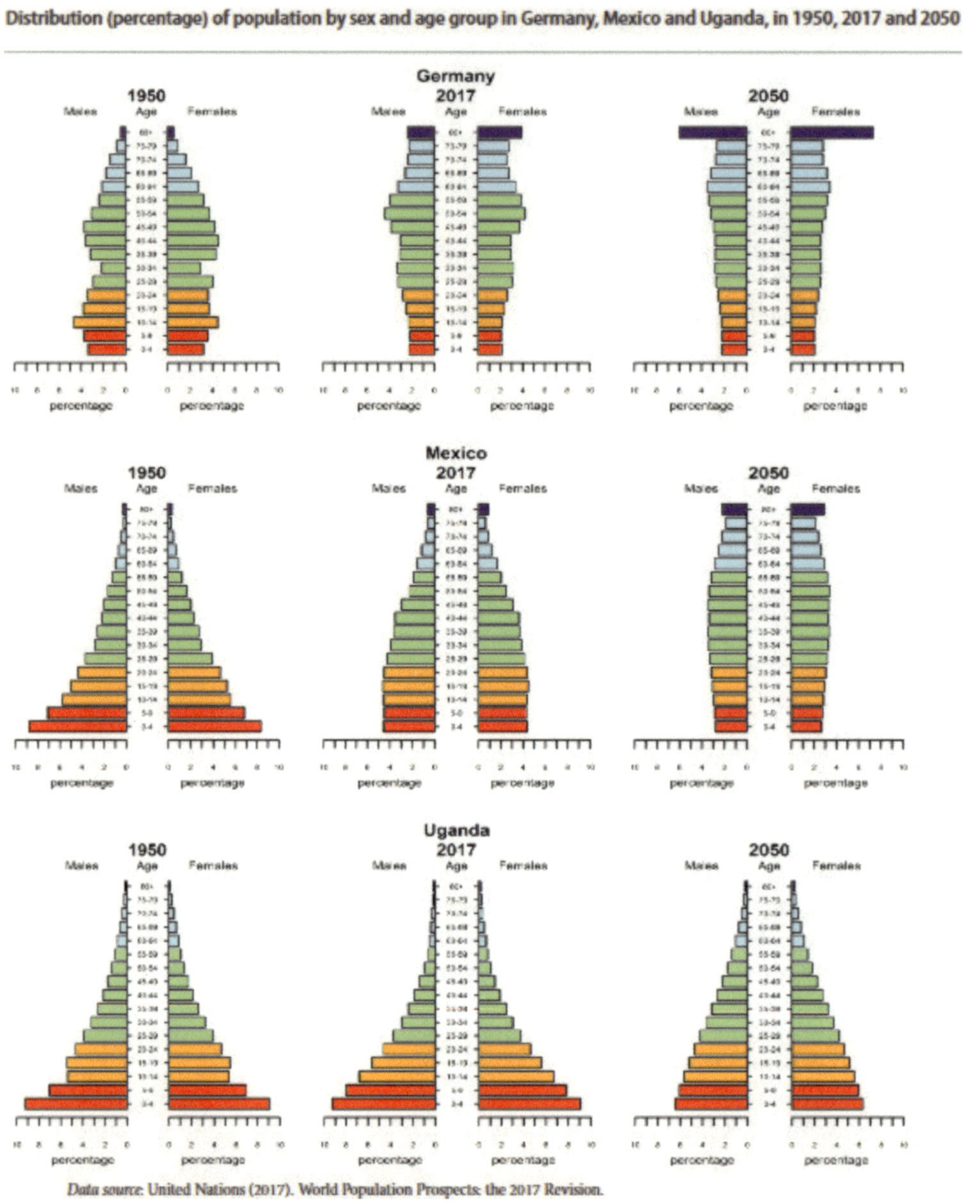

Figure 1.
Projection on demographic change LMIC versus HIC (from 1950 to 2050) [source [19]].

In addition, chronological or "official" definitions of aging can differ widely from traditional or community definitions of when a person is older. We will follow the lead of the developed worlds, for better or worse, and use the pensionable age limit often used by governments to set a standard for the definition [18].

According to the United Nations projection, about 79% of the world elder population aged 60 years or over will live in LMIC by 2050 [19]. Therefore, 20% of them will live in HIC.

Analyzing the population distribution (Figure 6 in [19]) reveals that in countries like Germany, population will count more aging people while an LMIC's population like Uganda's population will remain young.

As a conclusion, it is worth noting that the needs of nursing and care homes are higher in HIC than in LMIC. Therefore, the chapter will more focus on the nursing situation in HIC.

2.2.1 What is age-related impairment

Age-related impairment mostly known as cognitive impairment is a group of diseases, which occurs with advancing age. Cognitive impairment can also occur in young people. Mostly age-related cognitive impairments are dementia, Alzheimer's, Parkinson's, loss of vision, hearing loss, depression, incontinence, etc.

Obviously, cognitive impairment progresses with advancing age. In [17], the authors found out that elderly need from 75+ years severe nursing care. Though nursing care shows the potential to improve the individual's quality of life (QoL), most cognitive impairments cannot be cured. The patient, therefore, needs more attention, for example, reminding him to take food and drink enough water, and bringing him to get socialized again.

In order to better understand why these patients need more nursing care than others, it is worth understanding the symptoms of some cognitive diseases as follows.

2.2.1.1 Dementia

Dementia is a progressive health condition mostly in elderly people. Dementia is a consequence of health conditions like Alzheimer and is characterized by cognitive impairment (loss of cognitive capabilities or abilities).

The Journal of the American Medical Association defines Dementia as

Dementia is diagnosed only when both memory and another cognitive function are each affected severely enough to interfere with a person's ability to carry out routine daily activities.

The free dictionary gives a similar definition as

Loss of cognitive abilities, including memory, concentration, communication, planning, and abstract thinking, resulting from brain injury or from a disease such as Alzheimer's disease or Parkinson's disease. It is sometimes accompanied by emotional disturbance and personality changes.

Regarding the characteristics of dementia, it is highly requested to assist round the clock people suffering from such health condition in order to protect them against any accident that can result from forgetfulness. On one hand, they need assistance, and on the other hand, they can be refused to being admitted to nursing or care home. Furthermore, keeping these people at home remains challenging. Family members caring for these people are mostly by day time at their own job. In this case, the only solution is to employ care/nursing personnel to care for them during the absence of all family members. It is reportedly known that most "care/nursing personnel" hired for homecare are poorly skilled and mostly come from a different cultural background as the patient. The question is can all these factors impact the patient's QoL? Especially, can the cultural differences contribute to QoL loss? Answering this question is out of the scope of the present study.

2.2.1.2 Parkinson's

Parkinson's disease is one of the best-known and most common diseases of the nervous system. It is a cognitive disease and mostly related to advancing age. James Parkinson, the British physician, described the typical symptoms of the disease for the first time in 1817 and gave his name to the disease. Like a most cognitive disease, is a slowly progressive neurological disease that affects certain areas of the brain. The main symptom of Parkinson's disease is the movement disorder.

People suffering from Parkinson's disease are, therefore, dependent on other people since they are limited in their movement. Furthermore, they can lose the sense of smelling and mostly suffer from Dementia, depression, and anxiety.

2.3 Technologies enabling home automation

The main role of home automation is to control and manage devices at the local network(s) in the house. It can enable remote interactions with the network in order to access some information or to set command. For example, one can remotely ask his fridge or the fridge can send him a grocery list. Many technologies are included in home automation. Technologies like wireless sensor networks, videos, and connected devices support smart home automation paradigm. In [20], Toschi et al. reviewed the technologies that enabled a machine-to-machine (M2M)-based house automation. According to the authors, home automation is tending beyond connecting autonomous toward smart process and devices.

In this section, two technologies are briefly presented. In prior, the term automation is defined.

2.3.1 What is smart home automation?

In [21], Vasseur and Dunkels defined home automation as follows:

Home automation is an area of multiple and diverse applications that include lighting control, security and access control, comfort and convenience, energy management, remote home management, and aging independently and assisted living.

In the context of nursing care, home automation (HA) is a network system and application that includes at the first place bio-signal monitoring, well-being control, and other medical means like medication intake, physical exercises, etc. Further, HA includes temperature management, patient-safety, and security by preventing dangerous actions like leaving furnace or gas on, going out without adequate wearing.

Figure 2 (Source Figure 23.1 in [21]) presents a sample of home control devices.

In [22], Pham et al. defined smart home automation as an environment context-related data for precise health monitoring. They write:

A smart home environment provides ample contextual data related to a resident's health, which allows more accurate health monitoring than only using physiological signals.

They further presented cloud-based home automation that collects bio-signals and location information in order to accurately monitor nursing home residents.

2.3.2 Internet of things and the common architecture

The Internet of things is a paradigm for autonomous data gathering and processing. In [23], Luigi Atzori et al. had defined the Internet of things as follows:

"The Internet of Things (IoT) is a novel paradigm that is rapidly gaining ground in the scenario of modern wireless telecommunications. The basic idea of this concept is the pervasive presence around us of a variety of things or objects, such as Radio-Frequency IDentification (RFID) tags, sensors, actuators, mobile phones, etc., which, through

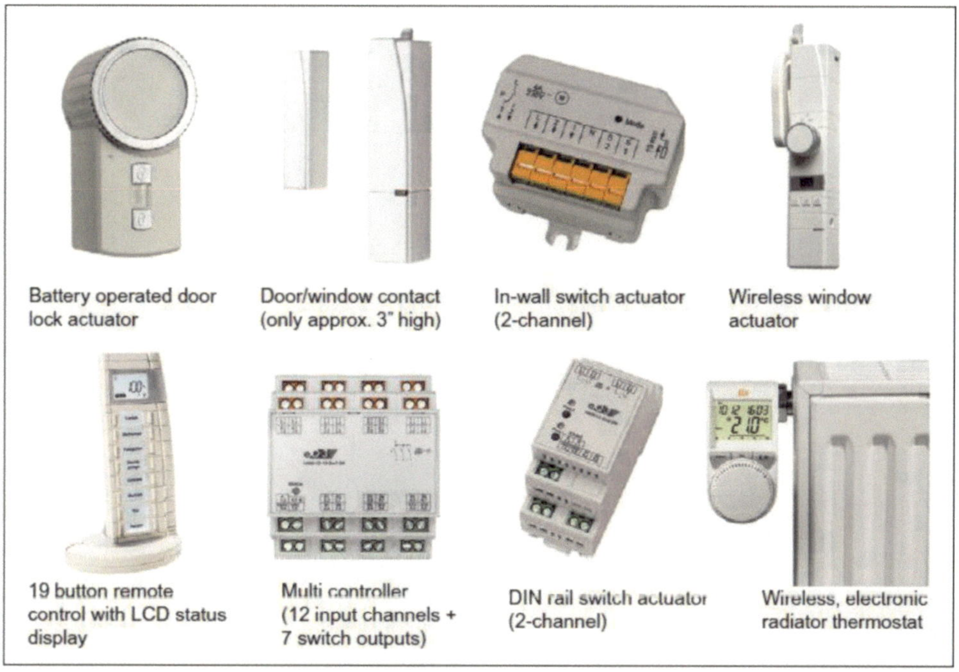

Figure 2.
Sample of home control devices (source: [21]/Figure 23.1).

unique addressing schemes, are able to interact with each other and cooperate with their neighbors to reach common goals."

Internet of health things (IoHT) is designed for medical data gathering and processing. IoHT connects unconnected health means with network connectivity ability. Digital and physical medical objects can thus network with each other in collaborating for data collection, processing, and storage. IoHT is a special case of the Internet of things (IoT) that combines health technologies and IoT and takes full advantage of IoT technology like the ability to initiate actions based on collected and analyzed data [24].

IoT finds its application already in the medical world as Istepanian et al. discussed in [25]. Williams et al. have defined the healthcare Internet of things (also called IoHT—Internet of health things) as

"*...the new embedded sensing capabilities of devices together with the availability of always being connected, to improve patient care whilst reducing costs* [26]."

The common architecture of IoT consists of sensors and actuators called things. Things are located at the data perception level. Behind the things are placed the IoT-gateways and data acquisition systems, followed by the edge IT and the data center (commonly on a remote server) and cloud. There are three (03) layers: (i) perception layer, (ii) gateway layer, and (iii) IoT platform layer.

IoT has the potential to enable home automation in collecting and processing data as well as to autonomously request actuators to execute some tasks for example temperature control by regulating the heater according to the set (for patient comfortable) room temperature.

IoT presents various domain-specific architectures that use various technologies and areas such as RFID, service-oriented architecture, wireless sensor network, supply chain management, industry, healthcare, smart city, logistics, connected living, big data, cloud computing, social computing, and security. **Figure 3** shows an IoT-enabled healthcare data perception system.

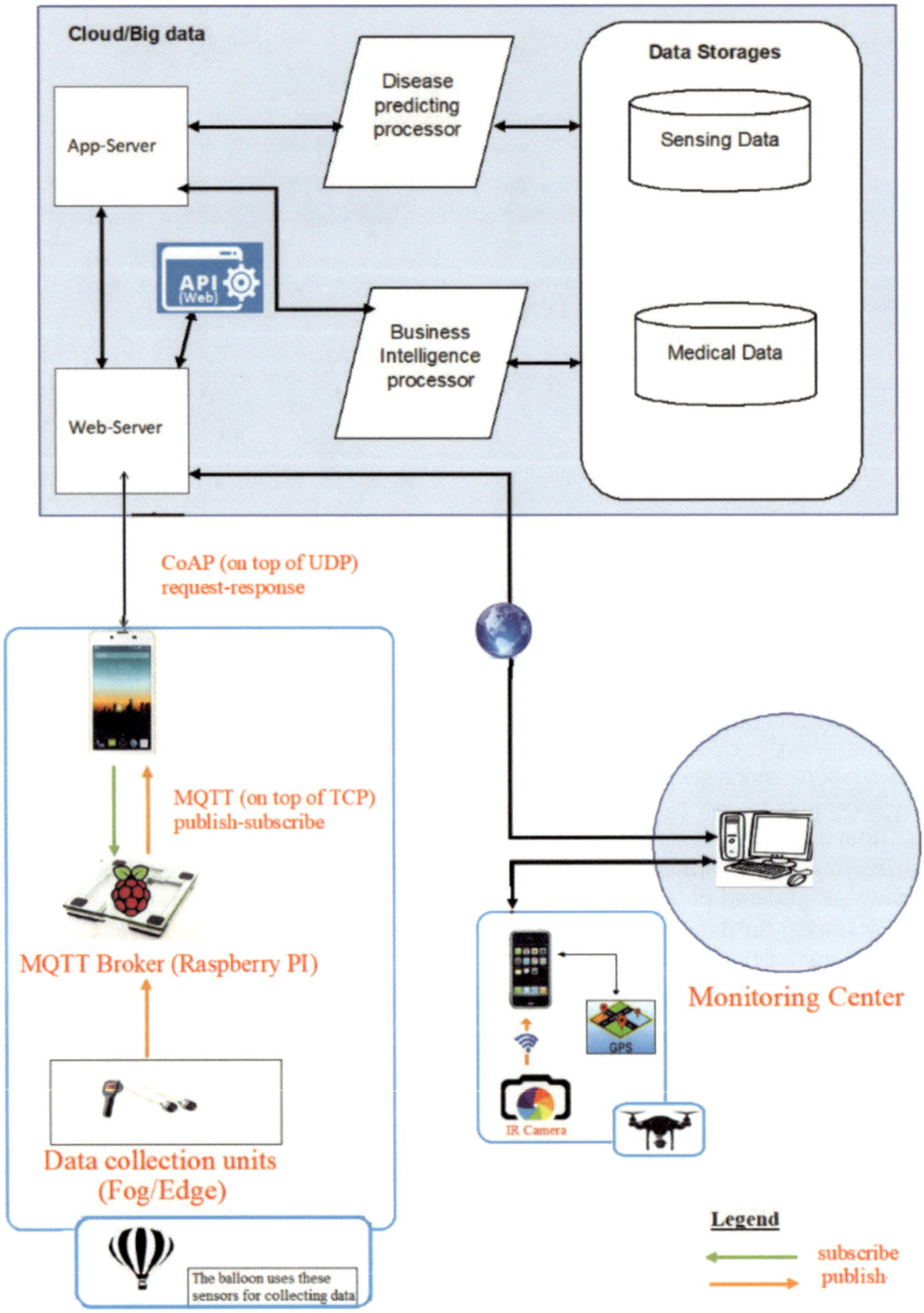

Figure 3.
Healthcare domain specific IoT architecture (source: [27]).

2.3.3 Wireless sensor networks (WSNs) and body area networks (BANs)

Wireless sensor networks (WSNs) find their use in smart home automation application since a while. They are used for medical application and devices to measure the patient's vital parameter. Bio-signals like body temperature, blood pressure, pulse oximetry, ECG, and breathing activity can autonomously and event-based automatically and seamless be measured.

Clinics, hospitals, and care/nursing homes can remotely use end-point devices like video and audio devices to assist family members to care for their sick member at home. Thus, home automation systems can be connected to medical emergency stations at clinics and hospitals close to the patient's residence place and regularly forward the patient's critical data gathered by WSMs and BANs. How this works is presented by Moghadam et al. in [28] where they have designed and implemented a communication system single and multi-antenna in a BAN. They wrote:

"an energy efficient data transmission technique for communication between a single-antenna medical sensor/microrobot inside the body to a multi-antenna receiver on the body surface through non-homogeneous propagation environment."

Transmitting over multiple spatial and temporal scales is challenging in advanced health informatics [29] though advancement achieved in the Internet of things (IoT) protocols like LoRaWAN with platforms like the things network (TTN) [https://www.thethingsnetwork.org/] eases transferring data today.

Wireless body area network (WBAN) is part of wireless sensor networks (WSNs) that can enable monitoring and collecting the patient bio-signal. This has been shown in a previous study [30] where a wireless sensor network system has been used at a cardiologic intensive care unit (CICU) for collecting and monitoring, round the clock, cardiologic activities in-patients. WSNs were connected to the patients and thus bio-signals have been collected in real time. This study has shown the feasibility of using WSNs and WBNs in home automation.

3. Research questions, hypotheses, and scopes

The main objectives behind the research questions are on one hand to investigate the tendency toward homecare regarding the nursing care homes and care workforce shortage and on the other hand to additionally investigate challenges and issues people are facing in homecare. Homecare is when family members care for their sick member at home. The needs in terms of appropriate solutions to overcome challenges and issues faced by caring for patients in homecare are assessed.

3.1 Questions

Q1: What is the tendency for homecare regarding the current nursing care crisis facing HIC?

Three nursing care options are noticed in Germany: (i) nursing care residences with 24 h registered care services and (ii) homecare with the assistance of ambulant nursing staff for a couple of hours per day. Many families hire care personnel from abroad (e.g., Yugoslavia, Budapest, and Ukraine), mostly with beginner's skills or no skills at all to care for their sick parents. Family members also care for the patient following medical instructions, (iii) regarding nursing care homes practiced fees, many families send their sick parents abroad in East European countries.

Q2: What challenges and issues are facing homecare?

Caring for a patient in homecare can be challenging for family members since many patients request round the clock nursing care. This is a full-time job. This study aims at investigating the challenges and issues that can be faced in such a situation.

Q3: Is it worth caring for dementia patients in homecare instead of at nursing care home?

Dementia patients are forgetful. They can forget to take food and drink water. They could forget to turn off a furnace or turn on a heater. Regarding these issues, it is worth assessing how to handle dementia patients.

Q4: What is the quality of life (QoL) of patients treated in homecare?

Many studies investigated the patient's QoL in nursing homes. Measuring or assessing the QoL level of patients in homecare is not achieved. This study aims at assessing it.

Q5: How to assess the factors impacting the health-related QoL for homecare?

There are well-established metrics for assessing the level of QoL in the nursing context though patient in homecare is exposed to additional environmental means. Therefore, it matters to investigate the impact of the QoL of other members on the QoL of the patient. Furthermore, can noise negatively impact the QoL in homecare? A grand-mutter (an elderly) will not be disturbed by a crying grand-child. It is, therefore, important to analyze which criteria are contributing to measuring the QoL in the case of homecare.

Q6: Can the technology assist to overcome homecare-related challenges and issues?

Round the clock care cannot be achieved by one person. It is a challenge. Previous studies have shown evidence for using the technology in healthcare to deliver care at remote, to monitor 24 h a day intensive care patients, etc. Many works have been achieved regarding mental health sensing and assessment, etc. In the present context, this study aims at investigating how home automation supported solution can assist in homecare and overcome challenges and issues faced.

3.2 Hypotheses

H1: The tendency to care for patients in homecare is on increase since the nursing crisis.

The study would like to verify if the nursing crisis has impacted the family member behavior.

H2: Smart automation home technology assists in homecare and impacts the QoL of both family members and the patient.

H3: Smart home automation enables to combine occupation (job) and caring adequately (efficiently and effectively) for a patient at home.

3.3 Scopes

Measuring the quality of services of nursing/care homes is out of the scope of this study, whereas only assessment of the patient's QoL in homecare before and after using the proposed solution constitutes the scope of the present work.

4. Methodology and data

This section presents the conducted literature review on the smart home automation for healthcare purposes. Additionally, data have been collected using semi-structured interview methodology with the objectives to answer the research question and verify the hypotheses.

4.1 Literature review

4.1.1 Paper sampling

In order to conduct a quantitative and qualitative literature review, papers have been sampled using snowball technology. Each found paper provides with numerous other papers through its references. Appropriate papers were thus found and used for the purpose of this study.

Papers were sought on three major bases: (i) home automation for medical applications, energy, security in the smart home, and trends in the smart home at cities. Beyond the technical part, papers dealing with the nursing care home, homecare, quality of life in nursing residences, and user satisfaction toward the nursing care are the main expressions used to find papers in the better academic literature database.

Table 1 summarizes the important papers reviewed.

Pos.	Title	Abstract	Year of publication	Reference
1	Design of an IoT smart home system	This paper basically deals with the design of an IoT smart home system (IoTSHS) which can provide the remote control to smart home through mobile, infrared (IR) remote control as well as with PC/laptop.	2018	[31]
2	A systematic review of the smart home literature: A user perspective	To facilitate the implementation and adoption of smart home technology, it is important to examine the user's perspective and the current state of smart homes. Given the fast pace with which the literature has been developing in this area, there is a strong need to revisit the literature. The aim of this paper is to systematically review the smart home literature and survey the current state of play from the users' perspective.	2019	[11]
3	Implementation of Smart home automation system on FPGA board using IoT	There has been a rapid introduction of network-enabled digital technologies in home automation. These technologies provide a lot of opportunities to improvise the connectivity of devices within the home. Internet helps to bring in with an immediate solution for many problems and also able to connect from any of the remote places which contribute to overall cost reduction and energy consumption. Intelligence based on microprocessors is used by home automation to incorporate electronic structures in the household.	2018	[32]
4	Smart home technologies in everyday life: do they address key energy challenges in households?	This paper interrogates their contribution to the ambitious carbon emission reduction efforts required under the 1.5 _C mitigation pathway set by the Paris Agreement and their suitability for energy poverty alleviation goals. In	2018	[32]

Pos.	Title	Abstract	Year of publication	Reference
		contrast to aspirational claims for a 'smart utopia' of greener, less energy-intensive, and more comfortable homes currently present in market and policy discourses, we argue that SHTs may reinforce unsustainable energy consumption patterns in the residential sector, which are not easily accessible by vulnerable consumers, and do little to help the 'energy poor' secure adequate and affordable access to energy at home.		
5	Environmental impacts and benefits of smart home automation: life cycle assessment of home energy management system	This paper discusses the life-cycle environmental impact of home energy management system (HEMS), in terms of its potential benefits and detrimental impacts. It is the expectation that adapting smart home automation (SHA) would lead to reduced electricity usage in the household and overall environmental advantages.	—	[33]
6	A review of smart homes—present state and future challenges	In the era of information technology, the elderly and disabled can be monitored with numerous intelligent devices. Sensors can be implanted into their home for continuous mobility assistance and nonobtrusive disease prevention. Modern sensor-embedded houses, or smart houses, cannot only assist people with reduced physical functions but help resolve the social isolation they face. They are capable of providing assistance without limiting or disturbing the resident's daily routine, giving him or her greater comfort, pleasure, and well-being. This article presents an international selection of leading smart home projects, as well as the associated technologies of wearable/implantable monitoring systems and assistive robotics. The latter are often designed as components of the larger smart home environment. The paper will conclude by discussing the future challenges of the domain.	2008	[10]
7	Home automation networks: A survey	Home automation networks provide a promising opportunity in designing smart home systems and applications. In this context, machine-to-machine (M2M) networks are emerging as an	2017	[20]

Pos.	Title	Abstract	Year of publication	Reference
		efficient means to provide automated communication among distributed ubiquitous devices in a standardized manner, but none have been adopted universally. In an effort to present the technologies used in the M2M and home integration environment, this paper presents the home area network elements and definitions and reviews the standards, architectures, and initiatives created to enable M2M communication and integration in several different environments, especially at the smart home domain. This paper points out the differences between them and identifies trends for the future.		

Table 1.
Selected literature among the sampling.

4.2 Data gathering and analysis

A semi-structured interview was conducted. Patients living at home as well as at nursing care home, care and nursing staffs, and people on the street were interviewed. The data collection was carried anonymously in accordance with the operative data privacy regulation in the country.

4.2.1 Data collection approach

4.2.1.1 Quantitative data collection approach

The data collection method has included questionnaires with a mixture of closed-ended (yes or no questions) and open-ended questions. Nursing home residents and patients in homecare were interviewed. Data were thus collected about nursing place tendencies and health-related as well as patient's quality of life with regard to the residence place: nursing home or homecare. No data on the quality of services in any nursing were collected.

4.2.1.2 Qualitative data collection approach

Quality of experience (QoE/QoX) or the satisfaction level is commonly based on a subjective appreciation of the quality of services. Patient's quality of life can be subjective somehow. For example, two distinct persons can differently appreciate noise or the presence of other people. Some elderly can feel uncomfortable when the nurse is a foreigner and ignore some elementary cultural rules. Therefore, nursing home residents were especially interviewed about their feeling, about what makes them feel uncomfortable in order to detect the impacts on their quality of life.

An important point was to determine their subjectivity level toward what makes them feel uncomfortable. Furthermore, test participants were asked about any discomfort the system has caused to them as well as if they feel observed or patronized.

4.2.2 Cohort sampling

4.2.2.1 Snowball sampling approach

This approach is the more appropriate method to sample the research cohort since sensible data were (anonymously) collected, and for this reason, precisely, it is difficult to find people willing to provide with their medical data. Participants have been selected on the basis of trust in the person who recruits them.

Table 2 summarizes the nursing home resident's cohort. A total of 33 patients were selected and classified per age range and gender. **Table 3** shows the structure of the patients in homecare. A total of 30 patients were selected. The two cohorts were interviewed for investigating the health-related QoL in nursing homes or in homecare as well as their preference in terms of staying at residence or living at home with their family.

Table 4 presents the cohort for investigating challenges and issues faced by homecare.

Table 5 presents an overview of the structure of the testing cohort. According to [17], elderly people aged 75+ years request severe nursing care. Based on this finding, the testing cohort is split into two groups: (i) < 65 year old participants and (ii) 65+ year old participants. All test participants are living at home. Participants living or having poor family support as well as participants with good family support have been selected. The objective was to verify to what extent the proposed solution can assist the patient even if he has no support. Furthermore, the limitations of the proposed system need to be tested in terms of to what extent the third person is needed so that they can fully assist the patient.

Age range	Total cohort size N = 33			
	N = 17		N = 16	
	Female		Male	
	Number	Health conditions	Number	Health conditions
< 50	2	1 Victim of road accident (outpatient)	4	1 Heart attack
		1 Physically disabled		1 Depression
				2 Schizophrenia
50–64	4	2 Dementia (early stage)	5	2 Depression
		1 Blindness + anxiety		2 Physically disabled
		1 Not diagnosed with a mental disorder		1 Anxiety
65–80	5	3 Alzheimer	4	2 Alzheimer
		2 Anxiety		2 Depression
>80	6	3 Parkinson's disease Advanced dementia	3	2 Alzheimer
		3 Alzheimer		1 Advanced dementia

Table 2.
Nursing home residents (cohort structure and diseases they are suffering from).

Age range	Total cohort size N = 30			
	N = 14		N = 16	
	Female		Male	
	Number	Health conditions	Number	Health conditions
< 50	4	3 Heart attack	7	2 Heart attack
		1 Physically disabled		3 Depression
				2 Blindness
50–64	5	2 Mental disorder (early stage)	5	2 Depression
		2 Anxiety		2 Physically Disabled
		1 Physically disabled		1 Anxiety
65–80	3	2 Alzheimer	3	Depression
		1 Anxiety		
> 80	2	1 Parkinson's disease	1	1 Alzheimer
		1 Alzheimer		

Table 3.
Homecare patients (cohort structure and diseases they are suffering from).

Assessment method	Total cohort size N = 515		
	N = 50	N = 68	N = 397
	Care staff*	Nursing staff*	Other individuals**
Paper-based questionnaires	45	60	385
1:1 semi-structural interview	5	8	12

***Only people that are caring or have cared for a family member were selected to participate.*
**Only staff involved in homecare.*

Table 4.
Nursing staff and people interviewed on the street.

Age range	Gender	Participants	Family support		Dementia		Other cognitive diseases		NCD/CD
			Good	Poor	Yes	No	Yes	No	
<65	Female	6	5	1	1	5	4	1	Diabetes
	Male	9	4	5	0	9	7	2	Heart diseases diabetes
>65	Female	11	8	3	3	8	8	3	
	Male	7	2	5	1	6	6	1	

**NCD, non-communicable diseases; CD, communicable diseases/infectious diseases. No data collected on CDs.*

Table 5.
The testing cohort.

4.2.3 Questionnaires for semi-structured interviews

This section summarizes the different questionnaires (**Tables 6–8**) used for the different surveys. At nursing homes and at participant's home (case of homecare), the questionnaires were used in 1:1 structured interview followed by a semi-structured interview. Distractor or control questions are inserted into the questionnaire in order to detect discrepancies in the responses and thus filter the biased responses (**Tables 6–8**).

4.2.4 Data analysis

Data analysis was made using IBM SPSS Statistics. Data were cleaned up; biased responses were not included in the analysis. Data dealing with a tendency for care at home as well as at nursing were accordingly classified. An AVG of the scores each category reaches was built. Before building the AVG, the different scores per category (stay at home or living at a nursing residence) obtained were compared with each other. The tendencies were plotted for visual analysis.

4.3 Testing design and methodology

Participants (**Table 5**) were selected using a snowball approach.

The action research methodology was applied for the testing. The system was adjusted according to the results in a phase and re-tested in the next phase. The test lasted one (01) week in the first phase. Data were collected and analyzed. The second phase took one (01) week again and findings from phase 1 were worked into phase 2.

Pos.	Questions	Observations
1	What do you most of all miss here?	Check how many patients prefer staying at home instead of living at the residence
2	Do your relatives visit you?	
3	How often do your parents visit you?	
4	Do you have any close friends here?	Socialization measurement
5	Are you missing your former friends?	Socialization, if he misses his former friends, this means he does not find a one here
6	Do you miss your parents, children, and grandchildren?	If yes, it means he does not receive enough or regular visits
7	Do you like living here?	
8	Do you have enough space for you?	
9	Are you missing your home?	
10	Do you receive enough and regularly food and water?	
11	What did you eat today?	Check if he is forgetful in order to consider or not the responses above
12	How do you feel today?	Assess the quality-of-life related to the patient's health state and care services he is provided with
13	Are the nurses nice to you?	
14	Which nurse is your best friend?	

Table 6.
Questionnaire for patients.

Pos.	Questions	Observations
1	Do you face any challenge during the admission process?	Assess how hard is it to get admitted to a nursing home
2	Would you prefer caring for your parent in homecare? If yes, why? If no, why?	Assess the tendency for homecare And find out why they have a tendency for one or other
3	Are you more confident to let care for your parent in a nursing home? If yes, what gives you that confidence? If no, why? Do you have no trust in nursing?	
4	Do you have a job? If yes, full-time or part-time?	Determine how one can manage both activities
5	If you respond to questions 2 and 4 by yes, then continue here; otherwise, go to the next question. How could you care for your parent in homecare and go to your job or on holidays?	
6	Have you ever experienced caring for a parent at home? If yes, how challenging was this?	Find out the real challenges people who experienced homecare are facing
7	Do you have any idea about which challenges and issues can be faced in homecare? If yes, which ones?	Challenges and issues in homecare. Home automation system should help to overcome these issues
8	Can modern information technology assist in homecare? If yes, how?	Determine the most needed functionality

Table 7.
Questionnaire for no-care staff to check their tendency for homecare or nursing care homes.

Pos.	Questions	Observations
1	Which challenges and issues are you facing daily?	
2	Do you have any technical assistance?	
3	How do you monitor the residents around the clock?	
4	Do you often assess the health-related quality of life of each patient?	
5	How many admissions do you register every year?	Assess the admission tendency
6	What is the admission tendency?	
7	How can you explain the tendency?	

Table 8.
Questionnaire for care-staff to investigate the trend toward the admission application.

At the end of each phase, a quantitative and qualitative analysis was performed. Patient's quality of life (QoL) and satisfaction level were measured in the light of the defined metrics (**Table 9**).

An important point was to involve participants living alone or having poor family support as well as those who have good family support. The objective to do so was to test if the system is well designed to assist people living alone too and how they are comfortable toward using the system (usability).

Quality of life measurement metrics	Description
Food and water intake	This metric verifies how many times the participant failed to take food and water.
Medication intake	Does the participant follow the medical instructions and take the medicine as prescribed?
Physical activities	Does the participant go out for physical activities or perform some at home?
Socialization	How many social contacts the patient has? Does he connect to other people or is he isolated?
Room temperature management	Does the system correctly learn from the participant preferences and set the temperature accordingly?
Noise and lighting control	Noise and light can make the individual feel uncomfortable
Familiarity	How familiar is the place to the individual
Accident rate	Does the system assist and prevent the participant from accident such as injury with a knife, fall down, etc.?
Emergency management	Does the system correctly detect emergency cases and thus manage the emergency?
Bio-signal gathering and data quality	

Table 9.
Quality of life measurement metrics.

4.3.1 Health-related quality of life (QoL) and user satisfaction (quality of experience) measurement metrics

In order to measure and asses the impacts of the proposed solution on the quality of life, a set of quality of life metrics were defined. The results of the experiment were analyzed in light of these metrics.

4.4 Ethical approval

Authorization and written informed participant consents were received from all major participants and their parents. An ad-hoc ethics committee at the involved clinics examined the request to conduct such an interview involving home's residents and approved it. Resident's parents also approved the study.

5. Findings and discussion

This section presents the study findings and discusses the results in light of data analytics.

5.1 Literature review on smart home automation for homecare

The literature review has pointed out that only a few previous types of research consider the *multidimensionality of the concept of the smart home.* Mostly the studies are focused on one aspect of smart home such as energy management [11].

A total of 656 abstracts and 239 full papers (journal and conference papers) were reviewed. Only 41 papers were retained having met the requirement of the present study. Unfortunately, only two papers have discussed many dimensions of smart homes. The rest mostly handle the topic of energy management at home. Smart home for elderly people is well considered in many papers, but the papers have failed to consider the multidimensionality of the concept of "aging at home".

Regarding the results, a novel solution considering the multidimensionality is therefore highly needed.

5.2 Interview findings

5.2.1 Patient's preferences

The interview with nursing home residents has revealed that elderly people prefer staying at home in their familiar and usual social environment (familiarity) and take care of their health by themselves as long as they are able to though only participants with good family support and those who have children, grandchildren, and good social contacts have the wish to stay at home as long as possible. However, alone living people, poor people, and people having no family support feel comfortable at the nursing residence.

Table 10 summarizes the results of the interview. Up to 91% of people living alone prefer residing in nursing homes, while more than 91% of people with good family support prefer staying at home with their family members.

5.2.2 Tendency toward homecare and nursing homes

Beyond the research questions, three (03) hypotheses were set. One hypothesis concerns the tendency for homecare as well as for nursing care home.

***H1:** The tendency to care for patients in homecare is on increase since the nursing crisis.*

The study verifies on the light of interview results the hypothesis H1. The survey was carried out to investigate the impact of the nursing crisis on the family member behavior toward the nursing care option for their patients.

The surveys point out the following results:

Category	Social status	Number	Preferences	
			Living at nursing care home	Staying at home
Nursing home residents (33 participants)	Poor family support Poor (financial) Have lived alone	21	17 (85%)	4 (15%)
	Rich Good family support Good social contact	12	01 (8.3%)	11 (91.7%)
Homecare participants (30 participants)	Living alone	23	21 (91.30%)	2 (8.70%)
	Living with family	7	0	7 (100%)

Table 10.
Participant's preferences toward living in nursing homes or staying at home.

There exist two categories of care: (i) stationary and (ii) ambulant nursing care [34].

People traditionally choose nursing residences for many reasons: (i) many people are living alone or have poor family support, (ii) the patient is at the end of life and needs severe intensive and palliative care, (iii) the care level (Ger. Pflegestufe).

A total of 118 healthcare staffs were interviewed. A total of 397 individuals on the street were also interviewed. A total of 56.78% of the interviewed care personnel admitted that the number of applicants for being admitted to a nursing home is

Assessment method	**N = 118**		
	Application for being admitted in nursing care residence		
	Decreasing	**Increasing**	**Stable**
Care personnel	67 (56.78%)	23 (19.50)	28 (23.72)

Table 11.
Tendency viewed by healthcare staff.

Assessment method	**N = 397**			
	Care experience	**Care for sick family members in homecare**		
		Prefer	**Do not prefer**	**Prefer going abroad**
People on the street	200 experienced with nursing homes	233 (58.69%)	97 (24.43%)	67 (16.87%)
	197 not experienced with nursing homes			

Table 12.
Preference of caring for a patient at home.

Pos.	**N = 397**			
	Challenges and issues		**Number (%)**	**Comments**
1	Nocturnal rest		397 (100%)	The family members have no rest. They can sleep well since assisted by the machine
2	Emergency issues		375 (99.5%)	Patient-centric data are collected.
3	Limited round the clock nursing		397 (100%)	x
4	Inaccurate collected data		40 (10.07%)	x
5	Combining job and care for a family member			The system shows potential to assist people in caring for their in parents aging in place
		Only part-time	317 (80.01%)	
		Stress	397 (100%)	
		Financial issue	298 (74.81%)	
6	Loss of quality of life		397 (100%)	
7	Limited social activities		290 (73.04%)	
8	Depression		15 (3.8%)	

Table 13.
Challenges and issues faced in homecare.

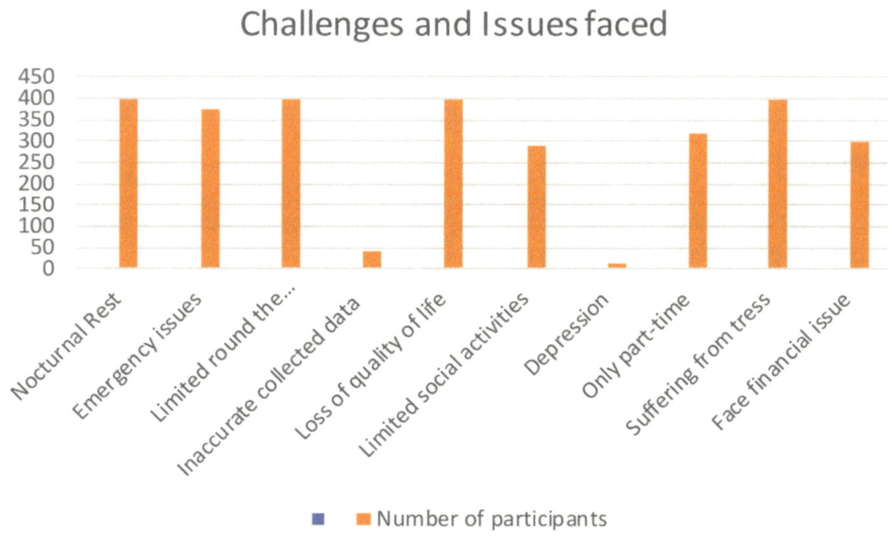

Diagram 1.
Number of participants facing challenges and issues in homecare.

Pos.	Metrics		N= 30 (Test last 2 weeks)			
			Number (%)			
			Before the test (AVG of 5 days observation)		After the test (Data provided by the system)	
			Have met the metrics			
			YES	NO	YES	NO
L1	Food and Water Intake		12 (40%)	18 (60%)	22 (73,33%)	8 (2,66%)
L2	Medication Intake		17 (56,6%)	13 (43,4%)	23 (76,6%)	7 (22,4%)
L3	Physical Activities		4 (13,3%)	26 (86,7%)	11 (36,6%)	19 (62,4%)
L4	Socialization		3 (10%)	27 (90%)	13 (43,33%)	17 (56,6%)
L5	Room Temperature management		3 (10%)	27 (90%)	26 (86,7%)	4 (13,3%)
L6	Noise and Lighting control		2 (6,6%)	28 (97%)	7 (22,4%)	23 (76,6%)
L7	Accident rate		1 (3,3%)	29 (26.7%)	1 (3,3%)	29 (26.7%)
L8	Good Emergency management		5 (16,6%)	25 (23,4%)	15	15
L9	Bio-signal gathering and data quality		1 (3,3%)	29 (26,7%)	18 (60%)	12 (40%)
L10	Results	Quality of life	Mediocre		Good++	
L11		Health outcome	Good-		Good+	

Legend (Comparing before and after test data)

Good++	
Bad	
Unchanged	

AVG	

Table 14.
Comparison of patient's quality of life before and after applying the proposed solution.

being slowly decreasing, while 58.69% of people interviewed on the street prefer to care for relatives in homecare.

Tables 11 and **12** show the tendencies of nursing care. The results obtained have confirmed the hypothesis H1.

5.2.3 Challenges and issues faced

The quantitative results regarding challenges and the number of people that reported these challenges and issues by caring for a family member are summarized in **Table 13** (**Diagram 1**). The quantitative data analysis reveals that very few people in home care are faced with data collection issues. This means data are rarely collected in home care. Thus, patients laying at home do not produce patient-centric data. The few data there produce is patient-centered. It is though known that patient-centered data are subjective, incomplete, and sometimes biased [27, 35].

5.3 Testing findings

H2: Smart automation home technology assists in homecare and impact the QoL of both family members and the patient.

The testing has confirmed the hypothesis (H2) regarding the user satisfaction's level and the quality of life (QoL) at both patient side and family side. **Table 14** (**Diagram 2**), **Table 15** (**Diagram 3**), and **Table 16** show detailed results.

H3: Smart home automation enables to combine job outside and adequately (efficiently and effectively) care for the patient in homecare.

The hypothesis is verified. Working family members can partially, full-time, work at home (home office), or go to the job and also care for a member.

5.3.1 Patients

Overall, broad satisfaction is noticed among the participants and their relations.

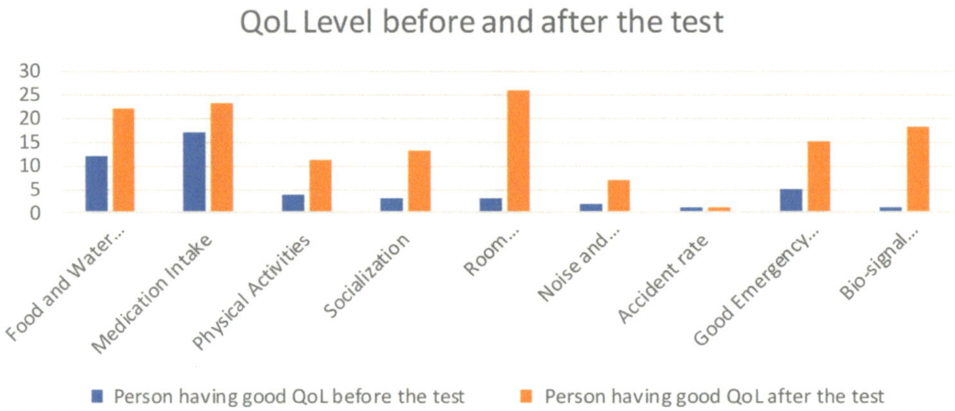

Diagram 2.
QoL level.

Metric [Link to Table 14 (column YES in after test)]	Cohort sampled after the test. It contains people having good metrics.		
	Total participants	Participants with good QoL and the family support level they received during the test	
		Poor	Good
Food and water intake [L1]	N = 22	7	15
Medication intake [L2]	N = 23	5	**18**
Physical activities [L3]	N = 11	2	9
Socialization [L4]	N = 13	3	10
Results	*About **7** have good QoL among 30 Patients after the test* *About **18** have good QoL among 30 Patients after the test*		

Table 15.
Impact of family support level on the patient's QoL.

The solution shows positive impacts on the quality of life (Good++, 36.6% started physical activities and 43.33% re-socialize). Due to the solution, 36.6% reconnect to physical activities, which means an increment of 23.3%. Nevertheless, about 62% remains without physical activities.

5.3.2 Family members

The solution has the pottential to assist people in combining full-time or parttime job with caring for a family member in home care. Since many people

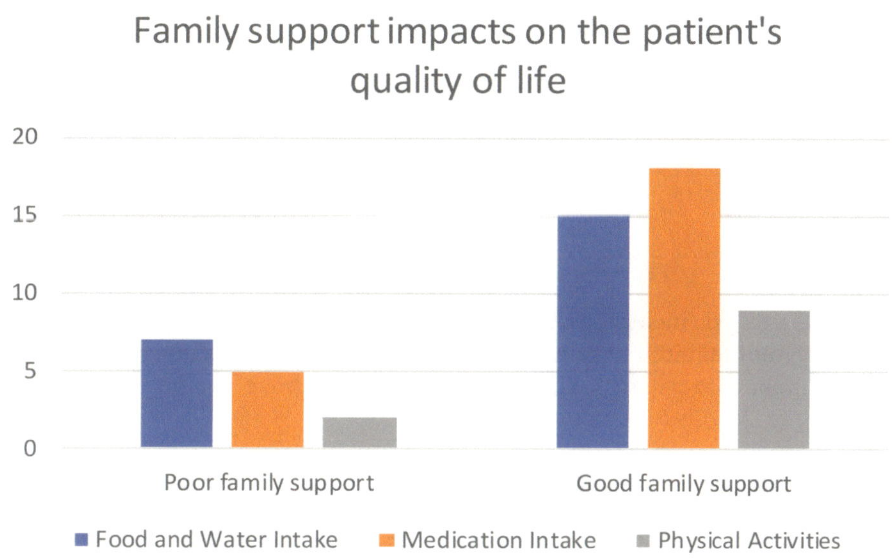

Diagram 3.
Impact of family support on the QoL.

N = 45 (patient's relatives)		
Metrics	**Before the test (Survey + 5 days observation)**	**After the test (Data provided by the system + observation)**
Quality of life	Average	Very good
Socialization	Few	Good
Nocturnal rest	Bad	Very good
Emergency management (quick medical assistance)	Bad	Good
Quality of communication with doctors	Bad	Improved
Bio-signal gathering and data quality	Worst	Improved (good)
Job situation	Worst (no job, part-time job)	Good (mostly full time)
Financial issues	Bad	Improved
Depression	Highly depressive	Less depressive

Table 16.
Comparison of family member's quality of life before and after applying the proposed solution.

aging in place still have the ability and capability to walk and can go out and back home alone, the system assists them and monitor their health condition in order to timely alerte parents and medical doctors in the case of emergency. The results had also shown evidence of improving the quality of life. An upcoming paper will report work conducted on this topic.

6. Health monitoring and daily personal assistance as approach

This section presents the concept of a *multidimensional smart home automation Internet of health things* for assisting dementia patients and elderly to "aging well at home". Additionally, the solution should assist the patient's family members to care for them and go to their occupation as usual.

The section presents the system requirements, features, concept, and architectural view.

6.1 System design

6.1.1 Requirements

The need analysis including the analysis of collected data leads to define the following system requirements and features, which the smart home solution for elderly and dementia patients will provide. As shown above, the health-related quality of life (QoL) is measurable by means of:

1. *The nutrition level (intake of food and water):* elderly and/or dementia patient is the most forgetful and could forget to take food and regularly drink water. This can cause severe health issues.

2. *Medication adherence:* medication adherence level influences patient health outcomes. Dementia patient who adheres to the prescribed medication could have comfortable days.

3. *Physical activities and socialization level:* both influence the patient's QoL.

4. *Family support:* Makes the patient feel more confident, secure, and safe. This is a factor impacting the patient's QoL.

5. *Space and comfort*: More space is a comfort that prevents anxiety in an individual since small space limits activities and movements.

Regarding the QoL measurement metrics, the following system requirements have been defined:

1. Qualitatively and quantitatively assess a patient's QoL level

2. Provide daily living assistance

3. Support patient empowerment and autonomy

4. Positively impact patient's health outcomes

5. Collect patient-centric data and information for accommodated and personalized health care services

6. Further, assist family members to efficiently and effectively care for their sick member at home

The main system-relevant requirement is to provide patients with a cheaper, simple, and better usability by considering their cognitive impairment like eye, hearing, and feeling impairment, restricted movement, etc. Additionally, the proposed solution should work online and offline.

6.1.2 System features

According to the system concept, the following features are provided to meet the requirements above.

6.1.2.1 Food and water intake monitoring

A designed water and food dispenser monitors the patient and can provide him with the food he needs. The system ensures that the patient drinks enough water so as to prevent him from feeling thirsty.

6.1.2.1.1 Food order process

1. Day menu presented.

2. The patient chooses a menu or the system selects 3 favorites based on historical data collected.

3. Food is ordered at the close restaurant and registered for the program.

4. Food is delivered.

6.1.2.1.2 Food dispenser

1. Food is stored in the special fridge (WaFoD).

2. At an appropriate time, the food is warmed.

3. The patient is served.

6.1.2.1.3 Reminder for family member

In case a family member is at home and wants to care for the patient, WaFoD sends an alert to the member.

In the case of ordering food, then food order process will run otherwise the food dispenser will run.

6.1.2.2 Medication intake monitoring

Similar to food intake, a drug dispenser is equipped with a high-resolution camera which logs the drug intake. A future extension will automatically perform anomaly detection on recorded films.

The medication intake is then logged. The logs are sent to the family member and the doctor.

6.1.2.3 Physical activities (in- and outdoor activities)

Special TV programs are displayed at certain times of the day to help the patient to train himself. The patient wears a body-area-networking (BAN) equipped with bio-sensors and accelerometer, which continually controls the position of the patient in order to detect if the patient is falling down or lying on the bed.

For dementia patients, no outdoor program is set.

6.1.2.4 Room temperature monitoring

Temperature control is a well-achieved domain application in smart home automation. Existing devices and systems are added to the network.

6.1.2.5 Noise and lighting control

This feature prevents any noise and controls the lighting.

6.1.2.6 Window and door monitoring

Doors and windows are controlled and closed when too noisy.

6.1.2.7 Reminder and assistance for indoor and outdoor

A smartphone-based application plays the role of a reminder and assistant. It follows the patient everywhere. Based on the patient calendar, this application can autonomously and automatically plan the whole day for the patient.

It can look for an appointment with the treating doctor for the next medical visit. The application is parametrizable.

6.1.3 Concept and architectural view

This section presents the concept of the proposed systems and gives an overview of its architecture.

The system features (i) a data perception unit, (ii) water, food, and medication management unit, and (iii) outside and inside activities.

6.1.3.1 Concept

6.1.3.1.1 Data gathering

IoT-enabled patient-monitoring systems present many advantages for the patient and for treating care personnel. Patient-centric data are collected. Personalized care can be based on these data. Actually, healthcare professionals base their treatment on patient-centered data, which can be biased since they are subjective. Further diagnoses are therefore needed or performed to verify the patient-centered data. Patient-centered data are data provided by the patient through narratives, while patient-centric data are data collected using modern information technologies like (wireless) body area network (W-BAN) or (wireless) sensor networks (WSNs).

Aging persons are often forgetful and thus provide mostly biased information when they are requested to report on their health conditions. Though in a smart home automation enabled healthcare solution for "aging well," collecting patient-centric data in an autonomous way is mandatory. In a previous study [35], various advantages of collecting patient-centric data were discussed. The healthcare personal gets a complete picture of the patient's health condition and can thus pose the right diagnosis.

Based on the requirement above, the proposed concept provides a patient-centric data collector in terms of sensors connected with the patient that fully collects any bio-signal as well as positions data and sends the data to a record system at the remote. A duplicated copy of the data is saved on the local server and serves as training data for a machine learning (ML) routing. Additionally, a set of networking capable video recorders are used to collect the patient's body expressions, behaviors, mimic, and any physical activities. These data are also used by the ML algorithm to predict patient's behaviors, expectations, and physiological needs (like thirst, hunger, going to the toilet, etc.).

Sensors (in a body area network) connect the patient to an IoT-gateway that transfers the collected data, using the MQTT protocol, to the local server. We talk of edge-computing that happens at the edge. Collected data are processed and stored on the local server. Using the CoApp protocol, data are sent to the cloud. Treating care/nursing homes or medical doctor as well as patient's family members can access the data and can send data to the local server, which would use received data to regulate some connected devices.

6.1.3.1.2 Food and water intake and control

For "food and water intake", a smart device is designed. This device combines microwaves and the fridge. The device called water and food dispenser (WaFoD) with networking ability is connected to the patient's smartphone and the local server, which in turn is connected to a remote server at the cloud that connects the home to the outside and can dispatch information and data in the whole network. WaFoD can learn from the individual's behaviors and preferences.

WaFoD is connected to the IoT gateway and can collect data, transfer data, and receive data from a remote unit (system or individual). Registered behaviors build the training data for a machine learning processor (ML) located on the local server, the master in the entire network. The ML processor predicts patient menus, proposes menus to the patient, and can order at the registered restaurant the selected menu. All proposed services to the patient are based on his behaviors and preferences.

WaFoD is designed to remind the patient to regularly drink water. It dispenses water or soft drinks. It can warm food and serve the food to the patient. The system logs each nutrition behavior and sends at the end of the day an activity journal, or in the case of emergency (that means the patient does not drink for a while or refuse to take food), it alerts the nursing home close to the patient's residence.

The patient is provided with a touchscreen that displays TV programs and can display the pictures of menus proposed by WaFoD.

The entire system is designed following the Internet of things (IoT) paradigm: (i) data collection unit(s) and (ii) IoT-gateway place between the local server. The local server is a light copy of the remote server at the cloud, which can perform complex and memory consuming computing activities; (iii) the IoT platform at the cloud.

A copy of data like room temperature, updated patient's preferences, etc. that are needed for any computing action are stored at the local server.

The patient is provided with a set of accelerometers (sensors to determine his position- fall down, laying , staying, seating, etc.). with the objective to detect, predict if the patient is falling down or will. Furthermore, other sensors like "Feuer alrm" have been used to monitor fire harzard.

6.1.3.1.3 Medication intake

A drug dispenser is provided. The dispenser is connected to the IoT-gateway via Bluetooth. It features an alarm and can remind the patient to take his medicine. The medication intake is logged and a protocol is stored on the network. Family members can be informed if the patient does not take the medicine on time, thus, action can be taken to help the patient to take the medicine. Care/nursing homes are also connected to the dispenser via the le cloud and can get alerted when the patient refuses to take the medicine.

6.1.3.1.4 Indoor activities

The local server is connected to a touchscreen TV. It can display physical activity programs, which can let the patient to also do so, for example, activities like a walk in the room, some light movements, etc.

6.1.3.1.5 Outdoor activities

Elderly people need real socialization. They need therefore to go out and meet other people. The solution proposed feature a smartphone-based application that manages and looks for senior-meeting close to residence place. This application integrates Google Maps that drives the patient to the meeting and takes him back home.

Similar is done with medical visit.

6.1.3.2 Architectural view

The architectural view presents 4 layers (**Figure 4**).

6.1.3.2.1 Network things

Bio-signals, behavior, preferences, room temperature, physical activities, food, water, and medication intake data are collected at this stage through sensors.

The data collected data are forwarded to the aggregation stage.

6.1.3.2.2 Data aggregation stage

At this stage, collected data are aggregated, filtered, cleaned up, processed, and pre-stored.

6.1.3.2.3 IT edge stage

Processed data from the prior stage are used here, but also forwarded the cloud.

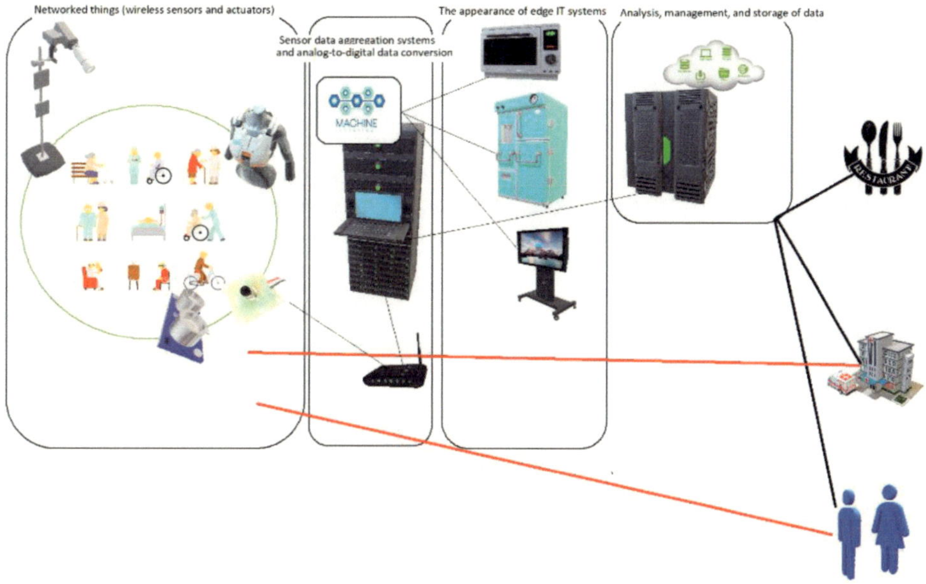

Figure 4.
Architectural view of the proposed system.

6.1.3.2.4 Analysis and other processes at the cloud

Stages A, B, and C happen in the local area (patient residence). In order to enable communication with the remote side, data are forwarded to the cloud. A communication line is, therefore, open between C and D.

Family members, restaurants, nursing homes, and all people authorized to deal with the stored data can access the data through the cloud.

6.1.3.3 Data security

This topic will be discussed in the upcoming paper.

7. Conclusion

This study has investigated the state-of-the-art of "ageing well at home." Many previous studies had archived interesting works on making the "home" comfortable and smarter for elderly and dementia patients. Though most of the previous works have failed in providing a complete solution of smart home automation (*multidimensionality of the concept of the smart home)* for people requesting homecare, this study covers this limitation and shows that smart home automation can impact the patient's and his family member's QoL in a positive way. People staying alone at home as well as those living in nursing care homes would take benefits from such a solution.

8. Upcoming works

The future works aim at launching a human hologram in the program to assist the patient. The patient would though see a family member and can receive from him any instructions or discuss with him. As **Table 15** shows, the presence of a family member has a great impact on the patient's health-related quality of life and thus on his health outcomes.

Setting and remotely regulating the room temperature is well achieved though patient temperature feeling also depends on the treatment he is under. Certain drug or after-physical activities make the patient feel warm or hot. There exists no system that can automatically and autonomously recognize that the patient's room temperature is not more appropriate. Therefore, we plan to design a wearable that can verify if the patient is feeling cold or hot and thus regulate the heater.

Author details

Thierry Edoh[1*] and Jules Degila[2]

1 Institut für Informatik, Technical University of Munich, Garching bei München, Germany

2 Institute of Mathematics and Physical Sciences, African Center of Excellence in Applied Mathematics, Dangbo, Benin

*Address all correspondence to: oscar.edoh@gmail.com

References

[1] Mazzola R, Hasseler M. Aktuelle herausforderungen an die profession pflege in der altenhilfe. In: Alter und Pflege im Sozialraum. Wiesbaden: Springer Fachmedien Wiesbaden; 2018. pp. 101-112

[2] Elke Peters KS. Versorgungsleistungen in der Pflege. In: Gesundheitswissenschaften Springer Ref. Pflege-Ther.-Gesundheit. Berlin: Springer; 2019. pp. 1-15

[3] Münnich R, Kopp J, Al E. Pflege an der Grenze. Wiesbaden: Springer Fachmedien Wiesbaden; 2019

[4] Hammer PDE. Wenn Pflege an Grenzen gerät-Gewalt im Pflegealltag. Alzheimer Gesellschaft Baden-Württenberg e. v., 2014. pp. 1–3

[5] Schmidt H, Golla H, Perrar KM, Eisenmann Y, Voltz R. Needs of people with advanced dementia in their final phase of life: A multi-perspective qualitative study in nursing homes. Palliative Medicine. 2017;**32**(3):657-667

[6] Marć M, Bartosiewicz A, Burzyńska J, Chmiel Z, Januszewicz P. A nursing shortage - a prospect of global and local policies. International Nursing Review. March 2019;**66**(1):9-16

[7] Burton A. How do we fix the shortage of neurologists? Lancet Neurology. 2018;**17**(6):502-503

[8] Palm R, Trutschel D, Sorg CGG, Dichter MN, Haastert B, Holle B. Quality of life in people with severe dementia and its association with the environment in nursing homes: An observational study. Gerontologist. 2018;**00**(00):1-10

[9] Kok JS, Nielen MMA, Scherder EJA. Quality of life in small-scaled homelike nursing homes: An 8-month controlled trial. Health and Quality of Life Outcomes. 2018;**16**(1):4-11

[10] Chan M, Estève D, Escriba C, Campo E. A review of smart homes-present state and future challenges. Computer Methods and Programs in Biomedicine. 2008;**91**(1):55-81

[11] Marikyan D, Papagiannidis S, Alamanos E. A systematic review of the smart home literature: A user perspective. Technological Forecasting and Social Change. 2019;**138**:139-154

[12] Simmons SF, Harrington C, M. Bates-Jensen B, Schnelle JF, Cadogan M, Garcia E. Relationship of nursing home staffing to quality of care. Health Services Research. 2004;**39**(2):225-250

[13] Sexton E, McLoughlin A, Williams DJ, Merriman NA, Donnelly N, Rohde D, et al. Systematic review and meta-analysis of the prevalence of cognitive impairment no dementia in the first year post-stroke. European Stroke Journal. 2019:1-12. https://doi.org/10.1177/2396987318825484

[14] Kamboh MI, Chang C-CH, Ganguli M, Berman SB, Snitz BE, Sullivan KJ, et al. Mild cognitive impairment that does not progress to dementia: A population-based study. Journal of the American Geriatrics Society. 2018; **67**(2):232-238

[15] Edoh TO. Advanced Systems for Improved Public Healthcare and Disease Prevention. USA: IGI Global, Medical Information Science Reference (an imprint of IGI Global); 2018

[16] Frid L et al. What technology can and cannot offer an ageing population: Current situation and future approach. In: Biswas P, Duarte C, Langdon P,

Almeida L, editors. A Multimodal End-2-End Approach to Accessible Computing. London: Springer; 2015

[17] Orimo H, Ito H, Suzuki T, Araki A, Hosoi T, Sawabe M. Reviewing the definition of elderly. Geriatrics & Gerontology International. 2006;**6**(3): 149-158

[18] World Health Organization. Proposed working definition of an older person in Africa for the MDS Project. Health statistics and information systems [Online]. 2002. Available: https://www.who.int/healthinfo/survey/ageingdefnolder/en/ [Accessed: March 23, 2019]

[19] United Nations. World Population Ageing 2017: Highlights; 2017

[20] Toschi GM, Campos LB, Cugnasca CE. Home automation networks: A survey. Computer Standards & Interfaces. 2017;**50**:42-54

[21] Vasseur J-P, Dunkels A. Home automation. In: Interconnecting Smart Objects with IP. 2010. pp. 353-360

[22] Pham M, Mengistu Y, Do H, Sheng W. Delivering home healthcare through a cloud-based smart home environment (CoSHE). Future Generation Computer Systems. 2018;**81**:129-140

[23] Atzori L, Iera A, Morabito G. The internet of things: A survey. Computer Networks. 2010;**54**(15):2787-2805

[24] Terry N. Will the internet of health things disrupt healthcare? Vanderbilt Journal of Entertainment & Technology Law. 2016;**19**(2):28-31

[25] Istepanian RSH, Hu S, Philip NY, Sungoor A. The potential of internet of m-health things "m-IoT" for non-invasive glucose level sensing. In: 2011 Annual International Conference of the IEEE Engineering in Medicine and Biology Society. 2011. pp. 5264-5266

[26] Williams PAH, McCauley V. Always connected: The security challenges of the healthcare internet of things. In: 2016 IEEE 3rd World Forum on Internet of Things (WF-IoT). 2016. pp. 30-35

[27] Edoh T. Risk prevention of spreading emerging infectious diseases using a hybridcrowdsensing paradigm, optical sensors, and smartphone. Journal of Medical Systems. 2018;**42**(5):91

[28] Moghadam NN, Farhadi H, Bengtsson M. An energy efficient communication technique for medical implants/micro robots. In: Xplore IEEE 2016 10th International Symposium on Medical Information and Communication Technology (ISMICT), Vol. 1, No. 1. 2016. pp. 1-5

[29] Yang G, Xie L, Mäntysalo M, Zhou X, Pang Z, Da Xu L, et al. A health-IoT platform based on the integration of intelligent packaging, unobtrusive bio-sensor, and intelligent medicine box. IEEE Transactions on Industrial Informatics. 2014;**10**(4):2180-2191

[30] Edoh TOC, Atchome A, Alahassa BRU, Pawar P. Evaluation of a multi-tier heterogeneous sensor network for patient monitoring—The case of Benin. In: MMHealth '16 Proceedings of the 2016 ACM Workshop on Multimedia for Personal Health and Health Care. 2016. pp. 23-29

[31] Khan A, Al-Zahrani A, Al-Harbi S, Al-Nashri S, Khan IA. Design of an IoT smart home system. In: 2018 15th Learn. Technol. Conf. (L&T 2018). 2018. pp. 1-5

[32] Tirado Herrero S, Nicholls L, Strengers Y. Smart home technologies in everyday life: Do they address key energy challenges in households? Current Opinion in Environment Sustainability. 2018;**31**:65-70

[33] Louis JN, Caló A, Leiviskä K, Pongrácz E. Environmental impacts and

benefits of smart home automation: Life cycle assessment of home energy management system. IFAC-PapersOnLine. 2015;**28**(1):880-885

[34] Augurzky B, Krolop S, Schmidt H, Schmitz H, Schwierz C. Pflegeversicherung, rating und Demographie Herausforderungen für deutsche Pflegeheime Heft 26. Essen/Germany, 2006. Retrieved from: http://www.rwi-essen.de/media/content/pages/publikationen/rwi-materialien/M_26_Pflegeheim-Rating-2006.pdf

[35] Edoh T. Internet of things in emergency medical care and services. In: Medical Internet of Things (m-IoT)—Enabling Technologies and Emerging Applications. Vol. 2. Rijeka, Croatia: IntechOpen; 2019. p. 64

Section 4

Artificial Intelligence and Other Applications

Chapter 6

Optimizing a Centralized Control Topology of an IoT Network Based on Hilbert Space

Jesús Jaime Moreno Escobar, Oswaldo Morales Matamoros, Hugo Quintana Espinosa, Ricardo Tejeida Padilla and Ana Gabriela Ramírez Gutiérrez

Abstract

An Internet of Things network (IoTN) is composed of many small devices or nodes located in homes and/or offices, to be operated through the Internet from anywhere, making these devices smarter and more efficient. For improving the efficiency of an IoTN, in this chapter an optimized fractal algorithm (OFA) was proposed for designing a centralized control topology of an IoTN, whose nodes are distributed according to the Hilbert space-filling fractal. We developed the OFA to find the best nodes where a smart home device can find the highly reliable link between its neighbors by a software-defined network (SDN) with a target coverage since OFA only considers reliable links among devices. Finally, through laboratory tests and computer simulations, we demonstrated the effectiveness of this proposal by using a large amount of IoT devices making them more efficient operating systems. The quality of service (QoS) is a challenge that guaranteed the level of service delivery to an IoTN, so that OFA takes less time to reach its destination after it is generated by its source, increasing the probability that the target node can recover the original packet before the lifetime expires.

Keywords: smart home, IoTN, centralized control topology, optimization, Hilbert fractal algorithm, quality of service

1. Introduction

In the coming fourth industrial revolution, several technological fields are impacting all disciplines, economies, and industries, such as artificial intelligence (AI) and the Internet of Things (IoT). Artificial intelligence permits machines to learn and adapt to different situations and problems. Due to its great versatility, AI can work in conjunction with the Internet of Things, allowing smart home devices or microcontroller units (MCUs) to create many Internet of Things networks (IoTNs) to be able to collect, process, and share data of different nature that flows through the network. It is expected that by 2025 26–30 million MCUs in homes and offices will be connected as networks to the Internet and equipped with sensors, processors, and embedded software.

As a network, IoTN behaves as a collective entity which does not depend on a central control or hub. Therefore, it is necessary to design IoTN topologies that are not subject to the performance of a single MCU or node for guarantying the quality of service (QoS) requirements due to an efficient connectivity based on a bandwidth that links each of MCUs with their immediate neighbors at 300 Mbps, in order to ensure that all of them really share their parameters when they are operating. This implies to design networks with shared parameters between their nodes and where the locality is preserved and limited the maximum average distance between the MCUs by using alternative tools, such as space-filling curves. The Hilbert fractal is a continuous space-filling curve whose locality-preserving behavior is better than that of Z-order curves, because the distance between each node in a Hilbert curve does not fluctuate, whereas that distance in a Z-order curve does fluctuate.

Accordingly, in this chapter an effective and a reliable optimized fractal algorithm (OFA) for extending the range of transmission of a given IoTN in an intelligent, adaptive, and dynamic way is proposed. The OFA finds a path from a source to a target to optimize the links by considering the quality of service constraints such as end-to-end reliability and delay. In this way, with OFA we pretend to achieve the best two MCUs where each smart home of the IoTN can transfer or share its parameters. Finally, through laboratory tests and computer simulations, we demonstrate the effectiveness of our approach by means of a fractal routing in IoTN, using from 16 to 64 smart home devices for real-time monitoring of different parameters, in order to make more efficient the Smart Home Automation by reducing the IoTN energy consumption.

The further sections of this chapter are organized as follows. First, in Section 2 we explain the relation between smart home devices distributed in networks and the energy efficiency. In Section 3 an explanation of the theoretical definition of the Hilbert space-filling fractal is given. OFA scheme is defined in Section 4, while in Section 5, the simulation and the comparison of the performances are announced, in order to verify the correction and feasibility of the proposal. Finally, the arguments and discussion of this chapter are analyzed in Section 6.

2. Merging smart home and software-defined networks

A house is smart when it has installed a series of electronic systems, sensors, and devices, so that any person can easily control it, even from a distance, and the house performs certain actions on their own.

An example of this would be the management of heating and air conditioning, so that our home always has a comfortable temperature, whatever the time of year.

This can be achieved by installing temperature sensors and connecting them to a computer or a dedicated device, which once we have programmed it, deciding which temperature is the best for the residents, sends orders to the heating or air-conditioning control devices. This automation is what is called home automation.

In this way, home automation has been around for a long time, as we can link various automations through domestic wiring, sometimes with cable inside the walls. The technological advance has given us wireless connectivity and miniaturization, which have allowed the extension of the concept of home automation to many other aspects, even outside the home.

Currently, and under the definition of the smart home, there are devices that allow to control almost everything imaginable, either automatically or facilitating human control.

In this sense, we can say that the smart home is the result of the intersection between architecture, interior design, and the most advanced technology, which allows the automation in an unattended way of certain domestic tasks, but also with the condition that there can be supervision or fixed objectives on the part of the residents.

Thus, a smart home seeks to:

- Increase our security cameras or motion sensors which are elements that warn us of intrusions and suspicious movements even when we are absent.

- Provide us more time. By automating certain tasks, we stop thinking about them. If we wake up every morning at a certain time, the same home (through its artificial intelligence) can raise the blinds and start to heat the coffee, so we can save this time to get to work a little earlier or simply to sleep 5 more minutes.

- Allow us remote control. Turning on the air-conditioning unit 10 min before we get home so that when we enter it.

- The energy saving with multiple sensors scattered around the home are more reliable than the human sensation, where exactly the energy needed to heat or cool the house is used.

The latter statement guides us to consider as one of the most important issues the way to save energy among sensors, namely, how efficient is the energy consumption. In this way we identify energy efficiency as the most critical challenges in IoTNs because the nodes or smart home devices in such networks have limited resources. In this way, software-defined networks (SDNs) are a way to approach the creation of networks in which control is detached from the hardware and given to a software application called a controller. So, SDNs, also known as programmable and automated networks, are presented as a proposal that provides greater speed, agile infrastructure, and better costs in cloud IT platforms; it is urgent to respond to the dynamism of the applications required by the user.

When a packet arrives at a switch in a conventional home network, the rules built into the proprietary firmware of the switch tell the switch where to transfer the packet. The switch or router sends each packet to the same destination on the same path and treats all packets in the exact same way. In this proposal, intelligent routers designed with specific MCUs are sophisticated enough to recognize different types of packages and treat them differently, but these router can be only the seed or the begging of the home network. Furthermore, in an SDN, a home network administrator can shape traffic from a centralized control console without having to touch individual smart devices, in this case sensors. The administrator can modify any rule of network switches when necessary—giving or removing priority or even blocking specific types of packets with a very detailed level of control.

This is especially useful in a multi-tenant architecture for cloud computing because it allows the administrator to handle traffic loads flexibly and more efficiently. Essentially, this allows the administrator to use fewer and more *intelligent routers* and have more control than ever over the flow of all the smart devices of the network traffic. Nowadays, the most popular specification for creating a software-defined network is an open standard called OpenFlow. OpenFlow allows network administrators to control routing tables remotely.

In addition, home networks have not changed about 20 years ago; unlike programs, we are facing a paradigm shift to which we must adapt. The development of

SDNs began in 1990, where programmable functions are included in the network; in 2001–2007, the control and data planes were improved, improving with this innovation. From 2007 to 2010, it was implemented. The OpenFlow API, is presented as an open interface, presents various ways the separation of the control plane and data to be scalable, practice where virtualization played an important role in this SDN evolution.

The progress and evolution of SDNs underlies because traditional home routers cannot respond to unpredictable traffic patterns much less to peak demand. In addition, there would be two alternatives to climb to a more expensive home network by spending more time in configuration or adapting to a dynamic home network. SDNs are suitable for people that require rapid changes in the short term, which happens in a conventional home where mobile devices are added and disappear arbitrarily over time. Currently a conventional home user uses social networks in many devices, which are in high demand or require sudden changes, for example, not only working on geographic traffic but environmental variables such as temperature or humidity, development of mobile devices, virtualization of network devices and sensors, and cloud services.

In this way it is important to point out that SDN is an architecture approach, not a specific product, which has traditionally been thought of as the virtualization of data center networks.

This means separating the management of the control plane from the home network devices from the underlying data plane that forwards the home network traffic. The use of a software-defined system to control this disaggregation provides many benefits, including greater network management flexibility and the ability to more easily implement high-precision security policies.

However, Kindness points out that smart home network operators think of SDN with a too narrow focus, although there has been an evolution in the SDN market in recent years, driven by the increase in demands on the network.

To meet these new challenges, the underlying technology that drives SDN has also been applied to other network areas such as the homes with a lot of sensor or smart devices.

SDNs emerged in early 2010 out of mere necessity. Many networks today were designed for client–server applications that run on a non-virtualized infrastructure. The purchase of Nicira by VMware in 2013 was considered a turning point for the SDN industry, as it turned the virtualization giant into a network provider. Today, VMware's NSX SDN product is based on that technology. Cisco Application Centric Infrastructure is the basis of its SDN offer. Many other companies, such as Juniper and Arista, also offer their solutions. IDC estimates that the SDN market has grown from an industry of 406 million dollars in 2013 to over 6600 million dollars in 2017. The consulting firm anticipates that the SDN market will continue to do so at a compound annual growth rate of 25.4%, reaching about 13,800 million dollars in 2021.

A survey conducted in 2017 by the Network World publication among 294 network professionals found that 49% of them are considering or actively piloting an SDN implementation; 18% already have an SDN installed.

Furthermore, IDC has identified a handful of major use cases for SDN today:

- To maximize investments in server virtualization and private cloud
- To enable network programming

Although so far many SDN deployments have focused on the data center network, Kindness, the Forrester analyst, states that the future of SDN will be defined by how this technology is used outside of the data center.

There are a variety of factors that will continue to pressure network operators, including the increasing use of public cloud computing, the attack of network traffic created by the Internet of Things, the proliferation of a mobile workforce, and an increasing number of distributed branches. Given this, SDN will play a role in configuring the next generation of networks for each of these use cases.

And there is already evidence of this in actual use: Software-Defined Wide Area Network (SD-WAN) is a software management platform to control access to the remote offices or branches of an organization.

In the past, home customers had only one connection to their branches; today SD-WAN allows companies to add several types of network connections to a branch and have a software management platform that allows high availability and automatically prioritize traffic.

SD-WANs can save a home customer's expense by installing expensive custom WAN acceleration hardware allowing them to run a software overlay on less expensive hardware.

Expects highlight that the SD-WAN will become a market of 6 billion by the year 2020.

In addition, software-defined IoTNs (SD-IoTNs) can be technically sorted in nine classifications:

1. Security: identifying the malicious user and their activities, namely, global overview of a device's status in the home network.

2. Routing: the network data and information is efficiently transferred.

3. Mobility: due to external forces, any node of the home network can be physically moved.

4. Reliability: the ability to be, almost always, in an operational state under adverse circumstances, usually after failures. Operating state is understood to be that in which the SDNs are capable of performing a specific operation.

5. Management: maintenance, network configuration, and provisioning.

6. Quality of service: it is a set of service requirements that SDN must accomplish when routing a data flow. It can be implemented in different situations, to manage congestion or to avoid it. It allows to control some significant characteristics of packet transmission. These characteristics can be specified in quantitative or statistical terms such as bandwidth, latency, jitter, and packet loss in the network, ensuring a pre-established degree of reliability that meets the traffic requirements, depending on the profile and bandwidth for a given data flow.

7. Wireless power transfer: energy sensor node can be transmitted to other nodes through an appropriate transmitter.

8. Localization: many applications of IoTNs need the information of each node.

9. Energy efficiency: sleep scheduling approaches are designed for switching the nodes into idle state if their functionality is not required.

In addition, this classification can be rewritten as follows:

- Coverage control: control activates or deactivates the sensor nodes to cover a network region.

- Clustering: nodes into clusters and a head node for each.
- Lifetime: possibility to utilize the node capabilities for a longer period of time.

Accordingly to SD-IoTN paradigm, no matter the sort of topology, any sensors inside a smart home network are connected individually to a central router, and they were almost never interconnected among them. Thus, the devices furthest from the router spend more energy than those that are closest to it. In addition, any communication between sensors must necessarily go through a centralized router, even if they are a foot away from each other.

In this sense, the main proposal of this chapter is to communicate with the nearest sensor using a non-centralized, rather a distributed and dynamic, topology that is reorganized according to a fractal function based on the Hilbert scanning.

3. Hilbert space-filling fractal

Mathematician B. Mandelbrot coined the fractal term in his pioneer work *The Fractal Geometry of Nature* [1], describing a fractal as an irregular or fragmented geometric structure that can be divided into parts: each of which is (or approximately) a smaller-size copy of the whole. He also pointed out that many fractals are found in the nature forming irregularly shaped objects or spatially nonstandardized phenomena in nature that cannot be attributed to Euclidean geometry, such as mountains or blood vessels, with fractional or non-integer dimension. From a mathematical point of view, fractals are a kind of composite geometric shapes which regularly display the property of self-similarity, such that a small segment of it can be reduced as a fractional scale replica of the whole [1].

All obtainable fractal objects in nature are generated from non-determined or random steps. Fractals generated by an iterative procedure, produced by consecutive dilations and conversions of a primary set, are deterministic. There are two properties attributed to fractals: self-similarity and space-filling. Self-similarity stands for a piece of the fractal geometry which seems to be like that of the total structure for all time, while the space-filling property means that a fractal outline can be packed in a limited region as the iteration increases without increasing the whole area.

Concerning the space-filling property, there are fractal curves that fill the plane that contains them in a specific order by continually changing direction or passing through each point that is in the defined space [2], such as the Hilbert fractal curve. The Hilbert space-filling curves are in a single layer that do not intersect; each point of them is at a constant distance unique to any other point, and these curves contain only one starting point and one stopping point in a single layer.

In 1878, mathematician G. Cantor demonstrated that there was a one-to-one correspondence between the unit interval $[0, 1]$ and the unit square (plane) curve. In 1879, Netto showed that any such mapping could not be continuous. In 1887, Jordan defined a (plane) curve as the set of points $(\phi(t), \psi(t))$ where ϕ and ψ are continuous functions on a closed interval $[0, 1]$, t is the time, and the curve is the path of a particle starting at $t = 0$ and ending at $t = 1$. In 1890, Peano discovered a space-filling curve. Thus the Peano curve must have multiple points, that is, points which are the images of two or more distinct values of t [3].

In 1891 Hilbert discovered another space-filling curve. Whereas Peano's curve was defined purely analytically, Hilbert's approach was geometric [2].

To construct Hilbert-type space-filling curves, lets denote the unit interval $[0, 1]$ as $I = \{t|0\} \leq t \leq 1$ and the unit square as $Q = \{(x; y)|0 \leq x \leq 1; 0 \leq y \leq 1\}$. For each positive integer n, the interval I is partitioned into 4^n subintervals of length 4^{-n} and the square Q into 4^n subsquares of side 2^{-n}. A one-to-one correspondence between the subintervals of I and the subsquares of Q subject to two conditions is constructed: (i) adjacent subintervals correspond to adjacent subsquares with an edge in common, *adjacency condition*, and (ii) if at the $n-th$ partition, the subinterval I_{nk} corresponds to a subsquare Q_{nk}, and then at the $(n+1)-st$ partition, the four subintervals of I_{nk} must correspond to the four subsquares of Q_{nk}, *nesting condition*.

For each n the 4^n subintervals are labeled in their natural order from left to right. The correspondence between the intervals and the squares amounts to numbering the squares so that the adjacency and nesting conditions are satisfied. Hilbert's enumeration of the squares is shown in **Figure 1** for $n = 1, 2, 3$. The first square is always in the lower left corner, and the last square is always in the lower right corner. This means that the Hilbert space-filling curve starts at $(\phi(0), \psi(0))$ at $t = 0$ and ends at $(\phi(1), \psi(1))$ at $t = 1$. With the first and last squares of each partition determined, there is only one enumeration of the squares that satisfies the adjacency and nesting conditions.

The construction of a Hilbert space-filling curve is presented in **Figure 1**, in which the dotted square shows the area to be filled by the curve. This square is divided into four squares.

On the other hand, many space-filling curves can be produced as the limit of a sequence of polygonal curves, where the curves are generated via an iterative process, for example, a Lindenmayer system (L-system), in order to visualize such sequences of curves. A small part of the curve at one step of the iteration is close to a corresponding part of the curve at the previous step, and so it is natural to add frames that continuously interpolate between the curves of the iteration.

The Hilbert fractal can be generated by using specific rewriting rules of the Lindenmayer systems [5].

Figure 1.
First four stages in the generation of Hilbert's space-filling curve, axiom = R (see Section 4.3) employed by Hilbert in [4].

The L-systems, developed by Lindenmayer in 1968, emerged as a way to discreetly describe the development and growth of multicellular organisms [6]. An L-system is a string rewriting system. We have an alphabet of symbols V; a string of symbols ω defining the start of the system, or *axiom*; and a set of production rules P determining how we produce a new string from an old string. The *L-system* grammar is the collection of these three parts, $G = (V, \omega, P)$. This generates a sequence of strings by repeatedly applying the rules P, starting with the axiom ω.

To construct the Hilbert space-filling curve, we proposed the following L-system grammar:

$$G_{Hilbert=}(\{F, -, +\}, R, \{R \rightarrow +RF - LFL - FR+\})$$

Then the output of the L-system is the following sequence of strings for the first three steps:

Step 1:	$-LF + RFR + FL-$
Step 2:	$- + RF - LFL - FR + F + -LF+$
	$RFR + FL - F - LF + RFR + FL$
	$- + F + RF - LFL - FR + -$
Step 3:	$- + -LF + RFR + FL - F - +RF - LFL - FR + F + RF - LFL-$
	$FR + -F - LF + RFR + FL - +$F + - + RF-LFL-FR + F + -
	$LF + RFR + FL - F - LF + RFR + FL - +F + RF - LFL - FR$
	$+ - F - +RF - LFL - FR + F + -LF + RFR + FL - F - LF+$
	$RFR + FL - +F + RF - LFL - FR + -F + -RL + RFR + FL-$
	$F - +FR - LFL - FR + F + RF - LFL - FR + -F - LF + RFR + FL - +-$

Given the L-system for the Hilbert space-filling curve, it is possible to view the produced strings as instructions for building these curves in a turtle graphic fashion [7]. That is, we start somewhere on the plane, with a specified forward-facing direction. We read the input string from left to right and interpret the symbols as actions to be performed. For the L-system for the Hilbert space-filling curve, we make the following interpretations:

1. F, move forward one unit

2. +, turn left (counterclockwise) 90°

3. −, turn right (clockwise) 90°

In this chapter it is proposed to use the symbols L and R to represent a kind of small deviation from a forward step (starts with a left turn) and to represent a small detour in the other direction (starts with a right turn), respectively.

4. Methodology

The proposed methodology in this chapter is defined by the following passes:

- Hub MCU

- $LRN_i(t)$ and $LSN_i(t)$

- Hilbert fractal scanning
- Initial topology

It is also important to consider the general parts of the environment of the proposal, so any configuration of IoTNs consists of a wireless access point (WAP) and a β number of IoTN nodes which can be increased or decreased in its amount. In this case β embedded systems or MCUs that composed the IoTN, which shares parameters among sensors inside a given room or floor plan; β ranges dynamically are $1 \geq \beta > 4^n$. In this way, we also define $IoTN_N(t) = \sum_{i\geq 1}^{\beta} MCU_i(t)$, where $t = \{t_0, t_1, t_2, \ldots t_j\}$, i.e., a specific period of time (t) when the IoTN is defined as a set of MCU_i with at least one member up to 4^n members. Whereas n is the Hilbert fractal level needed to go from the first to the last MCU_i and it is expressed by $n = \lceil \log_4(\beta) \rceil$, here i represents the index inside the IoTN (t) of a given MCU, namely, MCU_1 is the first MCU, MCU_2 the second one, and so on.

It is important to emphasize that the IoTN has β MCUs in the time frame (t_j), but it can have up to $4^n - 1$ devices. If $\beta > 4^n$, the value of n is recalculated, and a new topology is reconfigured; otherwise the algorithm just adds a new node in the network, with the purpose that each node of the same network is linked or indexed by the curves that make up the Hilbert fractal. Subsequently, the effectiveness of the methodology exchange parameter throughout the network is measured. Finally, the point-to-point algorithms of Wi-Fi (P2P) and SoftAP are configured.

4.1 Hub MCU

It is important to establish that $\beta(t_j)$ IoTN devices are randomly distributed along a certain floor plan which forms a matrix γ with $2^n \times 2^n$ dimension. Namely, our proposal estimates a certain topology according to the β devices in the room in a certain period of time t_j. Then, the main WAP or hub is placed in the center of the main room, although it can be placed anywhere within the room, to have connections at any direction. That is, every connection to the WAP will be the mainstream star topology. In this first generation of MNIs, a consecutive number will be assigned according to the speed of its link with the WAP. Thus, certain MCU_i has the best link so this device is labeled as $i = 1$ or MCU_1. Any floor plan had some walls, which attenuate the signal; that is why certain devices with shorter linear distance have a slower maximum speed than others that are further away from the WAP.

4.2 $LRN_i(t)$ and $LSN_i(t)$

$MCU_{i=1}(t)$ is identified as the main node in the IoTN. In this way, all the embedded systems both $IoTNe(t)$ and $MCU_1(t)$, enable Wi-Fi Direct mode and a full table with the bandwidth of the nodes are next to them, in order to generate a list of reliable nodes in a certain period of time t $(LRN_i(t))$. Every MCU generates a particular $LRN_i(t)$; then the proposed IoTN can generate 4^n LNRs at the instant t. In addition, every $LRN_i(t)$ contains the bandwidth of all $MCU_i(t)$ with which it establishes connection. In order to belong to the wireless sensor network, every $MCU_i(t)$ must connect to at least one link to another $MCU_i(t)$. All LRNs are shared, and $\sum_{i=1}^{4^n} MCU_i(t)$ knows the way to any node in the network, namely, everyone knows the topology of the network. In this way it is important to estimate the $LRN_i(t)$, but not all of these nodes are significant to be considered as the best option to establish a link, in which manner we define $LSN_i(t)$ as the list of significant nodes

(LSN) which is a vector that contains the best bandwidths of a certain $MCU_i(t)$. Algorithm 1 shows the methodology to estimate the $LSN_i(t)$ that needs i^{th} MCU and its $LRN_i(t)$. At this moment the IoTN can be considered as a scarce network, since only the $\frac{\rho_i(t)}{\beta}$ percent of the reliable links is connected.

Algorithm 1: Generator function to estimate the list of significant nodes (LSN) and the number of significant nodes ($\rho_i(t)$).

Input: i,$LNR_i(t)$
Output: $\rho_i(t)$,$LSN_i(t)$
j $\Longleftarrow$ size of the $LNR_i(t)$ for the $MCU_i(t)$
Sort all $LNR_i(t)$ regarding their bandwidth in $S_{LNR_i}(t)$
Estimate $B_{thr} = \mu_i(t) - \sigma_i(t)$
Initialize the counter $label = 1$ and $LSN = \{0, 0, 0\}$
for $label \leftarrow 1$ **to** j **do**
  Calculate $B_{LNR_k}(t)$: the bandwidth in Mbps of the k element in the $SLR_{LNR_k}(t)$
  if $B_{LNR_k}(t) \geq B_{thr}$ AND $k \leq 3$ **then**
    $LSN(k) = MCU_{label}(t)$
    $k \Longleftarrow k + 1$
  $LSN(k) = \{$ the three best links according B_{thr}, if possible $\}$
  $\rho_i(t)$ is the number of elements $\neq 0$

4.3 Hilbert fractal scanning

The construction of the Hilbert space-filling fractal, under the paper's graphic interpretation framework, was reduced to the replacements of the smallest elements (symbols) or interior replacements. This path can be optimized by means of the correlation of the strength of the signal or the bandwidth of the link by Eq. (1):

$$\phi_{xy} = \frac{\sum_{i=1}^{n}(x_i - \overline{x})(y_i - \overline{y})}{\sqrt{\sum_{i=1}^{n}(x_i - \overline{x})^2}\sqrt{\sum_{i=1}^{n}(y_i - \overline{y})^2}} \tag{1}$$

where ϕ_{xy} is the sample correlation coefficient, n is the stage 0 level of Hilbert curve (**Figure 2a**), and x_i and y_i are, respectively, the best and the second best individual bandwidths in a certain MCU_i indexed inside the IoTN as the ith element, while $\overline{x}$ and $\overline{y}$ are, respectively, the sample mean of the best and the second best individual bandwidths in a certain MCU_i. Thus, we can estimate the best link to a particular MCU_i.

All $\sum_{i=1}^{4^n} MCU_i(t)$ are distributed randomly giving as a result a shape as shown in **Figure 2b**. When γ is reordered as vector, it gives as a result the Hilbert curve indexing $\overrightarrow{\gamma}$ which contains the order of the $MCU_i(t)$ (see **Figure 2c**).

4.4 Initial topology

Once the Hilbert curve is defined as L-system, we adapt the production rules of the original work by Hilbert [4], who proposed an axiom with a $\mathcal{D}$ trajectory, while we propose to start with an $\mathcal{U}$ trajectory. Our proposal is based on the fact that most of the energy is concentrated in the nearest MCU, namely, at the right or left. In this way the production rules of the Hilbert curve are defined by:

- $\mathcal{U}$ is changed by $\mathcal{LUUR}$
- $\mathcal{L}$ by $\mathcal{ULLD}$

1	4	5	6	59	60	61	64
2	3	8	7	58	57	62	63
15	14	9	10	55	56	51	50
16	13	12	11	54	53	52	49
17	18	31	32	33	34	47	48
20	19	30	29	36	35	46	45
21	24	25	28	37	40	41	44
22	23	26	27	38	39	42	43

(a)

64	2	3	61	60	6	7	57
9	55	54	12	13	51	50	16
17	47	46	20	21	43	42	24
40	26	27	37	36	30	31	33
32	34	35	29	28	38	39	25
41	23	22	44	45	19	18	48
49	15	14	52	53	11	10	56
8	58	59	5	4	62	63	1

(b)

64	9	55	2	3	61	12	54	....	51	13	60	6	7	50	16	57
1	*2*	*3*	*4*	*5*	*6*	*7*	*8*	....	*57*	*58*	*59*	*60*	*61*	*62*	*63*	*64*

(c)

Figure 2.
(a) First three stages of the Hilbert curve, (b) matrix of dispersion or γ *of* $\sum_{i=1}^{4^n} MCU_i(t)$ *inside a convectional room, and (c) matrix* $\vec{\gamma}$ *ordered as a vector.*

- $\mathcal{R}$ by $\mathcal{DRRU}$
- $\mathcal{D}$ by $\mathcal{RDDL}$

The Hilbert curve has the property of remaining in an area as long as possible before moving to a neighboring spatial region. Hence, the correlation between neighbors $MCU_i(t)$ is maximized, which is an important property in the optimization of any system. The higher the correlation for estimating the final topology, the more efficient the data parameter sharing.

5. Experimental results

In this section, first we approach the basic concepts of the network simulator (NS) and then use it to compare the most current SDN topologies, which helps us to identify the effectiveness of the proposed algorithm.

5.1 Network simulator

Network simulator is a software to simulate discrete events, and it was designed to aid in the research of telematic networks. NS provides support for the simulation of a multitude of protocols of the application layers (http, ftp, cbr, etc.), transport (TCP, UDP, RTP, SRM), unicast and multicast routing protocols, etc., both for networks wired as local or satellite non-wired and with complex topologies with a large number of traffic generators. NS began in 1989 as a variant to the existing REAL Network Simulator and has evolved substantially in recent years, having been developed by DARPA with the help of several network research institutions, such as LBL, Xerox PARC, UCB, USC/ISI, etc.

NS is basically an object-oriented simulator, written in C++, whose user interface is presented as an object-oriented Tcl language interpreter or, in other words, OTcl language. The simulator supports a hierarchy of classes written in C++, also

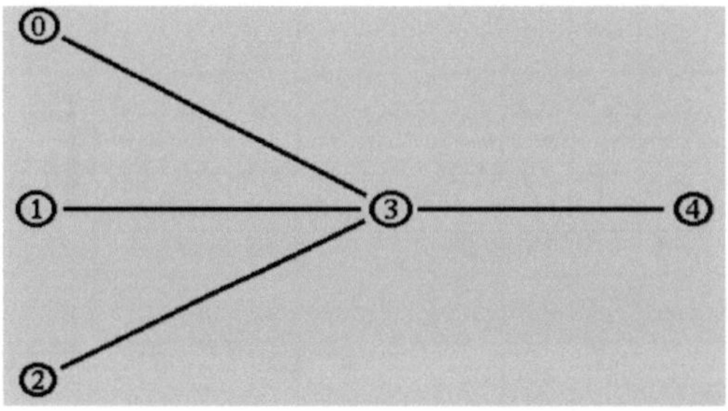

Figure 3.
Example of network topology.

called a compiled hierarchy, and another hierarchy of classes similar to the previous one but within the interpreter that is presented to the programmer in OTcl, and which is also known as an interpreted hierarchy. These two hierarchies are closely related to each other, so that, from the user's perspective, there is a one-to-one correspondence between one class in the interpreted hierarchy and one in the compiled hierarchy. The root of this hierarchy is a class called TclObject. When the user creates a simulator object from the interpreter, which usually starts a script, this object is created within the interpreter, and a close relationship is created with another identical object but within the compiled hierarchy. The interpreted class hierarchy is established automatically through methods defined within the class called TclClass. In addition, we can create other hierarchies to our liking in the scripts, written in OTcl, that are not split in the compiled hierarchy.

To create the topology, we just show a simple example for purposes of understanding the use of this simulator. For this example we use the connection depicted by **Figure 3**.

Then the nodes are created in the same way as in **Figure 3**, as follows:

```
set n0 [$ns node]
set n1 [$ns node]
set n2 [$ns node]
set n3 [$ns node]
set n4 [$ns node]

$ns duplex-link $n0 $n3 1Mb 100ms DropTail
$ns duplex-link $n1 $n3 1Mb 100ms DropTail
$ns duplex-link $n2 $n3 1Mb 100ms DropTail
$ns duplex-link $n3 $n4 1Mb 100ms DropTail
```

Then, traffic sources are attached to nodes n0, n1, and n2:

```
proc attach-expoo-traffic { node sink size burst idle rate } {
   #Get an instance of the simulator
   set ns [Simulator instance]

   #Create a UDP agent and attach it to the node
   set source [new Agent/UDP]
   $ns attach-agent $node $source

   #Create an Expoo traffic agent and set its configuration parameters
   set traffic [new Application/Traffic/Exponential]
   $traffic set packet-size $size
   $traffic set burst-time $burst
```

```
$traffic set idle-time $idle
$traffic set rate $rate
# Attach traffic source to the traffic generator
$traffic attach-agent $source
#Connect the source and the sink
$ns connect $source $sink
return $traffic
```

First, the procedure creates a traffic source and attaches it to the node. Then, it creates a Traffic/Expo object, sets its configuration parameters, and attaches it to the traffic source. Before, the source and sink have been connected. Now, the traffic sources (with different peak rates) are attached to n0, n1, and n2, and then they are connected to the three traffic sinks in node n4, which has been previously created:

```
set sink0 [new Agent/LossMonitor]
set sink1 [new Agent/LossMonitor]
set sink2 [new Agent/LossMonitor]
$ns attach-agent $n4 $sink0
$ns attach-agent $n4 $sink1
$ns attach-agent $n4 $sink2

set source0 [attach-expoo-traffic $n0 $sink0 200 2s 1s 100k]
set source1 [attach-expoo-traffic $n1 $sink1 200 2s 1s 200k]
set source2 [attach-expoo-traffic $n2 $sink2 200 2s 1s 300k]
```

In this example, Agent/Loss Monitor objects are used as traffic sinks, since they store the amount of data received, which can be used to calculate the bandwidth. In the same way, we construct all the nodes and interconnections for the simulation of the next section.

5.2 Comparing SDN performance

Once NS is defined, we estimate the performance of this chapter by the simulation of the effectiveness of the project node and the quality of service. Moreover, the duration of the OFA was evaluated by the end of the delay (E2ED) and the packet delivery ratio (PDR). So we compared the performance of OFA to newer algorithms similar to the latest technology:

1. Reliable and energy-efficient routing (REER) proposed by Li et al. in [8]

2. Delay-energy trade-off in wireless sensor networks with reliable communication (DETR) proposed by Liu et al. in [9]

3. Reliable routing with distributed learning automaton (RRDLA) proposed by Mostafaei in [10]

We use the network parameters recommended by Zorzi and Rao in [11] and Karp and Kung in our experiments [12].

In this way, it is important to define E2ED as the time it takes a packet to reach its destination after it is generated by its source, it is important to set the time for the packet to be reached because its next delay and delivery delays will be targeted at PDR probability of a total of [13] knot; the life span may be reset to original packages before the expiration of [14]. According to this definition, the correlation between the maximal and minimal end-to-delinquency values of the η is determined by the change between maximal and minimal packet delivery ratios. If the delay ends are stable, η will be equal to 0, and if the packet delivery rate is constant $\eta = \infty$, a delay will appear:

$$\eta(\varphi) = \frac{E2ED(\varphi)_{\max} - E2ED(\varphi)_{\min}}{PDR(\varphi)_{\max} - PDR(\varphi)_{\min}} \tag{2}$$

where φ is the effect of node density or the requirements for reliability of QoS. According to the models proposed by Niu et al. [15], it is based on the experience described for Mostafaei in [10] and Zeng et al. in [16].

In this way, **Figure 4a** and **b** show the comparison between PDR and E2ED, respectively, in terms of node intensity and reliability. We can say that better algorithms are reached with the higher E2ED and the lower PDR. Thus, both algorithms get the best results and try to minimize delays when sending target data which can restore original packages. OFA is slightly better in both cases, but the differences are not higher than 1.97%. Additionally, the delay in RRDLA and OFA is noticeably changed by increasing the intensity of the network or QoS node. DETR and REER also have worst results than the two best algorithms.

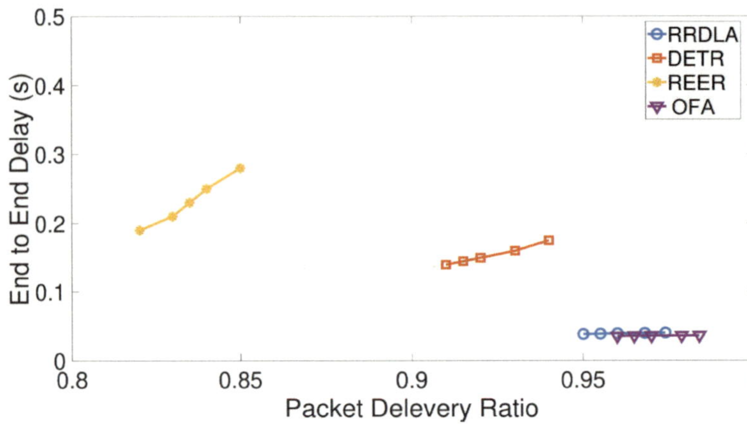

(a)

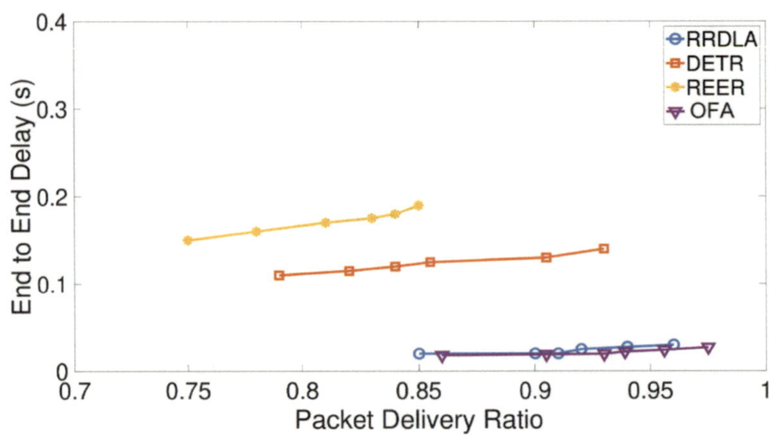

(b)

Figure 4.
The simulation of the impact of the node density and quality of service, case end-to-end delay vs. packet delivery ratio. (a) Node Density, and (b) Reliability Requirements.

Algorithm	η (node density)	η (quality of service)
RRDLA	*0.0833*	*0.0909*
DETR	1.1667	0.2143
REER	3.000	0.4000
OFA	**0.0386**	**0.0705**

Table 1.
$\eta(\varphi)$ for the impact of node density and quality of service.

Table 1 shows $\eta(\varphi)$ for node intensity and service quality, both in the OFA or best results (bold) to 0 and RDDLA will have the second (emphasized).

If **Table 1** is analyzed, we can realize that OFA obtains the better result when node density and quality of service are compared. It is important to mention that using the regression techniques of a variable of E2ED on a PDR variable, we look for a function that is a good approximation of a point cloud (x_i, y_i); this regression algorithm reduces in an important way the time of processing. The simple linear regression model has the following expression:

$$E2ED = \alpha + \beta PDR + \varepsilon \tag{3}$$

where α is the ordinate at the origin (the value that E2ED takes when PDR is 0), β is the slope of the line (and indicates how E2ED changes when increasing PDR in a unit), and ε is a variable that includes a large set of factors, each of which influences the response only in small magnitude, which we will call error. PDR and E2ED are random variables, so you cannot establish an exact linear relationship between them.

As a result, a package sent by OFA will spend less time to achieve its target, because the target node will increase the chance of recovering the original packet before its expiration.

6. Conclusions

Smart home technology, as part of IoT, is the future of residential-related technology which is designed to deliver and distribute services inside/outside the house via networked smart devices sharing information by means of a broadband Internet connection. In this chapter we developed an OFA for optimizing a central IoTN topology, increasing the range transmission of a given Wi-Fi network in an adaptive and dynamic way when sharing parameters, in order to achieve an energy-efficient IoT-based smart home.

With the IoTN topology proposed in this chapter, based on the L-system for the Hilbert space-filling fractal and focused on the efficient connection of the smart home devices or MCUs, the energy consumption of the connected devices is reduced. Most of this energy consumed by MCUs can be saved through inclusion of new nodes or devices connected in the same IoTN, since most of the information detected throughout the network is redundant due to geographically placed sensors.

Thus, our proposal optimizes the links among MCUs or smart home devices, since in a dense network we have 2016 possible links but with only 64 links we can share parameters of sensor, namely with only the 3.05% or we are reducing the 96.95% the number of links in the presented IoTN. In addition, we can manage scarce IoTNs in order to obtain more dense networks, since by means of a backbone, many WAPs can be interconnected to form more complex or dense topologies, which can grow using higher levels of Hilbert fractal.

Finally, we propose a software-defined network (SDN) with strong mobility since it can be reconfigured depending on the amount of nodes; also we employ a target coverage because OFA only considers reliable links among smart home devices. In terms of reliability, our proposal can share parameters such as battery, radio, hardware, or operating systems. By itself, the quality of service is a challenge that guaranteed the level of service delivery to an IoTN; OFA takes less time to reach its destination after it is generated by its source increasing the probability that the destination smart home device can recover the original packet before the life-time expires.

Acknowledgements

This article is supported by the National Polytechnic Institute (Instituto Poliécnico Nacional) of Mexico by means of Projects Nos. 20190046 and 20195208. The research described in this work was carried out at the Superior School of Mechanical and Electrical Engineering (Escuela Superior de Ingeniera Mecánica y Eléctrica), Campus Zacatenco.

Author details

Jesús Jaime Moreno Escobar[1*†], Oswaldo Morales Matamoros[1†], Hugo Quintana Espinosa[1†], Ricardo Tejeida Padilla[2†] and Ana Gabriela Ramŕez Gutiérrez[3†]

1 Escuela Superior de Ingeniería Mecánica y Eléctrica, Instituto Politécnico Nacional, Mexico City, México

2 Escuela Superior de Turismo, Instituto Politécnico Nacional, Mexico City, México

3 Escuela de Administración de Instituciones, Universidad Panamericana, Guadalajara, México

*Address all correspondence to: jemoreno@esimez.mx

† All the authors contributed equally.

References

[1] Mandelbrot BB. The Fractal Geometry of Nature. San Francisco, USA: Freeman; 1982. Chapter 7

[2] Sagan H. Space-Filling Curves. New York: Springer-Verlag; 1994

[3] Nicholas Rose J. Hilbert-Type Space-Filling Curves. Florida, USA: Bethune Cookman University; 2010

[4] Hilbert D. Über die stetige abbildung einer linie auf ein flächenstück. Mathematische Annalen. 1891;**38**(3): 459-460

[5] Prusinkiewicz P. Modeling and visualization of biological structures. In: Proceeding of Graphics Interface '93. Toronto, Canada; 1993. pp. 128-137

[6] Lindenmayer A. Mathematical models for cellular interaction in development, Parts I and II. Journal of Theoretical Biology. 1968;**18**:300-315

[7] Papert S. Mindstorms: Children, Computers, and Powerful Ideas. New York, USA: Basic Books; 1980

[8] Li X, Wang Y, Chen H, Chu X, Wu Y, Qi Y. Reliable and energy-efficient routing for static wireless ad hoc networks with unreliable links. IEEE Transactions on Parallel and Distributed Systems. 2009;**20**(10):1408-1421

[9] Liu Z, Dai L, Xue L, Guan X, Hua C. Reliability considered routing protocol in wireless sensor networks. In: Proceedings of the 30th Chinese Control Conference; 2011. pp. 5011-5016

[10] Mostafaei H. Energy-efficient algorithm for reliable routing of wireless sensor networks. IEEE Transactions on Industrial Electronics. 2019;**66**(7): 5567-5575

[11] Zorzi M, Rao RR. Geographic random forwarding (geraf) for ad hoc and sensor networks: Energy and latency performance. IEEE Transactions on Mobile Computing. 2003;**2**(4): 349-365

[12] Karp B, Kung HT. GPSR: Greedy perimeter stateless routing for wireless networks. In: Proceedings of the 6th Annual International Conference on Mobile Computing and Networking (MobiCom '00); ACM: New York, NY, USA; 2000. pp. 243-254

[13] Liu J, Sheng M, Xu Y, Li J, Jiang X. End-to-end delay modeling in buffer-limited manets: A general theoretical framework. IEEE Transactions on Wireless Communications. 2016;**15**(1): 498-511

[14] Yang B, Chen Y, Cai Y, Jiang X. Packet delivery ratio/cost in manets with erasure coding and packet replication. IEEE Transactions on Vehicular Technology. 2015;**64**(5): 2062-2070

[15] Niu J, Cheng L, Gu Y, Shu L, Das SK. R3e: Reliable reactive routing enhancement for wireless sensor networks. IEEE Transactions on Industrial Informatics. 2014;**10**(1): 784-794

[16] Zeng K, Lou W, Yang J, Brown DR III. On throughput efficiency of geographic opportunistic routing in multihop wireless networks. Mobile Networks and Applications. 2007;**12**(5): 347-357

Chapter 7

Living Function-Resilient Society in the Centenarian Era: Living Safety Technology Based on Connective, Artificial Intelligence

Yoshifumi Nishida and Koji Kitamura

Abstract

In the centenarian era, it has become even more imperative to address the physical and cognitive changes faced by children, the elderly, and disabled persons. We need a "living function resilient society," which ensures they enjoy safe living environments in ways that allow them to maintain active social participation levels despite the changes. To build such a society, more attention should be paid to problems: diversity in life function and intervention needs, gap between efficacy and effectiveness, fragmentation of living data and support service and variety in privacy exposure. To deal with these issues, this chapter describes a new approach referred to as "connective AI" that allows individual lives to be connected with each other and efficacy to be scaled to effectiveness by computerizing places of living in accordance with the private policy of individual facilities and connecting them with each other through a network. As a concrete example of connective AI, this chapter introduces smart living labs that are developed by the National Institute of Advanced Industrial Science and Technology (AIST) in cooperation with children's hospitals, rehabilitation hospitals, intensive care homes for the elderly, and private homes.

Keywords: living function, health monitoring, living laboratory, handrail, wearable device, RGBD camera, social participation, nursing care support

1. Introduction

A society that fosters resilience to changes in living function (a "living function-resilient society") is required now. We need social and industrial systems that ensure safety and high-level community involvement when we experience changes in cognitive function, physical function, or family function as we move toward an aging society (a dementia society). **Figure 1** shows the current age distribution of the Japanese population and that of accident rates, birth, and caregiving. Rapid, mental, and physical developments occur during infancy. Women experience significant changes during pregnancy and around birth. There are times when they or their child or parent needs nursing care. Changes in their physical and mental functions or in being able to care for someone in

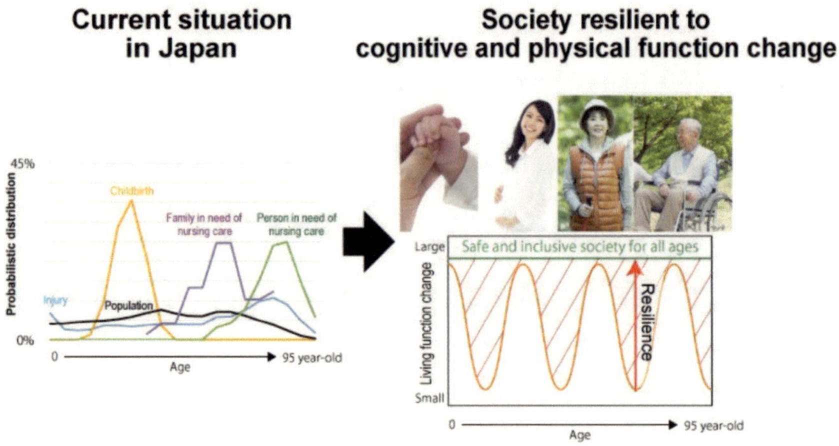

Figure 1.
Changes in living function and a living function-resilient society.

the family occur rather frequently in the family context. When viewed in terms of changes in living function, this means that a society will emerge in which we have to address changes in living function over generations, from a parenting generation that deals with development in children to a caregiver generation that deals with the declining living function of the elderly. The UN's 2030 Agenda for Sustainable Development, adopted in 2015, indicates the need to ensure the safety of people of all ages and the physically and intellectually challenged, ensure access to services, and implement urban design with consideration of safety and accessibility [1].

Figure 1 shows a snapshot of the current Japanese society. Significant changes will occur in a short period of 10 years or so: Japan's aging population will continue to grow until 2025 as the baby boomer generation grows old. As a result, our society will have a rapidly growing population that will need support in daily living. Compared to 2015, the elderly people aged 75 or over will increase by 5,000,000, and the number of dementia patients will increase by 2,000,000. China, as a driving force in the global economy, is also faced with an aging population: the proportion of its population aged 65 and over will be 16% by 2030. So China is likely to have the same issues as Japan, incurring enormous social costs. Today, Japan's social costs associated with dementia are huge, at US$ 127 billion. The costs are projected to reach US$ 181 billion in 2025 in Japan and US$ 2 trillion in 2030 globally. Thus, the issue of how society can develop adaptability to changes in living function will be increasingly important and is a key issue leading to a new growth strategy to develop adaptability (resilience) to changes in living function as a social mechanism separate from the issue of individual efforts.

In recent years, low-cost sensors, storage devices, and cloud computing services have become widely available. Artificial intelligence (AI) based on big data is rapidly advancing. These developments will make it possible to build a society that fosters resilience to changes in living function (a "living function-resilient society") that can adapt flexibly to changes in diverse physical and cognitive functions of children, women, the elderly, and the physically and intellectually challenged and allow people to exploit their potential to the fullest. There is an expectation that in the next 10 years, a new industry that helps translate the need for living function resilience into innovation will grow significantly.

This report identifies issues in building living function resilience and discusses, on the basis of our research, the potential for AI and the Internet of things (IoT) in building a living function-resilient society.

2. Issues in building a living function-resilient society

2.1 Variety in living function and intervention needs (one-size-fits-all issue)

The variety in living function needs to be acknowledged. A one-size-fits-all or universal intervention strategy is not necessarily effective: we need a mixed strategy combining universal, selective, and individual strategies [2]. We must have a basic understanding of what issues are involved in designing interventions based on a mixed strategy and where they arise. However, we do not fully understand them. We need to reveal the whole picture of issues associated with changes in the living function of the elderly and to base intervention design (precision intervention [2]) on that picture. To this end, we must have a system that collects data on changes of living functions and issues stemmed from the changes.

In the area of nursing care, providing one-to-one care services using human labor is considered ideal. In reality, it is difficult to provide services in terms of social costs. On the other hand, a universal strategy does not allow the variety in living function to be addressed. Precision care, which is defined as intermediate between universal and one-to-one strategies, is important. The use of AI technology may make it possible to align the variety in living function with individual adaptability by properly dividing the variety in living function into segments and selecting services that match each segment.

2.2 Gap between efficacy and effectiveness

Various intervention approaches have been proposed. An approach that is shown to have efficacy under laboratory conditions or under specific circumstances may have no effectiveness in other situations, such as the local community or society more broadly [3]. This is an issue of a lack of understanding of a local community or an individual life as a complex system and of error arising from the incorporation of a simplified model used in laboratory research into a complex system [4]. It is suggested that researchers act to hear user requests with a serious mind, for example, by conducting a complete interview survey. The issue is often discussed in terms of attitude or mindset in conducting research. However, we think that this discussion misses the point and that it is not a matter of mindset and effort but is a scientific issue arising from lack of science and technology and methodology for dealing with complex systems in field settings. Recently, it has been pointed out that the incremental approach (incrementalism) has a limit and design should be implemented in a manner that allows scaling (big change). We need to put in place in individual lives and facilities a system to evaluate the effectiveness of interventions and make continuous improvements. We need a method of designing and evaluating effective interventions in complex life systems that are present in individual lives and in facilities and in communities.

2.3 Fragmented living data and support services

Data on life are fragmented by facility and life situations. For example, data on illness, data on living function change, data on daily activities, and data on accidents/incidents exist in different facilities. It is necessary to use the data in an integrated manner and thereby to evaluate the variety in living function among children and elderly people, identify issues associated with the variety, and evaluate potential solutions. More importantly, we need a method of detecting their changes in daily life.

In addition to data fragmentation, services as solutions are fragmented. A collective impact model has been proposed to achieve effectiveness by collectively using stakeholders and social resources with a common purpose [5]. There is a need for a system for implementing the collective impact model based on the data.

2.4 Variety in privacy exposure

The definition of personal information has been changing. Besides that, there are a variety of ideas about privacy exposure. This means that there is a "one-size-fits-all" issue also in privacy policy; we need a system and technology to control information according to the variety of ideas of individuals and facilities about privacy, instead of developing a privacy policy common to all people and facilities. For example, there are facilities that are positive about installing cameras to prevent abuse of the elderly.

3. Problem solving by connective AI

3.1 Description of connective AI

To address issues in the variety in living function, intervention needs, data fragmentation, and variety in privacy exposure, we believe in the importance of an approach referred to as "connective AI" that allows individual lives to be connected with each other and efficacy to be scaled to effectiveness by computerizing places of living in accordance with the private policy of individual facilities and connecting them with each other through a network. We believe that connective AI is essential to building a living function-resilient society.

While our lives and living environments have individuality and are different from each other, there are many similar phenomena and environments. Skillful processing of information should make it possible to share information and convert it into knowledge. As pointed out by Herbert Simon, a physical phenomenon is essentially nonlinear when viewed hierarchically. We can develop a science based on the assumption that a physical phenomenon in the target layer can be modeled by associating it with feature quantities in the sublayers.

3.2 Smart living lab coevolving with connective AI

As **Figure 2** shows, to work with connective AI in concrete terms, we at the National Institute of Advanced Industrial Science and Technology (AIST) have developed a smart living lab in cooperation with children's hospitals, rehabilitation hospitals, intensive care homes for the elderly, and private homes. The term "smart living lab" here means (1) a place where we identify needs in field settings with user participation and adaptively explore whether new proposals to meet the needs are acceptable to the users in these "living labs" and (2) a place where we collect data, using AI and sensors, on activities of daily living of users (including children as non-main users) with a variety in living function (a smart field).

A system that allows daily life data fragmented in these places to be shared is essential to understanding the variety in living function. We need a new approach for information processing (AI for reality) to clarify real conditions. A system to

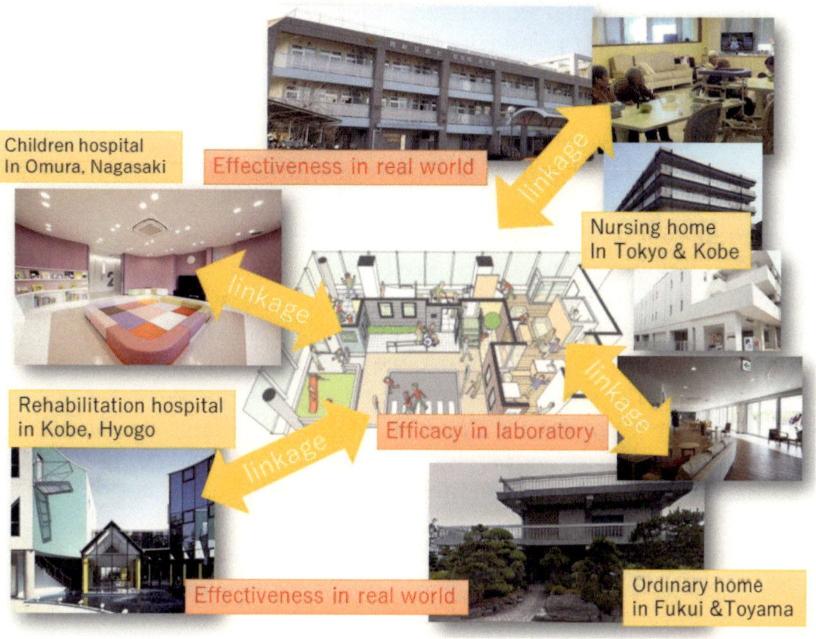

Figure 2.
Smart living lab developed by AIST.

link the verification of efficacy at the laboratory level to verification of effectiveness in facilities and communities is essential to developing solutions to support daily life. There is a strong need for a direction (reality for AI) for support technology to be put in place. We need to develop a living ecosystem suitable for individuals by combining stakeholders involved and social resources. We need to understand a field system as a complex system from a system science perspective and develop solutions that scale to different field settings. The next section describes connective AI technology being developed at AIST to build a living function-resilient society and smart living lab activities using the technology.

4. Attempt to demonstrate safety in daily life using AI and IoT technology

4.1 Raising awareness using data scattered across multiple organizations in an integrated manner (issue identification)

A recent text mining technique allows us to process quantities of data that are too big for humans to process. It has the potential for social function, which can be referred to as awareness (issue identification). Using data on medical treatment costs and situations resulting in injury from the Japan Sport Council, we at AIST are developing a technique that automatically analyzes situations that may result in severe injury. This technique identifies phrases unique to situations resulting in severe injury from free descriptive text, based on the assumption that treatment costs increase with increasing severity of injury. It is called "severity cliff analysis [6]." As shown in **Figure 3**, when similar situations are plotted in descending order of treatment costs, a cliff appears where the treatment costs change sharply. The technique allows us to identify the inflection point of the cliff and analyze what causes the severity of injury to increase.

Using this new technique, we analyzed injuries in school environments. Among injuries that involve running and tripping, the severity of injury is higher for hurdle running than for rope jumping and running on flat ground, because it involves hurdles. We investigated measures to prevent hurdle injury and found a hurdle with a top bar that opens like a double door when struck by the foot, which is in use at high schools in Miyagi Prefecture. Thus, severe injuries can be significantly reduced by taking effective preventive measures like this.

The technique allows us to understand in detail situations that result in severe injury by compiling incident data scattered across multiple organizations into big data for analysis by AI. Consequently, we can develop new prevention measures and associate them with existing measures. With data available only at some organizations (some schools, care facilities, etc.), we cannot know the overall occurrence and extent of severe injuries. As a result, known preventive measures remain isolated, and their widespread introduction is delayed. A new approach to improving situations in real-life settings by identifying problems and connecting them with solutions will be increasingly important in the future.

4.2 Monitoring changes in living function using IoT technology

Elderly people typically lose cognitive and motor functions with age and experience increasing challenges in daily life. There is a need for IoT sensors to detect changes in the living function of individual elderly people as they occur and to call for appropriate interventions. A recent projection for the next 10 years holds that smart homes will provide a market for sensors to support not only home security but also healthcare and safety in the home.

We developed a sensor to make it possible to measure how fast the elderly can walk and how well they can walk unaided. The sensor is designed to be built into an object used in daily life, in this case a handrail [7]. It collects only relevant data (maintaining privacy) and does not need to be attached and detached. We verified the basic functions of the sensor in the living lab at AIST and then installed it in the home of an 88-year-old woman who lives alone. Our study will verify the efficacy of the sensor through long-term monitoring.

Figure 4 shows how the sensor works and its installation. The sensor comprises two strain gauges fixed above and below a steel bracket secured to the wall. When the subject puts her hand on the handrail, the downward load is detected by the strain gauges.

We conducted a verification test of the sensor in a real-world setting. We installed several sensors on a handrail in the hallway in the subject's house (**Figure 4**, right). **Figure 5** (left) shows a sequence of images of the subject walking while holding the handrail. We collected data continuously for 24 months and plotted the subject's walking speed by using the installation's position-estimation capability (**Figure 5**, right).

We plotted the monthly median walking speed to reveal any changes in the walking pace from January 2016 to December 2017. As **Figure 6** shows, it changed substantially over the period: it decreased from February to August as physical strength declined, increased again from September to November, but declined again from January to March. The subject told us that she initially lost physical strength but regained it from September, but knee pain caused increasing difficulty in walking from January 2017. In May 2017, she broke her thighbone and was admitted to a hospital. In August 2017, she discharged from the hospital. Our results show that the sensor can detect some problems in daily life, although not the cause.

The walking pace of the elderly decreases with advancing age, along with walking patterns such as stride length, walking pace, and lower limb muscle strength.

Such declines increase the need for in-home support services for people who retain a strong need to remain in their own home. Low-cost monitoring of health and mobility would allow quick identification of risks to safety such as by falls. Such monitoring of individual elderly people would allow the timely implementation of appropriate interventions as a form of precision care or individualized care. Continued advances in AI and IoT will support this.

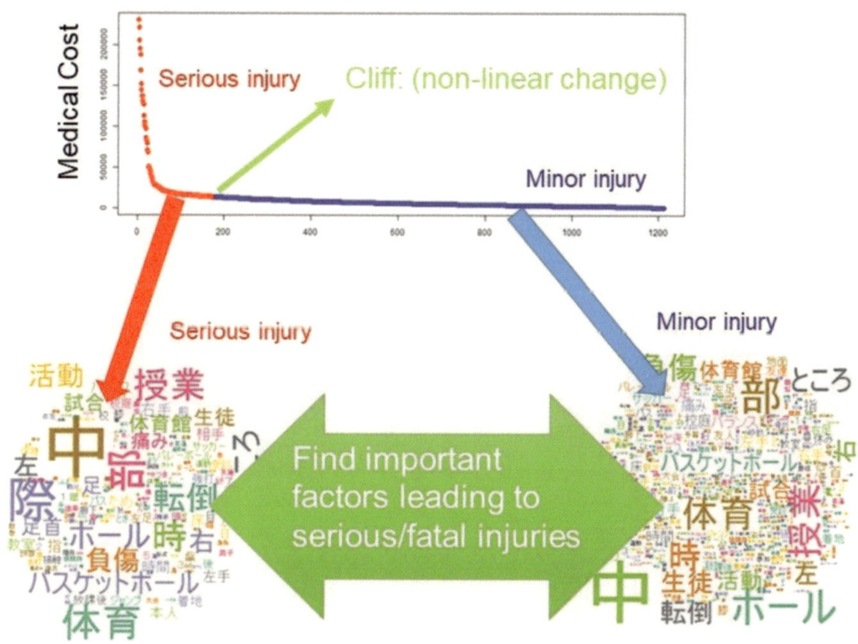

Figure 3.
Severity cliff analysis, using big data, to identify factors involved in situations resulting in severe injury.

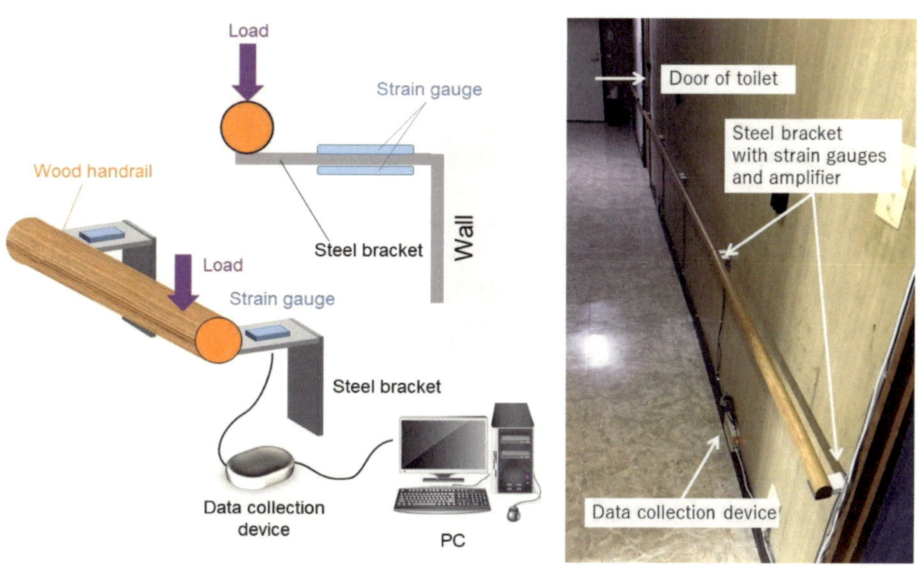

Figure 4.
Handrail-type IoT sensor (left, configuration; right, picture).

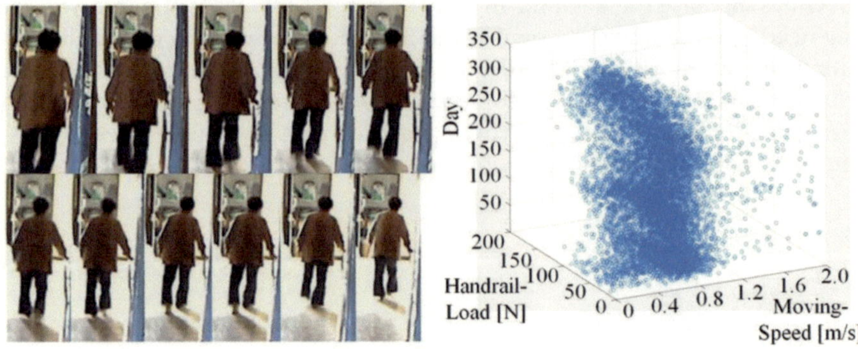

Figure 5.
The subject using the handrail (left) and motion data obtained (right).

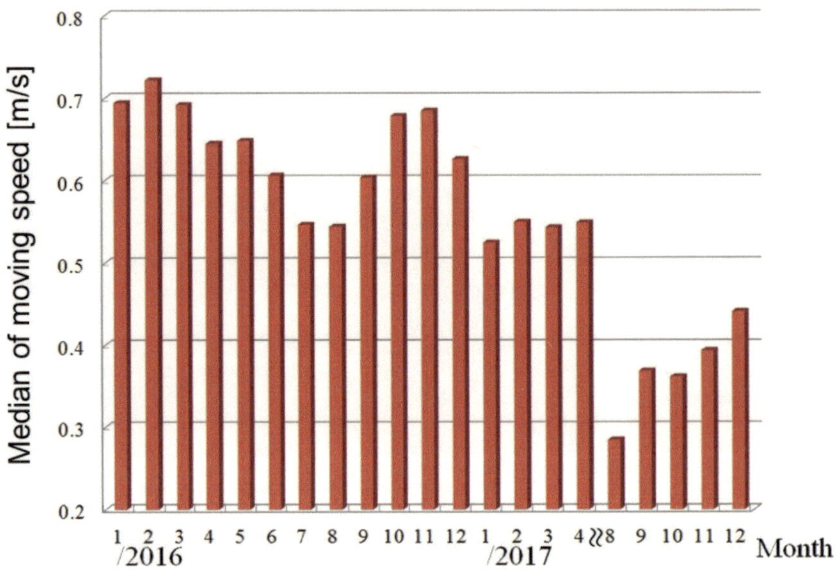

Figure 6.
Results of 15-month monitoring of walking pace as a health indicator.

4.3 Supporting community involvement of people with different living functions by using a daily life database

If we can detect changes in walking pace and other changes in daily life, what is the best way to use this information? One way is to provide services that support community involvement according to changes in living function.

To understand the living conditions and living function of the elderly, we interviewed elderly participants at home and collected data on their living conditions as a technological element for providing tailored support for community involvement. We developed a system to describe daily life data in terms of relationships between elements such as community involvement, experience, emotions, people, things, and activities related to community involvement [8]. We included elements and experiences used to describe daily life in the International Classification of Functioning, Disability and Health. **Figure** 7 shows an example of graphically represented life data of an elderly person at one time point. The plot represents the overall life structure. Such graphic representation allows an understanding of the entire life structure, including the relationships between individual elements, the use of graphic structure analysis, and numerical representation of the degree of similarity between individual graphic structures and searching on life data. Using this method, we have developed a life database of more than 70 elderly people.

The use of graphic representation of daily life allows calculation of the degree of similarity between individual graph structures and identification of those who have a similar life structure. **Figure 8** shows the life structures of 20 elderly people and plots the degree of similarity between them. It reveals groups concentrated at the top left corner of the graph, where the life structures have a degree of similarity, along with one person at the right edge (G15) and one at the bottom edge (G10), both substantially different from the others. The placement of G15 indicates that that person mostly feels happy about community involvement but sometimes feels sad, angry, or worried about it. The placement of G10 indicates that the person has mostly negative feelings such as loneliness, sadness, and anger. Such graphical representation allows us to identify elderly people with a different life structure from the majority who might therefore require interventions to support them in changing their living conditions.

For example, when we design an intervention to improve living conditions, by using a life structure distance space (or life structure manifold), we can identify people with similar life structures and encourage them to become involved in community activities, instead of putting together people with very different life structures. If a person's life structure later changes, we can again encourage community involvement with people with more similar life structures. This is a scientific approach to changing people's life structure step-by-step to bring it closer to what they want or by revising goals. We think that this approach will lead to a data-based scientific approach (life design methodology) to process design to achieve a desirable life structure.

Figure 9 shows the output of software that has read the life structure data of an 80-year-old woman and has searched for elderly persons with similar life structures

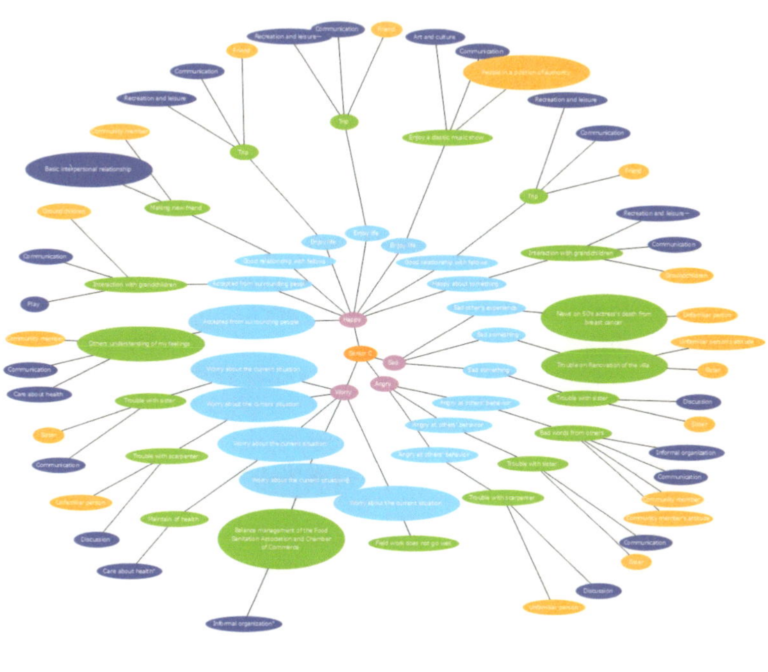

Figure 7.
Example of graphical representation of life structure data.

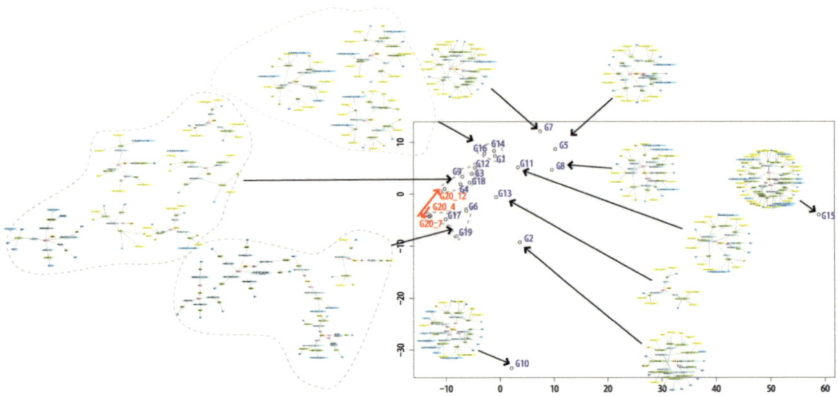

Figure 8.
Visualization of life structure patterns (life structure distance space or life structure manifold).

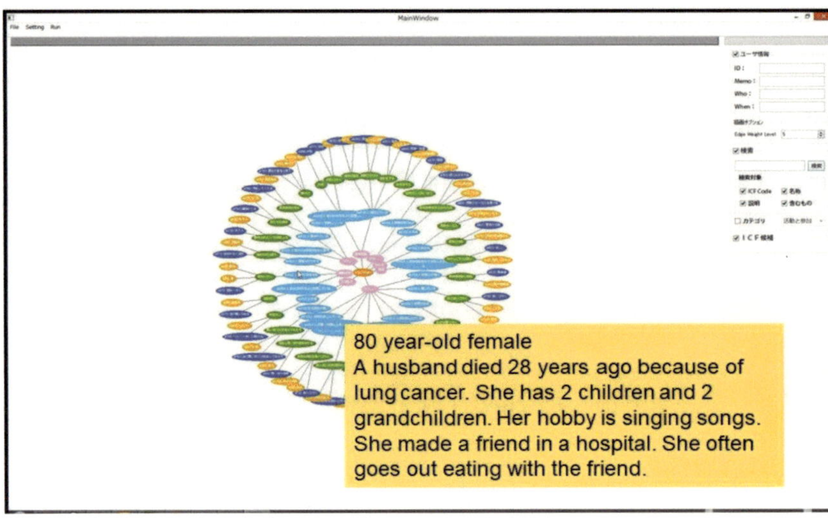

Figure 9.
Software to support life design based on an enormous amount of life data and life geometric operations (digital crystal ball).

and for things that make her happy. **Figure 10** shows a group of elderly people mapping the locations of little-known community involvement events. Currently, we are working with a community association and a local elderly care management center to provide the participants with advanced support, tailored to their individual living conditions, in community involvement, by combining life design support technology and local maps and making good use of resources available in the local community.

4.4 Trial of precision care at IoT-based care facilities

In collaboration with care homes, we are undertaking a project to develop monitoring technology, tailored to individual elderly persons, to prevent accidents and detect early changes in behavior in anticipation of the time when an elderly person with declining living function needs care or support. **Figure 11** shows a system to measure the location of an elderly person with dementia and monitor his

Figure 10.
Working with a local elderly care management center and a community association to create a map to support community involvement.

or her behavior, using a beacon embedded in the sole of the person's shoe [9]. Using sensors like this, we can monitor changes in walking patterns. **Figure 12** shows a case in which monitoring over 45 days revealed a change in the walking pattern of an elderly person with dementia: the distance walked decreased greatly about halfway through the monitoring period. Later, we found that the decrease was due to a broken bone caused by a fall. This case shows how the use of sensors allows us to quantify changes in individual persons' behaviors and to accurately detect changes that can be missed.

Both wearable sensors and smartphones can collect information on individuals, but they use battery power. This is a major hindrance, because devices that require frequent battery changes are not acceptable in real-life settings. At the same time, the use of AI technology not just to find people but also to identify them has made tremendous improvements. Such "non-wearable" has begun to appear. The combination of mounted RGBD cameras and face identification software can allow unintrusive long-term monitoring of individuals, as the monitors are not worn [10]. Some facilities have started to use it. **Figures 13** and **14** show a RGBD camera and a plot of a person's walking posture captured by RGBD camera. This person's walking pace tended to be slow in the morning and to vary greatly.

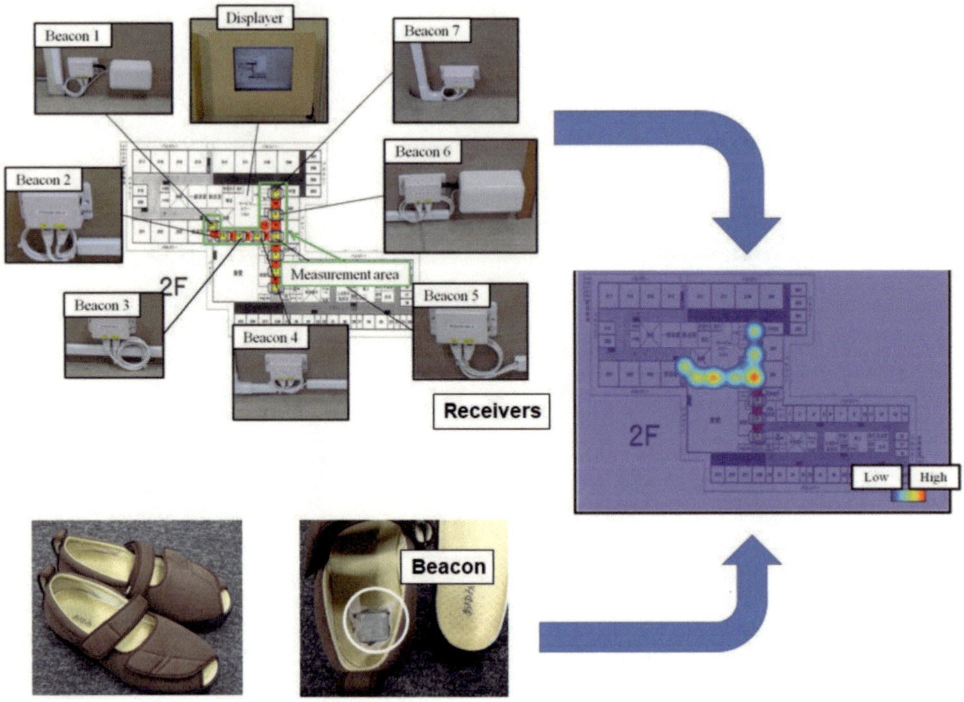

Figure 11.
Shoe-embedded location sensor for monitoring of the elderly.

The facility staff made the following comments on individualized monitoring:

1. The video of events that may not be accurately communicated by humans is recorded. This allows information on events to be shared accurately (video can be used, e.g., when passing information onto another staff member).

2. By watching the video of actual positioning of things and people at a care facility, instead of reading textbooks, staff awareness is raised, and crisis management training can be improved.

3. By watching the video, staff can know what happens when they are not on hand and can take action immediately.

4. Staff can monitor daily changes in the walking pace of elderly people and can associate the changes with medications, mental state (dementia), excretion, and pain. Many medications, notably sleeping pills, can cause falls. However, support tools for personal health management have not been available.

5. By associating the profile of an elderly person with risks, staff can know at a glance what risks the person faces, group people with similar needs for better management, and provide better care.

6. Being able to identify daily changes in individuals and their long-term trend, staff can determine the need for intervention and evaluate the effectiveness of intervention. The monitoring function can also be used as tool for nursing care.

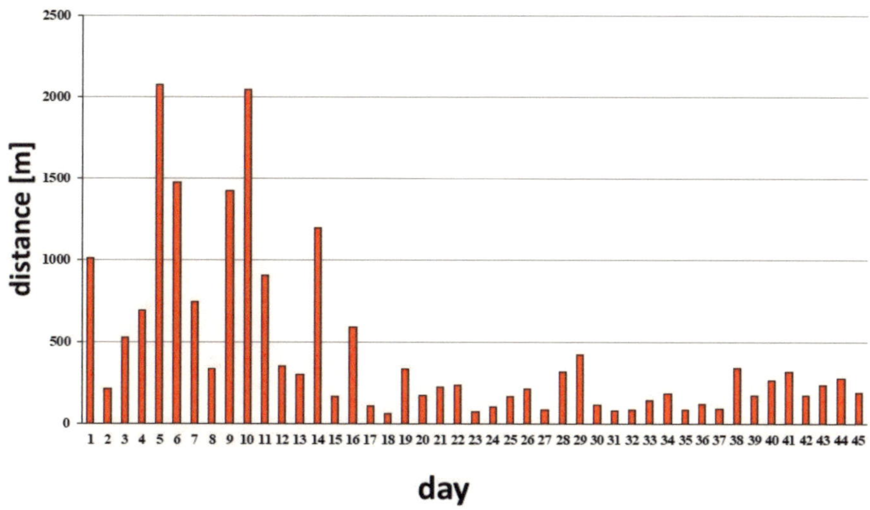

Figure 12.
Results of long-term monitoring with a shoe-embedded location sensor.

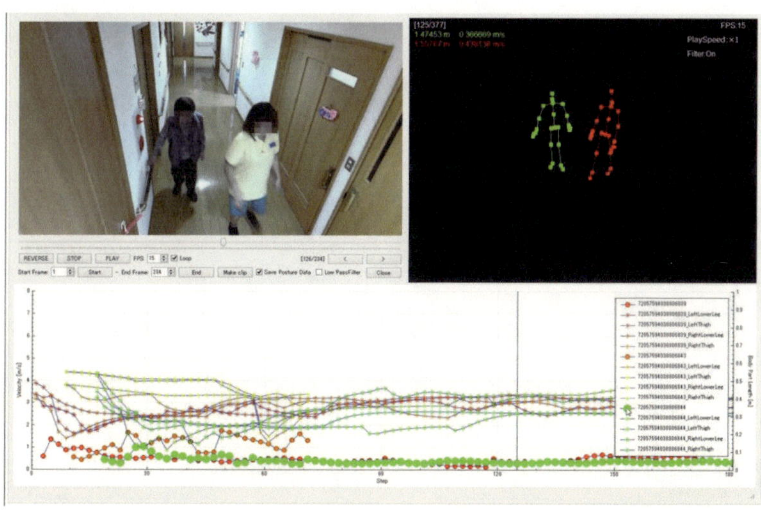

Figure 13.
Camera image (left) and measurement of walking posture (right).

Amid increasing reports of elder abuse, more facilities and users favor the use of sensors. Staff alone will not be able to monitor residents in the level of detail that this will entail. While acknowledging the need for privacy, we need to identify what services can be made possible by what sensing technology (with attendant risks to privacy). It is important to provide levels of services that suit users' needs best by preparing a variety of options for such services.

4.5 Elderly behavior library for searching for product usage by those with changing living function

Changes in living function vary among the elderly. Unlike in the case of children, this makes it difficult to classify events by age, because living function varies significantly among people of the same age. We can specify "a bed for babies up to

Figure 14.
Plot of walking pace monitored by mounted camera in a home.

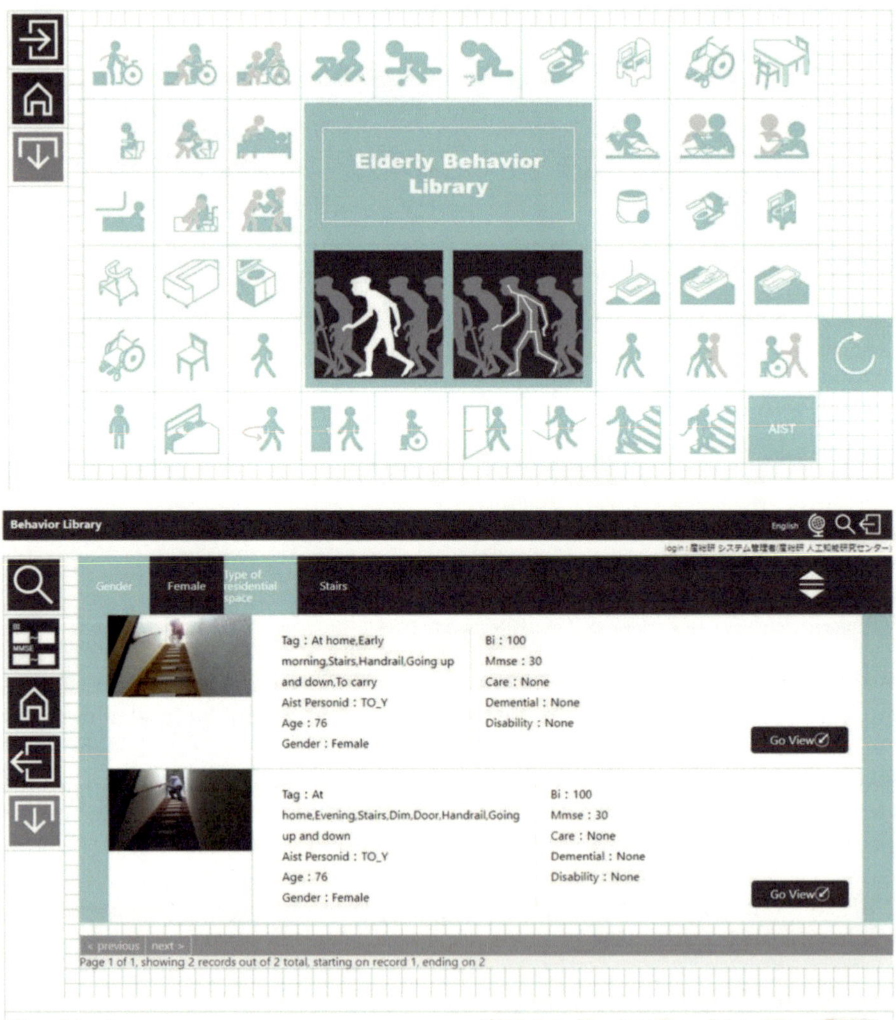

Figure 15.
Searchable library of elderly behaviors for identifying changes in living function.

24 month old," but not "a chair for persons 75 years old and older". As a means to understanding life dimensions in the elderly, we created a library of videos showing how different products (beds, chairs, wheelchairs, canes, doors, kitchen tools, handrails) are used by the elderly, depending not only on age but also on cognitive ability, physical ability, and level of care needed [11]. **Figure 15** shows snapshots of the developed library. The library was launched in March 2018. Its use requires registration. The library is a pioneering attempt to show how elderly use items, such as to identify handrails that are easy to use. We intend that it will be used to identify problems and develop solutions. If you are interested, please contact us at the address provided on the website [12].

5. Conclusions

In this report, based on our research, we describe "a living function-resilient society" as a desirable society and show the potential of AI and IoT technology to identify problems and resolve them. In terms of intelligence, to resolve issues associated with the variety in living function, we need to find innovative solutions using the variety in intelligence. Today, a society is emerging in which problems, data, and intelligence are ubiquitous. Sensing and recording technology is advancing and spreading throughout the society. Various organizations now store big data. In recent years, AI, such as data analysis technology, has been advancing. Human resources spread among universities, administrative organizations, care homes, regions, and companies can be linked to create an intelligence-ubiquitous society. New social issues emerge constantly. The type of innovation required today is the transformation of the society into an advanced interconnected society by using the ubiquity of data and intelligence in modern society to address newly emerging issues.

Author details

Yoshifumi Nishida* and Koji Kitamura
Artificial Intelligence Research Center, National Institute of Advanced Industrial Science and Technology, Tokyo, Japan

*Address all correspondence to: y.nishida@aist.go.jp

References

[1] United Nations Information Centre. Transforming Our World: The 2030 Agenda for Sustainable Development. 2015

[2] Winston FK et al. Precision prevention. Injury Prevention. 2016;**22**(2):87-91

[3] Green LW. From research to "best practices" in other settings and populations. American Journal of Health Behavior. 2001;**35**:165-178

[4] Hanson D, Allegrante JP, Sleet DA, Finch CF. Research alone is not sufficient to prevent sports injury. British Journal of Sports Medicine. 2012;**48**(8):682-684

[5] Kania J, Kramer M. Collective impact. Stanford Social Innovation Review. 2011;**9**(1):36-41

[6] Imai K, Kitamura K, Nishida Y, Yoneyama N, Takemura H, Nakayama T. Analysis of big data on sport injuries using a severity cliff analysis technique. 17th SICE System Integration Workshop. 2016;**2016**:2903-2907

[7] Takahashi Y, Nishida Y, Kitamura K, Mizoguchi H. Handrail IoT sensor for precision healthcare of elderly people in smart homes. In: Proc. of the IEEE 5th International Symposium on Robotics and Intelligent Sensors (IEEE IRIS 2017). 2017

[8] Nishida Y. Scientifically design living: Configuration of life function using a life structure database. Information Processing. 2013;**54**(8):772-778

[9] Nishimura T, Kitamura K, Nishida Y, Mizoguchi H. Development of a nursing care support system that seamlessly monitors both bedside and indoor location. In: Proceedings of the 6th International Conference on Applied Human Factors and Ergonomics. 2015. pp. 3808-3815

[10] Murata E, Kitamura K, Oono M, Shirato Y, Nishida Y. Behavior monitoring with non-wearable sensors for precision nursing. In: Proceedings of the 8th International Conference on Applied Human Factors and Ergonomics. 2017. pp. 384-392

[11] National Institute of Advanced Industrial Science and Technology. Report on the Project to Obtain Behavior Data of the Elderly Toward a Vintage Society. 2018

[12] https://www.airc.aist.go.jp/lirt/vintagelibrary.html (National Institute of Advanced Industrial Science and Technology, Elderly Behavior Library)

Supporting open minds since 2005

Antimicrobials, Antibiotic Resistance, Antibiofilm Strategies and Activity Methods
http://dx.doi.org/10.5772/intechopen.78751
Edited by Sahra Kırmusaoğlu

Contributors
Muhammad Usman Qamar, Muhammad Hidayat Rasool, Shah Jahan, Muhammad Shafique, Bilal Aslam, Laura Castrillon Rivera, Alejandro Palma Ramos, Jorge Ismael Castañeda-Sánchez, Maria Elisa Drago-Serrano, Eng Hwa Wong, Mansab Ali Saleemi, Navindra Kumari Palanisamy, Christian Agyare, Cynthia Amaning Danquah, Yaw Duah Boakye, Francis Adu, Newman Osafo, Angel Manteca, Paula Yagüe, Sahra Kırmusaoğlu

First published in London, United Kingdom, 2019 by IntechOpen
IntechOpen is the global imprint of INTECHOPEN LIMITED, registered in England and Wales, registration number: 11086078, The Shard, 25th floor, 32 London Bridge Street
London, SE19SG - United Kingdom
Printed in Croatia

British Library Cataloguing-in-Publication Data
A catalogue record for this book is available from the British Library

Additional hard copies can be obtained from orders@intechopen.com

Antimicrobials, Antibiotic Resistance, Antibiofilm Strategies and Activity Methods
Edited by Sahra Kırmusaoğlu
p. cm.
Print ISBN 978-1-78985-789-4
Online ISBN 978-1-78985-790-0

Meet the editor

Dr. Kırmusaoğlu, PhD, is an Assistant Professor of microbiology at the Department of Molecular Biology and Genetics, T.C. Haliç University, where she had previously held an administrative position as vice head of the same department. She specialized in microbiology at Abant Izzet Baysal University (Biology Department), Turkey. Her previous experience includes the position of laboratory manager at microbiology laboratories in several research and private hospitals. Throughout her career, she collaborated with academicians/researchers from AIBU, METU and Istanbul University Cerrahpaşa Faculty of Medicine, and participated in various research projects. She also worked as a lecturer in the Medical Laboratory Techniques Program held at the Department of Medical Services and Techniques, T.C. Istanbul Kavram Vocational Training School, before transitioning to T.C. Haliç University as an Assist.Prof. Dr. Kırmusaoğlu's research interests include pathogenic bacteria, bacterial biofilms, antibiofilm and antimicrobial agents, antibacterial resistance, synergistic activities of antimicrobials in combination with new agents such as chemical and natural bioactive compounds against pathogenic bacteria such as antibiotic-resistant and biofilm-producing microorganisms, regulation of antibiotic resistance, and virulence genes. She is also interested in research on antibiofilm medical devices for preventing biofilm infections associated with indwelling devices that could lead to systemic infections. She has published several international research articles, book chapters, and congress proceedings. More specifically, Dr. Kırmusaoğlu participated as an author in several books, e.g. *Microbial Biofilms: Importance and Applications and The Rise of Virulence and Antibiotic Resistance in Staphylococcus aureus* with her chapters "Staphylococcal biofilms: pathogenicity, mechanism and regulation of biofilm formation by quorum sensing system and antibiotic resistance mechanisms of biofilm embedded microorganisms" and "The mechanism of methicillin resistance and the influence of methicillin resistance on biofilm phenotype of Staphylococcus aureus." The chapter entitled "The methods for detection of biofilm and screening antibiofilm activity of agents" was written by Dr. Kırmusaoğlu. She is also the editor of *Disinfection* and *Bacterial Pathogenesis and Antibacterial Control*. In addition to these, she wrote a book *Genel Biyoloji Laboratuvar Kılavuzu (General Biology Laboratory Manual)* published by Hipokrat Publisher. She has been also contributing to an ongoing book project titled *Staphylococcus and Streptococcus* as an editor, and has contributed to a chapter translation of the book *Sherris Medical Microbiology by Ryan et al.* as one of the translation authors of *Sherris Tıbbi Mikrobiyoloji*, which is a Turkish translated book edited by Prof. Dr. Dürdal Us and Prof. Dr. Ahmet Başustaoğlu.

Contents

Preface

To prevent bacterial adherence caused by virulence factors such as biofilms, invasion, and infection, newly developed antimicrobials in combination with certain agents or bioactive compounds present in natural products, such as pharmacological nutrients and plant extracts as well as certain novel technologies, are being used and vastly researched nowadays.

Several factors such as natural selection, mutations in genes, the presence of efflux pumps, impermeability of the cell wall, structural changes in enzymes and receptors, biofilm formation, and quorum sensing cause microorganisms to develop resistance against antimicrobials. Multidrug-resistant pathogens that are the main causes of morbidity and mortality in the worldwide have the ability to synthesize a number of enzymes such as extended spectrum-β-lactamases, induced β-lactamases, carbapenamases, metallo-β-lactamases, New Delhi metallo-β-lactamases, and many others. Due to the wide range of antimicrobial resistance multidrug-resistant pathogens such as methicillin-resistant *Staphylococcus aureus* (MRSA), *Escherichia coli*, *Klebsiella pneumoniae*, *Enterobacter cloacae*, and *Pseudomonas aureginosa* have emerged throughout the worldwide, and mechanisms of resistance must be further investigated to treat untreatable infections caused by resistant pathogens. To overcome antimicrobial resistance, new strategies must be developed, and minimum doses of antimicrobials in combination with other agents such as synergistic compounds and nutrient extracts can be used.

Determining virulence factors such as biofilms and the level of antimicrobial activities of antimicrobial agents alone and in combination with appropriate doses against microorganisms is very important for the diagnosis, inhibition, and prevention of microbial infection, especially for recurrent ones.

This book contains seven chapters from valued experts in Turkey, Mexico, Spain, Netherlands, Pakistan, Malaysia, and Kumasi-Ghana. The goal of this book is to provide information on all the topics mentioned. It can be useful for researchers who are interested in the study of microbial responses against antimicrobials, and the development of new strategies for the treatment of infections caused by MDR and biofilm producer microorganisms.

I would like to thank all the authors who contributed to this book with their chapters. Especially, I would like to express my special thanks to Prof. Dr. Bekir Kocazeybek for his valued support. Finally, I would also like to thank Ms. Maja Bozicevic who assisted me in this project as an Author Service Manager and IntechOpen Publisher for their concern and encouragement in publishing this book.

Assist. Prof. Sahra Kırmusaoğlu
Haliç University, Faculty of Arts and Sciences,
Department of Molecular Biology and Genetics,
Beyoğlu-Istanbul, Turkey

Section 1

Antimicrobials and Antimicrobial Resistance

Chapter 1

Introductory Chapter: The Action Mechanisms of Antibiotics and Antibiotic Resistance

Sahra Kırmusaoğlu, Nesrin Gareayaghi and Bekir S. Kocazeybek

1. Antibiotics

Pathogenic microorganisms can infect tissues of human by destroying cellular functions. Microorganisms themselves or their toxins can damage host cells. Microbial infections are treated with antimicrobials by either inhibiting the microbial growth or killing the microorganism. Antibiotics are widely being used not only in the treatment of acute and chronic infections, but also in the prophylactic treatment [1]. Targets of antimicrobials are cell membrane, cell wall, protein synthesis, nucleic acid synthesis, and biological metabolic compound synthesis (**Figure 1**) [2]. Over usage of antibiotics, mutations in the genes, carrying resistance genes in chromosomes and plasmids, gaining resistance genes carried by transposons, insertion sequences (IS) and conjugation from the same or other species of microorganisms cause bacteria develop resistance to antimicrobials [3].

2. The action mechanisms of antibiotics and antibiotic resistance

2.1 Injury to cell membrane

Plasma membrane of microorganism that has selective permeability contributes active transport to gain energy as ATP. Cytoplasmic content and gradient such as micro and macromolecules and ions are controlled by active transportation via integral transporter proteins. When selective membrane permeability is disrupted by antimicrobials, ions are lost and cellular ion gradient is distorted, so, the organism undergoes cellular damage and death [4].

Plasma membranes of bacteria are constructed by fatty acids that can be synthesized in cell or taken from environment as building blocks. Targets of antimicrobials are metabolic steps of fatty acid synthesis and membrane phospholipids. Polymyxin B, that is a bactericidal antibiotic, has been used as one of very few drugs in the treatment of Gram-negative bacteria such as *Pseudomonas* [2]. Polymyxin B that has detergent-like peptides having lipophilic and hydrophilic groups disrupts phosphatidylethanolamine of membrane. Valinomycin, that is an ionophore, disrupts cellular membrane potential that contributes oxidative phosphorylation by forming pores in cellular membrane. Daphtomycin that is widely used in bloodstream, wound, and soft skin infections caused by β-lactam especially vancomycin-resistant *Staphylococcus aureus* disrupts membrane potential by depolarization, that means potassium ions are released from cytoplasm to extracellular matrix [4].

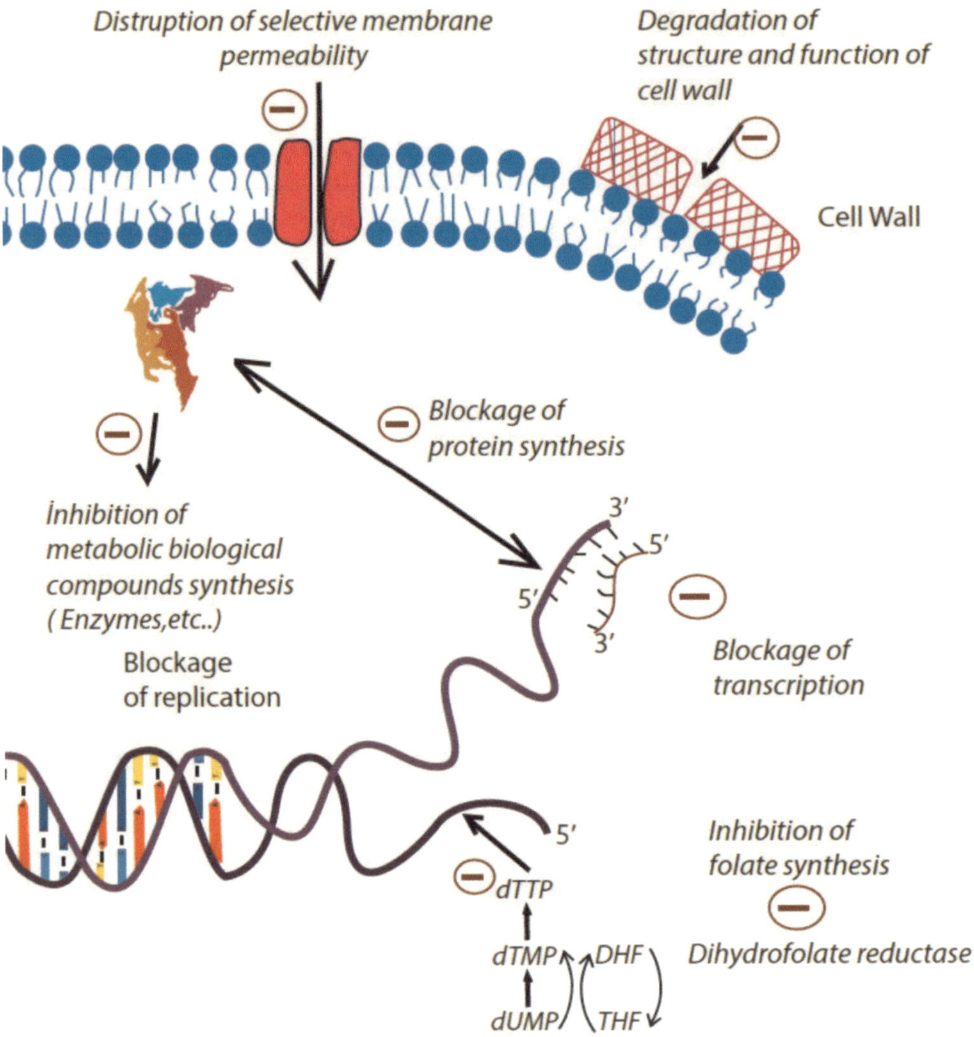

Figure 1.
Targets of antimicrobials.

Daptomycin, amphotericin B, colistin, imidazoles, and triazoles also act as inhibitors of cell membrane [2, 4].

2.2 The effect of antibiotics against cell wall

2.2.1 Cell wall synthesis

Cell walls of microorganisms are constructed by peptidoglycan. Glycan polysaccharide strands are linked by crosslink that bind polypeptides bound to N-acetyl muromic acid (NAM) of each polysaccharide strands.

After bactoprenol, which is a membrane-bound acceptor, transfers UDP-NAM-pentapeptide and UDP-NAG from cytoplasm to outer site of cell membrane, transglycosylation, and transpeptidation reactions are catalyzed by penicillin-binding proteins (PBPs) bound to cell membrane as a DD-peptidases to construct peptidoglycan [3].

2.2.2 Degradation of structure and function of cell wall

Certain antibiotics such as β-lactam antibiotics react with PBPs having high affinity to β-lactams by binding to PBPs as a substrate. These drugs are structural analogs of acyl-D-alanyl-D-alanine that binds to active site of PBP as a substrate of PBP during transpeptidation reaction. Transpeptidation reaction is blocked by these antibiotics inactivating transpeptidase domain of PBPs. Microorganisms are killed by these cell wall inhibitors that inhibit peptidoglycan biosynthesis [3].

2.2.3 β-lactamases and the mechanisms of β-lactam resistance

β-lactamase, that is an enzyme synthesized by many species of Gram-positive and Gram-negative bacteria, inactivates β-lactams degrading amide bond of β-lactam ring of β-lactam antibiotics [4] (bnm MRSA and MSSA). β-lactamases can be mediated by either plasmids or chromosomes, whereas penicillinases of *Staphylococcus aureus* are plasmid mediated, many Gram-negative bacteria are chromosomally mediated. β-lactamases that are plasmid mediated have tendency to be transferred between distinct species of bacteria. Chromosomally mediated β-lactamases can be either produced constructively as in the species of *Bacteroides* and *Acinetobacter* or induced as in the species of *Enterobacter*, *Citrobacter*, and *Pseudomonas.* Extended-spectrum β-lactamases (ESBLs) that are one class of β-lactamases having distinct ability to hydrolyze β-lactam rings of cefotaxime, ceftazidime, and aztreonam can be seen in a few species of Gram-negative bacteria such as *Klebsiella pneumoniae* and *Escherichia coli* [4].

β-lactamase is synthesized by *bla*Z gene and regulated by *bla*I and *bla*RI located in plasmid or transposon. When β-lactam is not used, by binding of BlaI synthesized by *bla*I gene to promoter-operator region of *bla*I-*bla*RI operon, β-lactamase is not transcriped by *bla*Z gene. But in the treatment with β-lactam antibiotics, active site of BlaRI which is a signal transducer integral protein of β-lactam is blocked by β-lactam. After this blockage of active site, intracellular zinc metalloprotease domain of BlaRI releases BlaI bound to *bla*I-*bla*RI operator. By upregulation of *bla*Z gene, β-lactamase enzyme is synthesized, so the microorganism develops resistance to β-lactam used [3].

2.3 Inhibition of metabolic biological compounds synthesis

Biological metabolic reactions are catabolized by enzymes that are activated by substrates. Synthesis of metabolic biological compounds can be inhibited by drugs as a competitive inhibition manner. Drugs that are structural analogs of substrates act as substrates for the enzymes used in metabolic reactions.

Para-aminobenzoic acid (PABA) is a substrate for folic acid synthesis that is a coenzyme in the reactions of purines, pyrimidine, and amino acids synthesis [2].

Sulfanilamide and 3,4,5-trimethoxybenzylpyrimidine are the examples of drugs inhibiting synthesis of metabolic biological compounds. Sulfonamides (Sulfa drugs) have been used in many infections such as urinary tract infections. Sulfonamide is widely used in combination with other compounds. Silver sulfadiazine, one of the combined drugs, is used in burn infections [2].

Trimethoprim sulfamethoxazole (TMP-SMZ) is another combined drug used widely because of its synergistic activity. Trimethoprim and sulfamethoxazole block distinct steps of DNA and RNA precursor synthesis, and protein. Sulfamethoxazole that is sulfonamides showing structurally analogy with PABA blocks the reaction

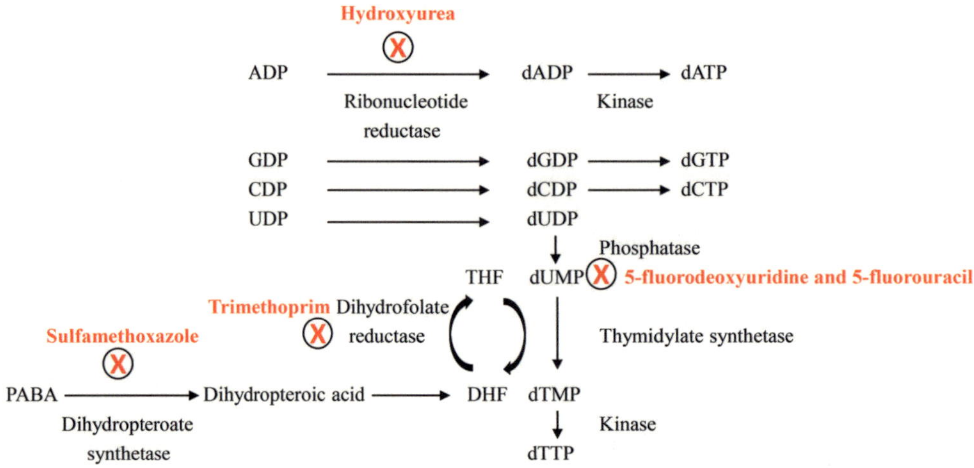

Figure 2.
The synthesis of deoxynucleoside triphosphate precursors (dNTP: dATP, dGTP, dCTP, and dTTP), and the inhibition of tetrahydrofolate (THF) and dNTP synthesis by antibiotics.

synthesizing dihydrofolic acid (DHF) from PABA, whereas trimethoprim that is sulfonamides showing structurally analogy with DHF blocks the reaction synthesizing tetrahydrofolic acid (THF) from DHF (**Figure 2**) [2].

2.4 Inhibition of nucleic acid synthesis

Antibiotics can inhibit replication, transcription, and folate synthesis of microorganisms.

2.4.1 Blockage of replication

2.4.1.1 Replication of DNA

Deoxyribonucleotide precursors are synthesized for the polymerization of deoxynucleotides. By kinase enzyme, deoxynucleoside triphosphates (dNTP: dGTP, dCTP, and dATP) are synthesized from deoxynucleoside diphosphates (dNDP: dGDP, dCDP and dADP) that are synthesized from ribonucleosides by ribonucleotide reductase. But deoxythymidine triphosphate (dTTP) is synthesized by different pathway. Deoxyuracil diphosphate (dUDP) that is synthesized from uracil diphosphate (UDP) by ribonucleotide reductase is converted to deoxyuracil monophosphate (dUMP). Thymidylate synthetase catalyzes a reaction that converts dUMP methylated by tetrahydrofolate (THF) to dTMP. In this step, tetrahydrofolate (THF) is synthesized from dihydrofolate (DHF) by dihydrofolate reductase. Finally, dTTP is synthesized from dTMP by kinase (**Figure 2**).

DNA gyrase opens DNA strands for the polymerization of deoxynucleotides by DNA polymerase according to the each circular template strand of chromosome [5].

2.4.1.2 Inhibition of replication and the mechanisms of resistance

Kinolons, such as nalidixic acid and ciprofloxacin that is used in the treatment of infections caused by *Pseudomonas* spp., prevent the formation of replication fork by inhibiting DNA gyrase, as a result of binding to *gyr*A subunit. Novobiocin and coumermycin prevent the formation of replication fork by inhibiting DNA gyrase, as a result of binding to *gyr*B subunit. When *gyr*A and *gyr*B genes of bacteria are mutated, bacteria develop resistant against these antibiotics [5]. Norfloxacin, gatifloxacin, gemifloxacin, and moxifloxacin are the other groups of floroquinolones having broader spectrum of activity. Floroquinolones are nontoxic. Ciprofloxacin, norfloxacin, gatifloxacin, and gemifloxacin are used to treat urinary tract infection and pneumonia [2].

Synthesis of deoxynucleotide precursors that are used in the replication of DNA can be blocked by trimethoprim, hydroxyurea, 5-fluorodeoxyuridine, and 5-fluorouracil.

Trimethoprim is an inhibitor of folate synthesis. Trimethoprim, that is a structural analog of DHF, prevents the synthesis of THF by inhibiting dihydrofolate reductase. So, dTMP that is a precursor of deoxynucleotide polymer is not synthesized. There are many mechanisms of trimethoprim resistance. If *thyA* gene of microorganism undergoes to mutation that inactivates thymidylate sentetaz dTMP and DHF are not synthesized, as a result of inability in the transfer of methyl group from THF to dUMP. Due to dihydrofolate reductase is not synthesized in the microorganism of which *thyA* gene is mutated, this microorganism is not inhibited by trimethoprim. If the microorganism changes the binding site of the dihydrofolate reductase, the trimethoprim cannot bind to dihydrofolate reductase. So, bacteria develop resistance to trimethoprim. Another mechanism of trimethoprim resistance is carrying gene causing resistance to trimethoprim [5].

Hydroxyurea inhibits ribonucleotide reductase that catalyzes deoxynucleoside diphosphate from ribonucleoside diphosphate (**Figure 2**). If microorganism of gene coding ribonucleotide reductase is mutated, microorganism develops resistance to hydroxyurea [5].

Precursor synthesis can be blocked by 5-fluorodeoxyuridine and 5-fluorouracil with competitive inhibition. Monophosphate forms of 5-fluorodeoxyuridine and 5-fluorouracil are structural analogs of dUMP that is the substrate of thymidylate synthase. They inhibit the synthesis of dTMP. If the microorganism of gene coding thymidylate synthase is mutated, the microorganism develops resistance to 5-fluorodeoxyuridine and 5-fluorouracil [5].

Dideoxynucleotides that are used as drugs are similar with deoxynucleotide precursors, except that the hydroxyl group is absent in their 3′ carbon. Dideoxynucleotides that mimic deoxynucleotide precursors corpore into DNA and then, stop replication, due to it cannot be bound by the next deoxynucleotide.

Mitomycin C blocks replication by binding guanine bases that are located in both template strands of DNA [5].

2.4.2 Blockage of transcription

2.4.2.1 Transcription

Genetic information is transcripted from DNA to RNA by RNA polymerase that catalyzes a reaction, binds ribonucleotides with phosphodiester bond. RNA polymerase is constructed by 2α, 1β, 1β′, 1ω, and 1σ subunit. Transcription that is initiated by σ subunit that binds to promoter elongates until it is terminated by termination protein P (Rho) that is a RNA-DNA helicase releasing transcript from template DNA by breaking hydrogen bonds produced between template DNA and transcript [5].

2.4.2.2 Inhibition of transcription

Rifampin, that is a derivative of rifamycine family of antibiotics, blocks initiation of transcription by binding to β subunit of RNA polymerase. Rifampin is used in the treatment of infections caused by *Mycobacterium tuberculosis*, *Mycobacterium leprae*, and bacteria for which treatment is hard is not toxic to humans, due to it does not inhibit eukaryotic RNA polymerase. Mutated gene that codes RNA polymerase containing a distinct β subunit structure causes the microorganism to resist against ripampin. Streptolydigin also blocks initiation of transcription by binding with the β subunit of RNA polymerase. Bicyclomycin, the target of which is termination protein P, prevents termination of transcription. Bleomycin produces nicks on DNA. Bleomycin is not suitable in the usage for humans, because they are not specific to bacteria and has high toxic effect against humans and animals; whereas, it is used for the experiments of transcription. Azaserin blocks transcription by inhibition of ribonucleoside triphosphate synthesis [5].

2.5 Inhibition of protein synthesis

2.5.1 Translation

Protein is translated from mRNA by tRNA in ribosome. Translation is initiated by the binding of formylmethionine tRNA-aminoacyl-tRNA (fMet-$tRNA_f^{Met}$), translation initiation region (TIR) of mRNA and initiation factor 2 (IF-2) to P site of 30S subunit of ribosome and the formation of 70S complex as a result of the release of IF-2. Translation continues with the binding of a new aminoacyl tRNA to A site, transferring the polypeptide from tRNA bond to P site to tRNA bond to A site by peptidyl transferase, and translocating of polypeptidyl tRNA from A to P site by elongation factor-G (EF-G), until translation is terminated by termination protein P (Rho) [5].

2.5.2 Inhibitors mimicking tRNA

Puromycin that mimics aminoacyl tRNA enters into ribosome and is added to polypeptide grown, but it is not translocated from A site to P site of ribosome. Polypeptide containing puromycin at the carboxyl terminal is released from ribosome and translation is terminated. Puromycin is toxic to humans and animals, as it inhibits translation of eukaryotes [5].

2.5.3 Inhibitors binding to 23S rRNA

2.5.3.1 Chloramphenicol

Chloramphenicol, that is a bacteriostatic agent and the inhibitor of 23S rRNA, inhibits transcription by preventing peptidyl transferase reaction, as a result of preventing the binding of aminoacyl tRNA to A site of ribosome. Due to its ability enter into blood-brain barrier, chloramphenicol is used in the treatment of many central nervous system infections such as bacterial meningitis. If the gene of ribosomal protein is mutated or bacteria has enzyme inactivating chloramphenicol, coded by *cat* gene of Tn9, bacteria resists to chloramphenicol. The product of *cat* gene that acetylates chloramphenicol inactivates chloramphenicol [5].

2.5.3.2 Erythromycin

Erythromycin, that belongs to macrolide class of antibiotics, inhibits translation by binding to 23S rRNA. As a result of the blockage of E site, that is the exit site for peptidyl-tRNA by erythromycin, premature polypeptide is released in translocation step. Macrolites, such as erythromycin, clarithromycin, azithromycin, and roxithromycin, are used in the treatment of Gram-positive and Gram-negative bacteria such as *Legionella*, *Mycoplasm*, and *Rickettsia*, due to usefulness of antibiotics. Mutational changes in 23S rRNA and efflux pumps, and conformational changes of 23S RNA that is caused by methylation of adenine localized in 23S rRNA by Erm methylase cause bacteria to resist against erythromycin [5].

2.5.3.3 Thiostrepton

Thiostrepton and other thiopeptide antibiotics block translation by binding to 23S RNA in the peptidyl transferase reaction and preventing the binding of EF-G that is a translocase translocating polypeptidyl tRNA from A to P site. Thiostrepton is used against Gram-positive bacteria. But the usage of thiostrepton is limited to veterinary and agriculture [5].

2.5.4 Inhibitors binding to A site of aminoacyl tRNA

2.5.4.1 Tetracyclin

Tetracyclin causes futile cycle to release aminoacyl tRNA from A site of ribosome by binding of release factors mimicking aminoacyl tRNA to A site. Tetracyclin is a broad spectrum antibiotic used to treat infections caused by Gram-positive and Gram-negative bacteria. *tet*M gene transferred from conjugative transposon Tn916 codes not only an enzyme that causes resistance by methylating certain bases of 16S rRNA, but also membrane proteins that pump tetracyclin to out of the cell, consequently resistance is developed. Ribosome protector proteins coded by *tet*O and *tet*Q gene bind to A site of ribosome by mimicking EF-G, consequently tetracyclin is released from A site [5].

2.5.5 Inhibitors of translocation

2.5.5.1 Aminoglycosides

Aminoglycosides that contain kanamycin, neomycin, gentamycin, streptomycin, amikacin, and tobramycin effect translocation by binding to A site. Aminoglycosides that have broad spectrum of activity cause false reading of mRNA and translation errors. Mutants that are resistant to aminoglycosides are seen rare. Aminoglycoside resistance is caused by the genes that inactivate aminoglycosides by phosphorylation, acetylation, and adenylation of them. *neo* gene that phosphorylates kanamycin and neomycin cause resistance of Gram-negative bacteria against kanamycin and neomycin [5].

2.5.5.2 Fusidic acid

Fusidic acid inhibits turnover of EF-G by preventing the release of EF-G from ribosome. Mutations that are occurred in *fus*A gene of *E. coli* cause resistance against fusidic acid. Certain acetyltransferases causing resistance against chloramphenicole can inhibit fusidic acid by binding to fusidic acid [5].

Author details

Sahra Kırmusaoğlu[1*], Nesrin Gareayaghi[2] and Bekir S. Kocazeybek[3]

1 Faculty of Arts and Sciences, Department of Molecular Biology and Genetics, Haliç University, Istanbul, Turkey

2 Istanbul Sisli Hamidiye Etfal Training and Research Hospital, Blood Center, University of Health Sciences, Istanbul, Turkey

3 Department of Medical Microbiology, Cerrahpasa Faculty of Medicine, Istanbul University Cerrahpasa, Istanbul, Turkey

*Address all correspondence to: kirmusaoglu_sahra@hotmail.com

References

[1] Sköld O. Antibiotics and Antibiotic Resistance. Hoboken, New Jersey: Wiley & Sons, Inc; 2011

[2] Tortora GJ, Funke BR, Case CL. Microbiology: An Introduction, Global Edition. 12th ed. London: Pearson Education, Inc; 2015

[3] Kırmusaoğlu S. MRSA and MSSA: The mechanism of methicillin resistance and the influence of meticillin resistance on biofilm phenotype of *Staphylococcus aureus*. In: Enany SME, Alexander LEC, editors. The Rise of Virulence and Antibiotic Resistance in *Staphylococcus aureus*. Croatia: InTech; 2016. pp. 25-41

[4] Brooks GF, Carroll KC, Butel JS, Morse SA, Mietzner TA. Jawetz, Melnick, & Adelberg's Medical Microbiology. 26th ed. New York; Chicago: McGraw Hill Education; 2013

[5] Snyder L, Champness W. Molecular Genetics of Bacteria. Washington: American Society of Microbiology Press; 2007

Chapter 2

Antimicrobial Resistance

Muhammad Usman Qamar, Muhammad Hidayat Rasool, Shah Jahan, Muhammad Shafique and Bilal Aslam

Abstract

Klebsiella pneumoniae (*K. pneumoniae*) pose an emerging threat to public health sector worldwide. They are one of the potent nosocomial pathogens and cause variety of infections including pneumonia, septicaemia, wound infections, urinary tract infections and catheter associated infections. From the last two decades, these pathogens are becoming more powerful due to the acquisition of resistomes on different types of plasmids and transposons. There are four main mechanisms of antibacterial resistance such as efflux pump, target alteration, membrane permeability and notably enzymes hydrolysis. *K. pneumoniae* produce different types of enzymes but most importantly extended spectrum-β-lactamase (ESBL), carbapenemase and metallo-β-lactamase (MBL). *K. pneumoniae* carbapenemases (KPCs) and New Delhi metallo-β-lactamase (NDM) producing isolates displayed resistance not only against the β-lactam drugs (penicillins, cephalosporins and carbapenems) but also to other classes of antibiotics (aminoglycosides and quinolones). Therapeutic options available to treat serious infections caused by these extensively drug-resistant pathogens are limited to colistin, tigecycline and fosfomycin. Hence, combination therapy has also been recommended to treat such bacteria with clinical side effects, therefore, new treatment regime must be required. Moreover, we are relying on conventional diagnostic tools, however, novel techniques must be required for robust identification of multi-drug-resistant bacteria.

Keywords: *Klebsiella pneumoniae*, antimicrobial resistant mechanisms, mortality and morbidity, economic burden

1. Introduction

Klebsiella pneumoniae is one of the notorious pathogens with increase number of severe infections and shortage of effective treatment regimes. They are extremely resilient pathogens because their success seems to follow the model of "the best defence for a pathogen is a good defence" rather than "the best defence for a pathogen is a good offense". *K. pneumoniae* is a Gram negative, non-motile, capsulated and an imperative member of *Enterobacteriaceae* family. They were firstly identified in late nineteenth century and given the name Friedlander's bacterium [1]. They are hermaphrodite and cause both community acquired infections (soil, surface water) and significantly hospital acquire infections (medical devices, catheters). Importantly, they can reside on different body mucosal surfaces of oropharynx and gastrointestinal tract and cause variety of infections including pneumonia, bacteraemia, septicaemia, gastritis, wound infections, urinary tract infections, and soft tissue surgical infections [2]. From the last 2 decades, *K. pneumoniae* strains acquired extra genetic traits and become more virulent and antibiotic resistant [3]. They develop resistance against a wide range of antibiotics including carbapenems. Therefore, they produce (or synthesise) extended spectrum-β-lactamases (ESBL), carbapenemases, metallo-β-lactamases (MBLs), New Delhi metallo-β-lactamases (NDM) and have many other resistant mechanisms such as efflux pump and membranous target alteration.

2. Antimicrobial resistance

Antimicrobial resistance (AMR) defined when a bacterial strain shows resistance against antibiotics that normally inhibit or stop their growth and allow them to withstand against drugs. Last 2 decades, there are various AMR pathogens have been emerged around the globe such an ESBL producing pathogens, multidrug resistant and extensively drug resistant *Mycobacterium tuberculosis* (MDR, XDR-MTB), MBL producing bacteria, methicillin resistance *Staphylococcus aureus* (MRSA) and the advent of NDM producing bacteria. These isolates are almost impossible to treat with the existent treatment regime [4]. AMR emerged even since the discovery of first antibiotic "penicillin" in 1940 due the natural process of bacteria. Certainly, the genes that confer resistance upon some strains of bacteria precede antibiotics by millions of years [5]. Currently, AMR is becoming a big health issue around the world due to the irrational, overused and misuse of antibiotics which can lead to extra financial burden, prolong hospital stay and even fatal consequences. Unfortunately, currently we deficiency of the novel effective antimicrobial compounds, peptides and pipelines antibiotics to combat with emerging superbugs which result us facing growing enemy with largely depleted armoury [6].

3. Mortality, morbidity and economic burden due to antimicrobial resistant bacteria

AMR pathogens are one of the main cause of morbidity and mortality in both developed and developing countries. According to a published data, around 700,000 people die due to infections caused by AMR pathogens every year worldwide and most of the data is still underreported because of poor reporting and surveillance studies [7]. On the other hand, as per The Centre of Disease and Control (CDC), above 2 million people get infected due to AMR bacteria with a mortality rate of 23,000 each year [8]. Similarly, an Indian study revealed that around 60,000 neonates die due to AMR infections per year [9]. Likewise, a study conducted in Pakistan in 2009 was also documented that 40% neonates die due to MDR and XDR pathogens (*Escherichia coli*, *K. pneumoniae*, *Enterobacter cloacae*) isolated from blood samples [10]. Another study from Pakistan testified that 37 out of 78 new-borns died due to infections caused by MDR-*Acinetobacter spp.*, [11]. Qamar et al., conducted a study in Islamabad, Pakistan also reported 4/9 children died due to NDM positive isolates mainly of *E. coli* and *K. pneumoniae* [12]. Recently, Khan et al., from Karachi, Pakistan also reported that 22 with <1 month of age and 10 were elderly patients died due to bacteraemia and septicaemia which was mainly caused by NDM positive *K. pneumoniae* [13]. Highest mortality rate is mainly reported in developing countries as compare to developed countries largely in immunocompromised persons. Major risk factors in public sectors hospitals of developing countries are, poor hygienic practices of health care workers, irrational use of antibiotics, poor sanitation and hygienic facilities, nosocomial pathogens, hospital and wards are overcrowded with patients and attended and lack of health care workers, floors and rooms are spilled with sputum and operation theatres are in deliberated condition and are poorly equipped, contaminated drinking water, more than one patients on single bed, presence of insects and flies and most importantly low health care budget by the Government [14].

Every year, millions of dollars are spent to tackle the burden of AMR bacterial infections around the globe. European Union invested around €1.5 billion annually to tackle the infections caused by AMR pathogens which are responsible for around 23,000 mortalities only in Europe [15]. However, United States of America also expenditures above $20 billion to treat the 20 million patients who get infected with AMR infection each year [16].

4. Mechanisms of antimicrobial resistance

Since 1990s, bacteria have been becoming gradually more lethal, due to acquisition of resistomes and develop resistance to different sets of antibiotics. Now a day, many bacterial pathogens notably *K. pneumoniae* are intrinsically resistant to various antibiotics or they acquire or encoded antimicrobial resistant genes mainly on plasmids as well on chromosomes [17]. There are different mechanisms such as conjugation, transformation and transduction in which susceptible strains can acquire resistance genes with transposons that eventually support different resistance genes to incorporate with host chromosome or plasmids. Presently, there are four fundamental mechanisms of bacterial drug resistance (1) target alteration (2) modification in membrane permeability (3) efflux pump and (4) antibiotic degradation by enzymes as shows in **Figure 1** [18]. In *K. pneumoniae* both the efflux pump and enzymatic degradation of different antibiotics are more efficient as compare to other two mechanisms.

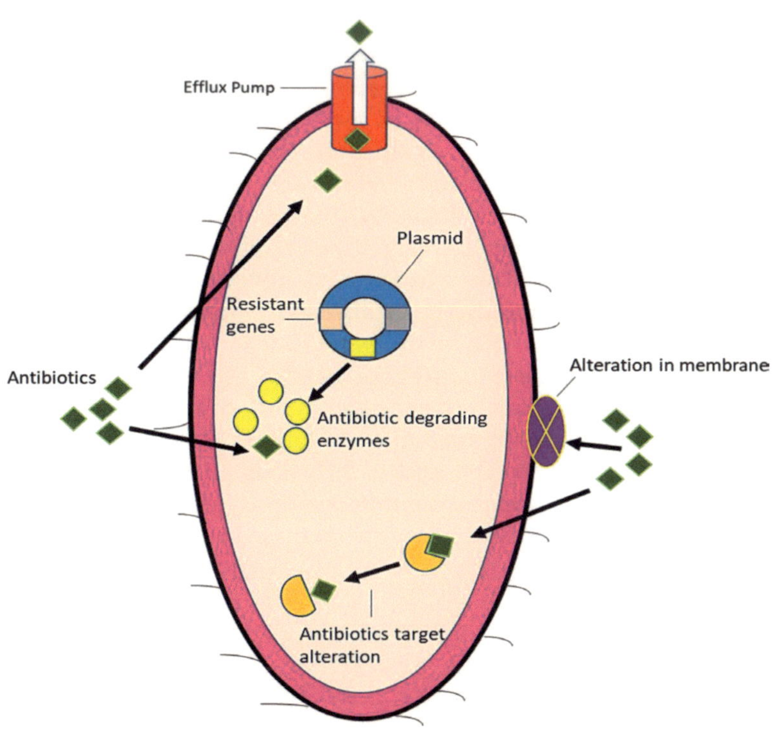

Figure 1.
Mechanism of antimicrobial resistance.

4.1 Target alteration of the antibiotics

In general, many antibiotics bind to different bacterial target sites with high affinity. So, most of the bacteria including *K. pneumoniae* alter the target sites and structure to block or inhibit the efficient binding sites [19]. In *K. pneumoniae*, antibiotic resistance occurs due to target modification in lipopolysaccharides (LPS), which is interceded by the activation of PmrA/PmrB two component regulatory system (TCRS). Moreover, another TCRS, PhoP/PhoQ contributes to produce polymyxin resistance by indirectly activating the PmrA/PmrB TCRS via PmrD linker protein. Similarly, inactivation of the PhoQ/PhoP negative regulator encoded by *mgrB* gene has also been identified to play a prominent role in polymyxin resistance in *K. pneumoniae* [20]. There are various pathogens that cause different infections and develop resistance even against single point mutation of resistant genes that encode an antibiotic target. These resistant strains then further produce numerous copies of bacteria [21].

4.2 Modification in membrane permeability

Outer membrane of the bacteria is first line barrier to protect the microorganisms against unfavourable environment including chemical and biological materials. In comparison, majority of the Gram-negative rods (GNR) are intrinsically less permeable to various antibiotics because of outer membrane barrier than Gram positive bacteria [22]. In the cell wall of *Enterobacteriaceae* family including *K. pneumoniae* contains various outer membrane proteins (Omps) such as OmpA, OmpF and OmpC and the hydrophilic antibiotics can cross through these Omps by diffusion. These proteins are involved in adhesion to other cell surfaces, maintenance of cell structure and regulation of transport of bactericidal agents and nutrition. OmpF expression is regulated at the posttranscriptional level via the *mic*F RNA, and many different environmental stresses impact the OmpC/OmpF ratio via changes in *mic*F levels. The expression of *mic*F is increased along with that of OmpF at high OmpR-P concentrations. The expression of *mic*F is also induced by toxic agents, such as paraquat and weak acids, by the SoxS, MarA, and Rob transcription factors, which bind to the same sequence element in the *mic*F promoter and they down regulate the synthesis of OmpF at transcriptional level (**Figure 2**) [23]. Therefore, the bacteria showed resistance by limiting the antibiotics entry in to the cell by the down regulation or replacement of porins with more selective channels. However, recent data showed that the loss of porins expression play a significant role to produce resistance to new drugs (cephalosporins and carbapenems), to which resistance is manly mediated by the enzyme degradation [24]. OmpA is one of the major protein of *K. pneumoniae* that prevent the activation of airway epithelial and loss of this protein make the bacteria susceptible. There are two important Omps in *K. pneumoniae*; OmpK35 and OmpK36 that allow the hydrophilic molecules such as nutrients and carbapenems to diffuse into bacteria. Moreover, *K. pneumoniae* also contain two alternative proteins; KpnO and OmpK26. Any mutation in OmpK36, OmpK26 and KpnO leads to increased resistance to cephalosporin and carbapenems [25].

4.3 Increased efflux pumps

Efflux pumps are the active transport mechanisms that expel out antibiotics from bacterial cell and play a vital role to develop intrinsic resistance against broad range of antibiotics in GNR [18]. Currently, there are several MDR efflux pumps are present in bacteria including FuaABC in *Stenotrophomonas*

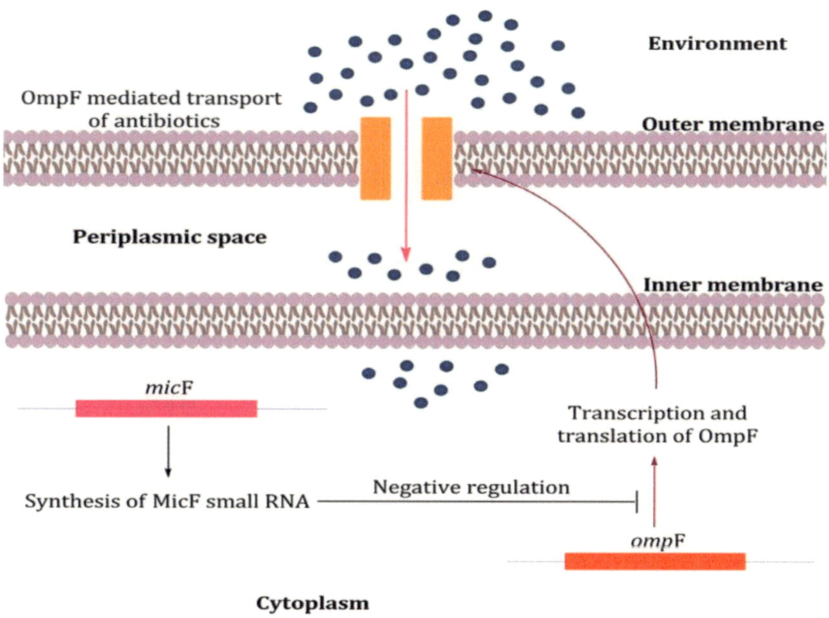

Figure 2.
Down regulation mechanism of outer membrane proteins (OMPs) in gram negative bacteria.

maltophilia, MedeA in *Streptococcus mutans* and AcrAB, KexD and more recently OqxAB efflux pumps in *K. pneumoniae* [26, 27]. In *K. pneumoniae*, most efficient antimicrobial resistant efflux pump is AcrAB, which is not only expel out the β-lactam, quinolones and chloramphenicol but also the host derived antimicrobial agents such as human antimicrobial peptides and human bronchoalveolar lavage fluid [25]. Recently, newly emerging OqxAB efflux pump has been recognised in different parts of the world that conferring resistance to quinolones and chloramphenicol [27].

4.4 Enzymatic hydrolysis of the antibiotics

In *K. pneumoniae*, the most potent and significant antimicrobial resistant mechanism is enzymatic hydrolysis of wide range of antibiotics including β-lactam, quinolones, aminoglycosides and macrolides [18]. β-lactamases are the enzymes that hydrolyze the effect of β-lactam drugs by attacking on their β-lactam ring. Currently, a total of 4037 β-lactam enzymes have been identified and mong these 1327 are class A, class B (n = 581), class C (n = 1359) and 660 belong to class D [28]. *K. pneumoniae* encode for many β-lactamases such as ESBLs, carbapenemases and MBLs. ESBLs and carbapenemases contain important subclasses of β-lactamase enzymes such as imipenemase (IMP), Verona integron encoded metallo-β-lactamase (VIM), *K. pneumoniae* carbapenemases (KPC), oxacillinase (OXA) and recently NDM [29].

4.4.1 Classification of β-lactamases

These enzymes are broadly classified into two main scheme one is Ambler Molecular classification scheme which is based on amino acid sequence and is widely acceptable and secondly, Bush Jacoby classification scheme which is based on biochemical properties [30]. Ambler class is further divided in to four major sub-classes A, B, C, and D based on conserved and distinguished amino acid motifs. Among these, class A, C, and D are the serine carbapenemases that hydrolyzed their substrate through active site by forming acyl groups and require serine as a cofactor for their activity. However, class B enzymes are the MBLs that utilised at least one zinc ion at their active site to hydrolyze the β-lactam antibiotics. MBLs are further classified into three super families B1, B2 and B3 [31]. However, Bush-Jacoby-Medeiros classification scheme is divided into three major groups; class C (Group 1 cephalosporinases), class A and D (Group 2 Serine β-lactamases) and class B (Group 3 MBLs). These are further divided in to several different subgroups as shows in **Table 1** [30].

4.4.2 Extended spectrum-β-lactamase producing K. pneumoniae

ESBL are the enzymes that hydrolyze the extended spectrum drugs mainly cephalosporins. ESBL producing *Enterobacteriaceae* mainly the *K. pneumoniae* and *E. coli* is becoming one of the serious threat to the public health sectors. According to World Health Organisation (WHO), around 50% of ESBL producing *K. pneumoniae* becomes the endemic in many parts of the world and it causes 30% resistance rate in the community [32]. In 1965, first β-lactamase enzyme Temoneria (TEM-1) and sulphydryl variable (SHV-1) has been identified. In 1983, first case SHV-2 and TEM-3 has been documented in Germany and France respectively. Similarly, in 1989, USA also reported ESBL producing *K. pneumoniae* and *E. coli*. Since then, incidence of these enzymes markedly sporadic from one region to another [33]. ESBL producing *K. pneumoniae* displayed resistance to clinically effective extended spectrum cephalosporins and monobactam and left limited treatment options. Moreover, ESBL producing *K. pneumoniae* has become major ESBL carrying bacteria associated with nosocomial infections and outbreak during 1990 to 2000. Later, these pathogens acquired variants of ESBL such as CTX-M, TEM and SHV. There are various types of ESBLs have been identified throughout the world. Up till now, TEM (n = 227) are the most predominant type of ESBLs followed by CTX-M (n = 207), SHV (n = 203), VEB (n = 19) and PER (n = 9) [28]. Majority of these ESBLs have been reported in *Enterobacteriaceae* notably *K. pneumoniae* and *E. coli*. ESBLs producing *K. pneumoniae* have also become an important hospital acquired pathogen. In China, around 50% of the hospital acquired infections are caused by CTX-M-14 and CTX-M-15 ESBL producing *K. pneumoniae* [34]. Recently, these ESBL producing *K. pneumoniae* were also reported from community such as water, soil and veterinary settings [35]. In Asia, SHV is the predominant genotype of ESBL and early SHV-5 and SHV-12 enzymes were documented in Korea and Japan. However, in Europe and south east Asia (India and Pakistan), CTX-M-2, CTXM-3, CTX-M-14 and CTX-M-15 producing *Enterobacteriaceae* at its rise. However, in Japan, SHV-12 and CTX-M-3 and CTXM-14 have been predominantly reported in *K. pneumoniae* and *E. coli*. In Indo-Pakistan region, most of the clinical isolates of *K. pneumoniae* encoded for TEM, CTX-M and SHV. Two highly populated countries in the world (China and India) represents the largest reservoir of CTX-M in the world [36].

4.4.3 AmpC producing K. pneumoniae

AmpC are the β-lactam enzymes that belong to the ambler class C of β-lactam. These enzymes hydrolyze cephalosporins antibiotics and produce resistance against penicillin, second and third generation cephalosporins and cephamycin. Moreover, these enzymes can also convey resistance to combination of these antibiotics along with β-lactam inhibitors. The versatility of the

Molecular class (subclass)	Bush Jacoby groups	Main substrate	Defining characteristic(s)	Representative	Bacteria
A	2a	Penicillins	Increased hydrolysis of benzylpenicillin	PC1	
A	2b	Penicillins and early Cephalosporins	Hydrolysis of the penicillins and cephalosporins	TEM and SHV	*Enterobacteriaceae* (*K. pneumoniae, E. coli, E. cloacae*)
A	2be	Extended spectrum-β-lactam drugs, monobactam	Hydrolysis of extended spectrum-β-lactam drugs	TEM, SHV and CTX-M	*Enterobacteriaceae* (*K. pneumoniae, E. coli, E. cloacae*)
A	2br	Penicillins	Resistant to β-lactam inhibitor	TEM and SHV	*Enterobacteriaceae* (*K. pneumoniae, E. coli, E. cloacae*)
A	2c	Carbenicillin	Hydrolysis of carbenicillin	CARB	*Enterobacteriaceae* (*K. pneumoniae, E. coli, E. cloacae*)
A	2ce	Carbenicillin and Cefepime	Hydrolysis of carbenicillin and cefepime	RTG	
B (B1)	3a	Carbapenems	Carbapenems but not aztreonam	IMP, VIM and NDM	*Enterobacteriaceae* (*K. pneumoniae, E. coli*), non-fermenters (*P. aeruginosa, A. baumannii*)
B (B2)	3b	Carbapenems	Prefer to hydrolyze carbapenems	CphA and Sfh	
C	1	Cephalosporins	More hydrolyze the cephalosporins	AmpC,	
C	1e	Cephalosporins	Highly hydrolyzed the ceftazidime	GC1, CMY-37	
D	2d	Oxacillin	Hydrolysis of oxacillin	OXA-1, OXA-10	*A. baumannii*
D	2de	Extended spectrum-β-lactam drugs	Hydrolysis of extended spectrum-β-lactam drugs	OXA-11, OXA-50	*A. baumannii*
D	2df	Carbapenems	Hydrolysis of oxacillin and carbapenems	OXA-23 and OXA-48	*A. baumannii*

Table 1.
Classification of the β-lactamases.

K. pneumoniae is to encode various β-lactam resistant genes on their transferable plasmids that able them to spread AmpC like resistomes in to other species [30]. Plasmid mediated AmpC gene was firstly identify during 1980s in

K. pneumoniae. The most common genes of AmpC families that have been reported in *K. pneumoniae* are DHA, FOX, CMY and MOX types in different parts of the world. Although they are plasmid mediated however, first case of chromosomal mediated AmpC *bla*CMY-2 containing *K. pneumoniae* was reported in 2009. AmpC producing *K. pneumoniae* along with porin loss and increase efflux pump display increase resistance against β-lactam drugs which can lead to the carbapenem resistance in future [37].

4.4.4 Class A carbapenemases producing K. pneumoniae

Since the sporadic spread of ESBL producing *K. pneumoniae* in different regions, carbapenems which is considering last resort to treat infections caused by such pathogens have widely used. The extensive usage of carbapenems have led the emergence of plasmid mediated carbapenemase genes such as IMP, VIM, KPC and NDM. Carbapenemases are the potent enzymes that hydrolyzed the carbapenem drugs and left no treatment option available [32]. There are various chromosomal mediated (SME, SFC and NMC-A) and plasmid mediated carbapenemases (KPC, GES, IMI) have been identified. It has determined that a carbapenemase has variants extending from KPC-1 to KPC-33 till now [28]. KPC gene is mainly located on different types of transmissible plasmids such as IncF, IncX, IncA/C, IncR and IncI2 which lead to the horizontal transfer of antimicrobial resistomes. These plasmids also acquired resistant genes of other antibiotics such as aminoglycosides, tetracycline and quinolones. Most of the KPC genes are associated with *Tn4401* transposons which jumped from one plasmid to other [38]. the endemic of KPC producing *K. pneumoniae* identified in different parts of the world including USA, UK, Japan, China, Greece, India, Spain, Germany and Italy. Moreover, out breaks of the KPC also documented in different countries like USA, France (travelling history from the endemic area), Spain, Netherlands, UK, Finland and Sweden [39].

4.4.5 Class B carbapenemase producing K. pneumoniae

Class B are the MBLs that require Zn^{+} as a cofactor for their activity. There are many types of MBLs have been identified such as IMP, VIM, and NDM in *Enterobacteriaceae*. These enzymes cause resistance to various antibiotics including carbapenems and left no treatment option [40]. NDM is one the potent enzyme of MBL that newly emerged in clinically isolates of *K. pneumoniae* and *E. coli* from a Swedish patient who had been hospitalised in New Delhi, India in 2008. NDM producing bacteria displayed resistance not only against β-lactam drugs but also to other 14 classes of antibiotics. NDM producing bacteria notable *K. pneumoniae* and *E. coli* are major cause of morbidity and mortality in developing countries. Qamar et al. reported 4/9 children died due to NDM producing bacteria [12]. Similarly, another study from Karachi, Pakistan also documented 57% mortality in neonates due to NDM producing bacteria mainly *K. pneumoniae* [13]. Up till now there are 19 (NDM-1 to NDM-19) variants have been identified in different parts of the world [41]. Among these NDM-1, NDM-5 and NDM-7 are most prevalent variants with a hyper variable region. NDM are mainly plasmid mediated as compared to chromosome and located on different size and number of plasmids such as IncF, IncX, IncA/C. NDM producing *K. pneumoniae* are belong to various sequence types such as ST11, ST147, ST258, ST340, ST15 and ST16. More than 50% of the NDM producing

K. pneumoniae have been identified from ST11 and ST147. NDM genes are mainly associated with IS*Aba125* transposons in *Enterobacteriaceae* family which hypothesised originated from the Tn*125* of *A. baumannii* [42]. Since the discovery of the NDM in Asia it is widely spread in different parts of the world including USA, UK, Middle East, Europe, Africa and Australia. Main reservoir for NDM are the Asian countries like India, Pakistan and China and the Balkan States [39].

4.4.6 Class D carbapenemases producing K. pneumoniae

These carbapenemase enzymes swiftly hydrolyzed the isoxazolylpenicillins drugs such as oxacillin, cloxacillins and dicloxacillin than benzylpenicillin. There are more than 574 variants have been identified around the globe but few of them have the carbapenemase activity. Therefore, class D are reclassified in to 12 main groups; Oxacillinase (OXA)-23, OXA-24, OXA-48, OXA-51, OXA-58, OXA-143, OXA134, OXA-211, OXA-213, OXA-214, OXA-229, OXA-235 [43]. Among these most potent variants is OXA-48 which was firstly identified in *K. pneumoniae* in Turkey in 2003. Later, this gene spread in different parts of the world such as Libya, Egypt, India, Spain Switzerland, France, Germany, Argentine, Israel, Saudi Arabia, Japan. The highest percentage of OXA-48 has been reported in France, Spain and Saudi Arabia (78%) followed by 88% in Lebanon and 56% in Pennsylvania [39]. Co-existence of OXA-48 like *K. pneumoniae* has been reported from different countries like Switzerland (OXA-48/NDM-1), Turkey (OXA-48/NDM-1) and India (OXA-181/VIM-5). OX-48 like genes are mainly present on transferable and conjugative group of plasmids (IncL/M and IncL) [39, 44].

5. Tacking of antibiotic resistance

AMR is becoming a menace to global health. According to United Nation (UN), in 2050, if we do not tackle the AMR now, more than 10 million people will die each year and the global income cost would be $100 trillion. So, in 2050, one person will die after every 3 seconds [7]. Firstly, it is need of the hour to give a public awareness through paper and electronic media to stop the irrational use of antibiotics. Secondly, people should improve their hygiene practices which will reduce the 60% antibiotics burden to reduce the diarrhoea. Thirdly, more than 70% world's antibiotics are being utilised in agriculture and live stocks, so overuse of antibiotics must be avoided in these settings. Fourthly, Global surveillance of AMR should be conducted and maintained throughout the world. Fifthly, there are many countries which do not have the rapid and effective diagnostics facilities of microbiology which allow the physician to prescribe the broad-spectrum empirical therapy. Sixthly, there is an urgent need to develop an alternative method to treat the AMR such as vaccines, phage therapy, probiotics, antibodies and lysins. Seventhly, there are shortages of microbiologists, infectious diseases specialists, infections control specialists, vegetarians, pharmacists and epidemiologists. Eighthly, A global funding must be required to support and encourage the less commercially attractive. Ninthly, the world gives better incentive to pharmaceutical companies to develop new and existing antibiotics. Finally, world leading organisations such as WHO, G20, UN must work together with coordination.

Author details

Muhammad Usman Qamar[1*], Muhammad Hidayat Rasool[1], Shah Jahan[2], Muhammad Shafique[1] and Bilal Aslam[1]

1 Department of Microbiology, Faculty of Life Sciences, Government College University, Faisalabad, Pakistan

2 Department of Immunology, University of Health Sciences, Lahore, Pakistan

*Address all correspondence to: musmanqamar@gcuf.edu.pk

References

[1] Merino S, Camprubi S, Alberti S, Benedi VJ, Tomas JM. Mechanisms of *Klebsiella pneumoniae* resistance to complement-mediated killing. Infection and Immunity. 1992;**60**:2529-2535

[2] Rock C, Thom KA, Masnick M, Johnson JK, Harris AD, Morgan DJ. Frequency of *Klebsiella pneumoniae* carbapenemase (KPC)-producing and non-KPC-producing *Klebsiella Species* contamination of healthcare workers and the environment. Infection Control. Hospital Epidemiology. 2016;**35**:426-429

[3] Paczosa MK, Mecsas J. *Klebsiella pneumoniae*: Going on the offense with a strong defense. Microbiology and Molecular Biology Reviews: MMBR. 2016;**80**:629-661

[4] Ventola CL. The antibiotic resistance crisis: Part 1: Causes and threats. P & T: A Peer-Reviewed Journal for Formulary Management. 2015;**40**:277-283

[5] Read AF, Woods RJ. Antibiotic resistance management. Evolution, Medicine, and Public Health. 2014;**2014**:147

[6] Viswanathan R, Singh AK, Basu S, Chatterjee S, Sardar S, Isaacs D. Multi-drug resistant gram negative bacilli causing early neonatal sepsis in India. Archives of Disease in Childhood Fetal and Neonatal Edition. 2012;**97**:182-187

[7] de Kraker MEA, Stewardson AJ, Harbarth S. Will 10 million people die a year due to antimicrobial resistance by 2050? PLOS Medicine. 2016;**13**:e1002184

[8] Lim C, Takahashi E, Hongsuwan M, Wuthiekanun V, Thamlikitkul V, Hinjoy S, et al. Epidemiology and burden of multidrug-resistant bacterial infection in a developing country. Life. 2016;**5**:e18082

[9] Laxminarayan R, Duse A, Wattal C, Zaidi AK, Wertheim HF, Sumpradit N, et al. Antibiotic resistance-the need for global solutions. The Lancet Infectious Diseases. 2013;**13**:1057-1098

[10] Hannan A, Qamar MU, Usman M, Waheed KAI, Rauf K. Multidrug resistant microorganisms causing neonatal septicaemia: In a tertiary care hospital Lahore, Pakistan. African Journal of Microbiology Research. 2013;**7**:1896-1902

[11] Saleem AF, Ahmed I, Mir F, Ali SR, Zaidi AK. Pan-resistant Acinetobacter infection in neonates in Karachi, Pakistan. Journal of Infection in Developing Countries. 2009;**4**:30-37

[12] Qamar MU, Nahid F, Walsh TR, Kamran R, Zahra R. Prevalence and clinical burden of NDM-1 positive infections in paediatric and neonatal patients in Pakistan. The Paediatric Infectious Disease Journal. 2015;**34**:452-454

[13] Khan E, Irfan S, Sultan BA, Nasir A, Hasan R. Dissemination and spread of New Delhi Metallo-beta-lactamase-1 superbugs in hospital settings. The Journal of the Pakistan Medical Association. 2016;**66**:999-1004

[14] Qamar MU, Hannan A, Arshad MU, Arshad M. Metallo-β-lactamase producing *Enterobacter cloacae*: An emerging threat in neonates. African Journal of Microbiology Research. 2014;**8**:525-528

[15] Davies SC, Fowler T, Watson J, Livermore DM, Walker D. Annual report of the chief medical officer: Infection and the rise of antimicrobial resistance. Lancet. 2013;**381**:1606-1609

[16] Hampton T. Report reveals scope of us antibiotic resistance threat. Journal

of the American Medical Association. 2013;**310**:1661-1663

[17] Tenover FC. Mechanisms of antimicrobial resistance in bacteria. American Journal of Infection Control. 2006;**34**:3-10; 64-73

[18] Blair JM, Webber MA, Baylay AJ, Ogbolu DO, Piddock LJ. Molecular mechanisms of antibiotic resistance. Nature Reviews Microbiology. 2015;**13**:42-51

[19] Billal DS, Feng J, Leprohon P, Légaré D, Ouellette M. Whole genome analysis of linezolid resistance in *Streptococcus pneumoniae* reveals resistance and compensatory mutations. BMC Genomics. 2011;**12**:512

[20] Haeili M, Javani A, Moradi J, Jafari Z, Feizabadi MM, Babaei E. MgrB alterations mediate colistin resistance in *Klebsiella pneumoniae* isolates from Iran. Frontiers in Microbiology. 2017;**8**:1-8

[21] Gao W, Chua K, Davies JK, Newton HJ, Seemann T, Harrison PF, et al. Two novel point mutations in clinical *Staphylococcus aureus* reduce linezolid susceptibility and switch on the stringent response to promote persistent infection. PLOS Pathogens. 2010;**6**:e1000944

[22] Kojima S, Nikaido H. Permeation rates of penicillins indicate that *Escherichia coli* porins function principally as nonspecific channels. Proceedings of the National Academy of Sciences of the United States of America. 2013;**110**:2629-2634

[23] Tran QT, Williams S, Farid R, Erdemli G, Pearlstein R. The translocation kinetics of antibiotics through porin OmpC: Insights from structure-based solvation mapping using WaterMap. Proteins. 2013;**81**:291-299

[24] Djahmi N, Dunyach-Remy C, Pantel A, Dekhil M, Sotto A, Lavigne J-P. Epidemiology of carbapenemase-producing *Enterobacteriaceae* and *Acinetobacter baumannii* in Mediterranean countries. BioMed Research International. 2014;**2014**:11

[25] Li B, Zhao Y, Liu C, Chen Z, Zhou D. Molecular pathogenesis of *Klebsiella pneumoniae*. Future Microbiology. 2014;**9**:1071-1081

[26] Hu R-M, Liao S-T, Huang C-C, Huang Y-W, Yang T-C. An inducible Fusaric acid tripartite efflux pump contributes to the fusaric acid resistance in *Stenotrophomonas maltophilia*. PLoS One. 2012;7:e51053

[27] Bialek-Davenet S, Lavigne J-P, Guyot K, Mayer N, Tournebize R, Brisse S, et al. Differential contribution of AcrAB and OqxAB efflux pumps to multidrug resistance and virulence in *Klebsiella pneumoniae*. Journal of Antimicrobial Chemotherapy. 2015;**70**:81-88

[28] Naas T, Oueslati S, Bonnin RA, Dabos ML, Zavala A, Dortet L, et al. Beta-lactamase database (BLDB)—Structure and function. Journal of Enzyme Inhibition and Medicinal Chemistry. 2017;**32**:917-919

[29] Livermore DM. Defining an extended-spectrum beta-lactamase. Clinical Microbiology and Infection: The Official Publication of the European Society of Clinical Microbiology and Infectious Diseases. 2008;**14**(Suppl 1):3-10

[30] Bush K, Jacoby GA. Updated functional classification of beta-lactamases. Antimicrobial agents and chemotherapy. 2010;**54**:969-976

[31] Öztürk H, Ozkirimli E, Özgür A. Classification of beta-lactamases and penicillin binding proteins using ligand-centric network models. PLoS One. 2015;**10**:e0117874

[32] Navon-Venezia S, Kondratyeva K, Carattoli A. *Klebsiella pneumoniae*: A major worldwide source and shuttle for antibiotic resistance. FEMS Microbiology Reviews. 2017;**41**:252-275

[33] Gupta A, Ampofo K, Rubenstein D, Saiman L. Extended spectrum beta lactamase-producing *Klebsiella pneumoniae* infections: A review of the literature. Journal of Perinatology: Official Journal of the California Perinatal Association. 2003;**23**:439-443

[34] Zhang J, Zhou K, Zheng B, Zhao L, Shen P, Ji J, et al. High prevalence of ESBL-producing *Klebsiella pneumoniae* causing community-onset infections in China. Frontiers in Microbiology. 2016;**7**:1830

[35] Toubiana J, Timsit S, Ferroni A, Grasseau M, Nassif X, Lortholary O, et al. Community-onset extended-spectrum β-lactamase-producing *Enterobacteriaceae* invasive infections in children in a University Hospital in France. Medicine. 2016;**95**:e3163

[36] Hawkey PM. Prevalence and clonality of extended-spectrum-beta-lactamases in Asia. Clinical Microbiology and Infection: The official Publication of the European Society of Clinical Microbiology and Infectious Diseases. 2008;**14**:159-165

[37] Zamorano L, Miro E, Juan C, Gomez L, Bou G, Gonzalez-Lopez JJ, et al. Mobile genetic elements related to the diffusion of plasmid-mediated AmpC beta-lactamases or carbapenemases from *Enterobacteriaceae*: Findings from a multicentre study in Spain. Antimicrobial Agents and Chemotherapy. 2015;**59**:5260-5266

[38] Naas T, Cuzon G, Villegas MV, Lartigue MF, Quinn JP, Nordmann P. Genetic structures at the origin of acquisition of the beta-lactamase blaKPC gene. Antimicrobial Agents and Chemotherapy. 2008;**52**:1257-1263

[39] Lee CR, Lee JH, Park KS, Kim YB, Jeong BC, Lee SH. Global dissemination of carbapenemase-producing *Klebsiella pneumoniae*: Epidemiology, genetic context, treatment options, and detection methods. Frontiers in Microbiology. 2016;**7**:895

[40] Qamar MU, Saleem S, Toleman MA, Saqalein M, Waseem M, Nisar MA, et al. In vitro and in vivo activity of Manuka honey against NDM-1-producing *Klebsiella pneumoniae* ST11. Future Microbiology. 2018;**13**:13-26

[41] Qamar MU, Walsh TR, Toleman MA, Saleem S, Jahan S. First identification of clinical isolate of a novel "NDM-4" producing *Escherichia coli* ST405 from urine sample in Pakistan. Brazilian Journal of Microbiology [Publication of the Brazilian Society for Microbiology]. 2018;**49**:949-950

[42] Toleman MA, Spencer J, Jones L, Walsh TR. blaNDM-1 is a chimera likely constructed in *Acinetobacter baumannii*. Antimicrobial Agents and Chemotherapy. 2012;**56**:2773-2776

[43] Jeong SH, Lee KM, Lee J, Bae IK, Kim JS, Kim HS, et al. Clonal and horizontal spread of the blaOXA-232 gene among *Enterobacteriaceae* in a Korean hospital. Diagnostic Microbiology and Infectious Disease. 2015;**82**:70-72

[44] Kilic A, Baysallar M. The first *Klebsiella pneumoniae* isolate co-producing OXA-48 and NDM-1 in Turkey. Annals of Laboratory Medicine. 2015;**35**:382-383

Section 2

Control Strategies of Microbial Biofilm

Chapter 3

Alternative Approaches to Combat Medicinally Important Biofilm-Forming Pathogens

Mansab Ali Saleemi, Navindra Kumari Palanisamy and Eng Hwa Wong

Abstract

Bacteria have developed the capability to produce structured communities (or cluster of cells) via adherence to surface to form biofilms that facilitate or prolong their survival under extreme environmental condition. Bacterial biomass adheres to inanimate and biotic surfaces in the hospital setting as well as in the environment. In the healthcare system, the biofilm formation on medical devices allows bacteria to sustain as a reservoir and becomes more resistant to antimicrobial agents. However, biofilm formation facilitates pathogens to sabotage the host defenses that are linked to long-term retention within the host cell. Therefore, in this review, we provide some steps leading to the formation of biofilm within the host and on inanimate surfaces, also emphasizing various medically significant pathogens and debate current developments on novel approaches that aimed to prevent biofilm formations and its dispersion to patients.

Keywords: biofilm, antibiotic resistant, distribution, control, therapeutic strategy

1. Introduction

Biofilm formation is structured accumulation of fastidious microorganisms attached on inanimate objects or compact surfaces that extensively have been examined in the past decades because they particularly cause infections and more often responsible for chronic infections [1–3]. They are predominantly problematic due to their antimicrobial resistant properties and their ability to evade host defense mechanisms, which substantially hinders disease treatment in the hospital [1–4]. Bacterial biofilms are ubiquitous in nature and harbor phenotypic adaptations in the environment with respect to broader perspective [1]. The nature of single cell organisms enables them to adhere to each other and form a "complex structure," which assists to survive under adverse environmental condition. The biofilm formation occurs from planktonic bacteria due to environmental changes and involves in conjugation gene transfer "multiple regulatory network" from one bacterium to another in response to environmental stress [5–9]. This type of cell-to-cell adhesion and gene transformation changes the expression of surface molecules, virulence factors, and nutrient utilization that enables their survival under unfavorable environmental condition [8, 10–17].

Bacteria are cocooned within the biofilm and form extracellular matrix, which represents 90% of the biomass [18]. The matrix as a stabilizing scaffold for the three-dimensional structure is composed of extracellular polymeric substance (EPS) along with extracellular DNS and carbohydrate binding protein [19–21]. Nutrients are trapped by the resident bacteria in the matrix and water is retained efficiently via H-bond interaction with hydrophilic polysaccharides [18, 22]. The composition of extracellular polymeric substance (EPS) is modified in response to alterations in nutrient availability [23, 24] by certain enzyme secretion of bacteria, thus tailoring biofilm formation to the more specific environment [23, 25]. Therefore, the skeletal components of the extracellular matrix are highly hydrated and provide high tensile strength that enables bacteria to exchange their DNA by conjugation and promote cell-to-cell interaction while defending the biomass from predation, radiation, desiccation, oxidizing molecules, and other dangerous agents [18, 26–28].

The multifaceted nature of biofilms that allow the bacteria to form a community, i.e., division of labor and express their virulence factors in response to local oxygen and nutrient availability, makes them resistant against different antimicrobial agents [29, 30]. Some studies have shown that there are presence of nondividing metabolically inactive recalcitrant bacteria within the biomass [29, 31], which play very crucial role to cause tolerance against broad-spectrum antimicrobial drugs. The matrix protein inside the host cell protects bacterial biofilm against innate immune defenses, i.e., phagocytosis and opsonization [32]. The spread of other virulence factors inside the host cell and drug resistance marker is due to the cell-to-cell interaction [15]. Thus, biofilm-forming pathogens retained and adhere to the infected surface and cause recalcitrant and chronic infection, i.e., upper respiratory tract infection (particularly, *Pseudomonas aeruginosa*) [33, 34], dental decay (mixed culture of *Streptococcus mutans,* and other pathogens) [35], ventilated-induced and other device-associated infections (*Escherichia coli*, *Klebsiella* spp., *Enterococcus faecalis*, *Staphylococcus aureus,* etc.) [36, 37], urinary tract infections [*Proteus* spp., uropathogenic *E. coli* (UPEC)] [38]. In particular, immunocompromised patients are the most common target to all these biofilm-forming pathogens, causing a devastating impact on patients, and in many cases, leading to death. Here, we analyze the formation of intracellular and extracellular biofilm which is the underlying factor for various medically important microorganisms. Given the recalcitrance and prevalence of infections caused by biofilm-forming pathogens, we discuss knowledge about the most current progresses in the advancement of novel strategies of biofilm.

2. Extracellular formation of biofilm

2.1 Bacterial attachment on surfaces and what does make it adhere to object surface?

Bacterial biofilm growth, subsequent maturation, and aggregation consist of irreversible and reversible stages, which involve various conserved and species-specific aspects. At the first stage, the bacteria are introduced on the surface; a process of at least a part of stochastic that is driven by gravitational forces and Brownian motion, and usually influenced by nearby hydrodynamic forces [39, 40]. Microorganisms encounter with repelling or attractive forces—within the niche that alter depending on ionic strength, pH, nutrient levels, and temperature. Bacterial cell wall composition, along with medium properties, affects direction and velocity toward or away by the contact surface of pathogens [39]. Motile bacteria utilize flagella in order to overcome repulsive and hydrodynamic forces, by having a competitive advantage. The main function of flagella is to provide motility and initial cell attachment to the surface for various pathogens, including *Listeria monocytogenes*, *E. coli*, *Vibrio cholerae*, and *P. aeruginosa* [41–45]. In some species of bacteria, chemotaxis plays very important role in direct attachment to nutrient composition, for instance, mutations arise in CheR1 methyltransferase, which have been observed to vary the response of amino acid of *P. aeruginosa* and impair maturation of bacterial biofilm and attachment [46]. Some earlier studies have been shown that chemotaxis in *E. coli* is dispensable [5]; moreover, current observations revealed that the disruption occurs in the chemotaxis methyl accepting protein II and informs biofilm defects particularly in uropathogenic *E. coli* cells [47]. With respect to intercepting surface, bacterial attachment is facilitated by additional secreted molecules such as adhesin protein and extracellular adhesive appendages.

Initially, the attachment is reversible and dynamic during which pathogens can separate and rejoin planktonic biomass if agitated through repulsive forces [48], hydrodynamic forces—detach bacteria off from the surface. Some bacteria attained irreversible attachment in order to maintain a firm grip on the cell surface. Serotypes of other *E. coli* and uropathogenic *E. coli* depend intensely on the type 1 pili [5, 40, 49–51]. Uropathogenic *E. coli* harbors several pili systems (means CUP system), which mediate adhering to a specific niche [38]. Attachment on the bacterial surface is facilitated by the adhesion protein (FimH), which identifies mannosylated moieties [50–52]. The adhesive protein (FimH) plays a critical role in the pathogenesis of uropathogenic *E. coli* because it facilitates adherence and causes invasion to epithelial cells of bladder in human, adheres to the human uroplakin and is also critical in preclinical murine cystitis model, which causes human disease [51, 53, 54]. FimH is much more consistent to play a critical role in the virulence of human disease under positive selection [52–56].

Furthermore, antigen 43, curli fibers, and type 1 pili have been observed to facilitate attachment and cell-to-cell interaction on inanimate surfaces [57]. Curli fiber also mediates attachment to the extracellular matrix components in eukaryotes such as plasminogen, fibronectin, and laminin [58]. *Pseudomonas aeruginosa*, for instance, uses various additional organelles, which assist in adherence to the surface, irreversibly. Contrary to UPEC and *P. aeruginosa*, Gram-positive bacteria (*Enterococci*) are lactose producing, nonmotile, and recently identified to contain nonadhesive (pili) that mediate attachment to the extracellular matrix components in eukaryotes. Examples of these include Ace (*E. faecalis*) and SagA (*E. faecium*), which attach to the collagen protein [59] and surface protein (Esp). This has been observed to stimulate abiotic formation of biofilm on the contact surface specifically in *E. faecalis* [60]. Current studies showed the existent of biofilm-associated pili (Ebp) and also confirmed their contribution toward urinary tract infections, endocarditis, and biofilm formation and attachment [61].

2.2 Maturation of biofilm

Cell-to-cell interaction triggers specific intrinsic responses that cause changes in the gene expression, upregulating factors favorable to sessility especially for those involved in extracellular matrix protein formation [40]. However, relatively very little information is obtained about the matrix constituents with respect to *E. coli* pathogen. Initially, cellulose was recognized as essential components in *E. coli* pellicle biofilms and later on expressed with curli fibers in gastrointestinal *E. coli* strains [62]. Curli fiber plays a critical role in pellicles, for instance, curli fiber (amyloid) that leads to the pellicle biofilm formation. It also acts as a curlicide to prevent pellicle formation, and some of them have deficient to form pellicles (known as curli mutants) [63]. Further studies revealed that colonic acid and polyglucosamine

(PGA) take part in biofilm architecture [64], while the PGA being predominant among the clinical strains, particularly in UPEC isolates. Thus, more detailed investigations are required for further characterization of extracellular matrix protein in *E. coli*. The composition of extracellular matrix protein has been extensively analyzed in *P. aeruginosa* and varies depending on external environmental conditions [65]. The primary components of EPS are Psl and Pel [25]. Psl enhances the attachment of *P. aeruginosa* to epithelial cells [66] and mucin, while the expression of Pel increased in small colony variants (SCV) isolated from the cystic fibrosis patients associated with *Pseudomonas* persistence in the airways of lung [67]. Moreover, intercellular interactions and biofilm stabilizations in *P. aeruginosa* are critical in response to environmental DNA (eDNA) [68].

Mature *P. aeruginosa* biofilm formations are more resistant to treatment with DNase as compared to young biofilms, demonstrating that eDNA remains stable because the components of EPS are not abundant during the initial stage of biofilm when the bacterial cells come to attach each other. In contrast, the concentration of eDNA increases during biofilm maturation stage due to the occurrence of bacterial cell lysis in response to quorum sensing mechanism of *Pseudomonas* quinolone signal (Pqs) [69]. In *Pseudomonas*, type IV pili play an essential role in the migrating pathogens to form aggregation in the area of high eDNA binding attraction [70]. The amount of eDNA to form biofilm structure has already been observed in *E. faecalis*. Some reports identified that biofilm formation in this organism is influenced by the affected autolysis of cells and intracellular release of DNA [71, 72]. Initial study reported that the mutant reduced the biofilm formation by 30% due to the lack of autolysin gene, Atn [59]. In another study investigated, it showed that specific stage of bacterial biofilm formation required temporal regulation by Atn for the release of DNA [73].

2.3 Matrix escape mechanisms

Bacterial mature biofilm provides a suitable living environment to the resident microorganisms for making compact surface adherence community, so as to share products and actively exchange their genetic materials by conjugation. Moreover, as biofilms mature, dispersal becomes a choice. In addition to passive dispersal caused by shear stress, the pathogen develops different ways to recognize environmental changes, which make it to stay within the biofilm. Bacterial biofilm dispersal occurs as a result of various clues such as oxygen fluctuations, modifications in nutrient availability, and increases in toxic products [74]. Biofilm dispersal is induced by the increase of extracellular iron in uropathogenic *E. coli* [75], while in *Pseudomonas* spp., it is due to the increased quantities of various nitrogen and carbon source [76]. The amounts of small molecules such as alterations in environment and changes in gene expression are monitored by various sensory systems [77]. Among various other signals, for instance, universal cyclic-di-GMP has been used in *P. aeruginosa* and *E. coli* causing implication in a shift between motility and sessility. Typically, an increase in the level of cyclic-di-GMP is favorable to sessility, while a reduction in cyclic-di-GMP induces upregulation of motility [78].

Recently, some results reported the factors responsible for such changes such as downregulation of extracellular polymeric substance, reduction of cyclic-di-GMP in bacterial biofilm communities, and upregulation of swarming and swimming motility [25]. Certain type of enzymes (such as alginate lyase) also participates in pathogen detachment from surface especially in *P. aeruginosa* [79], whereas in *E. coli,* the enzyme (CsrA) is responsible to repress the synthesis of PGA [80]. Along with that downregulation of EPS, certain molecules of surfactant are produced causing a reduction in cell-to-cell interaction. Moreover, studies identified that flagellated populations within the biofilms of *P. aeruginosa* migrate to other void surface in order to make colonies [65]. Initially, these colonies loosely attach to compact surface, but after maturation process, they make a hard shell in the surrounding and use the infected surface as a source of nutrient. Sometimes, live cells use dead cells as a source of carbon. When bacteria become dead, then live cells accumulate on it, bind to each other by sharing their genetic materials and form a compact layer that is usually very hard to break. Dead cells are also responsible for creating cavity within the bacterial biomass. The bacteria within the biofilm can be scattered by applying dispersal mechanism.

Due to dispersing nature of bacteria, they may have the ability to restart the biofilm formation process after encountering a favorable environmental condition [81]. This is another sophisticated mechanism of dispersal revealed by using *B. subtilis,* which could be prevalent among the bacterial species. Researchers reported that the pathogen (*B. subtilis*) lost its cellular integrity within 5–8 days and also found that disassembly of biofilm is associated with a mixture of different amino acids (D-tyrosine, D-methionine, etc.) that are formed during bacterial stationary growth phase [82]. These D-types of amino acids interfere with bacterial attachment to cell surface and perturbation to fiber dissociation, without influencing matrix component expression or bacterial growth [83]. In *B. subtilis*, the performance of biofilm is disrupted by the addition of D-type amino acid mixture [83]. Further studies showed that another factor such as norspermidine, which is produced by *B. subtilis,* works together with D-type amino acid leading to biofilm disassembly [84]. So, this type of association—norspermidine/D-type amino acid—is essential for the eradication of bacterial biofilm and makes them vulnerable to antimicrobial agents used in the hospital.

3. Bacterial intracellular biofilms

Gathering evidence have showed that numerous bacterial pathogenic species formerly considered as extracellular can retain within the host cell by adapting intracellular bacterial lifestyle that includes the bacterial communities having biofilm-like properties. First, a murine model of infection was used to assess the bacterial communities for UPEC [85]. Type 1 pili in uropathogenic *E. coli* bind to the receptor on superficial bladder cells [86], triggering to induce bacterial internalization. Toll-like receptor-4 (TLR-4)-dependent process used to expel out from inside the UPEC [87], but certain bacteria elude exocytic procedure and leave out from the cytoplasm of host cell, where they duplicate into intracellular bacterial communities (IBCs) [85]. Several developmental stages lead to the process of IBCs that indicate distinct morphological features [85]. After passing first 6 h ensuing bladder inoculation, UPEC rapidly divides (replication time 30–35 min) causing small clusters associated with loosely attached rods (during early IBCs), having a coccoid shape and an average bacterial length of about 0.7 mm. The bacterial exponential growth rate dramatically drops between 6 and 8 h, exceeding replication time to 60 min. This is the second stage where bacteria accumulate and are tightly packed within the biofilm and organized a compact sphere-shaped structure (mature-stage IBCs) (**Figure 1**).

The amount of IBCs is found between 3 and 700 in an infected patient's bladder—IBCs are composed of 10^4–10^5 bacterial cells [88]. There are numerous fibers surrounded on IBC bacteria that originate from the surface of pathogen and enclose pathogens in individualized sections. One of the main components present on the surface of IBCs called polysaccharide (sialic acid) that provides protection from the attack of immune system and environmental stress. The heterogeneous nature

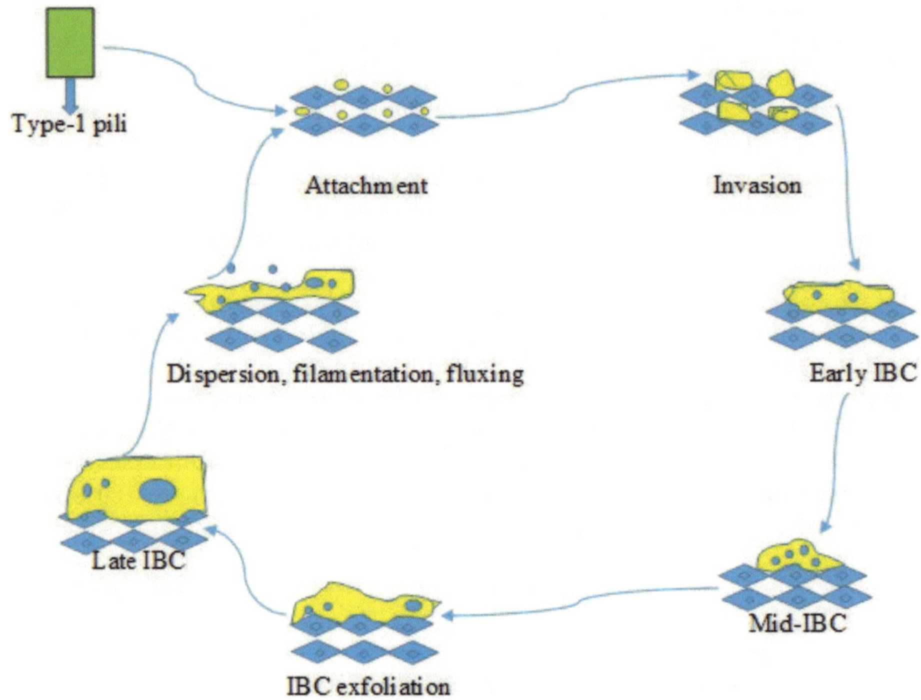

Figure 1.
Schematic diagram of the development of IBC cascade in uropathogenic E. coli (UPEC), taken by scanning electron microscope (SEM) images indicating different structural changes from attachment to dispersion and fluxing.

of IBCs, such as extracellular bacterial biofilm, composed of different subpopulation having distinct gene expression systems [89]. As IBCs expand, they induce the bacterial biofilm to cause interruption against cell membrane of host, producing a pod-like structure on the infected cell surface. Ultimately, UPEC detaches as filament or single rod at the IBC boundary and the infected cells are flux out into the lumen of bladder where can invade epithelial cells and restart the process through binding [85]. The inhibitor (SulA) of cell division has been observed to be crucial for dispersal and filamentation of UPEC from the bacterial biofilm. The patients suffered from urinary tract infections (UTIs) are more likely observed with the UPEC filaments in their urine, but not in comparison with healthy controls [90].

The formation of IBC is prevented by intense molecular blockages and during acute infection—development of chronic cystitis—the IBC numbers are higher, representing the significance of intracellular pathways in the pathogenesis of UTIs [88]. The cycle of IBC is dependent on FimH, causing interruption in the expression of type-1 pili after invasion to host cell, and disrupts normal development of IBC due to attenuation of UPEC [54]. The two-component system (QseBC) is a key factor influencing curli expression, formation of IBC and type-1 pili. Some studies indicated that the intracellular pathway of UPEC is necessary for the TCA cycle completion [47]. The techniques such as qPCR and DNA microarray analyses interpreting the UPEC expression patterns within IBC pathogen exposed that acquisition of iron in bacteria is upregulated, representing the significance of system biomass formation [91]. While in clinical isolates of UPEC, the iron acquisition patterns are prevalent [92]. Moreover, the pathogen *Klebsiella pneumoniae* is more commonly seen in community- and hospital-acquired infection. About up to 5% forms intracellular communities and is more predominant in hospitalized diabetic patients [93]. Likewise to UPEC, the *Klebsiella pneumoniae* invasion is mediated by type-1 pili and formation of IBC, although the differences occur in the expression kinetic of pili and filaments [90]. The ability to occupy an intracellular niche and persist within the host cell through transitioning from single microbial cell to the multicellular community is not confined to uropathogens. Researchers showed that by using different animal models and cell line of acute lung infection, the cluster formation occurs inside the lung airways due to *P. aeruginosa*, morphology similar to *Klebsiella* and UPEC (IBCs) [94]. The biofilm formation ability could be evolutionary adaptation of pathogens that enable the bacteria to persist within the host cell. All these findings represent the formation of IBC, a process that enables the bacteria to rapidly expand inside the host cell and take part in bacterial persistence.

4. Postantibiotic period: treatment strategy for biofilm

Broad-spectrum antibiotics are the drug of choice for the treatment of bacterial infections. Conventional antibiotics act as either killing the bacterial cell (bactericidal) or inhibiting the cell division (bacteriostatic). Numerous evidence shows that the use of antibiotics extensively causes damage to the host microbiota, producing a condition where invading bacteria can prevail and enhance the selective pressure against drug resistance [95]. Furthermore, surgery proceeded by administering antibiotics is highly successful in order to minimize the infection prophylactically. In certain cases, the perfect treatment of choice for foreign material associated with biofilm infections is the removal of infectious device. In some cases like pacemakers, cardiac implants and implantable prostheses, device removal is difficult [37]. Biofilm formation nature of bacteria that make them recalcitrant against different antimicrobial drugs is a result of prolonged treatment. There is a need for the irradiation or complete removal of these kinds of pathogens. Antibiotic resistance is not only due to increased resistance markers transmitted within the bacterial biofilm community, but also due to high metal ion concentration, low pH, and the presence of persistent cells that are metabolically inactive and inactivate the antibiotics

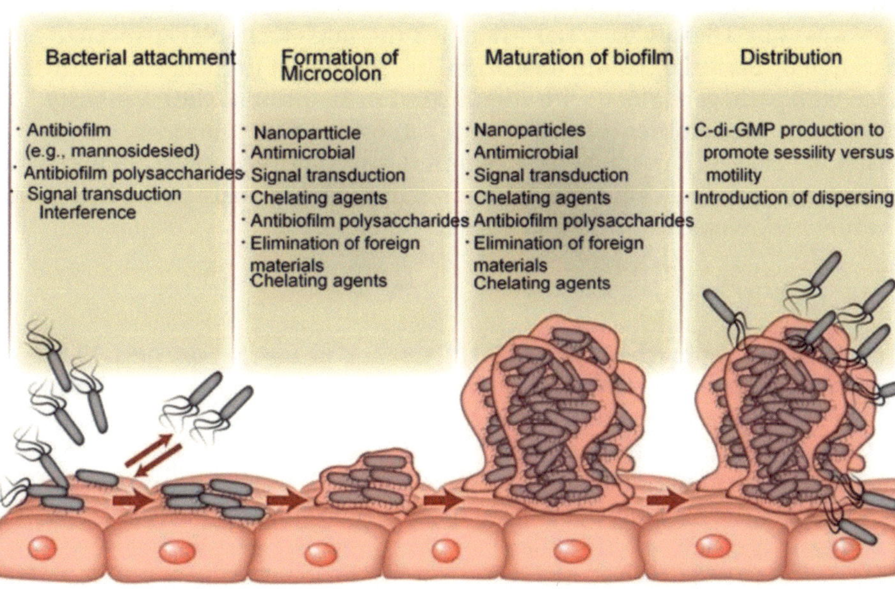

Figure 2.
Schematic diagram about the different stages in the development of biofilm and indicating the strategies to preventing and damaging the bacterial biofilm production at particular stages.

[31]. All these characteristics make bacterial biofilm more tolerant/resistant to antimicrobial drugs up to 1000-fold more when compared to planktonic bacterial cells [96]. Therefore, an alternative strategy must be investigated to combat the antibiotic resistant strains and make them vulnerable to antimicrobial drugs. Here below, we have mentioned some of the recent developments in strategies that are considered to prevent formation of biofilm by bactericidal mechanism or targeting distinct developmental stages of biofilm (**Figure 2**).

5. Bacterial killing strategies

5.1 Elimination of foreign material (indwelling devices) and abscess

There are studies that have reported that the presence of any foreign body (indwelling medical devices such as implants or prostheses or catheters) in low inoculums of *Staphylococcus aureus* (10^2 CFU/ml) in animal tissues was sufficient to form abscesses in the patients (95%) despite significant existence of leukocytes. In fact, this could be associated with the existence of any foreign material considerably intracellular bactericidal effects of body immune cells (leukocytes) and down-regulated the mechanism of phagocytosis [97]. The polymorphonuclear leukocytes cannot perform well in the presence of any foreign body because it provides a surface ideal for the bacterial attachment. Therefore, the existence of any foreign material considerably increases the chances of bacterial biofilm formation. This leads to the pathogen becoming more persistent and resistant against conventional antibiotics. Thus, potential therapeutic strategy is required for the elimination of such type of bacterial biofilm formations. Certain precautionary measures could be employed, for instance, to replace the infected devices used for medical purposes in the patients with a new one. Otherwise, it would be hard to overcome the problem regardless of applying various effective antimicrobial drugs in response to fastidious pathogens. Changing dialysis catheter if it is infected by the pathogens is another measure that could be taken. When pathogen forms biomass on the catheter, it could be the source of bacterial colonization leading to bacteremia which may be caused by a deadly bacterial strain. For the cure of catheter-associated infections caused by bacterial biofilm formation, it is important to change the catheter infected with pathogens along with administration of antibiotic intravenously during a short time in order to eradicate the pathogen before it invades into the bloodstream. However, in some cases, it is hard to change the catheter temporarily; therefore, antimicrobial drugs and other alternative therapy may be the best option for the minimal release of pathogens from the infected site.

5.2 Phage therapy

An alternative approach to antibiotic treatment is phage therapy [98]. Phages are present in a wide range in the environment. It can be isolated easily and ubiquitous in nature. Their host ranges from specific to narrow, they are able to self-replicate, and therefore, a small dosage may be sufficient to disturb the host microorganisms. Furthermore, high mutation rate of phage facilitates adaptation as conforming bacterial host aggregate mutations to fix in a specific environment. Phage therapy has various advantages during lytic cycle phage that does not enter prophage cycle and rarely transfers or contains a virulence gene, thus causing destruction of bacterial cell rapidly. Many phages are associated with EPS degrading protein [99] or spread during stationary growth phase; these features allow to persist inside the bacterial biofilm [100].

5.3 Antimicrobial peptides

This is another alternative approach used for the improvement of new type of antimicrobial drug, usually produced by innate immune response mechanism [101]. Contrary to that, their mechanism of action and antimicrobial spectrum activity must be defined more accurately before applying as a therapeutic strategy. Cathelicidin, for instance, possesses most essential type of antibacterial peptides. The biofilm formation of multidrug-resistant *Pseudomonas* strains, isolated from cystic fibrosis (CF) patients, is reduced considerably by BMAP-28, BMAP-27, and BMAP-29 [102]. According to a recent study by Pompilio et al. [102], antimicrobial activity of tobramycin against multidrug-resistant strains is less than cathelicidin peptides. This study indicates that the multidrug-resistant strains are vulnerable to cathelicidins due to antibiofilm agents. Another important group that can be used to assess the inhibitory effects is called lytic peptides. These peptides assist in attachment of lipopolysaccharides (LPS) to the cell membrane of pathogen and cause cell membrane disruption. The study on *Staphylococcus aureus* indicated that *in vitro* formation of biofilm is prohibited by the lytic peptide (PTP-7) and easily penetrates the bacterial biofilm causing death of the bacteria at a rate of 99.9%. This peptide has the capacity to bear extreme acidic environment and inhibit the biofilm formation of *Staphylococcus aureus* [103].

5.4 Silver nanoparticles

Many researchers have done research on the antimicrobial property of silver nanoparticles. Fey [37] found that the silver nanoparticles are the best alternative strategy to combat the bacterial biofilms. For example, antimicrobial agents (silver nanoparticles) have been incorporated with medical devices and have showed to inhibit the device-associated bacterial biofilms. Silver was frequently used as an antimicrobial agent for different pathogens over a 100 years; for instance, during World War 1, it was extensively used to sterilize the wound infections [104]. The antimicrobial activity of silver nanoparticles depends on the positively charged ions of metal and electrostatic interactions between negatively charged cell membrane of bacteria [105]. The thiol group in silver is the main cause of death in bacteria that play an important role in the inactivation of enzyme [106]. This is the reason why silver nanoparticles are increasingly used in response to various bacterial infections. The antimicrobial agents contain different properties such as high aspect ratios, nonimmunogenic, biocompatible, nonbiodegradable, ultralight weight, and easy cell membrane penetration. Due to such remarkable properties, we can apply silver nanoparticles in various applications such as infection therapy, gene therapy, and as antioxidants. The size of silver nanoparticles is typically smaller than 100 nm. The mechanism of action of silver nanoparticle is to interrupt the cell membrane of bacteria, generate the reactive oxygen species (ROS), interrupt the metabolic pathway, prevent the replication of DNA, disrupt the bacterial electron transport chain (ETC) [106], and release the toxic ions outside the bacterial cells that lead to the death of bacteria. There are large numbers of studies conducted regarding toxicity mechanism of silver nanoparticles in rabbits. There is a study that showed that silver nanoparticles inhibited bacterial biofilm formation against *Staphylococcus aureus,* without accumulating inside the host tissue [106, 107].

5.5 Polysaccharides

Bacterial cell-to-cell interaction mediated by the exopolysaccharides is a serious threat to the formation of biofilm and stabilization. Mutants incapable to export or synthesize such exopolysaccharides are usually deficient in the formation of biofilm and adherence and hence are extremely sensitive to killing through host immune defenses and antimicrobial drugs [108]. Recent studies showed that certain bacterial exopolysaccharides destabilize or prevent biofilm formation by some pathogenic species. For instance, the existence of *Pseudomonas aeruginosa* prevented biofilm formation of *S. epidermidis* in in vitro experiments [109]. Polysaccharides along with nonbactericidal antibiofilm characteristics have been separated from acellular biofilm (or biomass) extracts of various species [108]. The antibiofilm properties of *Pseudomonas aeruginosa* have the ability to act as signaling molecules that effect the expression of genes in susceptible pathogens, change the physical features of isolated bacterial cells, and prevent the protein-carbohydrate interactions. Most polysaccharides with antibiofilm properties allow a broad-spectrum inhibition of biofilm, while some are proficient of scattering preformed biofilms. So far, there are evidence suggests that polysaccharide with antibiofilm features acts as a surfactant molecule that alters the physical properties of abiotic surfaces and bacterial cells. Some results also show that polysaccharides might modulate the expression of genes of the recipient pathogenic bacteria by acting as signaling molecules [110]. Another potential mode of action of polysaccharide is to prevent competitively the multivalent protein-carbohydrate interactions [66]. As a result, polysaccharides with antibiofilm properties might block tip adhesins of pili and fimbriae, or block sugar or lectin-binding proteins that are present on the outer surface of pathogens. In pathogen *P. aeruginosa*, for instance, lectin-dependent adhesion to human cell is proficiently repressed by galactomannans [111]. This kind of polysaccharides that inhibit the biofilm could be a prominent strategy appropriate for the prevention of bacterial infections. Some scientist showed that antibiofilm polysaccharides can be used as an adjuvant because of enhancing antibiotic drug functions [108].

5.6 Interference with signal transference

Many studies have been carried out on biofilm inhibition caused by interruption of the pathogen signaling cascades. This is possible provided that the two-component systems in bacteria establish a dominant means of translating and intercepting the environmental changes. Signal transduction inhibition system plays a critical role in response to antimicrobial therapy because of this type of signaling cascade interruption. Not only does it kill the pathogen, but it also interferes with the gene expression. Two-component system (QseBC) is the best alternative candidate for targeting the drugs, particularly in Gram-negative biofilm-forming pathogens [112]. QseC/QseB establishes a significant association between the bacterial environmental signaling and the host stress response. The pathogen (*E. coli*) responds to autoinducer-3 in the intestine that is formed by the human stress hormones (such as epinephrine and norepinephrine) and gut flora. The cascade of signaling transduction comprises chemotaxis by activation of QseC and by using the serine receptor Tsr. In the quest for novel antimicrobial drugs and therapeutic targets, two-component system (QseBC) can play an important role to inhibit biofilm formation by blocking the binding of epinephrine or norepinephrine to QseC, as a result to reduce QseB/QseC signaling and decrease virulence and motility [113]. Studies have also suggested that the removal of QseC in EHEC and UPEC causes an excessive activation of response regulator QseB, owing to particular QseC phosphatase activity required for deactivation of QseB. The optimal strategy behind targeting the phosphate activity is to interfere with common gene expression in QseC containing pathogens [47]. Some other studies focused on the FsrATC/FsrA inhibitors in *E. faecalis*. The expression of gelE-sprE and FsrBDC control by the FsrC/FsrA leads to increase in the production of serine protease and gelatinase, both are crucial for the proper eDNA production [71].

5.7 Antimatrix agents

Apart from that, extracellular matrix with disrupting components is also very important to target the bacterial aggregates. Various observations exploited the inhibiting enzymes potentially involved in the modification or synthesis of cell wall-secreted or associated with EPS components. In these studies, use of engineered or naturally occurring enzyme and use of phage therapy as an enzyme delivery vehicle or to interrupt with matrix integrity by taking benefits from metal chelators have been recommended.

5.8 Chelating agents

Metal cations such as iron, magnesium, and calcium have been associated with stabilizing the matrix integrity [114]. Chelating agents indicated to cause interruption in the bacterial cell membrane stability besides disrupting the bacterial biomass structure [39]. *In vitro* study showed that biofilm formation was inhibited in various *Staphylococcus* species by sodium citrate [115]. Furthermore, eradication of bacterial biofilms in *in vitro* experiments is also facilitated by tetrasodium EDTA, while disodium EDTA only reduced the bacterial biofilm formations in *P. aeruginosa* and *Staphylococcus* species [116]. Current reports suggested that the solution of minocycline-EDTA was used to inhibit indwelling catheter-associated infections especially in children. There were no adverse side effects observed in patients treated with the solution of minocycline-EDTA but only a limited number (21%) of untreated group (control) developed infections [117]. Moreover, in hemodialysis patients, catheter-associated bloodstream infections were observed after applying minocycline-EDTA [118].

5.9 Enzyme

The main mechanism of active dispersal of bacterial biofilm is through the formation of extracellular enzymes (proteins) that act on several structural components (such as exopolysaccharides, surface proteins, and extracellular DNA) of the extracellular polymeric substances. These enzymes play an important role in the cell separation from the bacterial biofilm colonies and facilitate their planktonic discharge into the environment [119]. Through purifying and isolating these enzymes, therapist can apparently add them to preformed bacterial biofilms exogenously at raised concentrations, in order to make biofilm-associated bacteria more susceptible to antimicrobials/antibiotics and to achieve interventional dispersal of biofilms. For this purpose, several classes of enzymes (specifically proteases, glycoside hydrolases, and deoxyribonucleases) have been explored for the eradication of bacterial biofilms [119]. The enzymes dispersin-B and DNase-I have gained greater attention as possible antibiofilm agents, especially in response to Gram-positive bacteria. The DNase effect depends on its capability to interrupt the eDNA that is established within the bacterial biomass structure [73]. The treatment of DNase prevents biofilm formation in *Enterococcus* and *Staphylococcus* and dispersed bacterial biofilm [73]. For the treatment of patients with cystic fibrosis (CF), a recombinant enzyme (pulmozyme) is used in some cases [37]. However, treatment with dispersin-B represented to be more effective in response to *S. aureus* and *S. epidermidis* [77]. *In vitro* studies indicated that engineered dispersin-B used bacteriophage machinery in order to replicate during the stationary phase of cell growth, hence causing disruption of complete *E. coli* biofilms [120].

6. Conclusion

Currently, the removal of bacterial biofilm is the most challenging task for the clinicians and microbiologists. Antibiotics are not the best choice for the treatment of infections caused by bacteria forming biofilm. Biofilm formation allows the pathogen to adhere to the host surface under extreme condition and is resistant against a wide range of antibiotics. The choice of drug depends on the characteristics of the biofilm such as composition, age, solidity, and type of pathogens. These are the major components influencing the microbial susceptibility. As the bacterial biofilm matures, it enhances the accumulation of exopolymeric substance (EPS), attaches with the oxygen and nutrient gradients that effect bacterial growth rates and metabolism of cells, becomes impermeable, and reduces the activity of antimicrobial agents. This leads to resistance to most antibiotic regime. Therefore, novel potential therapeutic strategies should be considered to curb bacterial biofilm formation at specific stage without harming the pathogen. Antiadhesion and antimatrix agents are exciting strategies that may be used pending further investigation.

List of abbreviations

EPS	extracellular polymeric substance
DNS	deoxyribonuclease
CBP	carbohydrate-binding protein
DNA	deoxyribonucleic acid
CUP	chaperone-usher pathway
UPEC	uropathogenic *E. coli*
PGA	polyglucosamine
SCV	small colony variants
eDNA	environmental deoxyribonucleic acid
PQS	*Pseudomonas* quinolone signal
c-di-GMP	cyclic di-GMP
AL	alginate lyase
BS	*Bacillus subtilis*
TLR-4	toll-like receptor-4
IBC	intracellular bacterial communities
UTIs	urinary tract infections
SEM	scanning electron microscope
TCA	tricarboxylic acid
qPCR	quantitative polymerase chain reaction
HAIs	hospital-acquired infections
CFU	colony forming unit
MDR	multidrug resistant
LPS	lipopolysaccharides
ETC	electron transport chain
ROS	reactive oxygen species
EHEC	enterohemorrhagic *E. coli*
EDTA	ethylene-diamine-tetra-acetic acid
CF	cystic fibrosis

Author details

Mansab Ali Saleemi[1], Navindra Kumari Palanisamy[2] and Eng Hwa Wong[3]*

1 School of Biosciences, Taylor's University Lakeside Campus, Subang Jaya, Selangor Malaysia

2 Department of Medical Microbiology and Parasitology, Faculty of Medicine, Universiti Teknologi MARA (UiTM), Sungai Buloh Campus, Sungai Buloh, Selangor, Malaysia

3 School of Medicine, Taylor's University Lakeside Campus, Subang Jaya, Selangor, Malaysia

*Address all correspondence to: enghwa.wong@taylors.edu.my

References

[1] Hall-Stoodley L, Costerton JW, Stoodley P. Bacterial biofilms: From the natural environment to infectious diseases. Nature Reviews. Microbiology. 2004;**2**:95-108

[2] Costerton JW, Stewart PS, Greenberg EP. Bacterial biofilms: A common cause of persistent infections. Science. 1999;**284**:1318-1322

[3] O'Toole G, Kaplan HB, Kolter R. Biofilm formation as microbial development. Annual Review of Microbiology. 2000;**54**:49-79

[4] Kolter R. Biofilms in lab and nature: A molecular geneticist's voyage to microbial ecology. International Microbiology. 2010;**13**:1-7

[5] Pratt LA, Kolter R. Genetic analysis of *Escherichia coli* biofilm formation: Roles of flagella, motility, chemotaxis and type I pili. Molecular Microbiology. 1998;**30**:285-293

[6] Prigent-Combaret C, Brombacher E, Vidal O, Ambert A, Lejeune P, Landini P, et al. Complex regulatory network controls initial adhesion and biofilm formation in *Escherichia coli* via regulation of the csgD gene. Journal of Bacteriology. 2001;**183**:7213-7223

[7] Parsek MR, Singh PK. Bacterial biofilms: An emerging link to disease pathogenesis. Annual Review of Microbiology. 2003;**57**:677-701

[8] Lenz AP, Williamson KS, Pitts B, Stewart PS, Franklin MJ. Localized gene expression in *Pseudomonas aeruginosa* biofilms. Applied and Environmental Microbiology. 2008;**74**:4463-4471

[9] Monds RD, O'Toole GA. The developmental model of microbial biofilms: Ten years of a paradigm up for review. Trends in Microbiology. 2009;**17**:73-87

[10] Whiteley M, Bangera MG, Bumgarner RE, Parsek MR, Teitzel GM, Lory S, et al. Gene expression in *Pseudomonas aeruginosa* biofilms. Nature. 2001;**413**:860-864

[11] Stanley NR, Britton RA, Grossman AD, Lazazzera BA. Identification of catabolite repression as a physiological regulator of biofilm formation by *Bacillus subtilis* by use of DNA microarrays. Journal of Bacteriology. 2003;**185**:1951-1957

[12] Schembri MA, Kjaergaard K, Klemm P. Global gene expression in *Escherichia coli* biofilms. Molecular Microbiology. 2003;**48**:253-267

[13] Bagge N, Hentzer M, Andersen JB, Ciofu O, Givskov M, Hoiby N. Dynamics and spatial distribution of beta lactamase expression in *Pseudomonas aeruginosa* biofilms. Antimicrobial Agents and Chemotherapy. 2004;**48**:1168-1174

[14] Beloin C, Valle J, Latour-Lambert P, Faure P, Kzreminski M, Balestrino D, et al. Global impact of mature biofilm lifestyle on *Escherichia coli* K-12 gene expression. Molecular Microbiology. 2004;**51**:659-674

[15] Vuong C, Voyich JM, Fischer ER, Braughton KR, Whitney AR, DeLeo FR, et al. Polysaccharide intercellular adhesin (PIA) protects *Staphylococcus epidermidis* against major components of the human innate immune system. Cellular Microbiology. 2004;**6**:269-275

[16] Zhang L, Mah TF. Involvement of a novel efflux system in biofilm-specific resistance to antibiotics. Journal of Bacteriology. 2008;**190**:4447-4452

[17] Klebensberger J, Birkenmaier A, Geffers R, Kjelleberg S, Philipp B. SiaA and SiaD are essential for

inducing auto aggregation as a specific response to detergent stress in *Pseudomonas aeruginosa*. Environmental Microbiology. 2009;**11**:3073-3086

[18] Flemming HC, Wingender J. The biofilm matrix. Nature Reviews. Microbiology. 2010;**8**:623-633

[19] Tielker D, Hacker S, Loris R, Strathmann M, Wingender J, Wilhelm S, et al. *Pseudomonas aeruginosa* lectin LecB is located in the outer membrane and is involved in biofilm formation. Microbiology. 2005;**151**:1313-1323

[20] Branda SS, Chu F, Kearns DB, Losick R, Kolter R. A major protein component of the *Bacillus subtilis* biofilm matrix. Molecular Microbiology. 2006;**59**:1229-1238

[21] Diggle SP, Stacey RE, Dodd C, Camara M, Williams P, Winzer K. The galactophilic lectin, LecA, contributes to biofilm development in *Pseudomonas aeruginosa*. Environmental Microbiology. 2006;**8**:1095-1104

[22] Conrad A, Suutari MK, Keinanen MM, Cadoret A, Faure P, Mansuy-Huault L, et al. Fatty acids of lipid fractions in extracellular polymeric substances of activated sludge flocs. Lipids. 2003;**38**:1093-1105

[23] Sauer K, Cullen MC, Rickard AH, Zeef LA, Davies DG, Gilbert P. Characterization of nutrient-induced dispersion in *Pseudomonas aeruginosa* PAO1 biofilm. Journal of Bacteriology. 2004;**186**:7312-7326

[24] Gjermansen M, Ragas P, Sternberg C, Molin S, Tolker Nielsen T. Characterization of starvation-induced dispersion in *Pseudomonas putida* biofilms. Environmental Microbiology. 2005;**7**:894-906

[25] Ma L, Conover M, Lu H, Parsek MR, Bayles K, Wozniak DJ. Assembly and development of the *Pseudomonas aeruginosa* biofilm matrix. PLoS Pathogens. 2009;**5**:e1000354

[26] Walters MC 3rd, Roe F, Bugnicourt A, Franklin MJ, Stewart PS. Contributions of antibiotic penetration, oxygen limitation, and low metabolic activity to tolerance of *Pseudomonas aeruginosa* biofilms to ciprofloxacin and tobramycin. Antimicrobial Agents and Chemotherapy. 2003;**47**:317-323

[27] Jefferson KK, Goldmann DA, Pier GB. Use of confocal microscopy to analyze the rate of vancomycin penetration through *Staphylococcus aureus* biofilms. Antimicrobial Agents and Chemotherapy. 2005;**49**:2467-2473

[28] Mai-Prochnow A, Lucas-Elio P, Egan S, Thomas T, Webb JS, Sanchez-Amat A, et al. Hydrogen peroxide linked to lysine oxidase activity facilitates biofilm differentiation and dispersal in several gram-negative bacteria. Journal of Bacteriology. 2008;**190**:5493-5501

[29] Lewis K. Persister cells and the riddle of biofilm survival. Biochemistry (Mosc). 2005;**70**:267-274

[30] Domka J, Lee J, Bansal T, Wood TK. Temporal gene expression in *Escherichia coli* K-12 biofilms. Environmental Microbiology. 2007;**9**:332-346

[31] Lewis K. Multidrug tolerance of biofilms and persister cells. Current Topics in Microbiology and Immunology. 2008;**322**:107-131

[32] Cerca N, Jefferson KK, Oliveira R, Pier GB, Azeredo J. Comparative antibody-mediated phagocytosis of *Staphylococcus epidermidis* cells grown in a biofilm or in the planktonic state. Infection and Immunity. 2006;**74**:4849-4855

[33] Koch C, Hoiby N. Pathogenesis of cystic fibrosis. Lancet. 1993;**341**:1065-1069

[34] Govan JR, Deretic V. Microbial pathogenesis in cystic fibrosis: Mucoid *Pseudomonas aeruginosa* and *Burkholderia cepacia*. Microbiological Reviews. 1996;**60**:539-574

[35] Kuramitsu HK, Wang BY. The whole is greater than the sum of its parts: Dental plaque bacterial interactions can affect the virulence properties of cariogenic *Streptococcus mutans*. American Journal of Dentistry. 2011;**24**:153-154

[36] Venditti M, Biavasco F, Varaldo PE, Macchiarelli A, De Biase L, Marino B, et al. Catheter-related endocarditis due to glycopeptide-resistant *Enterococcus faecalis* in a transplanted heart. Clinical Infectious Diseases. 1993;**17**:524-525

[37] Fey PD. Modality of bacterial growth presents unique targets: How do we treat biofilm-mediated infections? Current Opinion in Microbiology. 2010;**13**:610-615

[38] Foxman B. The epidemiology of urinary tract infection. Nature Reviews. Urology. 2010;**7**:653-660

[39] Donlan RM. Biofilms: Microbial life on surfaces. Emerging Infectious Diseases. 2002;**8**:881-890

[40] Beloin C, Roux A, Ghigo JM. *Escherichia coli* biofilms. Current Topics in Microbiology and Immunology. 2008;**322**:249-289

[41] Toutain CM, Caizza NC, Zegans ME, O'Toole GA. Roles for flagellar stators in Biofilm formation by *Pseudomonas aeruginosa*. Research in Microbiology. 2007;**158**:471-477

[42] Lemon KP, Higgins DE, Kolter R. Flagellar motility is critical for *Listeria monocytogenes* biofilm formation. Journal of Bacteriology. 2007;**189**:4418-4424

[43] Klausen M, Aaes-Jorgensen A, Molin S, Tolker-Nielsen T. Involvement of bacterial migration in the development of complex multicellular structures in *Pseudomonas aeruginosa* biofilms. Molecular Microbiology. 2003a;**50**:61-68

[44] Klausen M, Heydorn A, Ragas P, Lambertsen L, Aaes-Jorgensen A, Molin S, et al. Biofilm formation by *Pseudomonas aeruginosa* wild type, flagella and type IV pili mutants. Molecular Microbiology. 2003b;**48**:1511-1524

[45] Watnick PI, Kolter R. Steps in the development of a *Vibrio cholerae* El Tor biofilm. Molecular Microbiology. 1999;**34**:586-595

[46] Schmidt J, Musken M, Becker T, Magnowska Z, Bertinetti D, Moller S, et al. The *Pseudomonas aeruginosa* chemotaxis methyltransferase CheR1 impacts on bacterial surface sampling. PLoS One. 2011;**6**:e18184

[47] Hadjifrangiskou M, Kostakioti M, Chen SL, Henderson JP, Greene SE, Hultgren SJ. A central metabolic circuit controlled by QseC in pathogenic *Escherichia coli*. Molecular Microbiology. 2011;**80**:1516-1529

[48] Dunne WM Jr. Bacterial adhesion: Seen any good biofilms lately? Clinical Microbiology Reviews. 2002;**15**:155-166

[49] Mulvey MA, Lopez-Boado YS, Wilson CL, Roth R, Parks WC, Heuser J, et al. Induction and evasion of host defenses by type-1-piliated uropathogenic *Escherichia coli*. Science. 1998;**282**:1494-1497

[50] Hung CS, Bouckaert J, Hung D, Pinkner J, Widberg C, DeFusco A, et al. Structural basis of tropism of *Escherichia coli* to the bladder during urinary tract infection. Molecular Microbiology. 2002;**44**:903-915

[51] Martinez JJ, Mulvey MA, Schilling JD, Pinkner JS, Hultgren SJ. Type 1 pilus-Mediated bacterial invasion of bladder epithelial cells. The EMBO Journal. 2000;**19**:2803-2812

[52] Nilsson LM, Yakovenko O, Tchesnokova V, Thomas WE, Schembri MA, Vogel V, et al. The cysteine bond in the *Escherichia coli* FimH adhesin is critical for adhesion under flow conditions. Molecular Microbiology. 2007;**65**:1158-1169

[53] Garofalo CK, Hooton TM, Martin SM, Stamm WE, Palermo JJ, Gordon JI, et al. *Escherichia coli* from urine of female patients with urinary tract infections is competent for intracellular bacterial community formation. Infection and Immunity. 2007;**75**:52-60

[54] Wright KJ, Seed PC, Hultgren SJ. Development of intracellular bacterial communities of uropathogenic *Escherichia coli* depends on type 1 pili. Cellular Microbiology. 2007;**9**:2230-2241

[55] Chen SL, Hung CS, Xu J, Reigstad CS, Magrini V, Sabo A, et al. Identification of genes subject to positive selection in uropathogenic strains of *Escherichia coli:* A comparative genomics approach. Proceedings of the National Academy of Sciences. 2006;**103**:5977-5982

[56] Chen SL, Hung CS, Pinkner JS, Walker JN, Cusumano CK, Li Z, et al. Positive selection identifies an in vivo role for FimH during urinary tract infection in addition to mannose binding. Proceedings of the National Academy of Sciences. 2009;**106**:22439-22444

[57] Hasman H, Chakraborty T, Klemm P. Antigen-43 mediated auto aggregation of *Escherichia coli* is blocked by fimbriation. Journal of Bacteriology. 1999;**181**:4834-4841

[58] Uhlich GA, Cooke PH, Solomon EB. Analyses of the red-dry-rough phenotype of an *Escherichia coli* O157:H7 strain and its role in biofilm formation and resistance to antibacterial agents. Applied and Environmental Microbiology. 2006;**72**:2564-2572

[59] Mohamed JA, Teng F, Nallapareddy SR, Murray BE. Pleiotrophic effects of 2 *Enterococcus faecalis* sagA-like genes, salA and salB, which encode proteins that are antigenic during human infection, on biofilm formation and binding to collagen type I and fibronectin. The Journal of Infectious Diseases. 2006;**193**:231-240

[60] Toledo-Arana A, Valle J, Solano C, Arrizubieta MJ, Cucarella C, Lamata M, et al. The enterococcal surface protein, Esp, is involved in *Enterococcus faecalis* biofilm formation. Applied and Environmental Microbiology. 2001;**67**:4538-4545

[61] Kemp KD, Singh KV, Nallapareddy SR, Murray BE. Relative contributions of *Enterococcus faecalis* OG1RF sortase-encoding genes, srtA and bps (srtC), to biofilm formation and a murine model of urinary tract infection. Infection and Immunity. 2007;**75**:5399-5404

[62] Kai-Larsen Y, Luthje P, Chromek M, Peters V, Wang X, Holm A, et al. Uropathogenic *Escherichia coli* modulates immune responses and its curli fimbriae interact with the antimicrobial peptide LL-37. PLoS Pathogens. 2010;**6**:e1001010

[63] Cegelski L, Pinkner JS, Hammer ND, Cusumano CK, Hung CS, Chorell E, et al. Small-molecule inhibitors target *Escherichia coli* amyloid biogenesis and biofilm formation. Nature Chemical Biology. 2009;**5**:913-919

[64] Wang X, Preston JF 3rd, Romeo T. The pgaABCD locus of *Escherichia coli* promotes the synthesis of a

polysaccharide adhesin required for biofilm formation. Journal of Bacteriology. 2004;**186**:2724-2734

[65] Harmsen M, Yang L, Pamp SJ, Tolker-Nielsen T. An update on *Pseudomonas aeruginosa* biofilm formation, tolerance, and dispersal. FEMS Immunology and Medical Microbiology. 2010;**59**:253-268

[66] Wittschier N, Lengsfeld C, Vorthems S, Stratmann U, Ernst JF, Verspohl EJ, et al. Large molecules as anti-adhesive compounds against pathogens. The Journal of Pharmacy and Pharmacology. 2007;**59**:777-786. [PubMed]

[67] Starkey M, Hickman JH, Ma L, Zhang N, DeLong S, Hinz A, et al. *Pseudomonas aeruginosa* rugose small-colony variants have adaptations that likely promote persistence in the cystic fibrosis lung. Journal of Bacteriology. 2009;**191**:3492-3503

[68] Yang L, Barken KB, Skindersoe ME, Christensen AB, Givskov M, Tolker-Nielsen T. Effects of iron on DNA release and biofilm development by *Pseudomonas aeruginosa*. Microbiology. 2007;**153**:1318-1328

[69] Allesen-Holm M, Barken KB, Yang L, Klausen M, Webb JS, Kjelleberg S, et al. A characterization of DNA release in *Pseudomonas aeruginosa* cultures and biofilms. Molecular Microbiology. 2006;**59**:1114-1128

[70] Barken KB, Pamp SJ, Yang L, Gjermansen M, Bertrand JJ, Klausen M, et al. Roles of type IV pili, flagellum-mediated motility and extracellular DNA in the formation of mature multicellular structures in *Pseudomonas aeruginosa* biofilms. Environmental Microbiology. 2008;**10**:2331-2343

[71] Thomas VC, Thurlow LR, Boyle D, Hancock LE. Regulation of autolysis-dependent extracellular DNA release by *Enterococcus faecalis* extracellular proteases influences biofilm development. Journal of Bacteriology. 2008;**190**:5690-5698

[72] Thomas VC, Hiromasa Y, Harms N, Thurlow L, Tomich J, Hancock LE. A fratricidal mechanism is responsible for eDNA release and contributes to biofilm development of *Enterococcus faecalis*. Molecular Microbiology. 2009;**72**:1022-1036

[73] Guiton PS, Hung CS, Kline KA, Roth R, Kau AL, Hayes E, et al. Contribution of autolysin and Sortase A during *Enterococcus faecalis* DNA-dependent biofilm development. Infection and Immunity. 2009;**77**:3626-3638

[74] Hong SH, Lee J, Wood TK. Engineering global regulator Hha of *Escherichia coli* to control biofilm dispersal. Microbial Biotechnology. 2010;**3**:717-728

[75] Rowe MC, Withers HL, Swift S. Uropathogenic *Escherichia coli* forms biofilm aggregates under iron restriction that disperse upon the supply of iron. FEMS Microbiology Letters. 2010;**307**:102-109

[76] Karatan E, Watnick P. Signals, regulatory networks, and materials that build and break bacterial biofilms. Microbiology and Molecular Biology Reviews. 2009;**73**:310-347

[77] Kaplan JB. Biofilm dispersal: Mechanisms, clinical implications, and potential therapeutic uses. Journal of Dental Research. 2010;**89**:205-218

[78] Wood TK, Hong SH, Ma Q. Engineering biofilm formation and dispersal. Trends in Biotechnology. 2010;**29**:87-94

[79] Boyd A, Chakrabarty AM. Role of alginate lyase in cell detachment of *Pseudomonas aeruginosa*. Applied

and Environmental Microbiology. 1994;**60**:2355-2359

[80] Wang X, Dubey AK, Suzuki K, Baker CS, Babitzke P, Romeo T. CsrA post-transcriptionally represses pgaABCD, responsible for synthesis of a biofilm polysaccharide adhesin of *Escherichia coli*. Molecular Microbiology. 2005;**56**:1648-1663

[81] Webb JS, Thompson LS, James S, Charlton T, Tolker-Nielsen T, Koch B, et al. Cell death in *Pseudomonas aeruginosa* biofilm development. Journal of Bacteriology. 2003;**185**:4585-4592

[82] Lam H, Oh DC, Cava F, Takacs CN, Clardy J, de Pedro MA, et al. D-amino acids govern stationary phase cell wall remodeling in bacteria. Science. 2009;**325**:1552-1555

[83] Kolodkin-Gal I, Romero D, Cao S, Clardy J, Kolter R, Losick R. D-amino acids trigger biofilm disassembly. Science. 2010;**328**:627-629

[84] Kolodkin-Gal I, Cao S, Chai L, Bottcher T, Kolter R, Clardy J, et al. A self-produced trigger for biofilm disassembly that targets exopolysaccharide. Cell. 2012;**149**:684-692

[85] Justice SS, Hung C, Theriot JA, Fletcher DA, Anderson GG, Footer MJ, et al. Differentiation and developmental pathways of uropathogenic *Escherichia coli* in urinary tract pathogenesis. Proceedings of the National Academy of Sciences. 2004;**101**:1333-1338

[86] Eto DS, Jones TA, Sundsbak JL, Mulvey MA. Integrin mediated host cell invasion by type 1-piliated uropathogenic *Escherichia coli*. PLoS Pathogens. 2007;**3**:e100

[87] Bishop BL, Duncan MJ, Song J, Li G, Zaas D, Abraham SN. Cyclic AMP-regulated exocytosis of *Escherichia coli* from infected bladder epithelial cells. Nature Medicine. 2007;**13**:625-630

[88] Schwartz DJ, Chen SL, Hultgren SJ, Seed PC. Population dynamics and niche distribution of uropathogenic *Escherichia coli* during acute and chronic urinary tract infection. Infection and Immunity. 2011;**79**:4250-4259

[89] Anderson GG, Martin SM, Hultgren SJ. Host subversion by formation of intracellular bacterial communities in the urinary tract. Microbes and Infection. 2004;**6**:1094-1101

[90] Rosen DA, Hooton TM, Stamm WE, Humphrey PA, Hultgren SJ. Detection of intracellular bacterial communities in human urinary tract infection. PLoS Medicine. 2007;**4**:e329

[91] Reigstad CS, Hultgren SJ, Gordon JI. Functional genomic studies of uropathogenic *Escherichia coli* and host urothelial cells when intracellular bacterial communities are assembled. The Journal of Biological Chemistry. 2007;**282**:21259-21267

[92] Henderson JP, Crowley JR, Pinkner JS, Walker JN, Tsukayama P, Stamm WE, et al. Quantitative metabolomics reveals an epigenetic blueprint for iron acquisition in uropathogenic *Escherichia coli*. PLoS Pathogens. 2009;**5**:e1000305

[93] Hansen DS, Gottschau A, Kolmos HJ. Epidemiology of *Klebsiella* bacteremia: A case control study using *Escherichia coli* bacteremia as control. The Journal of Hospital Infection. 1998;**38**:119-132

[94] Garcia-Medina R, Dunne WM, Singh PK, Brody SL. *Pseudomonas aeruginosa* acquires biofilm-like properties within airway epithelial cells. Infection and Immunity. 2005;**73**:8298-8305

[95] Ubeda C, Taur Y, Jenq RR, Equinda MJ, Son T, Samstein M, et al.

Vancomycin-resistant *Enterococcus* domination of intestinal microbiota is enabled by antibiotic treatment in mice and precedes bloodstream invasion in humans. The Journal of Clinical Investigation. 2010;**120**:4332-4341

[96] Hoiby N, Bjarnsholt T, Givskov M, Molin S, Ciofu O. Antibiotic resistance of bacterial biofilms. International Journal of Antimicrobial Agents. 2010;**35**:322-332

[97] Zimmerli W, Lew PD, Waldvogel FA. Pathogenesis of foreign body infection. Evidence for a local granulocyte defect. The Journal of Clinical Investigation. 1984;**73**(4):1191-1200

[98] Donlan RM. Preventing biofilms of clinically relevant organisms using bacteriophage. Trends in Microbiology. 2009;**17**:66-72

[99] Sutherland IW, Hughes KA, Skillman LC, Tait K. The interaction of phage and biofilms. FEMS Microbiology Letters. 2004;**232**:1-6

[100] Burrowes B, Harper DR, Anderson J, McConville M, Enright MC. Bacteriophage therapy: Potential uses in the control of antibiotic-resistant pathogens. Expert Review of Anti-Infective Therapy. 2011;**9**:775-785

[101] Yang D, Biragyn A, Kwak LW, Oppenheim JJ. Mammalian defensins in immunity: More than just microbicidal. Trends in Immunology. 2002;**23**:291-296

[102] Pompilio A, Scocchi M, Pomponio S, Guida F, Di Primio A, Fiscarelli E, et al. Antibacterial and anti-biofilm effects of cathelicidin peptides against pathogens isolated from cystic fibrosis patients. Peptides. 2011;**32**:1807-1814

[103] Kharidia R, Liang JF. The activity of a small lytic peptide PTP-7 on Staphylococcus aureus biofilms. Journal of Microbiology. 2011;**49**:663-668

[104] Chen X, Schluesener HJ. Nanosilver: A nanoproduct in medical application. Toxicology Letters. 2008;**176**:1-12

[105] Kim JS, Kuk E, Yu KN, Kim JH, Park SJ, Lee HJ, et al. Antimicrobial effects of silver nanoparticles. Nanomedicine. 2007;**3**:95-101

[106] Feng QL, Wu J, Chen GQ, Cui FZ, Kim TN, Kim JO. A mechanistic study of the anti-bacterial effect of silver ions on Escherichia coli and Staphylococcus aureus. Journal of Biomedical Materials Research. 2000;**52**:662-668

[107] Secinti KD, Ozalp H, Attar A, Sargon MF. Nanoparticle silver ion coatings inhibit biofilm formation on titanium implants. Journal of Clinical Neuroscience. 2011;**18**:391-395

[108] Rendueles O, Kaplan JB, Ghigo JM. Antibiofilm polysaccharides. Environmental Microbiology. 2012;**15**(2):334-346. DOI: 10.1111/j.1462 2920.2012.02810.x.

[109] Pihl M, Davies JR, Chavez de Paz LE, Svensater G. Differential effects of Pseudomonas aeruginosa on biofilm formation by different strains of Staphylococcus epidermidis. FEMS Immunology and Medical Microbiology. 2010;**59**:439-446

[110] Kim Y, Oh S, Kim SH. Released exopolysaccharide (r-EPS) produced from probiotic bacteria reduce biofilm formation of enterohemorrhagic Escherichia coli O157:H7. Biochemical and Biophysical Research Communications. 2009;**379**:324-329. [PubMed]

[111] Zinger-Yosovich KD, Gilboa-Garber N. Blocking of *Pseudomonas aeruginosa* and *Ralstonia solanacearum* lectins by plant and microbial branched polysaccharides used as food additives. Journal of Agricultural and Food Chemistry. 2009;**57**:6908-6913. [PubMed]

[112] Wang X, Wang Q, Yang M, Xiao J, Liu Q, Wu H, et al. QseBC controls flagellar motility, fimbrial hemagglutination and intracellular virulence in fish pathogen Edwardsiella tarda. Fish & Shellfish Immunology. 2011;**30**:944-953

[113] Prüß BM. Involvement of Two-Component Signaling on Bacterial Motility and Biofilm Development. Journal of Bacteriology. 2017;**199**(18):e00259-e00217

[114] Raad II, Fang X, Keutgen XM, Jiang Y, Sherertz R, Hachem R. The role of chelators in preventing biofilm formation and catheter-related bloodstream infections. Current Opinion in Infectious Diseases. 2008;**21**:385-392

[115] Shanks RM, Sargent JL, Martinez RM, Graber ML, O'Toole GA. Catheter lock solutions influence staphylococcal biofilm formation on abiotic surfaces. Nephrology, Dialysis, Transplantation. 2006;**21**:2247-2255

[116] Bookstaver PB, Williamson JC, Tucker BK, Raad II, Sherertz RJ. Activity of novel antibiotic lock solutions in a model against isolates of catheter-related bloodstream infections. The Annals of Pharmacotherapy. 2009;**43**:210-219

[117] Chatzinikolaou I, Zipf TF, Hanna H, Umphrey J, Roberts WM, Sherertz R, et al. Minocycline-ethylene-diaminetetraacetate lock solution for the prevention of implantable port infections in children with cancer. Clinical Infectious Diseases. 2003;**36**:116-119

[118] Bleyer AJ, Mason L, Russell G, Raad II, Sherertz RJ. A randomized, controlled trial of a new vascular catheter flush solution (minocycline-EDTA) in temporary hemodialysis access. Infection Control and Hospital Epidemiology. 2005;**26**:520-524

[119] Fleming D, Rumbaugh KP. Approaches to dispersing medical biofilm. Microorganisms. 2017;**5**:15. DOI: 10.3390/microorganisms5020015

[120] Lu TK, Collins JJ. Dispersing biofilms with engineered enzymatic bacteriophage. Proceedings of the National Academy of Sciences. 2007;**104**:11197-11202

Chapter 4

Origin and Control Strategies of Biofilms in the Cultural Heritage

Laura E. Castrillón Rivera, Alejandro Palma Ramos, Jorge I. Castañeda Sánchez and María Elisa Drago Serrano

Abstract

Biodeterioration is defined as the undesirable change in the properties of materials caused by the activity of biological agents. This process is complex and involves alterations in the physicochemical and mechanical properties by the action of organisms and depends on the microorganisms involved, type of substrate, and environmental conditions. The biodeterioration of cultural heritage is the physical or chemical damage caused by microorganisms on objects, monuments, or buildings that belong to the cultural heritage. Among the main materials that can be affected are: stone, metal, ceramic, polymers, and other materials. Among the main undesirable effects to these materials are: discoloration, dissolution, rupture, and efflorescence among others. Biofilms represent the usual form of growth of bacteria and consist of communities of microorganisms that grow attached to an inert surface or a living tissue, surrounded by an extracellular matrix that they themselves synthesize. The importance of biodeterioration by biofilms is mainly related to changes in pH values, ionic concentrations, oxide-reduction reactions in the biofilm thickness, and in the interface with the substrate and enzymatic degradation. This chapter presents evidence of the participation of biofilms and associated mechanisms in biodeterioration as well as the main prevention and control strategies.

Keywords: biofilms, cultural heritage, biodeterioration, art biodeterioration

1. Cultural heritage

The concept of cultural heritage refers to the cultural inheritance that corresponds to a given community and as such is protected and communicated to both the present and the future generations. This concept is subjective and dynamic, and it does not depend on the objects or goods, but on the values that society in general attributes to them at each moment of history and that determine which goods are those that must be protected and preserved for posterity. Therefore, the identification, protection, conservation, and dissemination of the world's cultural heritage are one of the most recognized tasks of the United Nations Organization for Education, Science and Culture [1–3].

Cultural heritage is made up of tangible or intangible assets that history has left to a country and those citizens in the present grant it a special and relevant historical, scientific, symbolic or esthetic importance. This inheritance left by our ancestors and received today is the clear testimony of their existence and vision of the world [4].

IntechOpen

The tangible heritage is also called material heritage, and there are movable and immovable property such as the objects of the artistic collections such as the collection of religious, ethnographic, technological, historical, artistic, and archeological and artisan objects. The real tangible heritage is made up of archeological sites, engineering works, places, buildings, and architectural ensembles.

The intangible cultural heritage includes the wealth of knowledge, also living expressions inherited from our ancestors and transmitted to our descendants, such as language, oral traditions, customs, performing arts, ways of life, rituals, festive events, knowledge, and practices related to nature and the universe, as well as knowledge and techniques linked to traditional crafts [5].

Cultural heritage is a nonrenewable resource with regard to its past, and that is why it manifests itself tangibly as an untouchable and irreplaceable resource of a people. This heritage is always linked to the human collective, since it is men and women who produce it, and therefore it is what gives identity, origin, and continuity to our people. Hence, it is the responsibility of all its conservation and restoration for which it is essential to know what are its main threats in order to prevent, delay its deterioration and, if necessary, restore this heritage.

The alteration of cultural assets is the characteristic of the continuous cycle of disintegration and reconstruction, and it is a natural condition since all matter follows a process of alteration, degradation, or decomposition which means that original physical, chemical, and optical qualities are lost and enter a process of instability promoted by factors or agents of deterioration that are of two types:

Intrinsic: it depends on the nature of the material, manufacturing technique, and procedures that were used to perform the work.

Extrinsic: It depends on the sources external to the object such as environmental factors (light, relative humidity, temperature, and air pollutants), anthropogenic factors (handling, use, consultation, vandalism, tourism, etc.), biological factors (microorganisms, plants, rodents, and insects), and catastrophic factors (floods, fires, etc.) [6–8].

Among the main mechanisms of deterioration, three processes are known:

Physical or mechanical processes where the behavior of the material is modified, where several mechanical forces participate (compression, traction, etc.). These change the behavior of a material without modifying its chemical composition.

Chemical processes: are those that compromise a chemical reaction that transforms the matter.

Biological processes: where living organisms, such as microorganisms, insects, rodents, plants, etc., can chemically attack the material or its mechanical resistance. This process is also known as biodeterioration, which has been defined as "undesirable changes in the properties of materials caused by the vital activity of organisms" [9].

2. Biofilms as biodeteriogens

It is expected that works of art last for a long period of time; however, these suffer deterioration, and previously it was believed that chemical and physical processes were the dominant factors in the degradation of materials. Since 1967 and in latter decades, dogma has changed, and today it is assumed that microorganisms only by their very presence can cause damage by esthetic destruction of the materials since they inhabit them and penetrate causing their loss due to their acid corrosion, enzymatic degradation, and mechanical attack [10]. These microorganisms can grow in nature in large, silty colonies known as biofilms where relationships and dependencies are established [11–13].

Biofilms are microbial monospecies or multispecies (consortium) communities that have demonstrated the most successful colonization among microorganisms are ubiquitous in nature and responsible for many diseases. They are considered growing communities of microorganisms embedded in a self-produced exopolysaccharides matrix and are attached to an inert surface or living tissue [14–16]. The microorganisms in biofilms have properties that are not shared by free organisms, and the requirements for the formation of biofilms are simple: surface, moisture, nutrients, and microorganisms. This complex microbial organization that can consist mainly of bacteria and fungi, offers several advantages for its survival, such as resistance to environmental stress through the formation of stable microcolonies, facilitates the exchange of genetic material, and there is accumulation of nutrients and water in its matrix that offers protection against toxic substances (biocides and antibiotics) and against desiccation as well as immune defenses in the case of the formation of biofilms in higher organisms [17].

The importance of biofilms in the biodeterioration of cultural heritage has been reported for several decades and is related to: (a) modifications in pH values and ionic concentrations, (b) reduction oxide conditions in the interface of biofilms and substrate, (c) covering the surface and masking its properties, (d) increasing the leaching of additives and monomers outside the polymer matrix by microbial degradation, (e) releasing enzymes that lead to embrittlement and loss of mechanical stability, (f) accumulating water that penetrates the matrix causing swelling and increased conductivity, and (g) excretion of lipophilic pigments among others [18, 19].

3. Biodeterioration of stone

The mineralogical nature of the stone, its surface properties, and environmental conditions act synergistically for its bioreceptivity (ability to be colonized by microorganisms), and its intensity will depend on the concentration of pollutants, microclimatic conditions, and anthropogenic eutrophication of the atmosphere [20].

The climatic conditions in which the monuments or architectural structures are exposed can be the wind that wears the rock eroding it, the solar radiation causing discoloration, the temperature, as well as the rain, snow and humidity that induce the process of physical and chemical wear. These factors affect the stability of the matrix of the stone or act through chemical corrosion forming minerals by oxidation and hydration reactions as well as by the dissolution of carbonates and solubilization of some elements of minerals with silicates [21, 22].

The microbial communities after the interaction with biotic and abiotic factors are developed using the stone as a substrate and are partially responsible for the chemical and physical deterioration of the same and alter the esthetic appearance and physical integrity of the material through different mechanisms (**Figure 1**). The effects of microbial activities on historic buildings may be: discoloration, water retention, growth stimulation of heterotrophic organisms and higher organisms, material breakage, disintegration of the material, formation of patinas, degradation (corrosion), wear and dissolution of the structure, alkaline dissolution, and alteration of stratified silicates [23].

The microorganisms that colonize the stone monuments can be distinguished according to their location in the stone. The so-called epilithic that are located on the surface of the rock and those that live inside the rock within fractures and cracks and pores in granites are known as endolithic [12]. The main microorganisms that play a potential role in biodeterioration are autotrophic and heterotrophic bacteria, fungi, algae, and lichens. Phototrophic microorganisms such as microalgae, cyanobacteria, and lichens are considered the pioneer colonizers of the surface of stone monuments.

Figure 1.
Biodeterioration on stone (photo: Laura Castillón).

Cyanobacteria and algae such as chlorophytes, chrysophytes, and diatoms are a morphologically diverse and widely distributed group endowed with remarkable adaptability to variable environmental conditions and effective protection mechanisms against various abiotic stresses that enable them to colonize almost all classes of extreme lithic habitats [24]. These microorganisms form pigmented scabs (patinas) and incrustations that affect the substrate esthetically and cause physical and chemical deterioration of the rock. The epilithic cyanobacteria play an important role in the dissolution of the limestone carbonate, being able to cause the detachment of parts of it, due to a decrease in the coherence of the crystals around the colonies [25, 26].

The external stones are an appropriate niche for the growth and development of pioneering microorganisms that include photoautotrophs, lithophiles, and chemolithotrophs. The colonization begins with cyanobacteria and algae, probably followed by lichens that synthesize extracellular organic matter, in addition, dead cells release their constituents that form sources of nutrients for the growth of heterotrophic microorganisms which are considered secondary colonizers [27]. The phototrophic metabolism of cyanobacteria and algae facilitates their growth in oligotrophic environments such as stone forming biofilms on rocky surfaces, and it is the characteristics of the substrate that determine the speed of their growth and therefore the intensity of biodeterioration [28]. Lichens are highly resistant to extreme temperatures and desiccation that allow their easy growth on the surface of the stone. The microbial populations present in the rocky substrate are the result of successive colonization by different heterotrophic microorganisms.

It is well known that stone surfaces are exposed to high levels of solar radiation, high temperature, and to prolonged periods of desiccation alternating with rainy and damp periods. Many cyanobacteria are known to tolerate environmental extremes like UV light and their resistance to desiccation and tolerance for high level of light intensities and UV radiation provide them a distinct advantage for their survival on exposed surfaces, and they synthesize UV sunscreen pigments including scytonemin, mycosporine like aminoacids and biopterin glucosides. There are several reports on the effect of UV radiation on nitrogenase activity as it relates to the role of cyanobacteria in the nitrogen economy of ecosystem [29].

Pigmentation as a mechanism of deterioration depends on the nature of the substrate, the presence of trace elements such as iron, zinc, etc., the acidity or basicity of the medium, and even environmental conditions. The microorganisms produce two types of pigments (a) endopigments: they are located inside the cell and only leave after the lysis of the same as photosynthetic pigments, such as chlorophyll and phycobilins and (b) exopigments emitted outside the cell as fungal pigments (black, violet, blue, green, and purple). The black pigment known as melanin protects fungi against environmental threats and cellular lysis. Moreover, mycosporines and carotenoids (ß-carotene, s-carotene, phytoene, torulene, and torularhodin) may protect fungi against excessive UV radiation, act as antioxidants, osmoprotectants, and provide desiccation tolerance [30, 31].

The wear of materials is accelerated by the presence of biofilms containing active and latent microorganisms and their metabolic products, such as corrosive organic and inorganic acids as well as polymeric materials. Polymers, usually polysaccharides, act as gums that trap dust and other particulate materials increasing the disfiguring effects of the biofilm [32].

Beyond the type of microorganisms, the formation of the biofilm is a biodeterioration factor. The exopolysaccharide matrix plays a crucial role in this phenomenon since it produces mechanical stress on the stone through the pores of the mineral structure because it modifies the circulation of water within the material, its sensitivity to temperature variations, and the cycles of swelling and contraction dependent on the concentration of water within the matrix [27, 33].

Biofilms are also associated with the degradation of buildings and mural paintings by a phenomenon known as salt efflorescence, involves secondary minerals produced through the reaction of anions from excreted acids with cations from the stone which is available in the wall by the biological process or simply due to comigration with the infiltrated water. The solubilization of the calcareous material is detected by the presence of hygroscopic salts including carbonates, chlorides, nitrates, sulfates, etc., can be found on the surfaces of decayed monuments caused by chemical reactions (chemical agents in the air) or by enzymatic reactions of certain microorganisms. The most frequently isolated genus was *Bacillus*, followed *Staphylococcus*, *Kocuria*, *Micrococcus*, *Paenibacillus,* and *Arthrobacter* (bacteria of the sulfur and nitrogen cycle) [34].

The precipitation of salts is due to changes in temperature or humidity, and the salts can precipitate on the exposed surfaces and cause a destructive effect. Some salts when hydrated occupy a large space causing additional pressure that eventually leads to the loss of material and destruction due to cracking and detachment of walls or calcareous structures. In studies conducted by electron microscopy of salt efflorescence zones on walls, biofilms have been reported by members of *Firmicutes, Actinobacteria,* and *Ascomycota* [13, 35, 36].

From the nutritional point of view, organisms show a wide range of metabolic modalities where they use different sources of carbon (organic or inorganic compounds) as well as light as an energy source, and they are classified as photoautotrophic, chemoautotrophs, chemoheterotrophs, chemoorganotrophs, and photosynthetic.

The main mechanisms of biodeterioration associated with the different types of stone-colonizing organisms according to their nutrition are presented in **Table 1** [20].

Microbial colonization on bare stone surface is thought to be initiated by pioneering microorganisms which includes photoautotrophs, lithophiles, and chemolithotrophs. These organisms may secrete carbohydrates and growth factors which help in the formation of biofilm (a three-dimensional structure regulating temperature and humidity) and support the growth of successive microbial communities that is predominated by heterotrophic bacteria and fungi [20].

Limestones are carbonate rocks composed of calcite, and their main uses are in construction, chemical products, smelting, agrochemicals, and glass. This material is highly porous and hydrophilic in nature, and it is highly susceptible to water (such as acid rain) and environmental contaminants. Water often penetrates the pores of the stones causing damage by corrosive ions such as chlorine and acids. Biofilms,

Mechanisms of Biodeterioration	
Organism	**Mechanism**
Cyanobacteria	Formation of biofilms, coloured patinas and crusts, bio-waste as a result of calcium uptake, calcium salt precipitation and mineral formation
Lichens	Extraction of nutrients from the surface of the stone, formation of oxalate, production of carbonic acid associated with biodeterioration, physical intrusions in small pores.
Algae	Biofilm formation, color alteration, black scabs
Mosses and liverworts	Discoloration, green-gray patches, extraction of minerals from the stone
Sulfur-oxidizing and nitrifying bacteria	Formation of black crusts, secreted acids Salt efflorescence
Heterotrophic bacteria	Scab formation, patina, exfoliation, color alteration Bio-waste by secreted acids
Actinomycetes	White powder - grayish, patinas, white salt efflorescence Biofilm formation
Fungus	Fungal diagenesis, color alteration, oxalate formation, bio-waste by secreted acids, chelating properties by secreted acids, physical intrusion of hyphal penetration, biofilm formation and destabilization of the texture of the stone
Sulfur reducing bacteria	Conversion of sulphate into sulfite that acts as a nutrient for sulfur-oxidizing bacteria, salt efflorescence
Superior plants	Intrusion of roots within the cracks and pores, collapse and detach the structure of the stone.

Table 1.
Mechanisms of biodeterioration.

industrial and persistent pollutants, particulate matter, ash and often smog are deposited on the stone, and as a result, its deterioration is accelerated [23, 37].

The wear of the rocks and monuments can also be a consequence of the removal and solubilization of cations present in the minerals of the stone in particular iron and manganese of the mineral network by the negatively charged exopolysaccharide (EPS) of the biofilms or by some microbial proteins called siderophores by organic transport complexes and metallic organic chelates. Under low iron stress, siderophores chelate iron and supply to bacterial and fungi cells by outer membrane receptors, and the role of these compounds is to scavenge iron from the environment and to make the mineral, which is almost always essential, available to the microbial cells [38, 39].

The ability to grow by the dissimilatory oxidation of inorganic electron donors (ferrous iron, hydrogen, sulfur, and reduced inorganic sulfur anions) is widespread among acidophilic prokaryotes. Both oxygen and ferric iron can act as electron acceptors from many species of chemolithotrophic acidophiles, enabling them to exploit anoxic as well as aerobic environments [40].

In aerobic conditions, electron donors may include ferrous ions or sulfur compounds which are oxidized into ferric iron and sulfuric acid, respectively, yielding high energy. However, during anaerobic conditions, ferric ions can replace oxygen as the electron acceptor with multiple substrates donating an electron. This pathway yields less energy than aerobic conditions, but energy can still be produced for growth. *A. ferrooxidans* is a chemolithoautotrophic bacterium which can use many different electron donors to support growth *Leptospirillum* spp. have been shown to use only ferrous iron as electron donor and are therefore (as a result of thermodynamic constraints) obligate aerobes.

The subsequent redox process is favored by the release of oxygen by photosynthetic bacteria, cyanobacteria as *Acidithiobacillus ferrooxidans, Bacillus* spp., *Leptospirillum* ssp., and chemoorganotrophic fungi such *Aureobasidium* spp. [24, 41, 42].

Mechanical damage to stone structures, monuments, and architecture is another type of biodeterioration mechanism which is due to the physical intrusion and penetration of bacteria, fungal hyphae in the gaps, pores, and fractures that destabilize the texture of the stone, causing mechanical deterioration or by the contraction, and expansion of the stem under fluctuations in humidity conditions. Also the mosses through the rhizoids can penetrate the rock causing holes (pitting) and the vascular plants through their mechanical deterioration through the growth of roots or chemically by the acidity and diverse exudates, alteration of the microclimatic parameters, increase of the risk of fires, and physical and visual obstruction [26, 43].

4. Biodeterioration of paintings

The chemistry of the manufacture of paintings and their function has now been transformed from art to science. The knowledge of the pictorial components allows to associate the type of microorganisms that can potentially colonize the paintings, and the different techniques used determines the final composition of the work.

The pictorial technique has multiple modalities: tempera, encaustic, fresco, mural, and oil among others. The first three techniques were used in the ancient Greco-Roman and Egyptian world. (a) Tempering consists basically of the mixture of earths or pigments with binder constituted by glue and water or by egg yolk and oils in any type of support, (b) encaustic or wax painting requires a previous preparation of the wall by means of wax, on which colors are applied to the tempera cast in wax using a hot palette, and (c) the fresco that consists essentially of painting on a surface of wet plaster, on the same wall, prepared by the artist himself [44]. This latter is the modality of frequent use and is performed on wet lime plaster that serves as a support for the various pigments dissolved in water which are chemically integrated into the wall, and so its durability is very high, and (d) oil is the best known pictorial technique and used on cloth or board. It consists of a mixture of colored pigments with oil, usually flax or walnut. Another technique related to painting on canvas is acrylic, which consists of a combination of acrylate molecules in emulsion with water [45].

The deterioration of a painting can have different origins such as: (a) alterations due to natural aging of the work that makes it more fragile, (b) defects inherent to the work such as low quality materials or bad techniques at the moment of being painted, and (c) influence of external conditions such as thermo-hygrometric conditions or other factors. The manifestations of the damage can be physical as lack of adhesion of the binder, or damage caused by the movements of the other, mechanical damage, etc. Chemical damage is manifested as a processes of gradual degradation and depolymerization or crosslinking of the materials of the work, damage caused by light, oxidation or biological attack [46–49].

The biodeteriorable character of the canvas is conditioned by the characteristics of the fabrics that are its support formed by cellulose fibers which is a polysaccharide whose constituent unit is D-glucose linked by glycosidic bonds β (1–4) forming linear chains, which in turn are links in parallel fibers called microfibrils. For cellulose, degradation involved different enzymes whose combined action allows obtaining glucose molecules that can be assimilated by microorganisms as a carbon source. The degree of polymerization and its orientation, the length of these chains, their crystallinity, and their orientation are detected by microorganisms and could be susceptible to biodeterioration. The susceptibility to biological attack depends on the percentage content of cellulose, lignin, and other organic components. The purest cellulose can hardly be attacked. *Alternaria, Aspergillus, Fusarium, Memnoniella, Myrothecium, Neurospora, Penicillium, Scopulariopsis, Stachybotrys, Stemphylium,* and *Chaetomium* are the main fungi associated to this process and as cellulolytic bacteria: *Cellvibrio, Sporocytophaga, Myxococcoides, Cellufalcicula,* and also *Clostridium* sp. as anaerobic bacteria has been reported [50]. This cellulolytic process is favored in conditions of relative humidity of high air or condensation water where the fiber of the fabric loses consistency and elasticity becoming brittle and falls apart.

The filmogenic substances act as a binder and as a vehicle for the pigments; they are mainly of organic origin; they are applied in liquid form, and with drying, they solidify forming a hard and flexible layer, with the passage of time and under certain adverse conditions, this layer loses its property of cohesion of pigments and causes dusty surfaces or the separation of the layers.

The most important component of the paintings on canvas is the pigments; they are either natural or synthetic origin, and have three main functions: they provide color, opacity, and brilliance, and protect the surface in which they are applied and protect the binder from its destruction by UV radiation.

The adhesives are a fundamental element for the final result of the work whose function is to facilitate a uniform distribution of color and prevent the paint layer from being absorbed by the fabric, and these adhesives have been changing in the course of history and have been classified depending on their origin in animal (gelatin, albumin, casein, and wax) and vegetal (starch, resins, gums, and gluten).

Gelatin is obtained from collagen which is an existing protein in the skin and cartilage, albumin (protein of egg or blood plasma), casein (protein of milk), and wax (secreted by bees composed of a mixtures of esters, hydrocarbons, and fatty acids). The starch is a polysaccharide of vegetable origin, which is formed predominantly of amylose and amylopectin. Vegetal resins are a mixture of organic compounds principally terpenes and derivatives. Gums consisting of mixtures of water soluble polysaccharides produced by exudation, usually from the stem of tree and gluten, refer to the proteins in cereal grains found in the endosperm plant embryos during germination (Coppen, 1995).

The organic composition of all these adhesives favors the growth of microorganisms due to their high nutritional content, and therefore they are easily attacked by them [30].

The use of varnishes is required to provide protection against environmental attacks. For them, natural or synthetic varnishes (resins) are used, which must be applied in such a way that they form a resistant, colorless, and transparent film. According to its chemical composition, the name of the natural resins (soft) depends on the number of isoprene units that contain such terpene molecules as monoterpenes, sesquiterpenes, diterpenes, and triterpenes. The deterioration of natural resins causes chemical changes such as polymerizations (crosslinking of polymer chains, hydrolysis of polymer chains, and oxidation of the main chain or side groups which causes the resin to become more insoluble, losing its resistance and changes in its coloring [51]. Synthetic (acrylic) resins have good adhesive properties and are currently widely used in preservation treatments, and the monomers of these resins are generated by the esterification of an acrylic acid with several alcohols. Its general formula is: $CH_2CR1COOR2$ and due to its hydrophobic nature are more resistant to microbial attack because they are not used as a source of nutrients [52, 53].

Among the forms of deterioration of pictorial heritage on canvas related to biodeterioration agents are [54]:

Alterations of the canvas: hydrolysis, colorations, loss of strength, loss of support, cracks, scales, and deformations.

Alterations of the binder/adhesive: enzymatic degradation, colorations, disintegration, and pulverulence.

Alterations of the varnish: yellowing, tiling, whitening, and peeling.

The biological origin of deterioration in paintings has been widely reported in wall and easel paintings since the 1980s in different parts of the world [49, 54], and in general, the main fungal species associated with the biodeterioration of painted walls are *Penicillium* sps., *A. niger, Rhizopus oryzae, Mucor, Trichophyton, Alternaria alternate,* and *Epidermophyton floccosum* [55, 56] and as biofilms *Acremonium*, *Cladosporium, Aspergillus, and Fusarium* [57]. In the case of canvases of oil paints, the bacterial strains of the phylum firmicutes such as *Bacillus* sp., *Micrococcus luteus,* and *Paenisporosarcina* sp. and nonculturable bacteria of the phylum *Proteobacteria such as Stenotrophomonas* sp. [58] as well as *Halobacillus* sp., *Halobacillus naozhouensis*, and *Nesterenkonia* sp. in wall paintings responsible for pigmentation by pink biofilms in Romanian monasteries [59].

The mere presence of microorganisms (colonization) in any type of surface does not determine their participation in the biodeterioration process, to be able to specify it, experimental strategies have been developed in the laboratory where traditional supports such as linen cloth prepared with layers enriched in gums of animal origin and linseed oil that are inoculated with suspensions of fungi and bacteria, later identifying the species that grow and deteriorate these materials, thus checking the postulates of Koch [60]. Another report associated with this proposal was made in the murals of the medieval church with the bacterium Arthrobacter responsible for the black spots as a result of the reaction of lead oxide of pigments and hydrogen sulfide produced by other bacteria responsible for spotting [61]. These models will allow to establish, under controlled conditions, which species colonize a given substrate and how the flora will change the substrate and how the substrate is modified by microbial colonization and how these modifications lead to the establishment of different microbial communities (microbial succession) [62].

There are several reports of the participation of biofilms as responsible for biodeterioration in the pictorial cultural heritage as in the case of the works of the Nerja and Treasure in Málaga [63], the church of St. Martins in Greene-Kreiensen, Germany [64], and the Mogao caves in Dunhuang, in Gansu Province of the People's Republic of China [65] among others. As an example of these works, mucilages have been analyzed in fragments detached from frescoes of the Santissima Annunziata Church in Siena Italy in damaged areas, and the presence of biofilms has been demonstrated where their growth is favored by external factors such as humidity, poor ventilation, and light which may be associated with biodeterioration, because their hydration retains particles and atmospheric pollutants that accelerate chemical corrosion by oxidation, reduction, and transformation of metal ions with the changes in the pigments in addition to the coexistence of saline efflorescence making the bioreceptives fresh and causing detachment of the paint layers [66].

Another threat has been reported to the cultural heritage of biodeterioration in frescoes caused by the presence of tourism such as the case of The Lascaux Cave, which is the most emblematic example of the damage that micro-organisms cause to mural paintings due to the amount of organic matter, respiration, and the sweat of visitors and workers that increased the concentration of CO2 favoring the growth of the fungus *Fusarium solani* and *Ochroconis lascauxensis* (black spots) and the alga *Bracteacoccus minor* forming a green patina (green disease) causing irreversible damage so it closed its access permanent visitors [62, 67].

At present, we must consider the use of commercial paints (canned) which, due to their composition, represent a carbon source for a large number of microorganisms and which can be a source of contamination and colonization for the surfaces in which they are applied. *Pseudomons, Flavobacterium, Escherichia, Bacillus, Enterobacter, Proteus, Micrococcus, Serratia, Aeromonas, or Stenotrophomonas maltophilia*, among others, have been reported in water-based paints, this contamination can occur during their production with the use of contaminated water or in the team [68–70].

5. Biodeterioration of textiles

Textiles are considered representative of cultural identity because they carry a significant value that transcends that of their materials and the work required for their manufacture. The desire of all cultures to express and communicate their social, esthetic and cultural values in their textile manifestations, and materials such as clothing or basketry (with ceremonial or ritual destination), are a unique cultural heritage, and the assignment of cultural value to a material object is the basis of conservation [71].

Textiles, such as clothing, fashion accessories, archeological objects, baskets, quilts, tapestries, embroideries, flags, funerary, and religious garb are often treasured for their artistic, technical, cultural and sentimental value, and for this reason, they are currently stored in collections in museums [72].

The textile heritage is very extensive, despite the loss to which it has been subjected throughout history, mainly due to the characteristics of its delicate materials and the interventions that have suffered and suffer from this type of pieces. The gradual deterioration of this material is very sensitive and can only be slowed down, the daily use of these fabrics, inadequate handling and bad storage conditions have caused the loss of unrepeatable textiles because the same materials and techniques are not available as well as the techniques that were used in their preparation that are already part of our past [73]. The state of preservation of textiles depends on the type of textile fiber, composition of the dye, age of the textile as well as its history of use and storage conditions.

It is called textile fiber to the set of filaments or strands susceptible to be used to form yarns (and of these fabrics). In the manufacture of the yarn for textiles, two types of fiber can be used: natural or synthetic. The natural fibers can be of animal or vegetable origin. In the first, they are generally of the protein type, such as wool from sheep's hair, goat, camelid, rabbit or another type of natural fiber such as silk from the silkworm. Among the natural fibers of plants include cotton, linen hemp, and jute among others. Synthetic fibers include polyester, polyamide polyurethane, polypropylene, polyacrylonitrile, and polyvinyl [74, 75].

In addition to the passage of time and the environmental characteristics of conservation as a possible source of deterioration, the development and presence of various types of organisms (microorganisms, rodents or insects) which are a threat to textiles and damage will depend on the type of fabric, its origin as well as storage conditions. The presence and permanence will depend on the availability of nutrients as well as light, humidity, and temperature conditions. The degradation of the materials that cause the damage by microorganisms is due to the processes of assimilation by fungi and bacteria that use these materials as a source of nutrients or to the degradation processes due to the effect of microbial metabolism.

The main manifestations of this process are the evident changes of the surface of these materials, discoloration, decrease in their resistance, changes in pH, and unpleasant odor. These damages can cause the total destruction of the material by the reduction of the degree of polymerization, decrease in its tensile strength, and elasticity. In general, natural fibers are more susceptible to microbial attack than synthetic fibers.

The main molecules responsible for the attack on textile fibers are organic acids, extracellular enzymes released or by exopigments of bacteria such as *Achromobacer* sp., *Bacillus* sp., *Brevibacterium* sp., *Corynebacterium* sp., *Pseudomonas* sp., *Rhodococcus* sp. *and Streptomyces* sp. or exopigments of fungi of the group *Aspergillus* sp., *Penicillium* sp., *Cryptococcus* sp., *Rhodotorula* sp., and others [76].

The biodeterioration mechanisms in textile objects will depend mainly on the type of fiber:

Cellulose fibers: The degradation of (1-4)-β-D-glucan or cellulose results from the activity of cellulolytic enzymes produced by several bacteria and especially fungi that hydrolyze cellulose to glucose by the enzymes: 1,4-β-D-glucan, cellobiohydrolase, endo-(1-4)-β-D-glucan glucanohydrolase, and glucohydrolase of β-D-glucosidases. These enzymes decrease the degree of polymerization of the long-chain cellulose molecules, resulting in a decrease in the strength of the fiber.

The presence of other components in fibers such as hemicellulose, pectins, other carbohydrates or substances added to fabrics (plasticizers) and even contaminants provide additional nutrients to microorganisms.

Among the genera of fungi associated with biodeterioration of cellulose are *Chaetomium, Myrothecium, Memnoniella, Stachybotrys, Verticillium, Alernaria, Trichoderma, Penicillium, Aspergillus, Aureobasidium, Cladosporium, Fusarium, Mucor, Paecilomyces, Rhizopus,* and *Trichothecium*. In the case of bacterial damage and with less significance, agents of degradation of cellulose are: *Arthrhobacter, Bacillus, Cellulomonas, Cellvibrio, Clostirdium, Cytophaga, Microbiospora, Nocardia, Pseudomonas, Sporocytophaga,* and *Streptomyces* [77].

Wool fibers: Keratin is the constituent protein of these fibers that form a polymer when disulfide bridges cross over this polymer. The mechanism of biodeterioration is by keratinolysis, sulfitolysis, proteolysis by peptidases, and deamination (metabolic processes with release of ammonia). The rate of degradation depends on the chemical composition, molecular structure, and degree of polymerization of the substrate and to a lesser degree on the structure of keratin [78].

Among the main biodeterioration agents are bacteria: *Arthrobacter, Bacillus (B. mesentericus, B. subtilis, B. cereus* and *B. mycoides), Cellulomonas, Cellvibrio, Clostridium, Cytophaga, Microbiospora, Nocardia, Pseudomonas, Sporocytophaga* and *Streptomyces* [79]. Degradation by fungi has been reported by the genera *Microsporum, Trichophyton, Fusarium, Rhizopus, Chaetomium, Aspergillus, Penicillium, Alternaria, Acremonium, Cephalothecium, Chrysosporium, Dematium, Oospora, Scopulariopsis, Stachybotrys, Trichoderma*, and *Ulocladium* [77, 80].

Silk fibers: They are produced by silkworms and are fibers of the fibroin protein that are joined to one another by rubber-like proteins, known as sericin that serves as protection from damage by light. This natural fiber is the most resistant to biodeterioration, and its decomposition depends on the proteolytic action on sericin and fibroin that are used as a carbon source by bacteria *Bacillus, Aeromonas, Arthobacter, Chyseomonas, Pseudomonas, Streptomyces, Serratia* and *Variovorax* and how biodeteriogenic fungal genera are: *Aspergillus, Chaetomium, Cladosporium, Penicillum* and *Rhizopus* [80].

There are very few works to which biofilms are directly associated as being responsible for biodeterioration in textile materials, and it may be the result that experimental designs have not been developed with these types of materials.

6. Biodeterioration of paper and parchments

The documentary production goes back in antiquity in different cultures that left numerous examples in different supports like tablets of mud, rolls of papyrus, parchment, sheet of amate and in more recent times the books. The invention of paper gave man a faithful support where the written memory will inhabit his journey through history, and thanks to its consistency and durability, the texts of our ancestors are still today, a faithful witness of his time [81].

The main components of paper are fiber or fibrous material (hemp, cotton, linen, bagasse, rice straw a,nd wood) and functional additives (sizing, optical brighteners, and consolidating agents such as gelatin, cellulose acetate and carboxymethylcellulose). In this chapter, cellulose fiber is the major component with a lower proportion of lignin, hemicellulose, and other macromolecules, its quality depends on the source of the raw material used, and the procedure applied to obtain the fiber. Its mechanical resistance depends on its degree of polymerization and its interfiber links.

The inks are an important component of the documents and consist of a liquid that is fixed to the support endowed with an intense, durable, odorless and variable pH, is composed of a pigment, a diluent and a binder. Among the oldest ones are ferrous ink, whose components are iron sulfate, gallotanic acid and a binder, usually gum arabic. Over time the components of plant and animal origin have been replaced by synthetic compounds [82].

The books are composed of a support (parchment or paper), supported elements (inks), binding elements (seams and adhesives), protective structures (covers), and each of them with particular chemical characteristics that can be elements of degradation [83].

The microorganisms that commonly appear in the documentary supports are bacteria and fungi (yeasts and filamentous fungi), which transport moisture and attract pests by modifying the nutritive environment of the substrate. Both colonize a susceptible medium when in a poorly ventilated place, with adequate pH and low illumination, where temperature higher than 25°C and ambient humidity greater than 65% with accumulation of dust and/or soot in the different types of surfaces.

Filamentous fungi are the most biodeteriogenic microorganisms because they have structures called hyphae that are vegetative and reproductive [14, 84]. Vegetative hyphae are intertwined in paper fibers and through enzymatic processes that degrade cellulose, absorb nutrients, produce acids, and affect the coloration of the support resulting in fragility of the paper and often its complete destruction (**Figure 2**). On the other hand, the reproductive structures (spores) are a potential threat because they can remain in a latent state, they can be airborne, and they accumulate in layers of dust as long as the environmental conditions for their germination are reached, such as the formation of condensation points and local microclimates due to poor ventilation and heterogeneous temperature on the surface of the material.

Among the fungi identified are the Ascomycetes as well as mitosporic xerophilic fungi (which grow with a small amount of water) such as those of the genus *Aspergillus, Paecilomyces, Chrysosporium, Penicillium and Cladosporium* [85] and among the cellulolytic bacteria *Cellvibrio* and *Cellfacicula* as well as *Cytophaga* (myxobacteria).

In the case of the parchments that are composed of collagen (protein), its degradation depends on the oxidative chemical deterioration of amino acid chains and hydrolytic breakdown of the peptide structure and the production of pigments and organic acids that modify this material. The bacteria that have been described in this process are those of the genera *Bacillus, Staphylococcus, Pseudomonas, Virgibacillus* and *Micromonospora* as well as some alkalophilic bacteria such as *Actinobacteria* and among the proteolytic fungi reported are the Ascomycetes: *Chaetomium and Gymnoascus* and the genus *Acremonium*, *Aspergillus, Aureobasidium, Epicoccum, Trichoderma* and *Verticillium* [13, 78].

Figure 2.
Paper biodeterioration. (Photo: Laura Castrillón).

7. Prevention, conservation, restoration, and control

To preserve the tangible cultural heritage, there are two ways of action: the prevention of deterioration (conservation) and the repair of damage (restoration). With the preventive conservation anticipates the damage generated by extrinsic causes, alien to the nature of the pieces to conserve, but that in more or less long term could degrade their cultural value.

Therefore, prevention methods and strategies are usually not directly applicable to the object to be treated, but are directed to the environment to control microclimatic conditions in order to eradicate harmful agents or elements that can temporarily or permanently influence degradation [82].

The prevention methods inhibit or slow down the biological growth modifying the factors that can condition or inhibit their presence (humidity, temperature, light, and ventilation), if these factors cannot be controlled as in the case of monuments or archeological zones can be modified eliminating dust, dirt and deposits of residues of plant or animal nature.

In contrast and in general terms, conservation can be defined as the set of operations that aim to prolong the life of the material, thanks to the anticipation of damage or the correction of deterioration.

In the field of the conservation of cultural goods, the purpose is to maintain the physical and cultural properties of what has reached the category of cultural property, with the purpose that its value does not diminish and lasts beyond a limited time segment. Preventive conservation, as a methodology aims to control the deterioration of works of art before they occur to reduce the need to intervene. The deterioration must be minimized and the optimum conditions of exhibition, transportation, handling, cleaning, and storage must be maintained [86]. Among the main measures applied for preventive conservation that have been incorporated in the facilities are: air conditioning free of biodeteriogens agents, environmental fumigations, humidity, and temperature control.

The restoration aims to recover the physical and functional integrity of the work, thanks to the correction of the alterations that it has suffered. Consequently, the curative methods are of direct application because they try to amend all the damages they have experienced through their own history, whenever these suppose mutilation or reduction of their documentary values. The restoration is more than an art is technical, thanks to the set of interdisciplinary scientific methods that give the work the authentic guarantee of the rigor of applied sciences to the field of conservation [87].

Restorers use the intervention techniques of biodeteriorated materials to eliminate the degradation products induced by microorganisms and if it is possible to delay their recurrence. The intervention treatment must be evaluated taking into account the identity of the biodeteriogens, degree and type of damage, safety of the treatment towards the materials of the object, risk for the worker, and possible environmental impacts. However, the growth of unwanted organisms will inevitably occur if the environmental conditions that favor their development persist.

Between the main methods of control of biodeteriogenic agents that grow as biofilms are:

Mechanical methods: they consist in the physical removal of fungi, bacteria, algae or any organism by shaving, abrasion, brushing, etc. Immediate but not lasting results are obtained, complete elimination is not achieved, and the results improve with the use of chemical agents.

Physical methods: modifying the temperature or pressure changes that are not suitable for the growth of organisms. Its biocidal effect depends on the denaturation or breaking of molecules of the organisms treated by breaking chemical bonds. Among these methods are electromagnetic radiation (microwave, ultraviolet rays, and gamma rays), anoxic treatments, and extreme temperatures. Its mechanism of action depends on its direct action with the genetic material or alteration of its structure and metabolic function.

Its main disadvantages are its high cost and the possible damage to the materials treated by its chemical alteration such as the pigmentation and hydrolysis of proteins and cellulose.

Chemical methods: these are the most commonly used intervention techniques through the use of biocides (disinfectants, bactericides or fungicides). Generally they are used in liquid or gas form, their mechanism of action is variable and they attack by disintegrating the bacterial or fungal membrane or by inhibiting their cellular processes, causing their death when they are used in the appropriate doses. Many products have been evaluated, however, due to their high risk and limited knowledge of the compatibility with the materials to which they apply their use has been limited. The selection of the biocide depends on the type of material, type of microorganism, and availability of the biocide. An additional problem is its long-term ineffectiveness [88].

A biocide can be a synthetic chemical, natural, of biological origin that is intended to destroy, counteract, neutralize, impede action or exercise control over any organism considered harmful to man. According to their action, they are divided into microbicides (bactericidal and fungicidal), growth inhibitors, and for the case of other organisms such as insects, rodents or birds, and there are also very toxic and lethal products such as pesticides, insecticides and/or repellents, acaricides, nematicides, avicides, rodenticides, etc.

A good biocide must have a broad spectrum of activity, be effective at low concentrations, be active over a wide range of pH, soluble in water, possess high persistence (effective over time), have low human and environmental toxicity, and have a low cost [89].

Among the main biocides used for the chemical treatment of tangible cultural goods are:

Thymol, orthophenylphenol (OFF), formaldehyde, pentachlorophenol, ethylene oxide, ethanol, etc., antibiotics and enzymes have also been used. For the best selection of the biocide, preliminary tests must be done to guarantee the innocuousness of the operation for the safety of the operator and absence of risk for the object not only immediately but in the long term. For this reason, the use of very strong and long retention solvents (glycols, formamide, turpentine, and butylamine) in the porous bodies has been ruled out.

The forms of application of the biocides are: sublimation, pulverization or fumigation, according to their possible solid, liquid or gaseous state. The solids in solution or dispersion have a longer time of action although less exterminating capacity. The sublimable solids have little penetrability and, unless they are applied in high concentrations, they become repellent air fresheners with little lethal efficacy.

Solvents that are unstable to light and have a tendency to yellow or polymerize are also eliminated (ethylenic compounds such as dipentene, turpentine, acetylacetone, N-methylpyrrolidone). These rules are applicable to all porous objects, paintings, polychrome sculptures, mural paintings, etc. In the area of textiles, the most suitable solvent is water. But there are fibers very altered or with very bad coloration that are too sensitive. Then we must resort to organic solvents but avoiding chlorinated solvents, taking into account an eventual acidity and the release of hydrochloric acid. Water-based solvents must be used with extreme prudence, since many materials can suffer deterioration [90, 91].

Because the control of pests that use chemicals that are generally expensive and have side effects in people and can deteriorate the material, the choice of a biocide is increasingly difficult, therefore alternatives substances with biocidal properties have been sought for many years such as the use of natural plant products for which more and more reports justify their use. Currently, there is already a database of the accumulated experience of a Spanish group of the use of natural extracts for disinfection and disinfestation of cultural goods [92], as well as the use of essential oils from medicinal plants such as *Mentha piperita, Thymus vulgaris, Origanum compactum, Salvia officinalis, Artemisa absinthium and Lanandula angustifolia*, among others [93–95].

The development of nanotechnology is currently an emerging field in the conservation of cultural heritage, consequently the FP7 NANOFORART project has arisen (nano-materials for the conservation and preservation of movable and immovable artworks) and as an example of its applications, there are reports of the use of zinc oxide nanoparticles to control fungal biofilms or nanosilver coated cotton fabrics application for antimicrobial textile finishing [96–99].

Biological methods: another option for the control and restoration of works of art has been the use of microorganisms in the processes of biocleaning and biomineralización that are presented below:

Biocleaning: The accumulation of organic material on surfaces either by deposition of atmospheric particles, traces of colonization of microorganisms and organic substances allow the growth of bacteria and fungi. This accumulation can cause damages to the art work in response to the growth of microbial and are therefore considered biodeteriorating agents, however, in recent years, bacteria have been used for the conservation and restoration processes for the elimination of these organic materials. This procedure is known as biocleaning.

One of the advantages of the use of microorganisms over physicochemical treatments (which are very drastic) is that they use substrate-specific enzymes that do not degrade complex substances and adapt easily to environmental conditions. The microorganisms selected should be nonpathogenic and nonsporulating so that it is not a risk for workers after application and not be able to produce forms of resistance (spores).

Examples of these treatments have been documented for the removal in stone of black scale (hydrated calcium sulfate and carbon residues) caused by sulfur dioxide, hydrocarbons, and particulate matter (soot) emitted by the exhaust pipes of vehicles, which were removed with the use of *Desulfovibrio desulfuricans* bacteria [100]. Good results have also been obtained for the elimination of nitrates in marble under anaerobic conditions with the use of *Pseudomonas denitrificans, Pseudomonas stuzeri, Pseudomonas pseudoalcaligenes* or *Paracoccus denitrificans* [101]. Biological cleaning of mural paints has also been explored to eliminate the remains of organic matter from old restorations or insoluble saline efflorescence with the use of *Pseudomonas stutzeri* [102].

Biomineralization: A modern and ecological alternative applicable to the restoration of historical monuments is the process known as biomineralization, specifically carbonatogenesis that can help in the restoration of cracks of statues or walls, since there are bacteria capable of mineralizing and filling these grooves when feeding them with means of culture that contain calcium salts in solution producing microcrystals of calcium carbonate that allow the restoration of damaged areas [103–105, 107, 108].

The carbonatogenesis or calcite production can occur either autotrophically or heterotrophically by the *Bacillus, Pseudomonas, Proteus, Myxococcus* and *Pantoea agglomerans* bacterial genera that allow obtaining a layer of a few millimeters thick by carbonatogenesis. Among the micro-environmental factors related to this process are: the concentration of dissolved inorganic carbon, the pH, the concentration of calcium ions and the presence of nucleation sites / or development of crystals for nucleation [23, 106].

8. Conclusion

In addition to the passage of time and environmental characteristics of the detriment of artistic and cultural heritage, this may also be the consequence of the microbial appearance favored by the enrichment of organic matter on the surfaces that colonize. The formation of biofilms is a strategy used by microorganisms to adapt to conditions that may be adverse for their growth. These biofilms are present in almost any type of surface, and the historical heritage is no exception. For the reason, it is important to know the characteristics that favor their formation, elimination and control to avoid being a threat to these objects.

As these biofilms represent a threat to this cultural heritage, the main challenges are to establish prevention measures for their appearance and to understand that if an intervention treatment is necessary, the resistance to treatment with biocides increases with this form of organization, which requires the use of high doses that compromise the physicochemical characteristics of the treated material.

The knowledge of the chemical composition of works of art made in different substrates such as stone, paintings, textile fibers, and paper composition allows us to understand what type of microbial colonization can be favored, and consequently be able to propose microbial growth as the responsible of biodeterioration. If these objects are valuable as a cultural heritage, it is the responsibility of all their conservation and restoration, for this reason the advanced techniques of identification and control of biofilms in cultural assets, currently, they are applied as strategies in the areas of preservation, restoration and control that will delay their natural deterioration and allow future generations to know this legacy.

Author details

Laura E. Castrillón Rivera*, Alejandro Palma Ramos, Jorge I. Castañeda Sánchez and María Elisa Drago Serrano
Departamento de Sistemas Biológicos, Universidad Autónoma Metropolitana, Mexico

*Address all correspondence to: lcrivera@correo.xoc.uam.mx

References

[1] García Cuetos P. El patrimonio cultural: Conceptos bàsicos. España: Ed. Prensas Universitarias de Zaragoza; 2011. ISBN: 978-84-15274-56-8

[2] Patrimonio Cultural: Instituto Andaluz del Patrimonio Histórico. Consejería de Cultura. Available from: www.iaph.es/web/canales/patrimonio-cultural [Accessed: Apr 5, 2018]

[3] UNESCO. Patrimonio Indicadores UNESCO de cultura para el Desarrollo. Available from: Vaillant CM. Materiales constituyentes de las colecciones de archivos y bibliotecas. In: Biodetrioro del patrimonio histórico documental: Alternativas para su erradicación y control. Río de Janeiro: Ed: Mast/FCRB; 2013

[4] Rivera CM. Patrimonio Cultural Tangible e Intangible. Available from: https://manuelrivera23.wordpress.com/blog/revista-ajayu/patrimonio-cultural-tangible-e-intangible/ [Accessed: May 7, 2018]

[5] ICOMOS, International Cultural Tourism Charter. Principles and guidelines for managing tourism at places of cultural and heritage significance. In: ICOMOS International Cultural Tourism Committee 2002. ISBN: 956-244-166-0

[6] Martianrena X. Conservación y Restauración. Cuadernos de Sección. Artes Plásticas y Documentales. 1992;**10**:177-224

[7] Moreno OM, Rogerio-Candelera MA, López NJT, Hernández JV. Estudio y conservación del patrimonio cultural. Coria Gráfica, SL: Actas. Ed.; 2015. ISBN: 978-84-608-2452-7

[8] Simmons EJ, Muñoz-Saba Y. Cuidado, manejjo y conservación de las colecciones biológicas. In: Conservación Internacional. Serie Manuales de campo 1. Bogotá Colombia. 2005. ISBN: 958-33-6969-1

[9] Hueck HJ. The biodeterioration of material as a part of hylobiology. Materials and Organisms. 1965;**1**:5-34

[10] Marijnissen RH. Degradation-conservation. Restauraction de 'Oeuvre D'Art. Ed Arcade; 1967. pp. 123-125

[11] Gacto M. Los microorganismos y el arte. Anales de Biologia. 2011;**33**:107-115

[12] Páramo AL, Narváez Z, De la Cruz E. Aislamiento e identificación de microorganismos en biopelículas provenientes del Castillo de Chapultepec, Ciudad de México. Nexo Revista Científica. 2011;**24**(2):83-91

[13] Sterflinger K, Piñar G. Microbial deterioration of cultural heritage and works of art-tilting at windmills? Applied Microbiology and Biotechnology. 2013;**97**:9637-9646

[14] Castrillón RLE, Palma RA, Padilla DMC. Biopelículas fúngicas. Dermatologia Revista Mexicana. 2013;**57**:350-361

[15] Castrillón RLE, Palma RA. Biofilms: A survival and resistance mechanism of microorganisms. In: Pana M, editor. Antibiotic Resistant Bacteria—A Continuous Challenge on the New Millennium. Rijeka . Ch 7: Intech Open Science; 2012. pp. 159-178

[16] Costerton JW, Cheng KJ, Geesey GG, Ladd TI, Nickel JC, Dasgupta M, Marrie TJ. Bacterial biofilms in nature and disease. Annual Review of Microbiology. 1987;**987**:435-464

[17] Castrillón RLE, Palma RA, Padilla DMC. Importancia de las biopelículas en la práctica médica. Dermatologia Revista Mexicana. 2010;**54**:14-24

[18] Morton LHG, Surman SB. Biofilms in biodetrioration—A review. International Biodeteriorarion & Biodegradation. 1994;**34**(3-4):203-221

[19] Videla HA. Manual of Biocorrosion. Boca Raton, PL, USA: Lewis Publishers/ CRC Press; 1996. p. 288

[20] Dakal TCH, Cameotra SS. Microbially induced deterioration of architectural heritages: Routes and mechanisms involved. Environmental Sciences Europe. 2012;**24**:36

[21] Tatis CRD, Barbosa LAL. Enfoque químico del deterioro y biodeterioro de rocas calcáreas conformantes de monumentos patrimoniales de importancia histórica y cultural. Luna Azul. 2013;**36**:247-284

[22] Warscheid T, Braams J. Biodeteriorarion of Stone: A review. International Biodeterioration & Biodegradation. 2000;**46**:343-368

[23] Cameotra SS, Dakal TC. Carbonatogenesis: Microbial contribution to the conservation of monuments and stone artwork. Conservation Science in Cultural Heritage. 2012;**12**:79-108

[24] Pandey VD. Rock-dwelling cyanobacteria: Survival strategies and biodeterioration of monuments. International Journal of Current Microbiology and Applied Sciences. 2013;**2**(2):519-524

[25] Macedo MF, Miller AZ, Dionisio A, Saiz-Jiménez C. Biodiversity of cyanobacteria an green algae on monuments in the mediterranean basin: An overview. Microbiology. 2009;**155**:3476-3490

[26] Videla AH, Guiamet SP, Gómez de Saravia SG. Biodeteriroro de materiales estructurales de sitios arqueológicos de la civilización maya. Revista Museo de la Plata. 2003;**44**:1-11

[27] Martino DP. What about biofilms on the surface of stone monuments? The Open Conference Proceedings Journal. 2016;(Suppl 1:M2):14-28

[28] Gómez de Saravia SG, Fontana JM, Guiamet SP. Estudio de biofilms fototróficos mediante microscopía óptica y microscopía electrónica de barrido. Revista Argentina de Microbiología. 2010;**42**:315

[29] Essays UK. Monuments and Caves: Effect of Cyanobacteria. 2013. Available from: https://www.ukessays.com/essays/biology/role-cyanobacteria-monuments-caves-3562.php?vref=1 [Accessed: Jun 8, 2018]

[30] López-Miras MM. Identificación y caracterización de comunidades microbianas presentas en pinturas sobre lienzo. Estudio de su capacidad como agentes de biodeterioro Tesis Doctoral Universidad de Granada. 2012. ISBN: 978-84-694-6935-4

[31] Salvatori O, Municchia CA. The role of fungi and lichens in the biodeterioration of stone monuments. The Open Conference Proceedings Journal. 2016;7(Suppl 1:M4):39-54

[32] Gaylarde CC, Morton GH. Deteriogenic biofilms on building and their control: A review. Biofouling. 1999;**14**(1):59-64

[33] Suihko ML, Alakomi HL, Gorbushina A, Fortune I, Marquardt J, Saarela M. Characterizaation of aerobic bacterial and fungal microbiota on surfact of historic Sctottish monuments. Systematic and Applied Microbiology. 2007;**30**:494-508

[34] Laiz L, Recio D, Hermosin B, Saiz-Jimenez C. Microbial communities in salt efflorescences. In: Ciferri O, Tiano P, Mastromei G, editors. Of Microbes and Art. Boston, MA: Springer; 2000

[35] Lepinay C, Mihajlovskia A, Seyer D, Touron S, Bousta F, Di Martino P. Biofilm communities survey at the areas of salt crystallization on the walls of a decorated shelter listed at UNESCO World cultural Heritage. International

Biodeterioration & Biodegradation. 2017;**122**:116-127

[36] Ortega-Morales BO, Nakamura S, Montejano-Zurita G, Camacho-Chab JC, Quintana P, De la Rosa-García SC. Implications of colonizing biofilms and microclimate on west stucco at North Acropolis, Tikal, Guatemala. Heritage Science. 2013;**1**:32

[37] Gómez-Cornelio S, Ortega-Morales O, Morón-Ríos A, Reyes-Estebanez M, de la Rosa-García S. Cambios en la composición de la comunidad fúngica de biopelículas sobre roca calcárea a través de una cronosecuencia en Campeche México. Acta Botánica Mexicana. 2016;**117**:59-77

[38] Neilands JB. Siderophores: Structure and function of microbial iron transport compounds. The Journal of Biological Chemistry. 1995;**270**(45):26723-26726

[39] Sajeed AS, Vidhale NN. Bacterial siderophore and their application: A review. International Journal of Current Microbiology and Applied Sciences. 2013;**2**(12):303-312

[40] Smith LS, Johnson BD. Growth of *Leptospirillum ferriphilum* in sulfur medium in co-culture with *Acidithiobacillus caldus*. Extremophiles. 2018;**22**:327-333

[41] Ortega-Morales O, Guezennec J, Hernandez-Duque G, Gaylarde CC, Gaylarde PM. Phototrophic biofilms on ancient Mayan buildings in Y ucatan, Mexico. Current Microbiology. 2000;**40**:81-85

[42] Schalk JI, Hannauer M, Braud A. New roles for bacterial siderophores in metal transport and tolerance. Environmental Microbiology. 2011;**13**(11):2844-2854

[43] Gutiérrez RJN, Nugkui TNM. Biodeterioro por plantas vasculares en el museo cementerio presbítero Matías Maestro, Lima-Perú. Revista Rebiol. 2016;**36**(2):40-53

[44] Sánchez FLP, Cardona ML. In: La pintura mural: Una pasión: análisis del mural fresco. San Diego, CA, USA. 2013. ISBN 13:9789588743479

[45] Triado JR, Subirana R. Las claves de la pintura. Barcelona: Ed. Planeta; 2002. 119 p

[46] Flemming CH. Relevance of biofilms for the biodeterioration of surfaces of polymeric materials. Polymer Degradation and Stability. 1998;**59**(1-3):309-315

[47] Garófano MI. Materiales orgánicos naturales presents en pinturas y policromías. Naturaleza, usos y composición química. Revista del Instituto Andaluz del Patrimonio Histórico. 2011;**80**:56-71

[48] Ravikumar HR, Rai SS, Karigar CS. Biodegradation of paints: A current status. Indian Journal of Science and Technology. 2012;**5**(11):1977-1987

[49] Villarquide Jevenois A. La pintura sobre tela II. Alteraciones, materiales y tratamientos de restauración. Ed. Nerea SA; 2005. ISBN:84-89569-50-9

[50] Caneva G, Nugari MP, Salvadori O. La Biología de la restauración. España: Ed: Nerea SA; 2002. ISBN: 8489569487

[51] Doménech-Carbó MT. Novel analytical methods for characterising binding media and protective coatings in artworks. Analytica Chimica Acta. 2008;**621**:109-139

[52] Nugari MP, Priori GF. Resistance of acrylic polymers to microorganisms-first part. In: Vth International Congress on Deterioration and Conservation of Stone; 25-27 September; Lausanne. 1985. pp. 685-693

[53] Pankhurst ES, Davies MJ, Blake HM. The ability of polymers or materials containing polymers to provide a source of carbon for selected microorganisms. In: Harry Walters A, Hueck-van der Plas EH, editors. Biodeterioration of Material. Vol. 2. Elsevier Science & Technology. pp. 76-90

[54] Poyatos JF. Procesos de biodeterioro en pinturas sobre lienzo del museo de Bellas Artes de Granada: Examen visual y gráfico. Tesis Doctoral Univesidad de Granada. 2007. ISBN: 978-84-338-4436-1

[55] Elumalai P, Elumalai EK, David E. Fungi associated with deteriorations of painted wall surfaces: Isolation and identification. European Journal of Academic Essays. 2014;**1**(3):48-50

[56] Ogu TC, Okaa AI, Ozokpo AC, Onochie CC. Microbial deterioration of paints. African Journal of Education, Science and Technology. 2016;**3**(1):190-194

[57] Gorbushina AA, Hayrman J, Dornieden T, González-Delvalle M, Krumbein EW, Laiz L, Peersen K, Saiz-Jiménez C, Swings J. Bacterial and fungal diversity and biodeterioration problems in mural painting environments of St. Martins church (Greene-Kreiensen, Germany). International Biodeterioration & Biodegradation. 2004;**53**:13-24

[58] López-Miras MM, Martin-Sánchez I, Yebra-Rodríguez A, Romero-Noguera J, Bolívar-Galiano F, Ettenaurer J, Sterflinger K, Piñar G. Contribution of microbial communities detected on an oil painting on canvas to its biodeterioration. PLoS One. 2013;**8**(11):e80198

[59] Gomoiu I, Mohanu D, Radvan R, Dumbravician M, Neagu SE, Cojoc LR, Eanche MI, Chelmus A, Mohanu I. Environment impact on biopigmentation of mural paintings. Acta Physica Polonica. 2017;**131**:48-51

[60] Seves A, Romano H, Scilione G, Maifreni T, Sora S. A laboratory investigation of the microbial degradation of cultural heritage (ICMC). In: Cifferri O, Tiano P, Mastromei G, editors. Of Microbes and Art. Boston, MA: Springer; 2000. pp. 121-133. ISBN 978-1-4613-6904-2

[61] Petushkova JP, Lyalikova NN. Microbiological degradation of lead-containing pigments in mural paintings. Studies in Conservation. 1986;**31**:65-69

[62] Ciferri O. Microbial degradation of paintings. Applied and Environmental Microbiology. 1999;**65**(3):879-885

[63] Del Rosal Y, Lobo JV, Hernández MM, Roldán MM, Sáiz-Jiménez C. Biofilms en cuevas turísticas: la Cueva de Nerja y la Cueva del Tesoro. In: Andreo B, Durán JJ, editors. El karst y el hombre: las cuevas como Patrimonio Mundial. Nerja (Málaga): Asociación de Cuevas Turíisticas Españolas; 2016. pp. 103-114. ISBN: 978-84-617-4702-3

[64] Anna A. Bacterial and fungal diversity and biodeterioration problems in mural painting environments of St. Martins church (Greene-Kraiensen, Germany). International Biodeterioration & Biodegradation. 2004;**53**:13-24

[65] Ma Y, Zhang H, Du Y, Tian T, Xiang T, Liu X, Wu X, An F, Wang W, Ji-Dong G, Feng H. The community distribution of bacteria and fungi on ancient wall paintings of the Magao Grottoes. Scientific Reports. 2015;**5**:7752

[66] Milanesi C, Baldi F, Borin S, Ciampolini F, Cresti M. Biodeterioration of fresco by biofilm forming bacteria. International Biodtererioration & Biodegradation. 2006;**57**:168-173

[67] Martín-Sánchez PM, Jurado V, Porca E, Bastian F, Lacanette D, Alabouvette C, Saiz-Jimènez C. Airborne microorganisms in Lascaux Cave (France). International Journal of Speleology. 2014;**4**(3):295-303

[68] Fazenda JMR. Tintas e vernizes: ciência e tecnologia. São Paulo: Ed. Abrafati; 1995. ISBN: 9788521203742

[69] Obidi OF, Aboaba OO, Makanjuola MS, SCU N. Microbial evaluation and deterioration of paints and paint-products. Journal of Environmental Biology. 2008;**30**(5):835-840

[70] Rochoel RF, Giese EC, DRF H, Pelayoi JS, Melo BA. Micobiological contamination of water-based paints from an industry in the state of Paraná, Brazil. Semina: Ciencias Exatas e da Terram Londrina. 2008;**29**:85-92

[71] Ruiz de Haro MI. Orígenes, evolución y contextos de la tecnología textile: La producción del tejido en la prehistory y la protohistoria. Arqueología y Territorio. 2012;**9**:133-145

[72] Ramos ENJ, Valverde GM, Aliaga AMT, Gimènez TA. Deteriroro causado por microorganismos en textil arquelógico y lienzos. Revista Boliviana de Quimica. 2012;**29**(2):170-176

[73] Olmedo PM. Conservación del patrimonio textil. "Guía de Buenas Prácticas" DL MA 995-2017. España. 2017

[74] Giraldo MJF. Manual Técnico Textil. 4th ed. Medellín, Colombia; 2015 Available from: http://www.microdenier.com.co/manualtecnico.pdf

[75] Szostak-Kotova J. Biodeterioration of textiles. International Biodeterioration & Biodegradation. 2004;**53**:165-170

[76] Joshi VK, Attri D, Bala A. Microbial pigments. Indian Journal of Biotechnology. 2003;**2**:362-369

[77] Pekhanov E, Neverov A, Kubika S, Zaikov G. Biodegradation and biodeterioration of some natural polymers. Chemistry & Chemical Technology. 2012;**6**(3):263-280

[78] Kunert J. Biochemical mechanism of keratin degradation by the actinomycete *Streptomyces fradiae* and the fungus *Microsporum gypseum:* A comparison. Journal of Basic Microbiology. 1989;**29**:597-604

[79] Ararwal PN, Puvathingal JM. Microbiological deterioration of woolen materials. Textile Research Journal. 1969;**38**:38-42

[80] Gutarowska B, Pietrzak K, Machnowski W, Milczarek MJ. Historical textiles—A review of microbial deterioration analysis and disinfection methods. Textile Research Journal. 2017;**87**(2):2388-2406

[81] Asenjo MJL, Hidalgo MMC. El papel: 2000 años de historia. 2010. Available from: http://www.interempresas.net/Graficas/Articulos/37870-El-papel-2000-anos-de-historia.html [Accessed: May 17, 2018]

[82] Vaillant Callol M. In: Biodeterioro del Patrimonio Histórico Documental: Alterantivas para su erradicación y Control. Río de Janeiro: Ed. MAST/FCRB; 2013

[83] Forniés MZ, Garcìa QR. Factores de degradación intrínseco en los libros: la naturaleza del material bibliográfico. 2014. Available from: http://www.bid.ub.edu/es/32/fornies2.htm

[84] Méndez TLJ., López MR., Hernández HF. Actualidades en Micología Médica. México: Ed. Sefirot SA de CV; 2012. ISBN: 978-60-7772-833-7

[85] Pinzari F, Montanari M. Mould growth on library materials stored in compactus-type shelving units

(Chapter 11 pp. 193-206). In: Abdul-Whahab Al-Sulaiman SA, editor. Sick Building Síndrome in Public Bulidings and Workplaces. London, New York: Springer, Heidelberg Dordrecth; 2011

[86] Reyes Valderrama L. Biodeterioro del patrimonio cultural: el sorprendente apetito de algunos organismos por el arte y la historia. Hipotesis. 2013;**15**:70-75

[87] Viñas V, Viñas R. Las técnicas tradicionales de restauración: Un estudio del RAMP. Programa general de Información y UISIST. UNESCO. PGI-88/WS/17. 1988

[88] Sequeira S, Cabrita EJ, Macedo MF. Antifungals on paper conservation: An overview. International Biodeterioration & Biodegradation. 2012;**74**:67-86

[89] Borrego-Alonso S. Los biocidas vegetales en el control del biodeterioro del patrimonio documental. Perspectivas e impacto. Revista CENIC Ciencias Biológicas. 2015;**46**(3):259-269

[90] López LP. Óxido de etileno, utilización como agente esterilizante y riesgos para la salud del personal sanitario. Revista CES Salud Pública. 2014;**5**(2):154-162

[91] Masschelein-Kleiner L. Los Solventes. Publicaciones Centro Nacional de Conservación y Restauración DIBAH; 2004

[92] Morales R, Blanco P, Lalana P, Pardo de Santayana M, Valentín N. La ciencia y el arte IV. Ciencias experimentales y conservación del patrimonio. España: Ed: Ministerio de Educación Cultura y Deporte, Subdirección General de Documentación y Publicaciones; 2013

[93] Matusiak K, Machnowski W, Wryosek H, Polak J, Rajkowska K, Smigielski K, Kunicka-Styczynska A. Application of *Cinnamomun zeylanicum* essential oil in vapour phase for heritage textiles disinfection. International Biodeterioration & Biodegradation. 2017;**xxx**:1-9

[94] Valdés-Pérez O, Borrego-Aloso S, Vivar-González I, Anaya-Villapanda M, Molina-Veloso A. Actividad antifúngica del aceite esencial de clavo de olor en el control del biodeterioro fúngico de documentos. Revista CENIC Ciencias Biológicas. 2016;**47**(2):78-85

[95] Walentowka J, Folsowicz-Flaczyk J. Thyme essential oil for antimicrobial protection of natural textiles. International Biodeterioration & Biodegradation. 2013;**84**:407-411

[96] Baglioni P, Gorgi R, Chelazzi D. Nano-materials for the conservation and preservation of movable an inmovable artworks. In: Progress in Cultural Heritage—EUROMED 2012. 2012. pp. 313-318 Available from: https://www.researchgate.net/publication/270365605

[97] Balakumaran MD, Ramachandran R, Jagadeeswari S, Kalaichelvan PT. *In vitro* biological and characterization of nanosilver coated cotton fabric—An application for antimicrobial textile finishing. International Biodeterioration & Biodegradation. 2016;**107**:48-55

[98] Di Salvo S. Nanotechnology for cultural heritage. Science, Technology and Society: An International Journal. 2014;**2**(2):28-32

[99] Gambino M, Ali Ahmed MAA, Villa F. Zinc oxide nanoparticles hinder fungal biofilm development in an acient Egyptian tomb. International Biodeterioration & Biodegradation. 2017;**122**:92-99

[100] Capitelli F, Zanardini E, Randalli G, Mello E, Deffonchio D, Sorlini C. Improved methodology for bioremoval of black crust on historical stone artwork by use of sulphate reducing bacteria. Applied

and Environmental Microbiology. 2006;**72**:3733-3737

[101] Cappitelli F, Ranalli G, Zanardini E, Mello E, Sorlini C. Biotreatment of salts in stone at Matera Cathedral. In: Conference on Heritage, Microbiology and Science. Portsmouth, UK: University of Portsmouth; 29 June–2 July 2005. p. 51

[102] Bosh RP, Regidor RJL, Soriano SP, Deménech CMT, Montes ER. Ensayos de biolimpieza con bacterias en pinturas murales. Vol. 4-5. Arché: Publicación del Instituto de Restauración del Patarimonio de la UPV; 2010. pp. 117-124

[103] Le Métayer-Levrel G, Castanier S, Orial G, Joubiére JF, Perthuisot JP. Application of bacterial carbonatogenesis to the protection and regeneration of limestones in buildings and historic patrimony. Sedimentary Geology. 1999;**126**:25-34

[104] Rinaldi A. Saving a fragil legacy. EMBO Reports. 2006;**11**(11):5-1079

[105] Zammit G, Sánchez-Moral S, Albertano P. Bacterially mediated mineralisation processes lead to biodeterirarion of artwork in Maltese catacombs. Science of the Total Environment. 2011;**409**:2773-2782

[106] Páramo ALA, Narváez ZJA, Ortega MBO. La bioprecipitación de carbonato de calcio por la biota nativa como método de restauración. Nexo Revista Científica. 2015;**28**(1):25-40

[107] Calvo A. Conservación y restauración de pintura sobre lienzo. In: De la A a la Z, editor. Conservación y Restauración: Materiales Técnicas y Procedimientos. Barcelona: Ed. Serbal; 1997. pp. 152-155. ISBN: 9788476281949

[108] Porqué es importante la conservación del patrimonio cultural? Available from: https://sabiendomsdenuestropatrimoniocultural.wordpress.com [Accessed: May 7, 2018]

Chapter 5

Antimicrobial Agents: Antibacterial Agents, Anti-biofilm Agents, Antibacterial Natural Compounds, and Antibacterial Chemicals

Yaw Duah Boakye, Newman Osafo, Cynthia Amaning Danquah, Francis Adu and Christian Agyare

Abstract

The surge in antimicrobial resistance coupled with the decline in the antimicrobial drug pipeline calls for the discovery and development of new agents to tackle antibiotic resistance and prevent a return to a post-antibiotic era. Several factors account for resistance of microbes; some are natural and others are acquired. Natural selection, presence of efflux pumps, impermeable cell wall, biofilm formation and quorum sensing are some of the factors. Though it is difficult to outwit the pathogens, the discovery and development of compounds with pleiotropic modes or mechanisms of action different from the conventional drugs currently being used can help us tackle antimicrobial resistance. Natural products have been known to be a rich source of bioactive compounds with diverse structures and functional group chirality. Various reports indicate medicinal plants with antibacterial, anti-biofilm, efflux pump inhibition, wound healing effects or properties and others used for upper respiratory and urinary tract infections. There is an urgent need to research into natural products particularly plants for antimicrobial agents including antibacterial agents, anti-biofilm agents, antibacterial natural compounds and antibacterial chemicals. This chapter throws more light on such antimicrobials.

Keywords: antimicrobials, biofilm inhibitors, natural products, antibacterials

1. Antibacterial agents

1.1 Introduction

The last decade has seen in a dramatic fashion, an accelerated microbiological evolution and resistance to antimicrobial agents. There is therefore the need to optimize appropriate stewardship of infection control in the light of an apparent stagnation in the development of novel antimicrobial agents. This chapter therefore considers current anti-infective agents of the various classes that are clinically used in treating infections.

1.2 Beta lactam derivatives

The beta lactam derivatives stand as the oldest class of antibiotics used. The beta lactam ring has proven to be the major weapon in the fight against bacterial infections. Several novel molecules modeled after Alexander Fleming's penicillin and its derivatives have been developed. These drug moieties share the common characteristic of the beta lactam ring being an integral part of the structural make up and its effectiveness. These include the cephalosporins, monobactams, cephamycins, and the carbapenems (imipenem and meropenem). Beta lactams are indicated for a varied number of bacterial infections ranging from respiratory and urinary tract infections, ear and eye infections and gonorrhea to more life-threatening conditions like meningitis, septicemia and pneumonia. It is also widely adopted for prophylactic use in bacterial endocarditis, surgical site infections and in immuno-compromised situations [1].

Beta lactam antibiotics are bactericidal in their action. They inhibit the building of bacterial cell wall by interfering with the synthesis of peptidoglycan. Penicillin binding proteins which are bacterial enzymes which are essentially for bacterial cell wall synthesis are usually the targets of beta lactams [1]. Beta lactam antibiotics are generally available for parenteral administration with some also showing good absorption from the gastrointestinal tract. In patients with intact renal function, most beta lactams have a serum half-life of 1–2 h. Ceftazidime and temocillin break off this usual norm with a half-life of 4–6 h and an even higher half-life of 8–10 h for ceftriaxone. Penicillins and cephalosporins are eliminated primarily through glomerular filtration with varying levels of active transport across the renal tubules as well as the hepatobiliary system [2].

Owing to the general abuse of antibiotics, the beta lactams have suffered the challenge of the development of resistance in target pathogenic organisms. The production of beta lactamases has been a major determinant in the resistance observed especially in Gram-negative pathogens. Alterations in the beta-lactam targets, the penicillin binding proteins, are also important in Gram-positive pathogens. Efflux mechanisms and/or exclusion of these agents also contribute more often in conjunction with the other two mechanisms [3]. New agents of the beta-lactam group that have been approved following development have come from the cephalosporin class.

1.3 Cephalosporins

Cephalosporins are usually classified based on spectrum, generation, chemical structure, clinical pharmacology and resistance to beta-lactamases. The first cephalosporins were assigned first-generation cephalosporins; later, more expanded spectrum cephalosporins were designated as second-generation cephalosporins. Each more current generation has altogether more prominent activity against Gram-negative bacteria than the preceding generation and much of the time with diminished action against Gram-positive bacteria. Fourth-generation cephalosporins, however, have true broad-spectrum activity. The recent addition to the block, the fifth generation cephalosporins has become very crucial due to its activity against multidrug-resistant *Staphylococcus aureus* (MRSA) [4].

The drug of choice in this latest generation is ceftaroline. It is the only beta-lactam with MRSA activity. Ceftaroline, which is available as a pro-drug, ceftaroline prosamil is again unique for its expanded and extensive Gram-positive activity beyond all presently available cephalosporins. Ceftaroline active against the

Gram-positive organisms (*Streptococcus pneumoniae*, *Staphylococcus aureus* and *Streptococcus pyogenes*) and Gram-negative species (*Haemophilus influenzae* and *Moraxella catarrhalis*), including resistant phenotypes. Data for its approval proved its effectiveness in the treatment of acute bacterial skin and skin structure infections and community acquired bacterial pneumonia [5]. Ceftaroline demonstrates dose-proportional pharmacokinetics similar to other renally excreted cephalosporins after intravenous administration. It is half-life after dose is 2.53 h with protein binding of approximately 20%. Ceftaroline also showed in clinical trials to have positive attributes with regards to antibacterial stewardship by having a low potential for development of resistance as well as favorable tolerability and safety profile [5].

Ceftolozane is a novel beta-lactam cephalosporin combined with beta-lactamase inhibitor, tazobactam for the management of complicated urinary tract and intra-abdominal infections. The peculiar chemistry and dosing accounts for its extensive coverage of Gram-negative organisms including multidrug-resistant *Pseudomonas aeruginosa* as well as extended-spectrum beta-lactamase producing organisms and some anaerobes. Its efficacy can be compared to levofloxacin in patients with complicated urinary tract infections, including pyelonephritis, and comparable to that of meropenem against complicated intra-abdominal infections. Ceftolozane-tazobactam has shown to be indispensable due to the lack of susceptibility to the usual mechanisms of resistance mostly by Gram-negative organisms such as the production of beta-lactamases, efflux pumps, alterations in penicillin binding proteins as well as porin loss [6].

1.4 Glycopeptides

Glycopeptide antibiotics are complex and rigid molecules that repress a late stage in bacterial cell wall peptidoglycan synthesis. The selective toxicity of glycopeptides is attributable to the fact that its 3D structure harbors a cleft into which peptides of a specific configuration found only in bacterial cell walls can fit. Glycopeptide has assumed a special role in the face of the general threat of antimicrobial activity resistance since its unique mechanism involving the attachment of a bulky inhibitor to a substrate with the goal that the active sites enzymes are unable to align themselves correctly, therefore renders resistance to glycopeptides more difficult to achieve than other antimicrobial agents [7].

A good number of glycopeptides have gone through development in recent years and have been approved for clinical use. Oritavancin, a lipoglycopeptide obtained from the naturally occurring chloroeremomycin of the eremomycin class of glycopeptides is very similar to vancomycin but possess two 4-epi-vancosamine monosaccharides, one supplanting vancosamine and the other connected to ring-6 via an amino acid residue. Based on the features of its pharmacophore and its stereochemistry, it has enhanced antimicrobial against Gram-positive organisms including those possessing both VanA- and VanB-mediated vancomycin-resistance [8]. Oritavancin is transcendently cleared by means of the reticuloendothelial system, accumulating most notably in macrophages of the liver (Kupffer cells), kidney, spleen and lungs, as well as in the intestinal mucosa, thymus, and lymph nodes. Subsequent release and elimination from these tissues does not occur readily and thus only trace amounts are recouped from urine and feces. Dosage adjustments are not required in hepatic and renal insufficiency [9].

Telavancin also derivative of vancomycin and a lipoglycopeptide has been shown to have a dual mechanism of action by causing an inhibition of the peptidoglycan synthesis and through membrane depolarization. Telavancin is reliably active against *Staphylococcus aureus*, including MRSA, vancomycin-intermediate-resistant *Staphylococcus aureus*, linezolid-resistant *Staphylococcus aureus*, and daptomycin-susceptible strains and therefore effective for the treatment of complicated skin and skin-structure infections. It has additionally proven effective in the treatment of

Gram-positive bacterial infections especially pneumonia. Its non-inferiority is compared with vancomycin, in the treatment of complicated skin and skin-structure infections and pneumonia. Telavancin is excreted by the kidneys, and thus, dosage adjustments are required in cases of renal failure. Telavancin is related with higher rates of renal events, changed taste, nausea and vomiting but however lesser rates of pruritus and infusion related events relative to vancomycin [10].

Dalbavancin has proven to be a valuable addition to the armamentarium of antimicrobial agents as it is the first once a week antibiotic with activity against a broad range of Gram-positive pathogens. Dalbavancin's uniqueness is its novel pharmacokinetic profile with a half-life of 170–210 h, which makes the once-weekly dosing optimal. Forty percent is eliminated via the renal route. Most of the drug is excreted as intact drug. Concentration was unchanged in patients with mild renal impairment. No adjustments are needed in hepatic insufficiency, as concentrations of the drug do not increase with severe hepatic impairment. It is still unknown if the drug penetrates the cerebrospinal fluid, or whether the drug is removed during hemodialysis. However, the high protein binding of dalbavancin would suggest both of these scenarios to be unlikely [11].

1.5 Oxazolidinones

Oxazolidinones are synthetic antimicrobial agent which inhibit bacterial protein synthesis. Linezolid, the first oxazolidinone to be approved for clinical use, has bacteriostatic activity against many important resistant pathogens including methicillin-resistant *Staphylococcus aureus*, vancomycin-resistant enterococci, and penicillin-resistant *Streptococcus pneumoniae* as seen from *in vitro* studies [12]. Tedizolid which is available as a prodrug (tedizolid phosphate) is the first in the class to be dosed once daily as it has a half-life of 12 h. It also has a profound oral bioavailability of about 90% and no dosage adjustment is required between intravenous and oral administration, nor is dosage adjustment needed based on hepatic or renal impairment. Its activity covers Gram-positive pathogens, including methicillin-resistant *Staphylococcus aureus.* It is indicated for the management of acute bacterial skin and skin structures infections. Tedizolid appears to depart from linezolid in the incidence of gastrointestinal and hematologic side effects and the apparent lack of drug interaction with selective serotonin reuptake inhibitors [13].

1.6 Fidaxomicin (macrocyclic antibiotics)

Fidaxomicin is the first in the new class of narrow spectrum macrocyclic antibiotics. It is derived from the organism *Dactylosporangium aurantiacum* as a fermentation product. Fidaxomicin's action leads to an inhibition transcription process by interfering with RNA polymerase. Fidaxomicin acts at a distinct site and step of RNA synthesis to that of the rifamycins and streptolydigin, and thus no overlapping antibiotic resistance has been identified. Its narrow spectrum of activity is against most Clostridial spp., including *Clostridium difficile*, and exhibits moderate activity against other Gram-positive organisms, such as staphylococci and enterococci. Systemic absorption is very limited with serum concentrations remaining generally low [14].

1.7 Aminoglycosides

Aminoglycosides are exceptionally potent, broad-spectrum antimicrobial agents with numerous alluring properties for the treatment of hazardous diseases. The antibacterial potency of the aminoglycosides is attributable to one or several aminated sugars linked by glycosidic bonds to a dibasic cyclitol in their chemical structures. Aminoglycosides act primarily by impairing bacterial protein synthesis through binding to prokaryotic ribosomes. Widespread resistance against these agents including semi-synthetic additions such as amikacin and netilmicin has prompted the need for development of alternative agents. The three main resistance mechanisms identified include a decreased cell permeability; alterations at the ribosomal binding sites; and production of aminoglycoside modifying enzymes [12]. Plazomicin is a next-generation semisynthetic aminoglycoside derived from sisomicin, a naturally occurring aminoglycoside antibiotic. Plazomicin is active against Gram-negative and selected Gram-positive bacteria and it is active against enterobacteriaceae, multidrug-resistant enterobacteriaceae (MDR-EC), aminoglycoside-resistant enterobacteriaceae (AR-EC), carbapenem-resistant Enterobacteriaceae (CR-EC), colistin-resistant enterobacteriaceae (CRE), tigecycline-resistant enterobacteriaceae (TR-EC). It has no nephrotoxic and ototoxic effects that characterize agents in this class. Its superior pharmacokinetic profile also supports a more convenient once daily IV dosing [15].

1.8 Plcuromutilins

Pleuromutilins are antimicrobial agents that selectively restrain bacterial translation and derivatives of the naturally occurring tricyclic diterpenoid pleuromutilin. The interest in pleuromutilins has resurged following the resistance in more prominent classes like the beta-lactams. Pleuromutilins inhibit bacterial protein synthesis by binding to the central part of domain V of the 50S ribosomal subunit at the peptidyl transferase center ultimately affecting peptide bond formation [16]. Retapamulin used as a topical agent for the treatment of impertigo has bacteriostatic effect against *Staphylococcus aureus* and *Streptococcus pyogenes*. Retapamulin is well tolerated with the most commonly reported adverse effect being pruritus at the application site. Although comparative efficacy has not been established with mupirocin, it is an effective alternative with dosing advantage of twice a day application.

1.9 Tetracyclines

The tetracyclines are another old class of natural product antibiotics. Until the development of doxycycline, they were used topically. However subsequent systemic application allowed for its use in the treatment in respiratory tract infections. Tetracyclines also inhibit protein synthesis by inhibiting acyl-tRNA transfer on the bacterial 30S ribosome. Tetracyclines have a fused linear tetracyclic structure and form chelation complexes with divalent cations such as calcium and thus its use has been limited to adults albeit with side effects. They usually serve as alternative to patients who are intolerant of macrolides or macrolide-resistant pathogens [12].

Tigecycline is the foremost drug in the new glycycline subclass of antibiotics. In spite of the fact that it is structurally identifies with minocycline, modifications to the molecule has brought about an expanded spectrum of activity and decreased susceptibility to the development of resistance when compared with other tetracycline antibiotics. Tigecycline has a broad spectrum of activity, including activity against drug-resistant Gram-positive organisms [17].

1.10 Macrolides

Macrolides which have erythromycin as its first member inhibit protein synthesis by binding to the 23S RNA of the bacterial 50S ribosomal subunit at the exit of the peptide synthesis tunnel. Macrolides are most widely adopted for respiratory tract infections as they have targeted activity against respiratory pathogens. Macrolide antibiotics achieve high tissue and intracellular concentrations, which helps to address bacteria that are intracellular. In addition, they have strong anti-inflammatory properties [12]. Telithromycin belongs to a class of drugs described as the ketolides which varies slightly from the existing class of the macrolides. The characteristic feature of the ketolides as opposed to other macrolides is the removal of the neutral sugar, L-cladinose from the three position of the macrolide ring and the subsequent oxidation of the 3-hydroxyl to a 3-keto functional group. Telithromycin is indicated for the treatment of upper respiratory tract infections such as community acquired pneumonia and sinusitis. Introduction of telithromycin was opportune due to the rise of microbial resistance in the existing macrolides as it appears to be effective against macrolide-resistant bacteria such as macrolide-resistant *Streptococcus pneumoniae* [18].

1.11 Quinolones

The quinolones although not obtained from microbial source, are derived indirectly from natural products, i.e., as a by-product of chloroquine synthesis, which is in itself an analogue of the plant alkaloid quinine. It has established itself as useful in the treatment of urinary tract infections. The introduction of fluorine group to the core structure, give rise to the fluoroquinolones like ciprofloxacin and moxifloxacin yielded products of an improved spectrum and better pharmacokinetic profiles. However, resistance to the class of antibiotics whose mechanism of action is to inhibit bacterial DNA gyrase and topoisomerase IV and even newer fluoroquinolones has prompted the need for the development of new and effective agents [12]. Delafloxacin is a new anionic fluoroquinolone used for the treatment of acute bacterial skin and skin structure infections caused by Gram-positive and Gram-negative organisms including MRSA and *Pseudomonas aeruginosa*. As with all fluoroquinolones, resistance is mediated through mutations in the target enzymes and drug efflux. However, delafloxacin has greater stability against target enzyme mutations in Gram-positive bacteria relative to other fluoroquinolones. Its availability in infusion and oral formulations, stability and enhanced antibacterial potency in acidic environments and overall tolerability gives a potentially better antimicrobial agent in the treatment of other infections [19].

2. Anti-biofilm agents

2.1 Biofilms

Microorganisms have a strong tendency to become associated with surfaces [20]. Bacteria, thus, live in communities, adhering to surfaces of implanted medical devices or damaged tissues. On these surfaces they encase themselves in a hydrated matrix of polysaccharide and protein forming a slimy layer known as biofilms [21]. Biofilm is a microbial culture which is identified as cells permanently bound to an interface or to other cells and are firmly attached to matrix consisting of polymers produced as a result of phenotypic alteration due to growth rate or transcription of genes. Bacteria tend to form biofilms in environments with rapid flow of matter. Planktonic bacteria can adhere to surfaces and initiate biofilm formation in the presence of shear forces that are higher than those of heart valves and exceed Reynolds numbers of 5000. The Reynolds number has no dimension and describes the turbulent flow of a liquid. If it is high, turbulent flow exists but if it is low then laminar flow conditions prevail. It is speculated that turbulent flow enhances bacterial adhesion to surfaces and biofilm formation by impinging the planktonic cells on the surface [22].

Biofilms do not form only between same species of microorganism but there can be inter-species adhesions. Interspecies binding outside the oral cavity have been described as well, most notably between pathogenic and commensal microorganisms of the urinary tract. Co-aggregation between aquatic bacteria has been reported, with *Micrococcus luteus* being mentioned as a bridging organism in the development of aquatic biofilms owing to its ability to co-aggregate with many aquatic heterotrophs [20, 23].

2.2 Bacterial biofilm formation

Genes are responsible for noted biochemistry of living things and biofilm production is no exception. Changes in microbial colony formation and organization may be due to mutations in one or more of certain genes. Mutations in a gene called *wspF*, which is part of a putative chemosensory signal-transduction operon result in cell aggregation and altered colony morphology. The phenotypic characteristics of *WspF* depend on the presence of *WspR*, which is a member of a family of signal transduction proteins known as response regulators. *WspR* contains a glycine-glycine-aspartic acid-glutamic acid-phenylalanine (GGDEF) domain known to catalyze formation of a cytoplasmic signaling molecule cyclic diguanylate (c-di-GMP). Mass sequencing of genomes in bacteria detected the highly abundant protein domains GGDEF and Glutamic acid-Alanine-Leucine (EAl) [24]. These two protein domains are involved in the turnover of c-diGMP *in vivo*.

The GGDEF domain stimulates c-diGMP production whereas EAL stimulates its degradation. Increased cellular levels of c-diGMP has been observed to correspond to increased biofilm formation in a *wspF* mutant while increased levels of EAL catalyze degradation of c-diGMP and reversed the phenotypes of a *WspF* mutant and inhibited biofilm initiation by wild-type cells, indicating that the presence of c-diGMP is necessary for biofilm formation. The *psl* and *pel* operons, which are involved in exopolysaccharide production and biofilm formation, were expressed at high levels in a *WspF* mutant [25].

2.3 Stages of biofilm formation

The process of biofilm formation is complex, but generally identified as consisting of five stages.

2.3.1 The conditioning film

The conditioning layer is the foundation on which a biofilm grows, and can be composed of many particles, organic or inorganic. Via gravitational pull or direction of flow of the bulk fluid, particles rest and become integral component of conditional layer. Hence, the conditioning layer which is basically made up of organic matter provides anchorage and nutrients for bacterial growth [26].

2.3.2 Reversible adhesion

When plankton bacteria come into close proximity with the conditioning layer, it attaches using bacteria appendages such as flagella, fimbriae, and pili. A fraction of the cells reaching the surface reversibly adheres. However, the ability of the plankton to adhere is dependent on factors such as available energy, bacteria orientation, temperature and pressure in the immediate environment. If repulsive forces are greater than the attractive forces, the bacteria will detach from the surface [26].

2.3.3 Irreversible adhesion

A number of the reversibly adsorbed cells remain immobilized and become irreversibly adsorbed. It has been reported that the physical appendages of bacteria overcome the physical repulsive forces. Henceforth, the attachments do come in close proximity with the bulk lattice of the conditioning layer culminating in oxidation and hydration which strengthen the bacteria–surface bond. There is a data supporting the assertion that microbial adhesion intensely relies on the hydrophobic–hydrophilic nature of interacting surfaces [26].

2.3.4 Cell growth

With the adhered cells undergoing binary division, the resulting clonally expanded cells spread from the point of adhesion leading to clusters formation. Ideally, such growth occurring in the biofilm yields a mushroom-like arrangement. Nutrient supply to the bacteria present within the biofilm is made possible through the utilization of this mushroom-like assemblage. As the cells increase, they produce polysaccharide intercellular adhesion polymers which facilitate stronger attraction between adjacent cells [26, 27].

2.3.5 Final stages of biofilm development

As the bioburden increases, organisms enter into the stationary phase of growth cycle where the rate of cell division equals the rate of cell death. At this level of high population density, the bacteria cells interact with each other via signaling mechanisms called quorum sensing, a good explanation for the role of auto inducers in the stimulation of the mechanical and enzymatic procedures through genetic expression.

Death phase entails the collapse of the formed biofilm. The bacteria culture produce enzymes which catalyze the catabolism of the structural polysaccharides within the biofilm leading to exposure of the bacteria within the matrix to colonize new substrates [26, 28].

2.4 Antibiotic resistance due to biofilm formation

Inside biofilms, organisms resist antibiotic action by multicellular strategies, rather than the known genetic processes that involve plasmids, transposons and mutations that make individual cells resistant [29]. Biofilms are the root cause of many persistent chronic infections due to bacteria. Several organisms have been known to produce biofilms. In *P. aeruginosa*, it has been shown that the gene (*alg*C) that controls phosphomannomutase involved in alginate (exopolysaccharide) synthesis is up-regulated within minutes of adhesion to a solid surface. It has also been shown that *alg*D, *alg*U, *rpo*S and the genes controlling polyphosphokinase synthesis are all up-regulated during biofilm formation [22]. Multi-system that includes poor antibiotic penetration, nutrient limitation and slow growth, adaptive stress responses, and formation of persister cells are hypothesized to constitute the organisms' resistance to antibiotics in biofilms [29]. Organisms that have shown resistance to antibiotics as a result of biofilm production include *Staphylococcus aureus*, *Pseudomonas aeruginosa*, *Escherichia coli*, *Pseudomonas pseudomallei*, and *Streptococcus sanguis* [22].

Susceptibility tests with *in vitro* biofilm models have shown the survival of bacterial biofilms after treatment with antibiotics at concentrations hundreds or even a thousand times the minimum inhibitory concentration of the bacteria measured in a suspension culture indicates a state of a high resistance to antibiotics [30]. Antibiotics used for the treatment of such bacterial infections suppress symptoms of infections by killing free-floating bacteria shed from the attached population, but fail to eradicate those bacterial cells still embedded in the biofilm. The biofilm, thus, offers protection to the organisms from the effects of the antibiotics. After the treatment course of the infection, the biofilm can act as a microbial repository for recurrence of infection. Biofilm infections can linger on for months, years, or even a lifetime as long as the colonized surface is not removed from the body [29].

In biofilms, the familiar mechanisms of antibiotic resistance such as efflux pumps, modifying enzymes, and target mutations do not account for the resistance of the organism and protection of bacteria in a biofilm. Even sensitive bacteria that do not have a known genetic basis for resistance can have profoundly reduced susceptibility when they form a biofilm. The fact that all these antibiotic resistance mechanisms are inherently multicellular helps to explain why bacteria dispersed from biofilms rapidly revert to a susceptible phenotype [31]. Although they are less common, fungal biofilms are also found on implanted medical devices. *Candida albicans* biofilms are drastically very resistant to most antifungal drugs, and are a major cause of morbidity in blood-stream infections [32, 33].

2.5 Other effects of biofilms

Biofilms can be both beneficial and detrimental. They are beneficial in the degradation of environmental hazardous substances in the soil, in a bioreactor and as bio-flocculants in the separation of coal particles from associated mineral matter. They are detrimental on food and slaughterhouse equipment, ship hulls, biomaterials implants, and in the oral cavity [20]. They are implicated in otitis media [34], otolaryngologic infections [35], osteomyelitis [36], bacterial endocarditis [37], cystic fibrosis and nonhealing wounds [38].

When organisms are in biofilms, they tend to cause infections with similar features although there could be significant variations in the causative organisms. Importantly, bacterial biofilms tend to bypass host defense mechanism and can withstand drug treatment irrespective of the competence of the individual's immune system. Actually, tissues adjacent to the biofilm might undergo collateral damage by immune complexes and invade the neutrophils [29]. Biofilms have also been documented as the major sources of infection by *Candida albicans*, specifically in view of the vast number of biomaterials that are now being used in the medical industry. Biomaterials serve as ideal substrates for microbial adhesion and eventual biofilm formation. Such materials include stents, catheters, and orthopedic joints [39].

2.6 Biofilm formation inhibitors

Prevention of biofilm formation or countering the resistance mechanisms due to biofilms may simplify the treatment of infections caused by biofilm producing organisms and bring back the usefulness of antibiotics that are out of use due to biofilm resistance [29]. Many substances, both natural and synthetic, have been found to inhibit biofilm formation. For example, silver nanoparticles (AgNPs) effectively prevent the formation of biofilms and kill bacteria in established biofilms produced by clinical strains including *Acinetobacter baumannii* and *Pseudomonas aeruginosa*, methicillin-resistant *Staphylococcus aureus* (MRSA) and *Streptococcus mutans*. Anti-biofilm formation activity of AgNPs was more pronounced on Gram-negative than Gram-positive bacteria although both groups exhibit equal antibacterial activity to the substance. AgNPs also inhibit biofilm production by *Candida albicans* [40]. The helical human cathelicidin LL-37 exhibits effective antimicrobial, anti-attachment, and anti-biofilm activity against *Staphylococcus aureus* [41]. Some plants have also been found to inhibit biofilm formation. Extracts of the sticks of *Salvadora persica* has been reported for its antibacterial and anti-biofilm activities against *Streptococcus mutans* [42]. Essential oils of lemon grass have biofilm inhibition activity against *S. aureus* [43] while the essential oils of Boswellia spp. (*B. papyrifera* and *B. rivae*) are active against staphylococcal and *C. albicans* biofilms [44].

2.7 Detection of biofilms

Various methods have been used to determine the presence of biofilms produced by bacteria. The methods include tissue culture plate method (TCP), tube method (TM), Congo red agar method (CRA), bioluminescent assay, piezoelectric sensors, and fluorescent microscopic examination. In the tube method a loopful of overnight culture of the test organism is inoculated into trypticase soy broth containing 1% glucose and incubated. The bacterial cultures are poured out of the solution and the tubes washed using phosphate buffer saline and dried. The dried tubes are stained with 0.1% crystal violet with the excess stained washed with deionized water. The tubes are dried while inverted with observation made for biofilm formation which confirmed by the presence of visible film lining the wall as well as the bottom of the tube. If there is an observed ring-like development at the liquid interface it does not indicate biofilm formation [45]. In the Congo red agar method, the test organism is grown in red Congo agar on Brain heart infusion broth containing Congo red dye (0.8 g/L). The appearance of black colonies with a dry crystalline consistency indicated biofilm formation [45, 46]. In the tissue culture plate method a 24-h broth culture of the test organism is inoculated into trypticase soy broth (TSB). This primary inoculum is then inoculated into TSB with 1% glucose prepared in different dilutions (1:20, 1:40, 1:80 and 1:100) and loaded into 96 wells flat bottom microtiter plate, which is an abiotic surface. The plate is then incubated at 37°C for 24 h. The content of the wells are decanted and washed with phosphate buffer, fixed with methanol for 15 min and decanted. They are then stained with crystal violet (0.5%) for 20 min, decanted and washed with distilled water. Glacial acetic acid is then added to extract the crystal violet and the optical density determined at 490 nm using an ELISA plate reader [45].

3. Antibacterial natural products

3.1 Introduction

Natural products play a significant role in the discovery of lead compounds for the development of drugs for the treatment of human diseases. The importance to medicine of natural product molecules lies not only in their pharmacological or chemotherapeutic effects but also in their role as template molecules for the production of new drug substances. Nature in one way or another continues to influence the design of small molecules and most of the antibacterial drugs in clinical usage are naturally derived.

Natural products have the ability to provide diversity, complexity, novelty and new scaffolds with various chiral centers, rings, bridges and functional groups in the molecule [47]. They differ from synthetic compounds by having more oxygen atoms and stereochemical elements such as polycycle (often bridged) carbon skeletons [48, 49]. Some of the most valuable products and promising leads in oncology were naturally derived or naturally inspired. For instance paclitaxel a chemically established drug came from natural sources. Doxorubicin, camptothecins, and tamoxifen derived from natural product leads, steroid hormones. Most of the promising pipeline candidates in oncology all arose from natural products screening followed by synthetic modifications.

3.2 Examples of antibacterial natural products

Antibacterial natural compounds include secondary metabolites isolated from plants, bacteria, fungi, marine organisms and algae. These compounds are categorized based on their chemical type as terpenes (sesquiterpenes, diterpenes, sesterterpenes, and triterpenes), steroids (sterols), alkaloids (indole, quinoline, pyridoacridone, and amine alkaloids), aromatics (flavonoids, chalcones, coumarins, lignans, xanthones, anthracenes, anthraquinones, naphthalene), polyketides (acetylenic fatty acids, polycyclic esters and quinones), and peptides. Sometimes, the categorization tends to put together structurally relevant natural products with low bioactivity and also those synthetic analogues with remarkable antibacterial activity [47].

Since there is a continual need for a pipeline of new agents to combat multidrug-resistant bacteria, it is important the search goes on especially from plant materials. Microbially derived products, of which there are many first class drug examples which can be readily fermented with few re-supply issues [50]. The value of natural products as a screening resource has recently been highlighted and it is likely the focus is on plants, microbes and marine organisms [51].

Examples of antibacterial natural products include cranberry juice *Vaccinium macrocarpon* (family: Ericaceae) which is used in the management of urinary tract infections and the prevention of recurrent cystitis. Berberine has antibacterial activity against various strains of methicillin resistant *Staphylococcus aureus* (MRSA). Thymol is used as an antiseptic. Bearberry, *Arctostaphylos uva-ursi* L. (family: Ericaceae) is an antimicrobial agent used particularly for urinary tract infections against several organisms including *Bacillus subtilis, Escherichia coli, Mycobacterium smegmatis, S. aureus* and *Shigella* spp. and the antibacterial activity is attributed to arbutin. Lemon balm, *Melissa officinalis* (family: Lamiaceae) and tea tree, *Melaleucae alternifolia* (family: Myrtaceae) are taken as herb teas and tea tree oil is applied as ointment [51]. Garlic, *Allium sativum* (family: Alliaceae) have antimicrobial and antiseptic properties and is used for respiratory tract infections [48, 49] .

Tannins and resins produced by plants have antimicrobial and wound-healing properties. The essential oil constituents of plants also possess antimicrobial activity. Plants produce phytoalexins in response to infections caused by fungi viruses and bacterial that may infect them. Resveratrol is an example of an antifungal phytoalexin, which has anticancer, antioxidant and cardioprotective benefits for humans. Plant metabolites with antibacterial properties include anti-staphylococcal activities of the acylphloroglucinols and terthiophenes [47].

The quest for newer and potent antimicrobial agents has ventured into studies on plants in this research driven direction. It, however, makes a cogent argument since plants are known to produce varied chemicals for defense purposes against microorganisms. Also they produce cytotoxic compounds some of which have been successfully utilized as chemotherapeutic agents and hence give a laudable reason to continue with search for new treatment protocols for man. There is also an ecological rationale for the production of natural products that modify microbial resistance. Plants may have evolved compounds which evade MDR mechanisms and that plant antimicrobials might be developed into broad-spectrum antibiotics in combination with inhibitors of MDR [52]. These MDR proteins are commonly found in nature as efflux pumps for foreign toxic substances, as they are in clinical isolates of resistant pathogens.

In the food industry, herbs, spices and essential oils are chiefly employed as preservatives of foods. A number of plants and spices are used as antimicrobial agents for killing and decreasing pathogenic bacterial load in foods hence improving its quality. These plant-based antimicrobials are produced from various extraction techniques using the flowers, buds, seeds, leaves, twigs, bark, herbs, wood, fruits and roots of various plants [50, 51].

4. Non-medicinal antibacterial chemicals

An ideal antibacterial agent (medicinal or non-medicinal) tends to have characteristics, which include ability to effectively inhibit or kill bacteria. Also it must have appreciable solubility either in water or any suitable solvent [53]. Antibacterial chemicals should be stored for a reasonable period of time with no significant loss of antibacterial action [54]. Such preparation must also be homogeneous with the active ingredients present in each application of the non-medicinal chemical against bacteria. An antibacterial chemical should be minimally inactivated when exposed to extraneous material such as proteins and other organic materials found in substances they are being utilized in treating. This decreases the amount of the agent the bacteria are exposed to [55, 56]. Antibacterial chemicals should not necessarily require elevated temperatures beyond that of the environments they are being used [57]. Again, such ideal antibacterial chemicals do possess cleansing properties, must be either odorless or have pleasant odor [58] and should be physically safe on surfaces they are used on.

The antimicrobial activities of these chemicals are often strongly influenced by the biocide affinity for the structural or molecular components of the cell, which can, in turn, depend on the attraction of dissimilar charges or hydrophobic interactions [53, 59]. It is henceforth convenient to consider the modes of action based on biocides target on the bacteria. Some probable drug target to be exploited include the cell wall, cell membrane and the cytoplasm. The precise mechanism of antimicrobial action of some of these agents will be discussed in this section as well as unraveling the concentration-dependent multiplicity in action of some of these agents [57].

4.1 Alcohols

Two water-soluble alcohols, i.e., ethyl alcohol and isopropyl alcohol, are normally employed as disinfectants due to their high germicidal activity [60]. They have rapid bactericidal and bacteriostatic activity against vegetative forms of bacteria. Although not active against bacterial spores, their tuberculocidal, fungicidal and virucidal activity are also apparent. Their cidal activity drops sharply when diluted below 50% concentration, and the optimum bactericidal concentration is 60–90% v/v solutions in water [60, 61]. Protein denaturation is the most likely cause of the alcohol action. This assertion is affirmed by the realization of the reduced bactericidal activity of absolute ethyl alcohol when compared with hydroalcohol due to the requirement of water to facilitate the protein denaturation process [62, 63]. This observation is also in conformation with the observed destruction of dehydrogenases of *Escherichia coli* by alcohol [64] and also the enhancement of the lag phase of *Enterobacter aerogenes* by alcohol [65] which is however reversed with some amino acids. Blockade of metabolites required for cell division has been linked to the bacteriostatic action of alcohol.

The role of hydrophobic alkyl groups of long-chain alcohols in antibacterial activity against *Staphylococcus aureus* and *Propionibacterium acnes* has been established [66]. The maximum activity was found to be dependent on the chain length from the hydrophilic hydroxyl group, and also the test bacteria. The antimycobacterial activities of alcohols with chain length ranging from C_5 to C_{13} against *Mycobacterium smegmatis* mc^2155 and *M. tuberculosis* $H_{37}R_v$ have been established, with best activity found with alcohol with C_{10} chain length [67]. This bactericidal activity is attributed to decanol's potential to harm the robust and complex cell envelope of *Mycobacteria* and its ability to reduce biofilm formation by *M. smegmatis* [67].

4.2 Aldehydes

The aqueous solutions of some aldehydes, such as formaldehyde and glutaraldehyde, have been found to be bactericidal, tuberculocidal, fungicidal, virucidal, and sporicidal [68–70]. Formaldehyde inactivates microorganisms by alkylating the amino and sulfhydryl groups of proteins and ring nitrogen atoms of purine bases [71]. The biocidal activity of glutaraldehyde results from its alkylation of sulfhydryl, hydroxyl, carboxyl, and amino groups of microorganisms, which alters RNA, DNA, and protein synthesis [72].

Studies have established that different concentrations of formaldehyde have deleterious effect on a lot of microorganisms. It has been realized that 8% formalin can inactivate poliovirus in 10 min although other viruses are inactivated by as low as 2% formalin [68]. Tuberculocidal potential can be seen with 4% formalin with the agent being able to inactivate 10^4 *Mycobacterium tuberculosis* cells with 2 min [69]. Also, about 10 million *Salmonella typhi* cells can be inactivated by 2.5% formaldehyde within 10 min even in the presence of organic matter [70]. However, formaldehyde has significantly reduced sporicidal activity than glutaraldehyde as seen from experimental studies [69].

Some aliphatic saturated and unsaturated aldehydes are produced by enzymatic cleavage of unsaturated fatty acids when plants undergo microbial attack; thus they may be one of the multichemical defense mechanisms used by several fruits to resist invasion by microorganisms [73]. In this direction, Trombetta and colleagues established the mechanism of the antimicrobial activity of aliphatic α,β-unsaturated aldehydes which included (E)-2-hexenal **(1)**, (E)-2-eptenal **(2)**, (E)-2-octenal **(3)**, (E)-2-nonenal **(4)**, (E)-2-decenal **(5)** and (E,E)-2,4-decadienal **(6)** (**Figure 1**). Their findings suggested that the 2E-alkenals tested elicit, very likely, a gross perturbation of the lipidic fraction of plasma membranes and are able to penetrate into bacterial cells.

The mechanisms of antimicrobial action of other aldehydes, such as *o*-phthalaldehyde **(7)**, are likely to involve interaction with the cytoplasmic membrane and increase in its permeability [74, 75]. *o*-Phthalaldehyde also appears to kill spores by blocking the spore germination process [76]. Although membrane functional proteins are generally supposed to be the potential targets toward which aldehydic antimicrobial agents are directed, other mechanisms of action/interaction can help explain their antimicrobial activity [73].

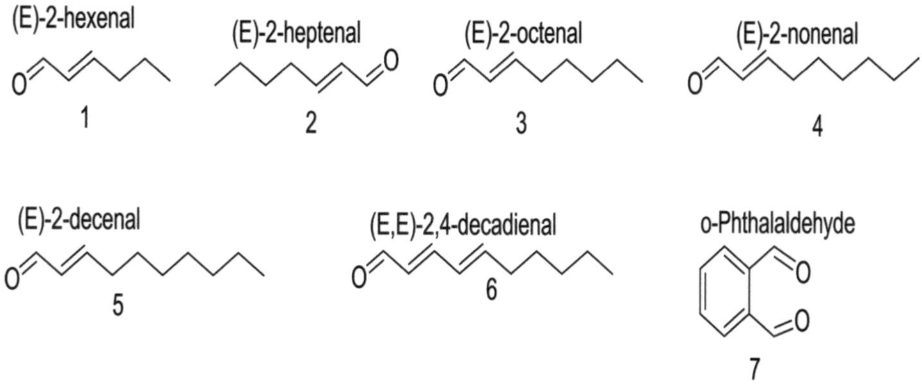

Figure 1.
Aldehydes with antibacterial properties.

4.3 Halogens

Over the years, tinctures and solutions of iodine have been used mainly as antiseptic agents on skin surfaces and other tissues. However, iodophors such as povidone iodine, are employed mainly as antiseptics and disinfectants. Such iodophors tend to possess germicidal ability but generally do not stain surfaces as well as do possess less toxic and irritant potential [77]. Iodine has the potential to penetrate the cell wall of microorganisms rapidly and tend to compromise proteins and nucleic acids as well as the synthesis of proteins.

Chlorine and products of chlorine do have broader spectrum of activity but they also leave behind toxic residues. They however are not affected by hardness of water, are less expensive and do possess faster onset of action [78]. They tend to possess activity on organisms fixed to surfaces and biofilms [79] with accompanied low risk of toxicity [80]. Its mechanism of antimicrobial action is associated with hypochlorous acid (HOCl) it yields. When HOCl dissociates, it produces hypochlorite ion (OCl^{-}) which has reduced antimicrobial activity and the process is pH-dependent [81]. It has not been mechanistically established how chlorine specifically kills microorganisms. However, chlorine tends to cause oxidation of sulfhydryl enzymes and amino acids leading to loss of intracellular contents with reduced nutrition uptake by cells. Also, chlorine tends to reduced adenosine triphosphate production as well as inhibition of DNA synthesis and increased DNA destruction [81]. A number of these process may account for the mechanism of antimicrobial action of chlorine [82].

4.4 Oxidizing agents

Peroxyacetic acid (**1**), chlorine dioxide (**2**) (**Figure 2**), and hydrogen peroxide have similar mechanism of antimicrobial action (chemical oxidation of cellular components), but they do vary greatly in their efficacy against microorganisms [83, 84]. Biochemically, there is significant variations in these agents with resultant variation in their outcomes on macromolecules. This attests to the variations in their biocidal activity, most especially between liquid and gas peroxide [84].

Administration of hydrogen peroxide results in the production of hydroxyl free radicals which attack lipid membranes, DNA and essential cell components leading to cell death. Aerobic organisms and facultative anaerobes tend to produce catalase and these enzymes offer protective benefits from hydrogen peroxide by catalyzing its conversion to water and oxygen. Such protective benefits are compromised by the concentrations of oxidizing agents administered as disinfectant [85]. Peracetic acid, which has rapid action on all microorganisms, does not yield any harmful products when it is broken down and is easily cleared from organic materials without leaving behind any residues [86]. In addition to its potent activity in the presence of organic matter, it retains sporicidal activity even at low temperatures.

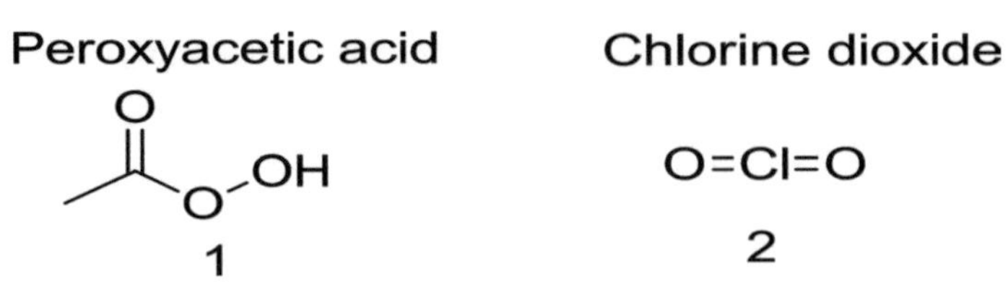

Figure 2.
Oxidizing agents with antibacterial activity.

DOI: http://dx.doi.org/10.5772/intechopen.82560

However, its specific mechanism of antimicrobial action has not been unraveled although postulated to be working in the same manner as other oxidizing agents. Hence possibly through oxidation of sulfhydryl and sulfur bonds in proteins, enzymes and other cellular metabolites [85].

4.5 Phenolic compounds

Phenolic compounds are produced from structural modification of phenol by addition of alkyl, phenyl, benzyl groups, and halogen as substituent on the aromatic ring. *o*-Phenylphenol **(1)** and *o*-benzyl-*p*-chlorophenol **(2)** are commonly employed phenol derivatives in disinfectants. Phenolic compounds tend to have enhanced antimicrobial potential than phenol [87].

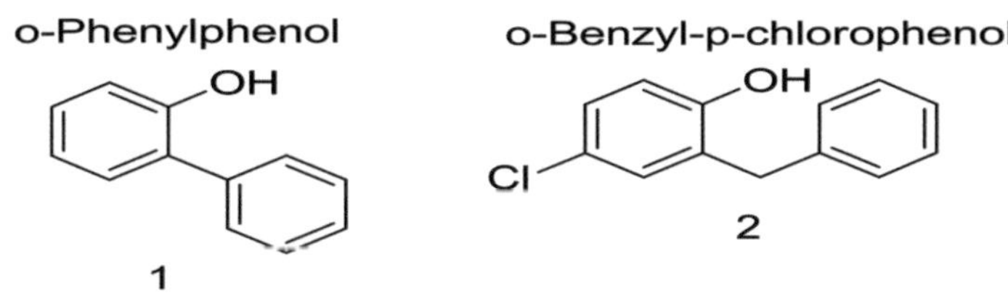

Figure 3.
Phenolic antibacterial chemicals.

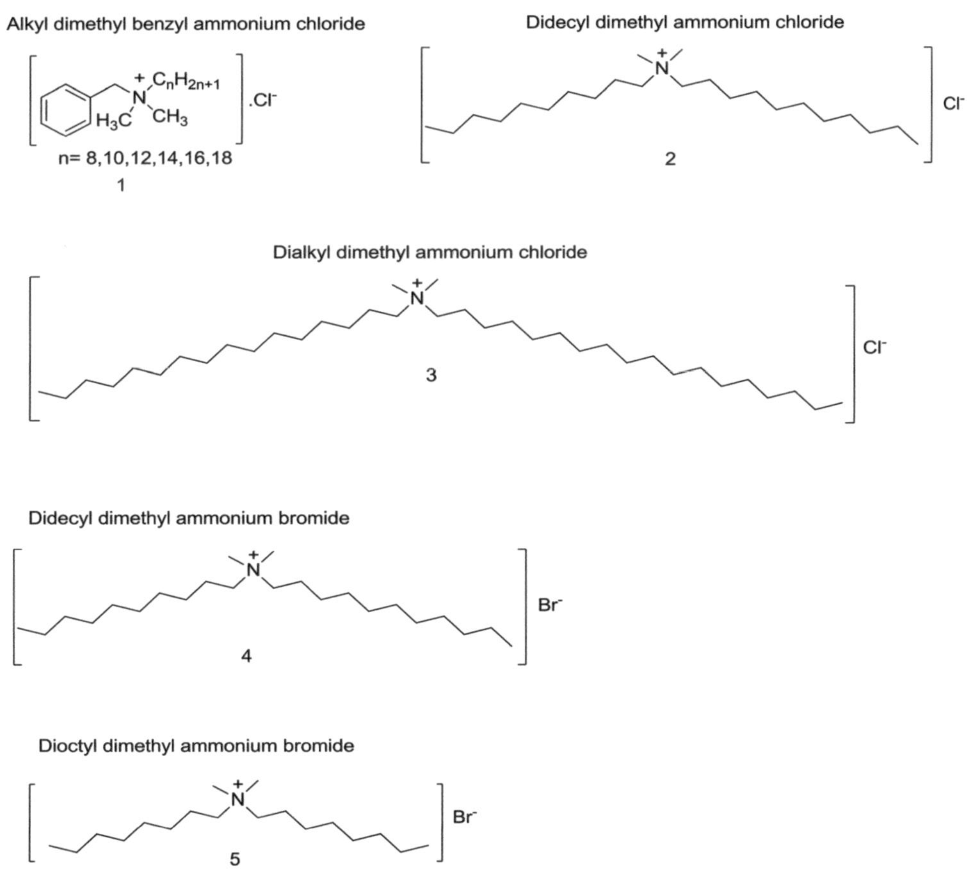

Figure 4.
Quaternary ammonium compounds with antibacterial activity.

Phenol acts as a penetrating and disrupting molecule on cell walls of organisms as well as increase precipitation in important proteins within cells leading to cell death at high concentrations. However, phenol and high molecular weight phenolic compounds tend to kill bacteria cells via enzyme inactivation and compromising of cell wall at lower concentration (**Figure 3**) [88].

4.6 Quaternary ammonium compounds

One of the most used disinfectants is the quaternary ammonium compounds. When contaminated quaternary ammonium compounds are employed in patient-care supplies, they contribute to significant proportion of its associated infections [89]. Gram-negative bacteria are documented to survive in this group of disinfectants [90].

The quaternary compounds are ammonium-derived moieties with nitrogen atom with a valence of 5. There are four alkyl or heterocyclic radical substitutes and a fifth halide, sulfate or similar radical substitute in the structure [91]. Some of these quaternaries include alkyl dimethyl benzyl ammonium chloride **(1)**, alkyl didecyl dimethyl ammonium chloride **(2)**, and dialkyl dimethyl ammonium chloride **(3)** (**Figure 4**). The innovative quaternary ammonium compounds (i.e., fourth generation), referred to as twin-chain or dialkyl quaternaries (e.g., didecyl dimethyl ammonium bromide **(4)** and dioctyl dimethyl ammonium bromide **(5)** (**Figure 4**)), have been reported to be active even in hard water and can withstand anionic residues [92]. Quaternaries are believed to be bactericidal due to their ability to inactivate energy-producing enzymes and also denature and disrupt cell proteins and membrane respectively [91, 92].

5. Conclusion

A multidisciplinary approach to antimicrobial drug discovery, involving the generation of novel molecular diversity from natural product sources, combined with total and combinatorial synthetic methodologies, and including the manipulation of biosynthetic pathways will continue to provide the best approach to antibiotic discovery and development and also overcome the challenges associated with antimicrobial resistance.

Conflict of interest

Authors declare no conflict of interest.

Author details

Yaw Duah Boakye[1], Newman Osafo[2], Cynthia Amaning Danquah[2], Francis Adu[1] and Christian Agyare[1]*

1 Department of Pharmaceutics, Faculty of Pharmacy and Pharmaceutical Sciences, Kwame Nkrumah University of Science and Technology, Kumasi, Ghana

2 Department of Pharmacology, Faculty of Pharmacy and Pharmaceutical Sciences, Kwame Nkrumah University of Science and Technology, Kumasi, Ghana

*Address all correspondence to: cagyare.pharm@knust.edu.gh

References

[1] Becker DE. Antimicrobial drugs. Anesthesia Progress. 2013;**60**(3):111-123

[2] Bergand T. Pharmacokinetics of beta lactam antibiotics. Scandinavian Journal of Infectious Diseases. Supplementum. 1984;**42**:83-98

[3] Kuriyama T, Karasawa T, Nakagawa K, Yamamoto E, Nakamura S. Incidence of β-lactamase production and antimicrobial susceptibility of anaerobic Gram-negative rods isolated from pus specimens of orofacial odontogenic infections. Oral Microbiology and Immunology. 2001;**16**(1):10-15

[4] Mehta D, Sharma AK. Cephalosporins: A review on imperative class of antibiotics. Inventi Rapid/Impact: Molecular Pharmacology. 2016;**1**:1-6

[5] Laudano JB. Ceftaroline fosamil: A new broad-spectrum cephalosporin. Journal of Antimicrobial Chemotherapy. 2011;**66**(3):11-8

[6] Cluck D, Lewis P, Stayer B, Spivey J, Moorman J. Ceftolozane-tazobactam: A new-generation cephalosporin. 2015;**72**:2135-2146

[7] Reynolds P. Structure biochemistry and mechanism of action of glycopeptide antibiotics. European Journal of Clinical Microbiology & Infectious Diseases. 1989;**8**(11):943-950

[8] Stein G. Oritavacin: A long-half life lipoglycopeptide. Clinical Infectious Diseases. 2015;**61**(4):627-632

[9] Karrine D, Ryback J. Oritavancin: A new lipoglycopeptide antibiotic in the treatment of Gram-positive infection. Infectious Disease and Therapy. 2016;**16**:1-15

[10] Saravolatz LD, Stein GE, Johnson LB. Telavancin: A novel lipoglycopeptide. Clinical Infectious Diseases. 2009;**49**(12):1908-1914

[11] Chen AY, Zervos MJ, Vazquez JA. Dalbavancin: A novel antimicrobial. International Journal of Clinical Practice. 2007;**5**(May):853-863

[12] Fernandes P, Martens E. Antibiotics in late clinical development. Biochemical Pharmacology. 2017;**133**:152-163

[13] Burdette SD, Trotman R. Tedizolid: The first once-daily oxazolidinone class antibiotic. Agents. 2018;**61**(June):1315-1321

[14] Mullane K, Gorbach S. Fidaxomicin: First-in-class macrocyclic antibiotic. Expert Review of Anti-Infective Therapy. 2011;**9**(7):767-778

[15] Gupta A. Plazomicin: A step toward next generation aminoglycosides. Review. Asian Journal of Research in Pharmaceutical Sciences. 2017;**7**(3):1-8

[16] Paukner S, Riedl R. Pleuromutilins: Potent drugs for resistant bugs—Mode of action and resistance. Cold Spring Harbor Perspectives in Medicine. 2017;**7**(1).pii:a027110

[17] Greer ND. Tigecycline (Tygacil): The first in the glycylcycline class of antibiotics. Proceedings (Baylor University, Medical Center). 2006;**19**(2):155-161

[18] Ackermann G, Rodloff AC. Drugs of the 21st century: Telithromycin (HMR 3647)—The first ketolide. The Journal of Antimicrobial Chemotherapy. 2003;**51**:497-511

[19] Mogle BT, Steele JM, Thomas SJ, Bohan KH, Kufel WD. Clinical review of delafloxacin: A novel anionic fluoroquinolone. The Journal

of Antimicrobial Chemotherapy. 2018;**73**(6):1439-1451

[20] Bos R, Van Der MHC, Busscher HJ. Physico-chemistry of initial microbial adhesive interactions its mechanisms and methods for study. FEMS Microbiology Reviews. 1999;**23**(2):179-230

[21] Stewart PS, Costerton JW. Antibiotic resistance of bacteria in biofilms. Lancet. 2001;**358**(9276):135-138

[22] Donlan RM, Costerton JW, Donlan RM, Costerton JW. Biofilms: Survival mechanisms of clinically relevant microorganisms. Clinical Microbiology Reviews. 2002;**15**(2):167-193

[23] Reid G, Cook L, McGroarty JA, Angotti R. Lactobacillus inhibitor production against *Escherichia coli* and coaggregation ability with uropathogens. Canadian Journal of Microbiology. 1988;**34**(3):344-351

[24] Simm R, Morr M, Kader A, Nimtz M, Römling U. GGDEF and EAL domains inversely regulate cyclic di-GMP levels and transition from sessility to motility. Molecular Microbiology. 2004;**53**(4):1123-1134

[25] Hickman JW, Tifrea DF, Harwood CS. A chemosensory system that regulates biofilm formation through modulation of cyclic diguanylate levels. Proceedings of the National Academy of Sciences of the United States of America. 2005;**102**(40):14422-14427

[26] Banerjee P, Singh M, Sharma V. Biofilm formation: A comprehensive review. 2015;**3**(2):556-560

[27] Hall-Stoodley P. Developmental regulation of microbial biofilms. Current Opinion in Biotechnology. 2002;**13**:228-233

[28] Dunne WM. Bacterial adhesion: Seen any good biofilms lately? 2002;**15**:155166

[29] Stewart PS. Mechanisms of antibiotic resistance in bacterial biofilms. International Journal of Medical Microbiology. 2002;**292**(2):107-113

[30] Ceri H, Olson ME, Stremick C, Read RR, Morck D. The calgary biofilm device: New technology for rapid determination of antibiotic susceptibilities of bacterial biofilms. Journal of Clinical Microbiology. 1999;**37**(6):1771-1776

[31] Walsh C. Molecular mechanisms that confer antibacterial drug resistance. Nature. 2000;**406**:775-781

[32] Douglas LJ. Candida biofilms and their role in infection. Trends in Microbiology. 2003;**11**(1):30-36

[33] Wesenberg-Ward KE, Tyler BJ, Sears JT. Adhesion and biofilm formation of Candida albicans on native and pluronic-treated polystyrene. Biofilms. 2005:63-71

[34] Bakaletz LO. Biofilms in otitis media: Evidence and relevance. The Pediatric Infectious Disease Journal. 2007;**26**(10):S17-S19

[35] Post JC, Stoodley P, Hall L, Ehrlich GD. The role of biofilms in otolaryngologic infections. Current Opinion in Otolaryngology & Head and Neck Surgery. 2004;**12**(3):185-190

[36] Brady RA, Leid JG, Calhoun JH, Costerton JW, Shirtliff ME. Osteomyelitis and the role of biofilms in chronic infection. FEMS Immunology and Medical Microbiology. 2008;**52**(1):13-22

[37] Høiby N. Understanding bacterial biofilms in patients with cystic fibrosis: Current and innovative approaches to potential therapies. Journal of Cystic Fibrosis. 2002;**1**:249-254

[38] James GA, Swogger E, Wolcott R, Pulcini E, Secor P, Sestrich J,

et al. Biofilms in chronic wounds. Wound Repair and Regeneration. 2008;**16**(1):37-44

[39] Ramage G, Saville SP, Wickes BL, López-ribot JL. Inhibition of *Candida albicans* biofilm formation by farnesol, a quorum-sensing molecule. Applied and Environmental Microbiology. 2002;**68**(11):5459-5463

[40] Martinez-Gutierrez F, Boegli L, Agostinho A, Sánchez EM, Bach H, Ruiz F, et al. Anti-biofilm activity of silver nanoparticles against different microorganisms. Biofouling. 2013;**29**(6):651-660

[41] Dean SN, Bishop BM, Van HML. Natural and synthetic cathelicidin peptides with anti-microbial and anti-biofilm activity against *Staphylococcus aureus*. BMC Microbiology. 2011;**11**:114

[42] Al-sohaibani S, Murugan K. Anti-biofilm activity of *Salvadora persica* on cariogenic isolates of *Streptococcus mutans*: *In vitro* and molecular docking studies. Biofouling. 2012;**28**(1):29-23

[43] Adukwu EC, Allen SCH, Phillips CA. The anti-biofilm activity of lemongrass (*Cymbopogon flexuosus*) and grapefruit (*Citrus paradisi*) essential oils against five strains of *Staphylococcus aureus*. Journal of Applied Microbiology 2012;**113**(5):1217-2

[44] Schillaci D, Arizza V, Dayton T, Camarda L, Stefano VD. *In vitro* anti-biofilm activity of Boswellia spp. oleogum resin essential oils. Letters in Applied Microbiology. 2008;**47**(5):433-438

[45] Deka N. Comparison of tissue culture plate method , tube method and congo red agar method for the detection of biofilm formation by coagulase negative Staphylococcus isolated from non-clinical isolates. International Journal of Current Microbiology and Applied Sciences. 2014;**3**(10):810-815

[46] Kaiser TD, Pereira EM, Dos Santos KR, Maciel EL, Schuenck RP, Nunes AP. Modification of the Congo red agar method to detect biofilm production by *Staphylococcus epidermidis*. Diagnostic Microbiology and Infectious Disease. 2013;**75**(3):235-239

[47] Gibbons S. Anti-staphylococcal plant natural products. Natural Product Reports. 2004;**2**:263-277

[48] Danquah CA, Kakagianni E, Khondkar P, Maitra A, Malkinson J, Bhakta S, et al. Analogues of disulfides from *Allium stipitatum* demonstrate potent anti-tubercular activities through drug efflux pump and biofilm inhibition. 2018, 1150;**8**(1):18

[49] O'Donnell G, Poeschl R, Zimhony O, Gunaratnam M, Moreira JBC, Neidle S, et al. Bioactive pyridine-N-oxide disulfides from *Allium stipitatum*. Journal of Natural Products. 2009;**72**(3):360-365

[50] Newman DJ, Cragg GM. Natural products as sources of new drugs over the 30 years from 1981 to 2010. Journal of Natural Products. 2012;**75**(3):311-335

[51] Heinrich M, Barnes J, Gibbons S, Williamson EM. Fundamentals of Pharmacognosy and Phytotherapy. Edinburgh: Churchill Livingstone/ Elsevier; 2012

[52] Tegos G, Mylonakis E, editors. Antimicrobial Drug Discover: Emerging Strategies. Wallingford: CABI; 2012

[53] McDonnell G, Russel AD. Antiseptics and disinfectants: Activity, action, and resistance. . 1999;**12**(1):147179

[54] Davies J, Davies D. Origins and evolution of antibiotic resistance. Microbiology and Molecular Biology Reviews. 2010;**74**(3):417-433

[55] Marriott NG. Principles of Food Sanitation. P. 143. Gaithersburg, MD: Aspen Publishers, Inc; 1999. p. 364

[56] Cords BR, Burnett SL, Hilgren J, Finley M, Magnuson J. Sanitizers: Halogens, surface-active agents, and peroxides. In: Davidson PM, Sofos J, Branen AL, editors. Antimicrobials in Food. 3rd ed. Boca Raton: Taylor and Francis; 2005

[57] Kahrilas GA, Blotevogel J, Stewart PS, Borch T. Biocides in hydraulic fracturing fluids: A critical review of their usage, mobility, degradation, and toxicity. Environmental Science & Technology. 2015;**49**(1):16-32

[58] Lambert D. Topical deodorant. In: Baran R, Maibach HI, editors. Textbook of Cosmetic Dermatology. 2nd ed. London, UK: Martin Dunitz Ltd; 1998. p. 172

[59] Denyer SP, Hodges NA, Gorman SP. Non-antibiotic antibacterial agents: Mode of action and resistance. In: Hugo and Russell's Pharmaceutical Microbiology. 7th ed. Oxford, UK: Blackwell Publishing Company; 2008. p. 309

[60] Spaulding EH. Alcohol as a surgical disinfectant: Pros and Cons of a much dicussed topic. AORN Journal. 1964;**2**:67-71

[61] Morton HE. The relationship of concentration and germicidal efficiency of ethyl alcohol. Annals of the New York Academy of Sciences. 1950;**53**(1):191-196

[62] Ali Y, Dolan MJ, Fendler EJ, Larson EL. Alcohols. In: Block SS, editor. Disinfection, Sterilization, and Preservation. Philadelphia: Lippincott Williams & Wilkins; 2001. pp. 229-254

[63] Morton HE. Alcohols. In: Block SS, editor. Disinfection, Sterilization, and Preservation. Philadelphia: Lea & Febiger; 1983. pp. 225-239

[64] Sykes G. The influence of germicides on the dehydrogenases of Bact. coli. Part I. The succinic acid dehydrogenase of Bact. coli. (London). 1939;**39**(4):463-469

[65] Dagley S, Dawes EA, Morrison GA. Inhibition of growth of Aerobacter aerogenes: The mode of action of phenols, alcohols, acetone and ethyl acetate. Journal of Bacteriology. 1950;**60**:369-378

[66] Kubo I, Muroi H, Himejima M, Kubo A. Antibacterial activity of long-chain alcohols: The role of hydrophobic alkyl groups. Bioorganic & Medicinal Chemistry Letters. 1993;**3**(6):1305-1308

[67] Mukherjee K, Tribedi P, Mukhopadhyay B, Sil AK. Antibacterial activity of long-chain fatty alcohols against mycobacteria. FEMS Microbiology Letters. 2013;**223**(2):177-183

[68] Klein M, DeForest A. The inactivation of viruses by germicides. Chemical Specialties Manufacturers Association Proceedings. 1963;**49**:116-118

[69] Rubbo SD, Gardner JF, Webb RL. Biocidal activities of glutaraldehyde and related compounds. Journal of Applied Microbiology. 1967;**30**:78-87

[70] McCulloch EC, Costigan S. A comparison of the efficiency of phenol, liquor cresolis, formaldehyde, sodium hypochlorite and sodium hydroxide against *Eberthella typhi* at various temperatures. The Journal of Infectious Diseases. 1936;**59**:281-284

[71] Favero MS, Bond WW. Chemical disinfection of medical and surgical materials. In: Block SS, editor. Disinfection, Sterilization, and

Preservation. Philadelphia: Lea & Febiger; 1991. pp. 617-641

[72] Scott EM, Gorman SP. Glutaraldehyde. In: Block SS, editor. Disinfection, Sterilization, and Preservation. Philadelphia: Lippincott Williams & Wilkins; 2001. pp. 361-381

[73] Trombetta D, Saija A, Bisignana G, Arena S, Caruso S, Mazzanti G, et al. Study on the mechanisms of the antibacterial action of some plant α,β-unsaturated aldehydes. Letters in Applied Microbiology. 2002;**35**:285-290

[74] Simons C, Walsh SE, Maillard JY, Russell AD. A note: Ortho-phthalaldehyde: Proposed mechanism of action of a new antimicrobial agent. Letters in Applied Microbiology. 2000;**31**:299-302

[75] Tsuchiya H. Biphasic effects of acetaldehyde-biogenic amine condensation products on membrane fluidity. The Journal of Pharmacy and Pharmacology. 2001;**53**:121-127

[76] Cabrera-Martinez RM, Setlow B, Setlow P. Studies on the mechanisms of the sporicidal action of ortho-phthalaldehyde. Journal of Applied Microbiology. 2002;**92**:675-680

[77] Gottardi W. Iodine and iodine compounds. In: Block SS, editor. Disinfection, Sterilization, and Preservation. Philadelphia: Lippincott Williams & Wilkins; 2001. pp. 159-184

[78] Rutala WA, Weber DJ. Uses of inorganic hypochlorite (bleach) in health-care facilities. Clinical Microbiology Reviews. 1997;**10**:597-610

[79] Merritt K, Hitchins VM, Brown SA. Safety and cleaning of medical materials and devices. 2000;**53**:131-136

[80] Heidemann SM, Goetting MG. Treatment of acute hypoxemic respiratory failure caused by chlorine exposure. 1991;**7**:87-88

[81] Dychdala GR. Chlorine and chlorine compounds. In: Block SS, editor. Disinfection, Sterilization, and Preservation. Philadelphia: Lippincott Williams & Wilkins; 2001. pp. 135-157

[82] Gerba CP, Rusin P. Relationship between the use of antiseptics/disinfectants and the development of antimicrobial resistance. In: Rutala WA, editor. Disinfection, Sterilization and Antisepsis: Principles and Practices in Healthcare Facilities. Washington, DC: Association for Professional in Infection Control and Epidemiology; 2001. pp. 187-194

[83] Denyer SP, Stewart GSAB. Mechanisms of action of disinfectants. International Biodet and Biodegrad. 1998;**41**:261-268

[84] Finnegan M, Linley E, Denyer SP, McDonnel G, Simons C, Maillard J-Y. Mode of action of hydrogen peroxide and other oxidizing agents: Differences between liquid and gas forms. The Journal of Antimicrobial Chemotherapy. 2010;**65**(10):2108-2115

[85] Block SS. Peroxygen compounds. In: Block SS, editor. Disinfection, Sterilization, and Preservation. Philadelphia: Lippincott Williams & Wilkins; 2001. pp. 185-204

[86] Tucker RC, Lestini BJ, Marchant RE. Surface analysis of clinically used expanded PTFE endoscopic tubing treated by the steris process. ASAIO Journal. 1996;**42**:306-313

[87] Kahn G. Depigmentation caused by phenolic detergent germicides. Archives of Dermatology. 1970;**102**:177-187

[88] Prindle RF. In: Block SS, editor. Phenolic compounds, Disinfection, Sterilization, and Preservation. Philadelphia: Lea & Febiger; 1983. pp. 197-224

[89] Ehrenkranz NJ, Bolyard EA, Wiener M, Cleary TJ. Antibiotic-sensitive *Serratia marcescens* infections complicating cardiopulmonary operations: contaminated disinfectant as a reservoir. Lancet. 1980;**2**:1289-1292

[90] Rutala WA, Cole EC. Antiseptics and disinfectants—Safe and effective? Infection Control. 1984;**5**:215-218

[91] Sykes G. Disinfection and Sterilization. London: E & FN Spon Ltd; 1965

[92] Merianos JJ. Surface-active agents. In: Block SS, editor. Disinfection, Sterilization, and Preservation. Philadelphia: Lippincott Williams & Wilkins; 2001. pp. 283-320

Chapter 6

The Methods for Detection of Biofilm and Screening Antibiofilm Activity of Agents

Sahra Kırmusaoğlu

Abstract

Biofilm producer microorganisms cause nosocomial and recurrent infections. Biofilm that is a sticky exopolysaccharide is the main virulence factor causing biofilm-related infections. Biofilm formation begins with attachment of bacteria to biotic surface such as host cell or abiotic surface such as prosthetic devices. After attachment, aggregation of bacteria is started by cell-cell adhesion. Aggregation continues with the maturation of biofilm. Dispersion is started by certain conditions such as phenol-soluble modulins (PSMs). By this way, sessile bacteria turn back into planktonic form. Bacteria embedded in biofilm (sessile form) are more resistant to antimicrobials than planktonic bacteria. So it is hard to treat biofilm-embedded bacteria than planktonic forms. For this reason, it is important to detect biofilm. There are a few biofilm detection and biofilm production methods on prosthetics, methods for screening antibacterial effect of agents against biofilm-embedded microorganism and antibiofilm effect of agents against biofilm production and mature biofilm. The aim of this chapter is to overview direct and indirect methods such as microscopy, fluorescent in situ hybridization, and Congo red agar, tube method, microtiter plate assay, checkerboard assay, plate counting, polymerase chain reaction, mass spectrometry, MALDI-TOF, and biological assays used by antibiofilm researches.

Keywords: mature biofilm, biofilm-embedded bacteria, antibiofilm methods, detection of biofilm gene expression, biofilm detection, MALDI-TOF, CLSM, checkerboard assay, microtiter plate assay, Congo red agar method

1. Introduction

1.1 The structure and pathogenesis of microbial biofilms

Microbial biofilms that are sticky exopolymeric substances (EPS) causing adherence of microorganism to biotic surfaces such as host cells or abiotic surfaces such as medical devices cause antimicrobial resistance, due to its molecular contents such as eDNA and exoenzymes (β-lactamase, toxins, etc.), limited diffusion of antimicrobials through the biofilm matrix, persister cell content, and limited nutrient and oxygen. Surface proteins and polysaccharide intercellular adhesions (PIA) play a role in the biofilm production and development. It is hard to treat biofilm-embedded bacteria than planktonic forms. Biofilm producer microorganism causes biofilm-related infections such as indwelling and medical device-related infections such as endocarditis, urinary tract infections, septic arthritis, chronic rhinosinusitis, ocular infections, wound infections, etc. The results of biofilm produced on indwelling medical devices are recurrent, untreatable infections and failure of medical device. To overcome chronic and recurrent infections, it is important to detect biofilms of microorganisms, maturation and dispersion, and determine antibiofilm and antibacterial activity of agents against biofilm and bacteria within biofilm, respectively [1]. Identification of genes involved in biofilm formation and measurement of gene expression as a result of antibiofilm and antibacterial activity of agents can be advantageous with carrying out high-throughput screens using microtiter plate assay system.

1.2 Techniques used to study biofilms and biofilm-embedded microorganisms

The standard antimicrobial susceptibility tests such as broth macrodilution and microdilution methods that are routinely used in laboratories and published by Clinical Laboratory Standards Institute (CLSI), National Committee for Clinical Laboratory Standards (NCCLS), and European Committee on Antimicrobial Susceptibility Testing (EUCAST) could never yield accurate results in biofilm producer microorganisms, due to being appropriate for the detection of antimicrobial activity of agents against planktonic microorganism [2].

There are several methods which have been used by clinical microbiologist for detection and measurement of microbial biofilms in response to agents (**Tables 1–3**). Several instruments as model system have been improved such as modified Robbins device, Calgary biofilm device, disk reactor, Centers for Disease Control (CDC) biofilm reactor, perfused biofilm fermenter, and model bladder. Model systems help to define susceptibility of antimicrobial agents against biofilm producer microorganisms by providing information about biofilm mechanisms. Substratums of modified Robbins device, Calgary biofilm device, disk reactor, CDC biofilm reactor, and perfused biofilm fermenter are silastic disks, plastic pegs, Teflon coupons, plastic needleless connectors, and cellulose acetate filters, respectively, whereas substratum of model bladder is urinary catheters (UCs). Medical devices of which dimensions are adjusted to appropriate sizes can also be used as a substratum (abiotic surfaces) for biofilm production by adapting and modifying to related methods by some biofilm researchers. The methods of modified Robbins device and Calgary biofilm device are based on viable counting. In Calgary biofilm device, pegs are sonicated before counting. The methods of disk reactor and CDC biofilm reactor based on direct and viable counting,

Method	Action of application	Aim
Roll plate	Extraluminal biofilm detection	Growth of biofilm-embedded bacteria
Sonication, vortex, and plate counting	Intraluminal and extraluminal biofilm detection	Growth of biofilm-embedded bacteria
Acridine orange staining	Extraluminal biofilm detection	Direct investigation of biofilm produced on catheter by microscopy
Streak plating of alginate swab	Investigation of biofilm produced on indwelling catheter	Growth of biofilm-embedded bacteria

Table 1.
The methods used for detection and measurement of biofilms produced on medical devices.

Method	Aim
Tube method (TM)	Qualitative detection by observing biofilm lined on bottom and walls of tube
Congo red agar (CRA)	Qualitative detection by observing colony color change
Microtiter plate (MtP)	Quantitative detection of biofilm by microplate reader (microELISA)
Real-time PCR	Detection of biofilm genes
Conventional PCR	
Multiplex PCR	

Table 2.
The methods used for detection of biofilm.

Method	Application	Target
Microtiter plate (MtP)	Measurement of biofilm produced on walls of wells in response to agent	Measures the effect of agents against biofilm production
Microtiter plate (MtP) (MBEC)	Measurement of biofilm remained on walls of wells in response to agent and detecting MBEC of agents	Measures the effect of agents against mature biofilm formed on walls of wells
Vortex and plate counting	Plate counting of biofilm-embedded bacteria and detecting bMBC of agents	Screens antimicrobial activity of agents against biofilm-embedded bacteria
Checkerboard assay	Plate counting of biofilm-embedded bacteria and FIC indexes are calculated	Screens antimicrobial activity of combination of agents
Sonication, vortex, and plate counting	Plate counting of biofilm-embedded bacteria and detecting bMBC of agents	Screens antimicrobial activity of agents against biofilm-embedded bacteria
Quantitative PCR	Measurement of specific biofilm gene expression	Monitors expression of biofilm genes in response to agents
Mass spectrometry (MS)	Measurement of exoenzymes located in biofilm matrix	Monitors expression of bacterial proteins in response to agents

Table 3.
The screening methods for antibiofilm and antimicrobial activity of agents against biofilm producer bacteria.

after substratums, are sonicated, vortexed, and homogenized. In perfused biofilm fermenter, viable counting is done, after filters are shaken in sterile distilled water, whereas in model bladder, UCs are examined directly by scanning electron microscopy (SEM) or transmission electron microscopy (TEM) or by chemical analysis [2]. Rate of biofilm formed on model system can be adjusted by parameters such as composition of medium that can contain glucose, iron, antimicrobial agents, multivalent cations such as Ca^{2+} and Mg^{2+} supporting adhesion of bacteria by cross-linking anionic groups on bacteria and substratum, shear force, retention time, flow rate, roughness, and chemistry of substratum and species of organisms (**Table 4**) [2, 3].

The aim of this chapter is to overview certain methods used by biofilm detection and antibacterial and antibiofilm researches such as tube method (TM), Congo red agar (CRA) method, microtiter plate (MtP) assay, plate counting of biofilm-embedded bacteria (sessile bacteria), PCR, mass spectrometry (MS), confocal laser scanning microscopy (CLSM), etc.

Instruments	Culture dynamics	Substratum	Method
Modified Robbins device	Batch culture	Silastic disks	Viable counting
Calgary biofilm device	Batch culture	Plastic polycarbonate pegs	Viable counting, after pegs are sonicated
Disk reactor	Batch culture	Teflon coupons	Direct or viable counting, after coupons are sonicated, vortexed, and homogenized
CDC biofilm reactor	Continuous culture	Plastic connectors	
Perfused biofilm fermenter	Continuous culture	Cellulose-acetate filters	Viable counting, after filters are shaken in sterile distilled water
Model bladder	Continuous culture	Urinary catheters	Examining directly by SEM or TEM or analyzing chemically
Flow cell	Continuous culture	Chambers with transparent surfaces	Examining by confocal laser scanning microscopy

CDC biofilm reactor, Centers for Disease Control biofilm reactor; SEM, scanning electron microscopy; TEM, transmission electron microscopy.

Table 4.
Instruments used to produce biofilm and examine biofilm process.

2. Detection of biofilm producer microorganism

2.1 Direct observation

Complexity and dynamics of biofilms can be observed by biofilm imagining optical technology including light microscopy, SEM, TEM, and CLSM. These techniques are used to visualize 3D structure and check the existence of biofilm [4].

2.1.1 Light microscope

Light microscopy is the easiest, cheapest, most simple, convenient and fastest method to quantitatively observe the morphology of microorganisms adhered to surfaces and to semiquantitatively estimate the amount of microorganism attached on surface (exist, absent, abundant, rare, etc.). Microorganisms including *Candida albicans*, *E. coli*, *Pseudomonas*, and *Staphylococcus epidermidis* adhered on acrylic sheets of polymethacrylate films, glass cover slips, and polystyrene petri dishes have been observed by light microscope, respectively. Observation with light microscopy that requires clear, transparent, and planar surfaces on which microorganisms attach does not create 3D vision of biofilm. Dyes can be used such as epifluorescence and fluorescent to enhance image clarity of microorganisms. The observation with the light microscope enables researchers to compare morphologies of sessile form and planktonic form of microorganism required by making smear and centrifuging of sample, respectively [3].

2.1.2 Transmission electron microscope (TEM)

Images of cells and cell structures such as protein and nucleic acid are obtained by electrons at high magnification and resolution. Monitoring of components of cell can be done directly in TEM by negative staining. Due to photons and electrons penetrating cells poorly, thin section of cell cut is stabilized and stained by certain chemicals with the treatment of osmic acid, permanganate, uranium, lanthanum, or lead salts. These

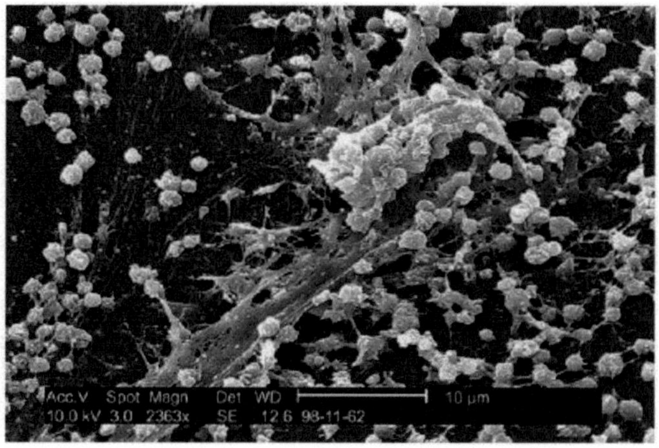

Figure 1.
The SEM image of S. aureus embedded in biofilm colonized on intravenous catheter [6].

stains contain high atomic weight. Due to stains having high atomic weight, contrast is accelerated by electron dispersion from sample. If observation of outer structure of cells will be done, it is not important whether the section of cell is thin or thick.

Due to inadequate stabilization of polysaccharides is done by the conventional fixatives such as aldehydes, glutaraldehyde, paraformaldehyde, and osmium tetroxide, water content of biofilm is eliminated by graded dehydration with alcohol after this postfixation step. After sample is infiltrated with resin, sample is embedded in gelatin capsule and headed for polymerization. Then, thin section taken is poststained with uranyl acetate and lead citrate.

Exopolysaccharide constituents are not observed with its own electron dense and staining poststains such as uranyl acetate and lead citrate with TEM, due to not only having high electron translucent, but also contrast is not developed by conventional poststains. According to the studies, glycocalyx of *Staphylococcus hominis* and *Staphylococcus epidermidis* can be stabilized by the usage of certain cationic reagent combinations including ruthenium red, alcian blue, lysine, lysine monohydrochloride, or lysine acetate and paraformaldehyde [5]. After all these steps are done, sample is observed in TEM (**Figure 1**).

2.1.3 Scanning electron microscope (SEM)

To visualize 3D images of cell sample is coated with heavy metals such as gold. Electrons released from metal coating of sample are caught by SEM for image

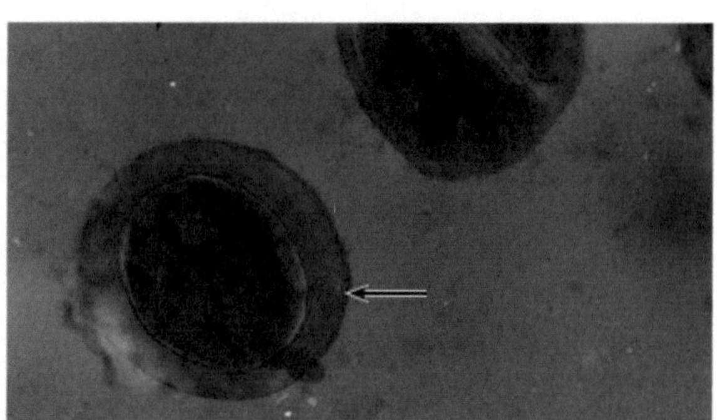

Figure 2.
The TEM image of Staphylococcus spp. surrounded by glycocalyx [6].

production. The procedure of SEM is similar to TEM except for some additional chemicals (gold), lacking infiltration, embedment in resin, polymerization, and thin section staining with lead citrate and uranyl acetate [5]. As in the steps of TEM method, postfixation and dehydration steps of SEM are similar to TEM. The step is applied after dehydration step is drying and coating sample with gold in the processing for SEM, rather than infiltration with resin, embedment in gelatin capsule, and staining with lead citrate and uranyl acetate in the processing for TEM. After dehydration process with graded alcohol, sample is dried and coated with gold palladium [5]. After all these steps are done, sample is observed in SEM (**Figure 2**).

2.1.4 Florescent tagging of biofilm

2.1.4.1 Confocal laser scanning microscopy (CLSM)

Biofilms formed on flow cells of which surface are transparent can be observed by confocal laser screening microscopy (CLSM). Three-dimensional (3D) morphology and physiology of biofilms can be screened by CLSM [2]. Thick samples such as biofilms and microorganisms localized in the depth such as biofilm-embedded microorganisms need to be observed by CLSM (**Figure 3**).

For observation of biofilm with confocal microscopy and related methods, biofilm must be fluorescent as a result of fluorescent molecules such as green fluorescent protein (GFP) that is fluorescent protein expressed by biofilm producer microorganism within biofilm (gene of cell interested is tagged by gene cassette encoding GFP) or staining components of heterogeneous mass of biofilm with fluorescence or fluorescence-labeled dyes [2]. Stains such as lectins target extracellular matrix, whereas certain fluorophores target extracellular DNA (eDNA) to visualize eDNA content of biofilm matrix [2, 7].

By scanning laser light across the sample, deep penetration of excitated energy is provided. As a result of fluorescence of biomolecules such as GFP or chlorophyll that are intrinsic fluorophores or molecules signed by exogenous probes such as fluorescent-labeled antibodies detected by photomultiplier, 3D digital image is formed. Observation of biofilms that are multilayered and have complex 3D structures requires additional resolution [2]. Images of each layers of biofilm obtained are combined by computer for construction of digital 3D images of whole biofilm.

Biofilm producer microorganisms can be manipulated genetically by tagging of microbial gene of interest by gene cassette encoding GFP as a reporter gene

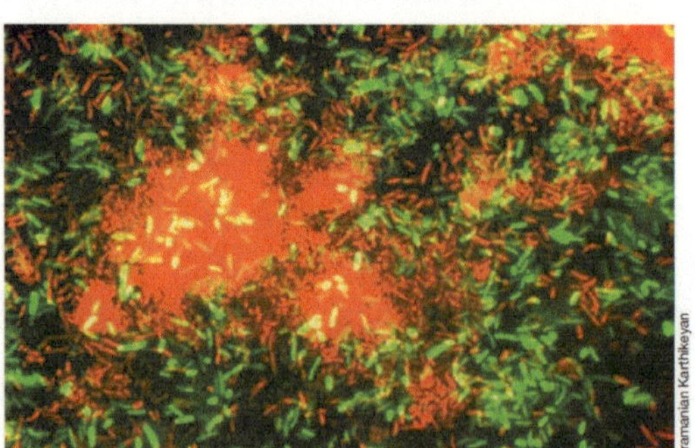

Figure 3.
Bacterial community embedded in a biofilm matrix visualized by CLSM. Each bacterium observed with a distinct color located at different depths of biofilm [11].

(*gfp* genes) to monitor gene expression and metabolic physiology activity in biofilms and determine location of microorganism within biofilm [2, 8, 9].

Idea about gene activity in biofilms is given by confocal microscopy applied to 3D localization of nonenzyme reporter systems such as GFP. Growth phase and activity of bacteria embedded in biofilms can be defined by promoter-reporter systems that is designed and fluoresced just in living dividing cells. In situ cellular growth activity of bacteria embedded in biofilms is determined by measuring ribosome-hybridization-signal intensity, due to synthesis rate and content of ribosome correlated with the growth rate (especially in exponential phase). Expression cassette that is active only in growing cells, labeled by GFP and controlled by rRNA promoter, can be constructed to monitor growth phase and activity of bacteria within biofilm [2, 10].

2.1.4.2 Fluorescent in situ hybridization (FISH)

Specific microorganisms present in a heterogeneous biofilm community can be identified by the probes of fluorescent in situ hybridization (FISH) method. GFP that is translated enables procedure not to require fixation or staining. Fluorescent-labeled microorganism within biofilm can also be examined by FISH. DNA probes designed to hybridize 16S rRNA of microorganism integrated to either fluorescent dye such as FITC or Rhodamine or enzyme such as horseradish peroxidase. The advantage of probes conjugated with horseradish peroxidase is not to destroy microorganism within biofilm. The growth rate of microorganism within biofilm can be determined by FISH method, due to the amounts of ribosomes existing in a microorganism that is directly proportional to growth activity of microorganism. Probe must be designed to label conserved region of only a single species (**Figure 4**) [2, 12].

2.2 Indirect observation

2.2.1 Roll plate method

Roll plate method is applied for the detection of possible microbial colonization having a potential to develop indwelling device-associated infection on the outer surface of cylindrical materials such as catheters and vascular grafts. Microorganism colonize on external surface of catheter is detected by roll plate method, instead of microorganism colonize on intraluminal site of catheter. Material is touched and rolled on the surface of medium [3].

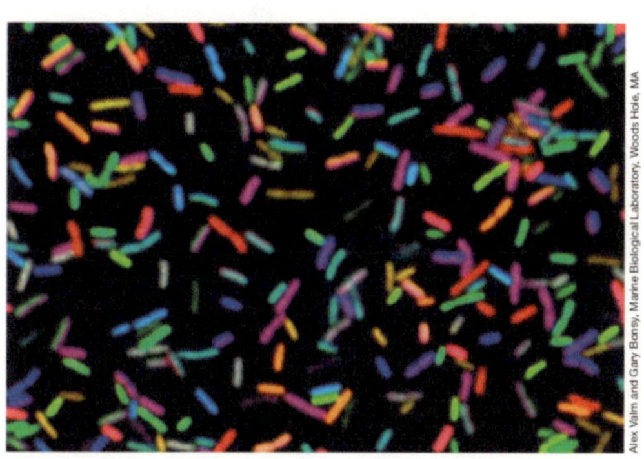

Figure 4.
Fluorescence image of 28 distinct E. coli strains labeled by fluorophore-conjugated oligonucleotides complementary to 16S rRNA of E. coli [11].

2.2.2 Congo red agar (CRA) method

Congo red agar (CRA) method that is a qualitative assay for detection of biofilm producer microorganism, as a result of color change of colonies inoculated on CRA medium, is described by Freeman et al. The CRA medium is constructed by mixing 0.8 g of Congo red and 36 g of sucrose to 37 g/L of Brain heart infusion (BHI) agar. After incubation period that was 24 h at 37°C, morphology of colonies that undergone to different colors is differentiated as biofilm producers or not. Black colonies with a dry crystalline consistency indicate biofilm producers, whereas colonies retained pink are non-biofilm producers (**Figure 5**) [13].

2.2.3 Tube method (TM)

Tube method (TM) that is a qualitative assay for detection of biofilm producer microorganism, as a result of the occurrence of visible film, is described by Christensen et al. [14]. Isolates are inoculated in polystyrene test tube which contained TSB and incubated at 24 h at 37°C. The sessile isolates of which biofilms formed on the walls of polystyrene test tube are stained with safranine for 1 h, after planktonic cells are discharged by rinsing twice with phosphate-buffered saline (PBS). Then, safranine-stained polystyrene test tube is rinsed twice with PBS to discharge stain. After air drying of test tube process, the occurrence of visible film lined the walls, and the bottom of the tube indicates biofilm production (**Figure 6**) [14].

2.2.4 Detection of biofilm production by microtiter plate assay

Microtiter plate (MtP) assay is a quantitative method to determine biofilm production by microplate reader. Bacterial suspension is prepared in MHB supplemented with 1% glucose and adjusted to 0.5 McFarland (1.10^8 cfu/ml). This bacterial suspension is 20-fold (1/20) diluted to reach 5.10^6 cfu/ml. Then 180 µl of Mueller-Hinton Broth (MHB) supplemented with 1% glucose [15] and 20 µl of bacterial suspensions are inoculated into 96-well flat-bottomed sterile polystyrene microplate to obtain 5.10^5 cfu/ml as a final concentration (tenfold dilution (1/10)). Microplates are incubated at 24 h at 37°C. The sessile isolates of which biofilms formed on the walls of wells of microplate are stained with only 150 µl of safranine for 15 min, after planktonic cells in wells of microplate are discharged by washing twice with phosphate-buffered saline (PBS) (pH 7.2) and wells are dried at 60°C for

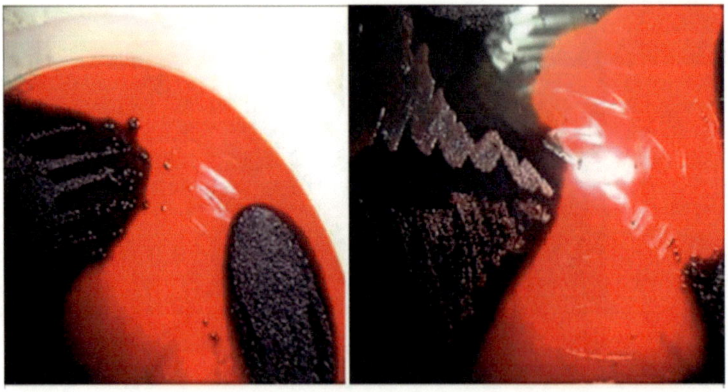

Figure 5.
CRA method applied on CRA medium. Black crystalline colonies of biofilm producer cell and pinkish-red colonies of biofilm nonproducer cell.

DOI: http://dx.doi.org/10.5772/intechopen.84411

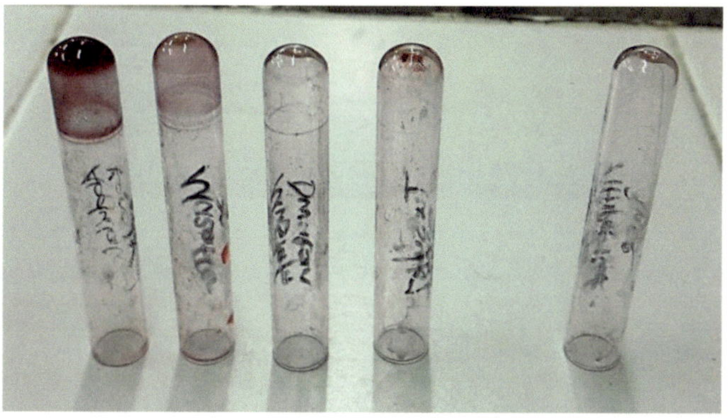

Figure 6.
Tube method. The first two polystyrene test tubes from the left indicate biofilm production. Other test tubes rather than the first two polystyrene test tubes from the left indicate lacking of biofilm production.

1 h [14]. Before staining with safranine, fixation of biofilms can be done by either subjecting to 150 μl of methanol for 20 min or drying at 60°C for 1 h. Then safranine stained wells of microplates are washed twice with PBS to discharge safranine stain. After air drying process of wells of microplate, dye of biofilms that lined the walls of the microplate is resolubilized by 150 μl of 95% ethanol or 33% glacial acetic acid or methanol. Then microplate is measured spectrophotometrically at 570 nm by a microplate reader [15, 16]. The studies are repeated in triplicates. Uninoculated wells containing sterile MHB supplemented with 1% glucose that are considered to be the negative controls are used as blanks. The blank absorbance values are used to identify whether biofilm formation of isolates exists or not. The wells of isolates of which OD values are higher than blank well are considered to be biofilm producers. Cut off value (ODc) can provide categorization of isolates as biofilm producer or not.

$$\text{Odc: Average OD of negative control} + (3 \times \text{standard deviation (SD) of negative control}) \tag{1}$$

$$OD_{\text{isolate}}\text{: Average OD of isolate} - \text{ODc} \tag{2}$$

Negative value obtained from this formula and represented as zero indicates lack of biofilm production, whereas positive value indicates biofilm production (**Figure 7**).

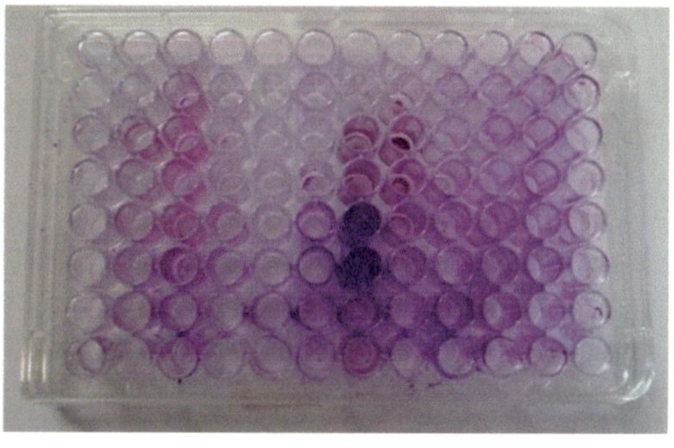

Figure 7.
Microtiter plate assay indicating biofilm production.

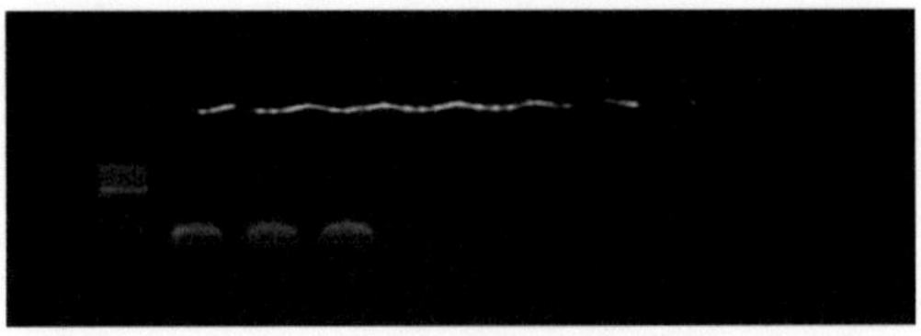

Figure 8.
Image of mecA gene on agarose gel. First sample is DNA size marker (ranging from 250 to 10,000 kb), second sample is ATCC 43300 methicillin-resistant Staphylococcus aureus (MRSA) positive control, third and fourth samples are mecA gene-positive isolates, and fifth sample is ATCC 29213 methicillin-sensitive Staphylococcus aureus (MSSA) as a negative control.

To interpret results, categorization can be done as no biofilm production (0), weak (+ or 1), moderate (++ or 2), and strong biofilm production (+++ or 3) by the calculation of cutoff value (ODc) shown below [15]:

OD ≤ ODc no biofilm production

ODc< OD ≤ 2 × ODc weak biofilm production

2 × ODc< OD ≤ 4 × ODc moderate biofilm production

4 × ODc< OD strong biofilm production.

2.2.5 Detection of biofilm-associated genes by PCR

PCR techniques is used for not only identification of pathogens by amplifying species-specific nucleic acid sequences but also detection of virulence factors by amplifying target virulence genes such as biofilm genes with the usage of gene-specific primers, even in the uncultured pathogen present in the sample.

Forward and reverse primers of biofilm-associated gene are designed. Firstly, multiple alignments were done in the NCBI to find oligonucleotide sequences specific to the species. Then primer pair of biofilm-associated gene is designed by using Primer3Plus verified by FASTA analysis checking specificity of primers for microbial sequences in the database [17, 18].

Genomic DNA of microorganism is extracted by extraction kits of which protocols can vary according to species and Gram-positivity or Gram-negativity of microorganisms. DNA of microorganism is measured spectrophotometrically by microplate spectrophotometry reader to determine the amount of DNA extracted as microgram per microliter.

Biofilm-associated gene is amplified by PCR such as qualitative real-time PCR, multiplex and conventional PCR that is used to detect whether biofilm-associated gene is present or not in microorganism. If conventional and multiplex PCR protocols are applied to detect biofilm gene, rather than qualitative real-time PCR, PCR product isolated is visualized on an agarose gel containing a DNA-intercalating dye such as ethidium bromide to confirm the presence of amplified gene (**Figure 8**). Only in qualitative real-time PCR, the amplicon is detected by fluorescence using a pair of specific hybridization probes labeled with fluorescence dye [11].

3. Detection of antimicrobial efficacy of agents against biofilm production

3.1 Detection by microtiter plate assay

Microtiter plate (MtP) assay is a qualitative assay to detect efficacy of agent against biofilm production by microplate reader.

Bacterial suspension is prepared in MHB supplemented with 1% glucose and adjusted to 0.5 McFarland (1.10^8 cfu/ml). This bacterial suspension is 20-fold (1/20) diluted to reach 5.10^6 cfu/ml.

A 180 µl of agent doses and 20 µl of bacterial suspension are dispersed to each well of microplate to obtain 5.10^5 cfu/ml as a final concentration (tenfold dilution (1/10)). After incubation at 37°C for 24 h, ongoing processes are done according to MtP assay as mentioned previously for the determination of effect of agent against biofilm production [14, 16].

4. Detection of antimicrobial efficacy of agents against mature biofilms

4.1 Determination of MBEC by microtiter plate assay

Biofilms remained after eradication by agent are measured by this technique. Biofilms of bacteria that line the walls of wells are formed according to MtP method.

After the content of microplate is discharged, 200 µl of each dose of agents is dispersed to each well of microplate of which the walls are lined with biofilm. A 200 µl of distilled water is added into a well of microplate of which the walls are lined with biofilm as a control. Then the effect of agent against mature biofilm is determined according to MtP assay as mentioned previously. Minimum concentration of agent eradicating mature biofilm that is named as minimum biofilm eradication concentration (MBEC) can be determined by this modified plate assay. MBEC50 and MBEC90 indicate the minimum concentrations of agents inhibiting 50 and 90% of mature biofilm formed.

5. Detection of antimicrobial efficacy of agents against biofilm-embedded bacteria

5.1 Determination of bMIC and bMBC by plate counting

In summary, biofilm formation process on abiotic surfaces by bacteria is done. Quantification of sessile biofilm-embedded bacteria lined on abiotic surface and sessile biofilm-embedded bacteria remained on abiotic surface after addition of agent on abiotic surface on which mature biofilms formed is determined by plate counting. Bacterial suspension is prepared and adjusted to 0.5 McFarland (1.10^8 cfu/mL) in Mueller-Hinton Broth (MHB) supplemented with 1% glucose [15]. This bacterial suspension is 200-fold (1/200) diluted to gain 5.10^5 cfu/mL. Kirschner wire orthopedic implants are placed into each test tube containing 5.10^5 cfu/mL isolate and incubated at 37°C for 24 h to lead bacteria to produce biofilm on Kirschner wire. After incubation, Kirschner wires on which biofilms are produced are discharged and rinsed with PBS (pH 7.2) and then transferred into each test tubes containing agent concentrations. After incubation at 37°C for 24 h, Kirschner wires are discharged and placed into test tubes containing 1 mL of sterile MHB and sonicated at 42 kHz for 2 min after vortexed for 5 min. Then 100 µl samples of each test tube sonicated and vortexed are inoculated on Mueller-Hinton agar (MHA) and incubated at 37°C for 24 h [19].

The lowest concentration of agent in which bacterial growth is below or equal to control is determined as biofilm minimum inhibitory concentration (bMIC) for biofilm. bMIC50 and bMIC90 indicate the minimum concentrations of agent inhibiting 50 and 90% of biofilm-embedded bacteria. After incubation, the lowest concentration of agent in which colonies of biofilm-embedded bacteria are not grown is determined as biofilm minimum bactericidal concentration (bMBC) of agent for biofilm [19].

5.2 Determination of combination effects of agents against biofilm

Checkerboard assay is used for the determination of combination effects of two different agents. A 250 μl twofold dilutions of each agent from the stock solutions are dispersed to each row and column to obtain final varying concentrations by starting at fourfold of zero MIC for each isolate. So each well contains distinct combination of concentrations of two agents. First wells of rows and columns are left behind for sole treatments of each dose of agents. One well is used for bacterial control (**Figure 9**). Kirschner wires on which bacterial biofilm is produced are dispersed to each well. This microplate is incubated at 37°C for 24 h. After incubation, Kirschner wires are discharged and sonicated at 42 kHz for 2 min after vortexed for 5 min. The lowest concentration of agent in which bacterial growth that is not observed is determined as biofilm minimum inhibitory concentration (bMIC) of agent for biofilm. Then 100 μl samples of each test tube sonicated and vortexed are inoculated on MHA and incubated at 37°C for 24 h. After incubation, the lowest concentration of agent in which colonies of biofilm-embedded bacteria are not grown is determined as biofilm minimum bactericidal concentration (bMBC) of agent for biofilm. For the determination of whether the synergism is present between agents or not, fractional inhibitory concentrations (FICs) index that are calculated for each agent are summed up according to formula written below:

$$\text{FIC (A): MIC A (A in the presence of B (A + B))/MIC A (A alone)} \quad (3)$$

$$\text{FIC (B): MIC B (B in the presence of A (A + B))/MIC B (B alone)} \quad (4)$$

$$\Sigma\text{FIC: FIC (A) + FIC (B)} \quad (5)$$

When ∑FIC is equal and lesser than 0.5, between 0.5 and 1, equal to 1, higher than 1 and equal and lesser than 4, and higher than 4, it is interpreted that the effect between agents in combination is synergistic, partial synergistic, additive, indifferent, and antagonistic, respectively [20].

The wells having the highest synergy rates of two agents that constitute the combinations are determined by taking the average and standard deviation of FIC indexes calculated of the wells with the lowest drug combination without bacterial growth in each row and column (**Figure 9**).

5.3 Measurement of biofilm-associated gene expression by quantitative real-time PCR

Measurements of biofilm genes repressed or induced by agent are done by quantitative real-time PCR (qPCR). So efficacy of agent against biofilm-associated genes can be detected by qPCR.

Complementary DNA (cDNA) is copied from RNA by enzyme reverse transcriptase. Gene expression in pathogen is monitored by qPCR copying cDNA from RNA of target gene. Amplified cDNA probed for identification. Fluorescent probes such as dye SYBR Green are used to indicate double-stranded DNA, consequently amplification. Accumulation of PCR amplicons labeled fluorescently is monitored through the qPCR processes (**Figure 10**). Visualization of amplicon on agarose gel is not needed to confirm amplification in qPCR [11].

Bacterial control	Agent X Dose 1	Agent X Dose 2	Agent X Dose 3	Agent X Dose 4	Agent X Dose 5	Agent X Dose 6	Agent X Dose 7
Agent Y Dose 1							
Agent Y Dose 2							
Agent Y Dose 3							
Agent Y Dose 4							
Agent Y Dose 5							
Agent Y Dose 6							
Agent Y Dose 7							

Figure 9.
The schematization of checkerboard assay.

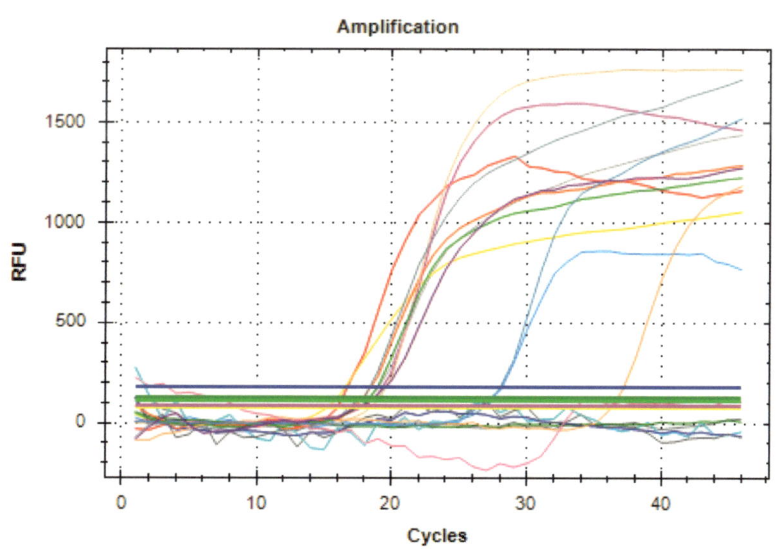

Figure 10.
Quantitative real-time polymerase chain reaction (qPCR) of atl (light blue), 16S RNA (red), mecA (purple), and nuc (light pink) genes of sample; icaA (gray), icaD (plato), atl (pink), mecA (blue), and nuc (green) genes of ATCC 43300 methicillin-resistant Staphylococcus aureus (MRSA); and icaA (orange), 16S RNA (yellow), and nuc (grayish blue) genes of ATCC 29213 methicillin-sensitive Staphylococcus aureus (MSSA). Expressions of all these genes are monitored by qPCR except nuc (light pink) gene sample. Lines below threshold monitored by qPCR indicate the negativity of genes such as icaA and icaD genes samples and icaD (turquoise), atl (plato), and mecA (light pink) genes of ATCC 29213 MSSA. RFI, relative fluorescence intensity.

Total RNA of microorganism is isolated according to protocols of RNA isolation kits. Kit protocols can vary according to the species of microorganism. Total RNA of microorganism is measured spectrophotometrically by microplate spectrophotometry reader to determine the amount of total RNA isolated as microgram per microliter. Then cDNA is synthesized from total RNA with qPCR using primer pair of the biofilm-associated gene, which is designed using Primer3 and verified by FASTA analysis, which controls the specificity of the primers for microbial sequences in the data system, after multiple alignments were done in the NCBI to find oligonucleotide sequences specific to the species [17, 18].

5.4 Mass spectrometry

Extracellular polymeric substances (EPS) not only contain polysaccharides but also contain proteins such as extracellular enzymes. These expressed proteins located in the matrix of EPS can be detected and characterized by mass spectrometry (MS) [1]. Large biologic molecules can be also detected and characterized in complex biologic structures such as EPS by MS. Chemicals involved in biofilm process are examined in detail by MS. Electrospray ionization (ESI) and matrix-assisted laser desorption ionization (MALDI) are the types of MS [2]. In time-of-flight (TOF) mass spectrometer, mass is analyzed by ions desorbed in vacuum chamber. These two technics are combined and called MALDI-TOF.

Sample is ionized and vaporized by laser. Ions generated from sample by laser pass through the column of MALDI-TOF device toward TOF detector by an electric field. Depending on the mass/charge ratio of molecule, measurements are done by TOF. If this ratio is smaller, ions move faster (**Figure 11**).

Bacteria are identified, expression of bacterial proteins such as surface proteins and exoenzymes like β-lactamase in response to antimicrobials can be monitored, and growth of bacteria is measured by applications of MALDI. MS has high sensitivity and requires minimum amount of sample [2].

5.5 Biological assay

Biofilm-embedded bacteria can be estimated by biologic assays that is an indirect assay. Biological assays that measure production of microbial product give an opinion about estimation of the number of microorganism within biofilm. Amount of biologic product is correlated with biofilm-embedded microorganism producing the product by standardization of planktonic microorganism. Biologic products produced by planktonic microorganism are similar to biologic products produced

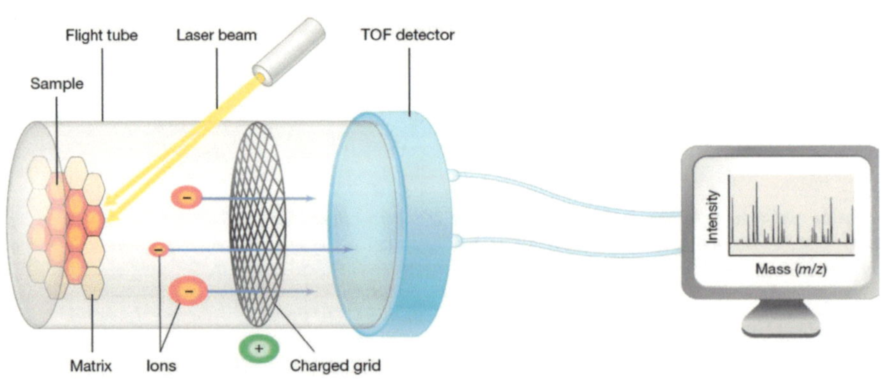

Figure 11.
MALDI-TOF mass spectrometry device [11].

by biofilm-embedded bacteria. Standardization curves of each microorganism tested need to be formed. Measurement of total protein at the absorbance is 550 or 950 nm; tryptophan fluorescence, endotoxin [2], ATP production via bioluminescence caused by luciferin and luciferase, urease production to estimate number of attached microorganism, and electron transport via the production of formazan are done by biological assays [3].

6. Conclusion

Biofilms cause resistance to many antimicrobial agents. The results of biofilm produced on indwelling medical devices are recurrent, untreatable infections and failure of medical device. To overcome chronic and recurrent infections, it is important to detect biofilms of microorganisms, determine antibiofilm activity of agents against biofilm, and determine antibacterial activity of agents against biofilm-embedded microorganism with the appropriate methods by clinical microbiologist and biofilm researcher microbiologist. Identification of genes involved in biofilm formation and measurement of gene expression as a result of antibiofilm and antibacterial activity of agents can be advantageous in biofilm studies.

Author details

Sahra Kırmusaoğlu
Department of Molecular Biology and Genetics, Faculty of Arts and Sciences, Haliç University, Istanbul, Turkey

*Address all correspondence to: kirmusaoglu_sahra@hotmail.com

References

[1] Kırmusaoğlu S. Staphylococcal biofilms: Pathogenicity, mechanism and regulation of biofilm formation by quorum sensing system and antibiotic resistance mechanisms of biofilm embedded microorganisms. In: Dhanasekaran D, Thajuddin N, editors. Microbial Biofilms—Importance and Applications. Croatia: Intech; 2016. pp. 189-209. DOI: 10.5772/62943

[2] Shunmugaperumal T. Biofilm Eradication and Prevention: A Pharmaceutical Approach to Medical Device Infections. New Jersey: John Wiley & Sons, Inc.; 2010. pp. 116-151

[3] Christensen GD, Simpson WA, Anglen JO, Gainor BJ. Methods for evaluating attached bacteria and biofilms. In: An YH, Friedman RJ, editors. Handbook of Bacterial Adhesion: Principles Methods, and Applications. Totowa: Humana Press; 2000. pp. 213-233

[4] Roy R, Tiwari M, Donelli G, Donelli V. Strategies for combating bacterial biofilms: A focus on anti-biofilm agents and their mechanisms of action. Virulence. 2018;**9**(1):522-554

[5] Fassel TA, Edmiston CE. Evaluating adherent bacteria and biofilm using electron microscopy. In: Yuehuei H, Friedman RJ, editors. Handbook of Bacterial Adhesion: Principles, Methods, and Applications. New Jersey: Humana Press Inc; 2000. pp. 235-248

[6] Willey JM, Sherwood LM, Woolverton CJ. Prescott, Harley, and Klein's Microbiology. 9th ed. New York: The McGraw-Hill Companies, Inc; 2013

[7] Allesen-Holm M, Barken KB, Yang L, Klausen M, Webb JS, Kjelleberg S, et al. A characterization of DNA release in *Pseudomonas aeruginosa* cultures and biofilms. Molecular Microbiology. 2006;**59**:1114-1128

[8] Lambertsen L, Sternberg C, Molin S. Mini-Tn7 transposons for site-specific tagging of bacteria with fluorescent proteins. Environmental Microbiology. 2004;**6**:726-732

[9] Sternberg C, Christensen BB, Johansen T, Na T, Andersen JB, Givskov M, et al. Distribution of bacterial growth activity in flow-chamber biofilms. Applied and Environmental Microbiology. 1999;**65**:4108-4117

[10] Lee N, Nielsen P, Andreasen K, Juretschko S, Nielsen J, Schleifer KH, et al. Combination of fluorescent in situ hybridization and microautoradiography, a new tool for structure-function analyses in microbial ecology. Applied and Environmental Microbiology. 1999;**65**:1289-1297

[11] Madigan MT, Martinko JM, Bender KS, Buckley DH, Stahl DA. Brock Biology of Microorganisms. 14th ed. The United States of America: Pearson Education, Inc; 2015

[12] Stahl DA, Amann R. Development and application of nucleic acid probes. In: Stackebrandt E, Goodfellow M, editors. Nucleic Acid Techniques in Bacterial Systematics. New York: John Wiley & Sons, Inc; 1991. pp. 205-248

[13] Freeman J, Falkiner FR, Keane CT. New method for detecting slime production by coagulase negative staphylococci. Journal of Clinical Pathology. 1989;**42**:872-874

[14] Christensen GD, Simpson WA, Younger JJ, Baddour LM, Barrett FF, Melton DM, et al. Adherence of coagulase-negative staphylococci to plastic tissue culture plates: A quantitative model for the adherence of staphylococci to medical devices. Journal of Clinical Microbiology. 1985;**22**:996-1006

[15] Stepanović S, Vuković D, Hola V, Bonaventura GD, Djukić S, Ćirković I, et al. Quantification of biofilm in microtiter plates: Overview of testing conditions and practical recommendations for assessment of biofilm production by staphylococci. APMIS. 2007;**115**(8):891-899

[16] Tang HJ, Chen CC, Ko WC, Yu WL, Chiang SR, Chuang YC. In vitro efficacy of antimicrobial agents against high-inoculum or biofilm-embedded methicillin-resistant *Staphylococcus aureus* with vancomycin minimal inhibitory concentrations equal to 2 μg/mL (VA2-MRSA). International Journal of Antimicrobial Agents. 2011;**38**:46-51

[17] http://www.bioinformatics.nl/cgi-bin/primer3plus/primer3plus.cgi. 2018 [Accessed: 28-11-2018]

[18] http://www.ncbi.nlm.nih.gov/GenBank [Accessed: 28-11-2018]

[19] Wayne PA. Performance standards for antimicrobial susceptibility testing. In: Clinical and Laboratory Standards Institute. Twenty-Third Informational Supplement CLSI Document M100-S23. 2013

[20] Moody J. Susceptibility tests to evaluate synergism. In: Isenberg HD, editor. Clinical Microbiology Procedures Handbook. Washington: ASM Press; 2004

Section 3

Antibiotic Production

Chapter 7

Streptomyces as a Source of Antimicrobials: Novel Approaches to Activate Cryptic Secondary Metabolite Pathways

Ángel Manteca and Paula Yagüe

Abstract

Streptomyces is the most important bacterial genus for bioactive compound production. These soil bacteria are characterized by a complex differentiation cycle. *Streptomyces* is extremely important in biotechnology, producing approximately two thirds of all antibiotics, as well as many compounds of medical and agricultural interest. Drug discovery from streptomycetes became challenging once the most common compounds were discovered, and the system was basically abandoned by industry. Simultaneously, antibiotic resistance is increasing dramatically, and new antibiotics are required. Screening from nature is being resumed (exploring new environments, looking for elicitors, metagenome, etc.). Secondary metabolism is conditioned by differentiation; although the relationship between both has long remained elusive, differentiation as a trigger for antibiotic production remains basically unexplored. Most cultures used in screening campaigns for new bioactive molecules have been performed empirically, and workflow was extremely productive during the so-called golden age of antibiotics; however, currently there is a bottleneck. *Streptomyces* is still the most important natural source of antibiotics, and it also harbors many cryptic secondary metabolite pathways not expressed under laboratory conditions. In this chapter, we review strategies based on differentiation, one of the keys improving secondary metabolite production and activating cryptic pathways to face the challenges of drug discovery.

Keywords: *Streptomyces*, screening, antibiotics, secondary metabolism, differentiation, elicitors, morphology, liquid cultures

1. Introduction

1.1 *Streptomyces* development

1.1.1 Solid cultures

The complex *Streptomyces* life cycle on solid-grown cultures is made up of different phases.

Microorganisms on soils are exposed to different stress factors: chemical, physical and/or biological. Of these variable conditions, sometimes nutrients can be limited, preventing growth [1]. A good survival strategy of some bacteria and fungi is the latency stage of spores. Under this stage, cells stop growing and their metabolism is mainly inactive [2]. Spores are very resistant to stress conditions. In Streptomycetes, they are formed by thick hydrophobic covers that preserve genetic information during unfavorable conditions. Spores are dispersed by wind, water or insects and remain in latency until germination in favorable situations [3]. Besides the thickness of the cell wall, their resistance to extreme temperatures and other physicochemical factors is because of their low water content [4]. In this dehydrated environment, molecules have inactive conformations (denaturalized proteins, etc.) that are stabilized and protected by the sugar trehalose [2]. From this null or limited metabolism, spores are able to germinate, as the macromolecules needed have already been synthesized before the latency stage [2]. For germination, optimal hydration conditions are required. However, it can be faster and more successful in the presence of nutrients and specific stimuli. Spores also need enough levels of intracellular trehalose [5], and polyphosphates [6] that allow metabolic activity to commence before cells are able to detect and assimilate an external source of energy [2].

Despite the fact that germination is a very important stage in the *Streptomyces* life cycle, it has remained largely unstudied, because it is not an industrially important stage. When environmental conditions are prosperous, spores on soils/laboratory conditions (petri plates with solid culture medium) germinate. Germination of *Streptomyces* spores is a sequential process, divided into at least four stages: darkening, swelling, emission of the germination tube [7] and reduced swelling [8].

Darkening is the stage in which spores allow the entry of water into the cell losing their impermeability and heat resistance. After a few minutes, spores change their optical characteristics and the cells are able to reactive their metabolism. Darkwelling requires bivalent cations such as Ca^{+2}, Mg^{+2}, Mn^{+2}, Zn^{+2} and Fe^{+2} as well as energy reserves of the spore [7, 9, 10]. Degradation of the cover leads to the reactivation of the cell wall hydrolases. These enzymes, like Rpf (Resuscitation-promoting factors), are involved in cell wall degradation, facilitating peptidoglycan remodeling by hydrolyzing glycosylic bonds B-1,4 between cell wall components, N-acetylglucosamine and N-acetylmuramic acid, which in turn, allow the access to external nutrients [2].

Swelling is the stage in which spores increase its volume due to the entrance of water. This increases the levels of glucose, as a result of trehalose hydrolysis [11]. This seems to be an essential step in germination since only when trehalose concentrations decrease and proteins able to recover their functional conformations [12, 13]. Ribosomes are inactive inside the spores [14, 15]. However, only few minutes after germination starts (30–60 min), they are fully functional and new proteins can be translated [6]. During this stage, spores are metabolically active and are able to use trehalose as main energy source [11].

Emission of the germination tube: This stage begins at the same time as the first DNA replication, when spores are able to detect external sources of nutrients and adjust their metabolic pathways as a result [2]. The accurate point from which the germination tube is emitted is indicated by the accumulation of the protein, SsgA [3]. SsgA is not essential for germination; nevertheless, its absence significantly reduces the number of germ tubes in each spore, and its overexpression has the opposite effect [3].

Reduced swelling: Once the germ tube emerges, the spore starts to gradually reduce its swelling. This stage depends on the quantity of peptidoglycan cross-link that involves a carboxypeptidase (which corresponds to the SCO4439 gene in the *Streptomyces coelicolor* genome www.strepdb.com). A deletion of this carboxypeptidase causes spores to continue swelling until 5 μm diameter (2.5 times more than the wild-type strain), after the emission of the germ tube [8]. The existence of this stage demonstrates that the regulation of water entering into the spores is complex and highly regulated.

During germination, several important morphological changes occur, including the degradation of the spore cover, which is mainly formed of peptidoglycan, and causing a huge increase of metabolic activity [2]. On the other hand, shortly after the emission of the germ tube, the expression of the protein DivIVA starts [6]. This protein plays a critical function in the vegetative growth and is located at the apex of the hyphae [16–18]. After germination, genes involved in cell division (like *ftsZ*) and proteins involved in cellular growth (like FilP) are expressed [2]. The expression of these genes is considered as the final of the germination phase [2, 6].

Curiously, during germination, three different secondary metabolites are produced: albaflavenone (a terpenoid) with an antibacterial effect against *Bacillus subtilis*, giving an important advantage in competitive environments, and two polyketides (germicidin A and chancone) with a germination inhibitory effect. These compounds are *de novo* generated during germination and are good examples of compounds produced during vegetative growth [19].

After germination, the *Streptomyces* mycelium starts to grow apically (regulated by *div*IVA gene) as a very transitory, multinucleated stage called first mycelium (MI), which is multinucleated [20]. This is a very transitory stage in which dead and living segments alternate in the same hyphae [21]. There are two kinds of septa separating segments during the MI stage: one of them is composed of a cell membrane and cell wall (with peptidoglycan) and the other, recently discovered, is a 1-μm spacing cross-membrane-based septa, without peptidoglycan [20]. *Fts*Z participates in this septation, but it is not essential, as cross-membranes were observed in the *Fts*Z null mutant. Later, the dead segments are completely disintegrated and living segments continue apical growth. In this phase, the growth is especially branched [18].

After this early stage of branching growth, there is a morphological differentiation into a second mycelium (MII), a sort of multinucleated mycelium with sporadic and apparently randomly positioned septa (formed of cell membrane and cell wall), which is the secondary metabolite producer [22]. This mycelium has two different populations: one of which remains as a branching mycelium growing on surfaces and is called substrate mycelium and the other develops hydrophobic covers and starts to grow into the air without branching (aerial mycelium) [18]. This is known as the "sky-pathway," which regulates the expression of genes encoding proteins forming hydrophobic proteins (rodlins, chaplins and ramS) that cover the aerial hyphae. Some of the so-called "bald" (*bld*) mutants among others (defective in aerial growth) regulate the activation of this pathway [23]. At the end of the cycle, aerial hyphae undergo extensive synchronous septation (cell division) to create up to 100 uninucleoid compartments, which finally differentiate to create chains of spores [24]. At the end of the cycle, the "white" (*whi*) genes are activated (whose mutants are defective in the formation of mature gray spores on the fluffy aerial mycelium) and the hyphae septate and sporulate [25].

1.1.2 Liquid cultures

Streptomyces differentiation in liquid cultures has hardly been studied, mainly due to the fact that most strains do not sporulate under these conditions. However, the fact that *Streptomyces venezuelae* is capable of sporulation in liquid cultures has made it a new study model in recent years [26].

The applications of novel methodologies (confocal microscopy, proteomics, transcriptomics) to the study of *Streptomyces* biology have extended the knowledge regarding the development during the phases preceding sporulation in liquid cultures [27–31].

In these conditions, in most *Streptomyces* strains, aerial mycelium formation and sporulation are blocked [28]; so, there is no hydrophobic cover formation, or sporulation, but there is a first exponential growth phase corresponding to the compartmentalized MI morphology, and then there is a growth arrest phase corresponding to programmed cell death (PCD). After that, the hyphae, which remain alive, continue growing in a second exponential growth phase in which the morphology of the hyphae becomes MII multinucleated mycelium, which corresponds to the antibiotic production [27, 30].

1.2 Correlation between *Streptomyces* life cycle and antibiotic production

Most industrial processes for secondary metabolite production are performed in liquid (flasks or bioreactors). This fact makes the knowledge of the *Streptomyces* behavior essential in these conditions. The optimization of *Streptomyces* liquid cultures has been empiric for decades; however, the end of the so-called "golden age of antibiotics," which meant fewer and fewer compound discoveries or the advances in genome sequences revealing the existence of cryptic biosynthetic pathways that are not expressed under laboratory conditions, calls the attention of the scientific community.

Work on *Streptomyces* differentiation in liquid cultures has largely focused on the analysis of mycelial macroscopic morphology (pellets/clumps formation), media composition and bioreactor design [31]. The differentiation of the hyphae during the life cycle is essential for secondary metabolism [30]; however, there is still some controversy over which is the best morphology for production. Some authors affirm that pellets are better for production, while others report about clumps or even disaggregated growth. Hence, these conditions seem to be strain dependent. For instance, in *Streptomyces olidensis* (retamycin) [32], *Streptomyces tendae* (nikkomycins) [33], *Streptomyces lividans* (hybrid antibiotics) [34] and *Streptomyces coelicolor* (Undecylprodigiosin, Actinorhodin) [30], pellet formation is essential for good production. However, in *Streptomyces noursei* (nystatin) [35] and *Streptomyces fradiae* (tylosin) [36], formation of pellets leads to worse antibiotic production than disaggregated growth. The lack of a reliable developmental model in streptomycetes liquid cultures has hindered the precise identification of reliable phenotypes for use in the analysis and optimization of industrial fermentations. It has been demonstrated that antibiotics are produced by the substrate mycelium at the end of the proliferation phase, i.e., second mycelium morphology (after PCD process) [27, 30, 37]. However, despite the fact that there is a distinctive mycelium producing antibiotics (MII) [27, 30] (hence there is not secondary metabolite production until the differentiation of MII), antibiotic production has additional regulations, and each *Streptomyces* strain does not display its entire potential secondary metabolism at a specific developmental condition.

1.3 Screening for new bioactive compounds and drug discovery approaches

Most antibiotics and other bioactive compounds (antitumorals, immunosuppressors, etc.) were discovered from actinomycetes, but this source of drugs became challenging once the most common compounds were discovered. New antibiotic scaffolds are required, and resuming screening from nature is mandatory.

During the past 30 years, only two new classes of antibiotics have been brought to the clinic for treatment systemic infections [38]. At the same time, microbial resistance to existing antibiotics has increased dramatically, rendering some microbial infections that are extremely hard to treat. The current scenario looks gloomy, and there is a concrete risk of humanity returning to the pre-antibiotic era, with a high mortality from routine surgical or chemotherapeutic procedures because of infections by antibiotic-resistant pathogens. There is a general consensus that new antibiotics are urgently required and are the best chance in the fight against resistance.

Only a fraction of secondary metabolite pathways is active in actinomycete laboratory cultures, and there are a large number of strains whose potential to produce bioactive compounds remains unexplored [39]. When whole genome sequencing became available at the start of the twenty-first century, the existence of many silent or cryptic biosynthetic gene clusters in actinomycetes was revealed, which may encode antibiotic-like substances but are not, or only poorly, expressed under laboratory growth conditions. Each *Streptomyces* strain harbors 20–50 biosynthetic pathways for natural products, for which only some products have been identified [39]. The big challenge is to find ways to activate these pathways so as to allow screening for new secondary metabolites from nature to resume.

2. Strategies in drug discovery

Once the most common antibiotics and bioactive compounds were discovered, drug discovery has become more challenging, and industry had mostly discontinued the search for new drugs, with *Streptomyces* as the source. Nevertheless, the urgent necessity of new antibiotics in the clinic has caused screening from nature to be resumed [39]. New environments are being explored and some new actinomycete strains have been discovered [40]; the potential of the marine environment has been highly explored during last decade, as well as some, *a priori* hostile soils, such as high mountains, deserts or icy places. The scientific community is also paying more attention to symbiotic relationships.

Thinking of natural conditions of growth is another important point of view. Laboratory/industrial conditions in which microorganisms are typically grown are very different conditions from nature, which imply the dormancy of most of the genetic pathways involved in antibiotic production, since these substances are related to stress conditions or defense against niche competitors. Furthermore, it is necessary to take into account that the majority of environmental bacteria are not easily culturable or are even nonculturable in laboratory conditions. This broad view is now possible, thanks to the access to the metagenome in different environments [41].

Another current strategy is to look for elicitors. Elicitors are small molecules, which are able to induce *Streptomyces* differentiation as well as trigger cryptic antibiotic pathways.

Additionally, secondary metabolism is conditioned by differentiation, so it is another aspect that requires special attention. As mentioned above, the life cycle stages and the mycelium morphology are keys for antibiotic production.

At present, different complementary strategies are being developed to improve secondary metabolite production and active cryptic biosynthetic gene clusters. One way to classify them is by unselective and selective methods [42], considering unselective methods are used in the improvement of screening of new activities and selective methods are used in the improvement of production of known molecules.

2.1 Methods for searching for new bacteria/compounds

2.1.1 Exploring new environments

The actinomycetes are widely distributed in the environment; they were discovered at the end of the nineteenth century as a group of soil-living microorganisms. However, their importance started in 1943 with the discovery of streptomycin, the first effective treatment against tuberculosis by the most important actinomycetes genus, *Streptomyces*. Researchers have isolated different *Streptomyces* as soil inhabitants during decades as the most important source of antibiotics and bioactive compounds in nature. Clavulanic acid (*Streptomyces clavuligerus* [43]), neomycin (*S. marinensis* [44]), chloramphenicol (*Streptomyces venezuelae* [45]), the insecticide avermectin (*Streptomyces avermitilis* [46]), the immunosuppressant tacrolimus (*Streptomyces tsukubaensis* [47]), kanamycin (*Streptomyces kanamyceticus* [48]) and potent antitumoral platenolides (*Streptomyces platensis* [49]) are only few examples of the 12,400 bioactive compounds produced by the genus *Streptomyces* that society is routinely using in the clinic or agriculture. The rest of the phylum produces 3600 bioactive compounds (being 3400 antibiotics) [50]. Until the 1980s, new bioactive compounds were discovered relatively easily; however, since 1985, only three new compounds have been discovered, with the last one, platensimycin (*S. platensis* [40]), in 2006.

Once the most common compounds are discovered and producer strains are isolated, one of the strategies for founding new activities is the exploration of new environments, under-exploited habitats, generally difficult to access. Marine environment are an important niche of new species of *Streptomyces* [51]. Besides marine actinomycetes, they have been isolated from Himalaya Mountains [52], several islands [53], Atacama Desert [54], Antarctica [55] and several extremophile habitats [56–58].

Promising results and new species are being discovered from these unexplored areas, in combination with next-generation genome sequencing, metagenomics and new ways of bacterial culture [59], being one of the most important approaches in the new age of bioactive compound research.

2.1.2 Symbiosis relationships

As mentioned above, *Streptomyces* are ubiquitous in soils. However, it has been discovered that they are not only free-living soil bacteria but many have also evolved to live in symbiosis with plants, fungi and animals [60]. Their secondary metabolism is a consequence of these complex interactions [60].

There are many and very diverse symbiotic relationships, which involve *Streptomyces*. One of the first discovered is its relationship with plant roots providing a natural defense against fungal infections [61]; it is very common to find bacteria of the *Streptomyces* genus in the rhizosphere. This fact raises the possibility, currently discussed, about its use as biofertilizers in crops [62, 63]. Furthermore, genome mining is now allowing us to know on one hand which genes are involved in this interaction and on the other hand the discovery of potential novel molecules produced only in these conditions [64].

Other important interactions of *Streptomyces* genera take place between bacteria and insects. For instance, in the case of Beewolf digger wasps, *Streptomyces* provides protection against pathogens to the larval cocoon producing a mixture of nine antibiotic compounds [65, 66].

In *Streptomyces* isolates from colonies of the South African termite *Macrotermes natalensis*, a number of interesting novel compounds including natalamycin A

[67] and the potent antitumor geldanamycin [67] have been discovered. Another *Streptomyces* isolate from the same termite produces a novel cyclic analogue of dentigerumycin [68], the antifungal from the fungus-growing ant system. Finally, a series of glycosylated isoflavonoid compounds, including termisoflavone A [69], were recovered from a third *Streptomyces* isolate.

In 2012, 30 *Streptomyces* strains were isolated from a different South African fungus-growing termite, *Microtermes* sp., and two novel polyketides, microtermolides A and B [69], were identified.

Two *Streptomyces* strains, isolated from *Dendroctonus frontalis* beetles, inhibit the antagonistic *Ophiostoma* fungus, but not the mutualistic fungus *Entomocorticium*, suggesting a defensive role that supports the beetle-fungus mutualism. The molecules responsible for this selective antifungal activity are the polyketide mycangimycin [70] and the polycyclic tetramate macrolactams, frontalamides A and B [71, 72].

Other important symbiotic relationship takes place between ants and *Streptomyces*. In recent years, new strains and compounds have been isolated from ant head, legs and nests [73–76].

2.1.3 Nonculturable microorganisms

Since the beginning of microbiology, it was known that many microorganisms do not grow under laboratory conditions and we cannot cultivate them. The new techniques of massive sequencing have revealed the existence of a huge amount of nonculturable microorganisms. Nonculturable microorganisms are, in fact, a majority, and they represent an important challenge in the screening for new secondary metabolites. Next generation sequencing (NGS) has revealed the big pharmacological potential of uncultured bacteria; hence, approaches to improve cultivation-based methods, such as isolation chip (iChip), which is a method of culturing bacteria within its soil environment, or co-cultures, culturing actinomycetes with other bacteria (generally species with which they naturally coexist), are being used successfully [70–72]. The combination of iChip and unexplored ecosystems or symbiotic relationships, mentioned above, is promising as well.

2.2 Methods for screening improvement in a known microorganism (unselective)

Unselective methods include classical strategies for improving the production of a certain compound by modifying the growth of the *Streptomyces* strain. This strategy used to be empiric.

2.2.1 Changing media components

Streptomyces grow well in rich culture mediums; nevertheless, antibiotic production often needs some specific nutrients. Sometimes defined medium can be better for determined compounds. Optimization of nutritional requirements has been one of the most useful changes for secondary metabolite activities [73–76].

2.2.2 Inducing stress response

Stressful situations are one of the reasons why microorganisms start to produce and export bioactive compounds. For this reason, the induction of cultures with several treatments of stress is one of the improvement strategies for synthesis of secondary metabolite compounds in bacteria. Heat and ethanol shock treatments, high salt conditions, increased hydrostatic pressure, acidic pH shock or feed limitations are some examples of this classical strategy (reviewed in [77]).

2.2.3 Random mutagenesis

There are different random mutagenesis methods used in *Streptomyces*: chemical mutagenesis, in which are used (e.g., by N-methyl-N′-nitro-N-nitrosoguanidine (MNNG), ethyl methanesulfonate (EMS), or nitrous acid (NA)) and physical mutagenesis (e.g., by ultraviolet (UV) light or X-rays)) that remain being robust methods to generate very high-producing strains [78–80].

On the other hand, there is transposon mutagenesis, a powerful tool for random mutagenesis of bacterial genomes and insertion of foreign DNA. A plasmid containing the enzyme transposase and the transposon (small fragment of DNA with transposase target regions) is extracted and inserted randomly into the host chromosome. Transposition creates single mutations that can be identified and mapped by plasmid rescue and DNA sequencing, generating a direct link between the higher yield (phenotype) and the transposon insertion (genotype) [81, 82].

2.2.4 Ribosomal engineering

The fact that some antibiotic resistances trigger some cryptic pathways of secondary metabolites has been recently used as another strategy in the search for new compounds.

It has been demonstrated that this occurs in strains affected by antibiotics, which target a ribosomal protein (for instance, streptomycin, paromomycin and gentamicin [42]), so the effects of the mutations at this level have had very good results, for example, in *Streptomyces coelicolor*, for improving secondary metabolite production [83]. The mechanism of this strategy is still unknown, but it has been proved that there is activation of the positive regulators like *act*II-*orf*4 gene at least in *S. coelicolor* [42]. Through this approach, the antibacterial piperidamycins family compounds have been discovered [84].

2.2.5 Elicitors

Elicitors are small diffusible molecules, which are being used as signals for improving secondary metabolite production or even activating cryptic antibiotic pathways [85].

Elicitors are also present in *Streptomyces* natural environments; hence one of the strategies for stimulating bioactive compound production is the co-culture of different bacteria [85, 86]. Co-cultures usually include species that have symbiotic relationships with *Streptomyces* in nature [87, 88] or pathogenic partners to activate the production of antimicrobial compounds against them [71, 89, 90].

Other kinds of elicitors are also useful, for instance, fungal elicitors (a complex mix of fungal cell walls and filtered fungi cultures) can positively affect the production of natamycin [91]; small molecules, such as GlcNac or phosphate, can activate differentiation and antibiotic production in *S. coelicolor* through the activation of *actII*-ORF4/*red*Z genes in some conditions [92].

At the end of the 1980s, the discovery of new compounds reached a bottleneck, and screening processes often led to the rediscovery of compounds already known. To avoid rediscovery, Pimentel-Elardo et al. [93] developed a workflow based on the use of chemical elicitors combined with activity-independent screening [93]. Activity-independent screening is based on the low-abundant compounds found in biological samples. The use of elicitors increases the production of these low-abundant compounds, distinguishing them from the most abundant ones. The elicitor "CI-ARC" has been identified as being responsible for triggering several cryptic biosynthetic genes [93].

2.2.6 Differentiation strategies based on morphology

2.2.6.1 Streptomyces behavior in liquid cultures

Large-scale antibiotic production is mostly performed in liquid cultures. Macroscopic morphology of the mycelium (pellets and clumps) correlates with the production of secondary metabolites [30]. The genes controlling mycelium aggregation have been characterized in the *S. coelicolor mat* gene cluster [94], and the *cslA, glxA* and *dtpA* genes [95–97] are responsible for pellet formation. This knowledge can be used for controlling and improving morphology in industrial fermentations.

Since spores are inoculated in liquid medium, there is an aggregation trend [98]. Germlings determine the macroscopic morphology (pellets and clumps) of the culture, which involves the highly conserved genes *matA, matB* [94, 98] and the *cslA/glxA* operon [98].

In liquid cultures, there are some strains able to sporulate [99], but even those which aren't able to sporulate in flasks can sporulate in bioreactors, beacuse of the stess conditions inside the fermenter, such as lack of oxygen, strong agitation, etc. [100]. When *Streptomyces* bacteria sporulate, all the metabolic efforts are focused on that, which means that the secondary metabolism stops by. Therefore, one of the main things for taking into account during a screening for new secondary metabolites is to avoid sporulation in the industrial fermentations. Hence, it is very important to look out the cultures for keeping secondary metabolism as much as possible [100].

Morphology in liquid cultures can be monitored by several tools. Laser diffraction has been used to measure pellet size, an improvement over image analysis, which is trickier and requires more time [101], and a technique based on flow cytometry has been used to establish the pellet size distribution of a culture population [102, 103]. Recently, a useful algorithm has been developed as a plugin for the open-source software ImageJ to characterize the morphology of submerged growing cells [104]. On the other hand, mathematical models have been performed to predict the behavior of *Streptomyces* liquid cultures based on pellet/clump morphology [105, 106].

Furthermore, it has been reported that the oxygen transfer rate (OTR) and oxygen uptake rate (OUR) are crucial for the development and production in liquid cultures [100]. Controlling these rates, which directly affects morphology, was described as being crucial for scaling up production to industrial conditions [107]. Biophysical parameters, including pH, viscosity, agitation, dissolved oxygen levels and surface tension, directly affect mycelium morphology. The optimization of these factors is another factor to consider [31].

2.2.6.2 Programmed cell death and MII differentiation

As mentioned in the introduction of this chapter, programmed cell death (PCD) is a key event, triggering the differentiation of antibiotic producer mycelium (MII) in liquid and solid cultures [30]. However, the specific signals derived from cell death are not yet known. The N-acetylglucosamine resulting from peptidoglycan dismantling accelerates the development and antibiotic production [108, 109] and might be one of the signals released during PCD. A simple methodology based on fluorimetric measures was designed to quantify PCD in liquid cultures [37], allowing prediction of the efficiency of antibiotic production based on the level of PCD.

PCD and MII differentiation can be modulated by modifying culture conditions and morphology. Modifying developmental conditions to enhance PCD and MII differentiation leads to the improvement of secondary metabolite production. This approach was recently applied in a study combining the heterologous expression of plant flavonoids in *Streptomyces albus* with a strategy to enhance PCD and MII differentiation [110]. The same strategy was successful at improving mTGase production from *Streptomyces mobaraensis* [111]. Therefore, PCD and MII differentiation were demonstrated to be crucial for the streamlining of secondary metabolite production in bioreactors [100].

2.3 Methods for production improvement in a known compound (selective)

Selective methods with a specific target include regulatory engineering, heterologous expression, genome mining and combinatorial biosynthesis.

2.3.1 Regulatory engineering

Each active compound, for example, antibiotic or antitumor, etc., has several molecules synthesized first in the pathway, and sometimes these molecules are a limited source for the compound of interest. In this sense, one strategy widely used is the addition of these precursors, previously synthetized from an external source. In the same way, it is possible to enhance the production of a specific compound through the overexpression of its positive regulators. Inversely, knocking out the suppressors (negative regulation) of the gene of interest has also had good results increasing the final compound production [112, 113].

2.3.2 Heterologous expression

In microbiology, this molecular genetic technique is based on the expression of a gene or a group of genes (gene cluster) in a host microorganism that does not have this particular gene or gene cluster in its own genome. This procedure, performed by recombinant DNA technology, is very useful when hosts are bacteria with simple developmental cycles, such as *E. coli* or *Bacillus* [114]. However, in the case of *Streptomyces,* due to its complex pathways to produce antibiotics, heterologous expression usually works only among *Streptomyces* strains [115, 116], not solving the problem of its tricky growth.

Nevertheless, good results have been achieved in activating cryptic metabolites, for example, by using the widely conserved *Streptomyces coelicolor* pleiotropic regulator *Afs*Q, which activates the production of new compounds in several streptomycetes [117]. Despite the fact that heterologous expression in *Streptomyces* is a challenge, this technique continues to be improved [118].

2.3.3 Genome mining

Genome mining is defined as the obtaining of information about an organism, based on genome analysis. In the postgenomic era, the combination of easier genome sequencing and bioinformatics analyses allows researchers to uncover the biosynthetic potential of the microorganisms, i.e., the gene clusters for producing secondary metabolites into actinomycetes strains that are in the genome, but silent. Genome mining is nowadays one of the more powerful tools in the screening for new antibiotics [119–122].

2.3.4 Combinatorial biosynthesis

Combinatorial biosynthesis can be defined as the application to modify biosynthetic pathways to natural products to produce new and altered structures using nature's biosynthetic machinery [123]. The chemical modification of existing chemical scaffolds is challenging and sometimes provides only a temporary solution against resistant organisms [124]. Combinatorial biosynthesis has been largely developed over the last 20 years, and it has been enhanced, thanks to genome mining and synthetic biology [125–127].

3. Conclusions

Humanity faces the great challenge of fighting antibiotic resistance, which is growing at a faster rate than our capacity to find new antimicrobials and new strategies to solve this problem.

The *Streptomyces* genus is still a huge source of natural bioactive compounds, but we need to elaborate new multidisciplinary strategies to avoid rediscovering the same compounds. Therefore, we need to invest efforts into developing alternative strategies to resume screening from natural actinomycetes.

There is not a "magic" methodology to activate cryptic pathways and improve the discovery of new bioactive compounds, but multidisciplinary approaches combining the methodologies discussed in this chapter will play a key role in the improvement of screening for new bioactive compounds from streptomycetes.

Acknowledgements

We thank the Spanish "Ministerio de Economía y Competitividad" (MINECO; BIO2015-65709-R) and the "Marie Curie cofund Clarin" Grant for financial support and Proof-Reading-Service.com for proofreading the final manuscript.

Conflict of interest

There is no conflict of interest in this work.

Author details

Ángel Manteca[1] and Paula Yagüe[2]*

1 Oviedo University, Oviedo, Spain

2 Leiden University, Leiden, Netherlands

*Address all correspondence to: p.yague.menendez@biology.leidenuniv.nl

References

[1] Challis GL, Hopwood DA. Synergy and contingency as driving forces for the evolution of multiple secondary metabolite production by *Streptomyces* species. Proceedings of the National Academy of Sciences of the United States of America. 2003;**100**(Suppl 2):14555-14561

[2] Bobek J, Smidova K, Cihak M. A waking review: Old and novel insights into the spore germination in *Streptomyces*. Frontiers in Microbiology. 2017;**8**:2205

[3] Noens EE, Mersinias V, Willemse J, Traag BA, Laing E, Chater KF, et al. Loss of the controlled localization of growth stage-specific cell-wall synthesis pleiotropically affects developmental gene expression in an ssgA mutant of *Streptomyces coelicolor*. Molecular Microbiology. 2007;**64**(5):1244-1259

[4] Kalakoutskii LV, Agre NS. Comparative aspects of development and differentiation in actinomycetes. Bacteriological Reviews. 1976;**40**(2):469-524

[5] Ranade N, Vining LC. Accumulation of intracellular carbon reserves in relation to chloramphenicol biosynthesis by *Streptomyces venezuelae*. Canadian Journal of Microbiology. 1993;**39**(4):377-383

[6] Strakova E, Bobek J, Zikova A, Vohradsky J. Global features of gene expression on the proteome and transcriptome levels in *S. coelicolor* during germination. PLoS One. 2013;**8**(9):e72842

[7] Hardisson C, Manzanal MB, Salas JA, Suarez JE. Fine structure, physiology and biochemistry of arthrospore germination in *Streptomyces antibioticus*. Journal of General Microbiology. 1978;**105**(2):203-214

[8] Rioseras B, Yague P, Lopez-Garcia MT, Gonzalez-Quinonez N, Binda E, Marinelli F, et al. Characterization of SCO4439, a D-alanyl-D-alanine carboxypeptidase involved in spore cell wall maturation, resistance, and germination in *Streptomyces coelicolor*. Scientific Reports. 2016;**6**:21659

[9] Grund AD, Ensign JC. Properties of the germination inhibitor of *Streptomyces viridochromogenes* spores. Journal of General Microbiology. 1985;**131**(4):833-847

[10] Salas JA, Guijarro JA, Hardisson C. High calcium content in *Streptomyces* spores and its release as an early event during spore germination. Journal of Bacteriology. 1983;**155**(3):1316-1323

[11] McBride MJ, Ensign JC. Effects of intracellular trehalose content on *Streptomyces griseus* spores. Journal of Bacteriology. 1987;**169**(11):4995-5001

[12] Nwaka S, Holzer H. Molecular biology of trehalose and the trehalases in the yeast *Saccharomyces cerevisiae*. Progress in Nucleic Acid Research and Molecular Biology. 1998;**58**:197-237

[13] Singer MA, Lindquist S. Multiple effects of trehalose on protein folding in vitro and in vivo. Molecular Cell. 1998;**1**(5):639-648

[14] Mikulik K, Janda I, Weiser J, Stastna J, Jiranova A. RNA and ribosomal protein patterns during aerial spore germination in *Streptomyces* granaticolor. European Journal of Biochemistry. 1984;**145**(2):381-388

[15] Bobek J, Halada P, Angelis J, Vohradsky J, Mikulik K. Activation and expression of proteins during synchronous germination of aerial spores of *Streptomyces* granaticolor. Proteomics. 2004;**4**(12):3864-3880

[16] Flardh K. Essential role of DivIVA in polar growth and morphogenesis in *Streptomyces coelicolor* A3(2). Molecular Microbiology. 2003;**49**(6):1523-1536

[17] Flardh K. Growth polarity and cell division in *Streptomyces*. Current Opinion in Microbiology. 2003;**6**(6):564-571

[18] Flardh K, Richards DM, Hempel AM, Howard M, Buttner MJ. Regulation of apical growth and hyphal branching in *Streptomyces*. Current Opinion in Microbiology. 2012;**15**(6):737-743

[19] Cihak M, Kamenik Z, Smidova K, Bergman N, Benada O, Kofronova O, et al. Secondary Metabolites Produced during the Germination of *Streptomyces coelicolor*. Frontiers in Microbiology. 2017;**8**:2495

[20] Yague P, Willemse J, Koning RI, Rioseras B, Lopez-Garcia MT, Gonzalez-Quinonez N, et al. Subcompartmentalization by cross-membranes during early growth of *Streptomyces hyphae*. Nature Communications. 2016;**7**:12467

[21] Manteca A, Fernandez M, Sanchez J. A death round affecting a young compartmentalized mycelium precedes aerial mycelium dismantling in confluent surface cultures of *Streptomyces antibioticus*. Microbiology. 2005;**151**(Pt 11):3689-3697

[22] Yague P, Rodriguez-Garcia A, Lopez-Garcia MT, Martin JF, Rioseras B, Sanchez J, et al. Transcriptomic analysis of *Streptomyces coelicolor* differentiation in solid sporulating cultures: First compartmentalized and second multinucleated mycelia have different and distinctive transcriptomes. PLoS One. 2013;**8**(3):e60665

[23] Claessen D, de Jong W, Dijkhuizen L, Wosten HA. Regulation of *Streptomyces* development: Reach for the sky! Trends in Microbiology. 2006;**14**(7):313-319

[24] Flardh K, Buttner MJ. *Streptomyces morphogenetics*: Dissecting differentiation in a filamentous bacterium. Nature Reviews. Microbiology. 2009;**7**(1):36-49

[25] Kaiser BK, Stoddard BL. DNA recognition and transcriptional regulation by the WhiA sporulation factor. Scientific Reports. 2011;**1**:156

[26] Schlimpert S, Flardh K, Buttner J. Fluorescence time-lapse imaging of the complete *S. venezuelae* life cycle using a microfluidic device. Journal of Visualized Experiments. 2016;**108**:53863

[27] Yague P, Rodriguez-Garcia A, Lopez-Garcia MT, Rioseras B, Martin JF, Sanchez J, et al. Transcriptomic analysis of liquid non-sporulating *Streptomyces coelicolor* cultures demonstrates the existence of a complex differentiation comparable to that occurring in solid sporulating cultures. PLoS One. 2014;**9**(1):e86296

[28] Manteca A, Jung HR, Schwammle V, Jensen ON, Sanchez J. Quantitative proteome analysis of *Streptomyces coelicolor* nonsporulating liquid cultures demonstrates a complex differentiation process comparable to that occurring in sporulating solid cultures. Journal of Proteome Research. 2010;**9**(9):4801-4811

[29] Rioseras B, Shliaha PV, Gorshkov V, Yague P, Lopez-Garcia MT, Gonzalez-Quinonez N, et al. Quantitative proteome and phosphoproteome analyses of *Streptomyces coelicolor* reveal proteins and phosphoproteins modulating differentiation and secondary metabolism. Molecular & Cellular Proteomics. 2018;**17**(8):1591-1611

[30] Manteca A, Alvarez R, Salazar N, Yague P, Sanchez J. Mycelium differentiation and antibiotic production in submerged cultures of *Streptomyces coelicolor*. Applied

and Environmental Microbiology. 2008;**74**(12):3877-3886

[31] van Dissel D, Claessen D, van Wezel GP. Morphogenesis of *Streptomyces* in submerged cultures. Advances in Applied Microbiology. 2014;**89**:1-45

[32] Giudici R, Pamboukian CR, Facciotti MC. Morphologically structured model for antitumoral retamycin production during batch and fed-batch cultivations of *Streptomyces olindensis*. Biotechnology and Bioengineering. 2004;**86**(4):414-424

[33] Vecht-Lifshitz SE, Sasson Y, Braun S. Nikkomycin production in pellets of *Streptomyces tendae*. The Journal of Applied Bacteriology. 1992;**72**(3):195-200

[34] Sarra M, Casas C, Godia F. Continuous production of a hybrid antibiotic by *Streptomyces lividans* TK21 pellets in a three-phase fluidized-bed bioreactor. Biotechnology and Bioengineering. 1997;**53**(6):601-610

[35] Jonsbu E, McIntyre M, Nielsen J. The influence of carbon sources and morphology on nystatin production by *Streptomyces noursei*. Journal of Biotechnology. 2002;**95**(2):133-144

[36] Stratigopoulos G, Gandecha AR, Cundliffe E. Regulation of tylosin production and morphological differentiation in *Streptomyces fradiae* by TylP, a deduced gamma-butyrolactone receptor. Molecular Microbiology. 2002;**45**(3):735-744

[37] Yague P, Manteca A, Simon A, Diaz-Garcia ME, Sanchez J. New method for monitoring programmed cell death and differentiation in submerged *Streptomyces* cultures. Applied and Environmental Microbiology. 2010;**76**(10):3401-3404

[38] Procopio RE, Silva IR, Martins MK, Azevedo JL, Araujo JM. Antibiotics produced by *Streptomyces*. The Brazilian Journal of Infectious Diseases. 2012;**16**(5):466-471

[39] Genilloud O. The re-emerging role of microbial natural products in antibiotic discovery. Antonie Van Leeuwenhoek. 2014;**106**(1):173-188

[40] Takahashi Y, Omura S. Isolation of new actinomycete strains for the screening of new bioactive compounds. The Journal of General and Applied Microbiology. 2003;**49**(3):141-154

[41] McMahon MD, Guan C, Handelsman J, Thomas MG. Metagenomic analysis of *Streptomyces lividans* reveals host-dependent functional expression. Applied and Environmental Microbiology. 2012;**78**(10):3622-3629

[42] Craney A, Ahmed S, Nodwell J. Towards a new science of secondary metabolism. Journal of Antibiotics (Tokyo). 2013;**66**(7):387-400

[43] Reading C, Cole M. Clavulanic acid: A beta-lactamase-inhiting beta-lactam from *Streptomyces clavuligerus*. Antimicrobial Agents and Chemotherapy. 1977;**11**(5):852-857

[44] Waksman SA, Lechevalier HA. Neomycin, a new antibiotic active against streptomycin-resistant bacteria, including tuberculosis organisms. Science. 1949;**109**(2830):305-307

[45] Shapiro S, Vining LC. Nitrogen metabolism and chloramphenicol production in *Streptomyces venezuelae*. Canadian Journal of Microbiology. 1983;**29**(12):1706-1714

[46] Ikeda H, Nonomiya T, Usami M, Ohta T, Omura S. Organization of the biosynthetic gene cluster for the polyketide anthelmintic macrolide avermectin in *Streptomyces avermitilis*. Proceedings of the National Academy

of Sciences of the United States of America. 1999;**96**(17):9509-9514

[47] Karlsson H, Truedsson L, Nassberger L. The immunosuppressive agent FK506 inhibits in vitro expression of membrane-bound and soluble interleukin-2 receptors on resting but not on activated human lymphocytes. Immunology Letters. 1991;**30**(1):129-132

[48] Kharel MK, Subba B, Basnet DB, Woo JS, Lee HC, Liou K, et al. A gene cluster for biosynthesis of kanamycin from *Streptomyces kanamyceticus*: Comparison with gentamicin biosynthetic gene cluster. Archives of Biochemistry and Biophysics. 2004;**429**(2):204-214

[49] Sakai T, Sameshima T, Matsufuji M, Kawamura N, Dobashi K, Mizui Y. Pladienolides, new substances from culture of *Streptomyces platensis* Mer-11107. I. Taxonomy, fermentation, isolation and screening. Journal of Antibiotics (Tokyo). 2004;**57**(3):173-179

[50] Hamedi J, Poorinmohammad N, Wink J. The role of actinobacteria in biotechnology. In: Wink J, Mohammadipanah F, Hamedi J, editors. Biology and Biotechnology of Actinobacteria. Cham: Springer; 2017. pp. 269-328. DOI: 10.1007/9783319603391

[51] Lam KS. Discovery of novel metabolites from marine actinomycetes. Current Opinion in Microbiology. 2006;**9**(3):245-251

[52] Bhattacharjee K, Banerjee S, Joshi SR. Diversity of *Streptomyces* spp. in eastern Himalayan region– computational RNomics approach to phylogeny. Bioinformation. 2012;**8**(12):548-554

[53] Charousova I, Medo J, Halenarova E, Javorekova S. Antimicrobial and enzymatic activity of actinomycetes isolated from soils of coastal islands. Journal of Advanced Pharmaceutical Technology & Research. 2017;**8**(2):46-51

[54] Idris H, Labeda DP, Nouioui I, Castro JF, Del Carmen Montero-Calasanz M, Bull AT, et al. *Streptomyces aridus* sp. nov., isolated from a high altitude Atacama Desert soil and emended description of *Streptomyces noboritoensis* Isono et al. 1957. Antonie Van Leeuwenhoek. 2017;**110**(5):705-717

[55] Sanyal A, Antony R, Samui G, Thamban M. Microbial communities and their potential for degradation of dissolved organic carbon in cryoconite hole environments of Himalaya and Antarctica. Microbiological Research. 2018;**208**:32-42

[56] Bonilla JO, Kurth DG, Cid FD, Ulacco JH, Gil RA, Villegas LB. Prokaryotic and eukaryotic community structure affected by the presence of an acid mine drainage from an abandoned gold mine. Extremophiles. 2018;**22**(5):699-711

[57] Ben Abdallah M, Karray F, Kallel N, Armougom F, Mhiri N, Quemeneur M, et al. Abundance and diversity of prokaryotes in ephemeral hypersaline lake Chott El Jerid using Illumina Miseq sequencing, DGGE and qPCR assays. Extremophiles. 2018;**22**(5):811-823

[58] Bull AT, Idris H, Sanderson R, Asenjo J, Andrews B, Goodfellow M. High altitude, hyper-arid soils of the Central-Andes harbor mega-diverse communities of actinobacteria. Extremophiles. 2018;**22**(1):47-57

[59] Stewart EJ. Growing unculturable bacteria. Journal of Bacteriology. 2012;**194**(16):4151-4160

[60] Seipke RF, Kaltenpoth M, Hutchings MI. *Streptomyces* as symbionts: An emerging and widespread theme?

FEMS Microbiology Reviews. 2012;**36**(4):862-876

[61] Schrey SD, Tarkka MT. Friends and foes: Streptomycetes as modulators of plant disease and symbiosis. Antonie Van Leeuwenhoek. 2008;**94**(1):11-19

[62] Rey T, Dumas B. Plenty is no plague: *Streptomyces* symbiosis with crops. Trends in Plant Science. 2017;**22**(1):30-37

[63] Vurukonda S, Giovanardi D, Stefani E. Plant growth promoting and biocontrol activity of *Streptomyces* spp. as endophytes. International Journal of Molecular Sciences. 2018;**19**(4)

[64] Ceapa CD, Vazquez-Hernandez M, Rodriguez-Luna SD, Cruz Vazquez AP, Jimenez Suarez V, Rodriguez-Sanoja R, et al. Genome mining of *Streptomyces scabrisporus* NF3 reveals symbiotic features including genes related to plant interactions. PLoS One. 2018;**13**(2):e0192618

[65] Kroiss J, Kaltenpoth M, Schneider B, Schwinger MG, Hertweck C, Maddula RK, et al. Symbiotic Streptomycetes provide antibiotic combination prophylaxis for wasp offspring. Nature Chemical Biology. 2010;**6**(4):261-263

[66] Engl T, Kroiss J, Kai M, Nechitaylo TY, Svatos A, Kaltenpoth M. Evolutionary stability of antibiotic protection in a defensive symbiosis. Proceedings of the National Academy of Sciences of the United States of America. 2018;**115**(9):E2020-E2029

[67] Kim KH, Ramadhar TR, Beemelmanns C, Cao S, Poulsen M, Currie CR, et al. Natalamycin A, an ansamycin from a termite-associated *Streptomyces* sp. Chemical Science. 2014;**5**(11):4333-4338

[68] Wyche TP, Ruzzini AC, Beemelmanns C, Kim KH, Klassen JL, Cao S, et al. Linear peptides are the major products of a biosynthetic pathway that encodes for cyclic depsipeptides. Organic Letters. 2017;**19**(7):1772-1775

[69] Kang HR, Lee D, Benndorf R, Jung WH, Beemelmanns C, Kang KS, et al. Termisoflavones A-C, isoflavonoid glycosides from termite-associated *Streptomyces* sp. RB1. Journal of Natural Products. 2016;**79**(12):3072-3078

[70] Kealey C, Creaven CA, Murphy CD, Brady CB. New approaches to antibiotic discovery. Biotechnology Letters. 2017;**39**(6):805-817

[71] Sung AA, Gromek SM, Balunas MJ. Upregulation and identification of antibiotic activity of a marine-derived *Streptomyces* sp. via co-cultures with human pathogens. Marine Drugs. 2017;**15**(8)

[72] Onaka H, Mori Y, Igarashi Y, Furumai T. Mycolic acid-containing bacteria induce natural-product biosynthesis in *Streptomyces* species. Applied and Environmental Microbiology. 2011;**77**(2):400-406

[73] Vastrad BM, Neelagund SE. Optimization of medium composition for the production of neomycin by *Streptomyces* fradiae NCIM 2418 in solid state fermentation. Biotechnology Research International. 2014;**2014**:674286

[74] Alduina R, Sosio M, Donadio S. Complex regulatory networks governing production of the glycopeptide A40926. Antibiotics (Basel). 2018;**7**(2)

[75] Jakeman DL, Graham CL, Young W, Vining LC. Culture conditions improving the production of jadomycin B. Journal of Industrial Microbiology & Biotechnology. 2006;**33**(9):767-772

[76] Saudagar PS, Singhal RS. Optimization of nutritional requirements and feeding strategies

for clavulanic acid production by *Streptomyces clavuligerus*. Bioresource Technology. 2007;**98**(10):2010-2017

[77] Yoon V, Nodwell JR. Activating secondary metabolism with stress and chemicals. Journal of Industrial Microbiology & Biotechnology. 2014;**41**(2):415-424

[78] Baltz RH. Genetic manipulation of secondary metabolite biosynthesis for improved production in *Streptomyces* and other actinomycetes. Journal of Industrial Microbiology & Biotechnology. 2016;**43**(2-3):343-370

[79] Khaliq S, Akhtar K, Afzal Ghauri M, Iqbal R, Mukhtar Khalid A, Muddassar M. Change in colony morphology and kinetics of tylosin production after UV and gamma irradiation mutagenesis of *Streptomyces fradiae* NRRL-2702. Microbiological Research. 2009;**164**(4):469-477

[80] Korbekandi H, Darkhal P, Hojati Z, Abedi D, Hamedi J, Pourhosein M. Overproduction of clavulanic acid by UV mutagenesis of *Streptomyces clavuligerus*. Iranian Journal of Pharmaceutical Research. 2010;**9**(2):177-181

[81] Fedashchin A, Cernota WH, Gonzalez MC, Leach BI, Kwan N, Wesley RK, et al. Random transposon mutagenesis of the *Saccharopolyspora erythraea* genome reveals additional genes influencing erythromycin biosynthesis. FEMS Microbiology Letters. 2015;**362**(22)

[82] Xu Z, Wang Y, Chater KF, Ou HY, Xu HH, Deng Z, et al. Large-scale transposition mutagenesis of *Streptomyces coelicolor* identifies hundreds of genes influencing antibiotic biosynthesis. Applied and Environmental Microbiology. 2017;**83**(6)

[83] Wang G, Hosaka T, Ochi K. Dramatic activation of antibiotic production in *Streptomyces coelicolor* by cumulative drug resistance mutations. Applied and Environmental Microbiology. 2008;**74**(9):2834-2840

[84] Hosaka T, Ohnishi-Kameyama M, Muramatsu H, Murakami K, Tsurumi Y, Kodani S, et al. Antibacterial discovery in actinomycetes strains with mutations in RNA polymerase or ribosomal protein S12. Nature Biotechnology. 2009;**27**(5):462-464

[85] Onaka H. Novel antibiotic screening methods to awaken silent or cryptic secondary metabolic pathways in actinomycetes. Journal of Antibiotics (Tokyo). 2017;**70**(8):865-870

[86] Marmann A, Aly AH, Lin W, Wang B, Proksch P. Co-cultivation--a powerful emerging tool for enhancing the chemical diversity of microorganisms. Marine Drugs. 2014;**12**(2):1043-1065

[87] Piel J. Metabolites from symbiotic bacteria. Natural Product Reports. 2004;**21**(4):519-538

[88] Yu J, Liu Q, Chen C, Qi X. Antifungal activity change of *Streptomyces rimosus* MY02 mediated by confront culture with other microorganism. Journal of Basic Microbiology. 2017;**57**(3):276-282

[89] Perez J, Munoz-Dorado J, Brana AF, Shimkets LJ, Sevillano L, Santamaria RI. *Myxococcus xanthus* induces actinorhodin overproduction and aerial mycelium formation by *Streptomyces coelicolor*. Microbial Biotechnology. 2011;**4**(2):175-183

[90] Wang Y, Wang L, Zhuang Y, Kong F, Zhang C, Zhu W. Phenolic polyketides from the co-cultivation of marine-derived *Penicillium* sp. WC-29-5 and *Streptomyces fradiae* 007. Marine Drugs. 2014;**12**(4):2079-2088

[91] Wang D, Yuan J, Gu S, Shi Q. Influence of fungal elicitors on biosynthesis of natamycin by *Streptomyces natalensis* HW-2. Applied Microbiology and Biotechnology. 2013;**97**(12):5527-5534

[92] Liu G, Chater KF, Chandra G, Niu G, Tan H. Molecular regulation of antibiotic biosynthesis in *Streptomyces*. Microbiology and Molecular Biology Reviews. 2013;**77**(1):112-143

[93] Pimentel-Elardo SM, Sorensen D, Ho L, Ziko M, Bueler SA, Lu S, et al. Activity-independent discovery of secondary metabolites using chemical elicitation and cheminformatic inference. ACS Chemical Biology. 2015;**10**(11):2616-2623

[94] van Dissel D, Claessen D, Roth M, van Wezel GP. A novel locus for mycelial aggregation forms a gateway to improved *Streptomyces* cell factories. Microbial Cell Factories. 2015;**14**:44

[95] Chaplin AK, Petrus ML, Mangiameli G, Hough MA, Svistunenko DA, Nicholls P, et al. GlxA is a new structural member of the radical copper oxidase family and is required for glycan deposition at hyphal tips and morphogenesis of *Streptomyces lividans*. The Biochemical Journal. 2015;**469**(3):433-444

[96] Petrus ML, Vijgenboom E, Chaplin AK, Worrall JA, van Wezel GP, Claessen D. The DyP-type peroxidase DtpA is a Tat-substrate required for GlxA maturation and morphogenesis in *Streptomyces*. Open Biology. 2016;**6**(1):150149

[97] Petrus ML, Claessen D. Pivotal roles for *Streptomyces* cell surface polymers in morphological differentiation, attachment and mycelial architecture. Antonie Van Leeuwenhoek. 2014;**106**(1):127-139

[98] Zacchetti B, Willemse J, Recter B, van Dissel D, van Wezel GP, Wosten HA, et al. Aggregation of germlings is a major contributing factor towards mycelial heterogeneity of *Streptomyces*. Scientific Reports. 2016;**6**:27045

[99] Geneviève Girard BAT, Sangal V, Mascini N, Hoskisson PA, Goodfellow M, van Wezel GP. A novel taxonomic marker that discriminates between morphologically complex actinomycetes. Open Biology. 2013;**3**(10):130073

[100] Rioseras B, Lopez-Garcia MT, Yague P, Sanchez J, Manteca A. Mycelium differentiation and development of *Streptomyces coelicolor* in lab-scale bioreactors: Programmed cell death, differentiation, and lysis are closely linked to undecylprodigiosin and actinorhodin production. Bioresource Technology. 2014;**151**:191-198

[101] Ronnest NP, Stocks SM, Lantz AE, Gernaey KV. Comparison of laser diffraction and image analysis for measurement of *Streptomyces coelicolor* cell clumps and pellets. Biotechnology Letters. 2012;**34**(8):1465-1473

[102] van Veluw GJ, Petrus ML, Gubbens J, de Graaf R, de Jong IP, van Wezel GP, et al. Analysis of two distinct mycelial populations in liquid-grown *Streptomyces* cultures using a flow cytometry-based proteomics approach. Applied Microbiology and Biotechnology. 2012;**96**(5):1301-1312

[103] Petrus ML, van Veluw GJ, Wosten HA, Claessen D. Sorting of *Streptomyces* cell pellets using a complex object parametric analyzer and sorter. Journal of Visualized Experiments. 2014;**84**:e51178

[104] Willemse J, Buke F, van Dissel D, Grevink S, Claessen D, van Wezel GP. SParticle, an algorithm for the analysis of filamentous microorganisms in submerged cultures. Antonie Van Leeuwenhoek. 2017;**111**(2):171-182

[105] Celler PC. K, van Loosdrecht MC, van Wezel GP., Structured

morphological modeling as a framework for rational strain design of *Streptomyces* species. Antonie Van Leeuwenhoek. 2012;**102**(3):409-423

[106] Nieminen L, Webb S, Smith MC, Hoskisson PA. A flexible mathematical model platform for studying branching networks: Experimentally validated using the model actinomycete, *Streptomyces coelicolor*. PLoS One. 2013;**8**(2):e54316

[107] Ha S, Lee KJ, Lee SI, Gwak HJ, Lee JH, Kim TW, et al. Optimization of Herbicidin A Production in Submerged Culture of *Streptomyces* scopuliridis M40. Journal of Microbiology and Biotechnology. 2017;**27**(5):947-955

[108] Rigali S, Nothaft H, Noens EE, Schlicht M, Colson S, Müller M, et al. The sugar phosphotransferase system of *Streptomyces coelicolor* is regulated by the GntR-family regulator DasR and links N-acetylglucosamine metabolism to the control of development. Molecular Microbiology. 2006;**61**(5):1237-1251

[109] Rigali S, Titgemeyer F, Barends S, Mulder S, Thomae AW, Hopwood DA, et al. Feast or famine: The global regulator DasR links nutrient stress to antibiotic production by *Streptomyces*. EMBO Reports. 2008;**9**(7):670-675

[110] Marin L, Gutierrez-Del-Rio I, Yague P, Manteca A, Villar CJ, Lombo F. De novo biosynthesis of apigenin, luteolin, and eriodictyol in the actinomycete *Streptomyces albus* and production improvement by feeding and spore conditioning. Frontiers in Microbiology. 2017;**8**:921

[111] LuciaTreppiccione AO, Luongo D, Maurano F, Manteca Á, FelipeLombó MR. Development of gluten with immunomodulatory properties using mTG-active food grade supernatants from *Streptomyces mobaraensis* cultures. Journal of Functional Foods. 2017;**34**

[112] Yi JS, Kim M, Kim EJ, Kim BG. Production of pikromycin using branched chain amino acid catabolism in *Streptomyces venezuelae* ATCC 15439. Journal of Industrial Microbiology & Biotechnology. 2018;**45**(5):293-303

[113] Razmilic V, Castro JF, Andrews B, Asenjo JA. Analysis of metabolic networks of *Streptomyces leeuwenhoekii* C34 by means of a genome scale model: Prediction of modifications that enhance the production of specialized metabolites. Biotechnology and Bioengineering. 2018;**115**(7):1815-1828

[114] Ece S, Lambertz C, Fischer R, Commandeur U. Heterologous expression of a *Streptomyces cyaneus* laccase for biomass modification applications. AMB Express. 2017;**7**(1):86

[115] Novakova R, Nunez LE, Homerova D, Knirschova R, Feckova L, Rezuchova B, et al. Increased heterologous production of the antitumoral polyketide mithramycin A by engineered *Streptomyces lividans* TK24 strains. Applied Microbiology and Biotechnology. 2018;**102**(2):857-869

[116] Kawahara T, Izumikawa M, Kozone I, Hashimoto J, Kagaya N, Koiwai H, et al. Neothioviridamide, a polythioamide compound produced by heterologous expression of a *Streptomyces* sp. cryptic RiPP biosynthetic gene cluster. Journal of Natural Products. 2018

[117] Daniel-Ivad M, Hameed N, Tan S, Dhanjal R, Socko D, Pak P, et al. An engineered allele of afsQ1 facilitates the discovery and investigation of cryptic natural products. ACS Chemical Biology. 2017;**12**(3):628-634

[118] Mevaere J, Goulard C, Schneider O, Sekurova ON, Ma H, Zirah S, et al. An orthogonal system for heterologous expression of actinobacterial lasso peptides in *Streptomyces* hosts. Scientific Reports. 2018;**8**(1):8232

[119] Yu Y, Tang B, Dai R, Zhang B, Chen L, Yang H, et al. Identification of the streptothricin and tunicamycin biosynthetic gene clusters by genome mining in *Streptomyces* sp. strain fd1-xmd. Applied Microbiology and Biotechnology. 2018;**102**(6):2621-2633

[120] Liu W, Sun F, Hu Y. Genome mining-mediated discovery of a new avermipeptin analogue in *Streptomyces actuosus* ATCC 25421. ChemistryOpen. 2018;7(7):558-561

[121] Gran-Scheuch A, Trajkovic M, Parra L, Fraaije MW. Mining the genome of *Streptomyces leeuwenhoekii*: Two new type I Baeyer-Villiger Monooxygenases from Atacama Desert. Frontiers in Microbiology. 2018;**9**:1609

[122] Ye S, Molloy B, Brana AF, Zabala D, Olano C, Cortes J, et al. Identification by genome mining of a type I polyketide gene cluster from *Streptomyces argillaceus* involved in the biosynthesis of pyridine and piperidine alkaloids argimycins P. Frontiers in Microbiology. 2017;**8**:194

[123] Floss HG. Combinatorial biosynthesis--potential and problems. Journal of Biotechnology. 2006;**124**(1):242-257

[124] Genilloud O. Actinomycetes: Still a source of novel antibiotics. Natural Product Reports. 2017;**34**(10):1203-1232

[125] Olano C, Mendez C, Salas JA. Post-PKS tailoring steps in natural product-producing actinomycetes from the perspective of combinatorial biosynthesis. Natural Product Reports. 2010;**27**(4):571-616

[126] González A, Rodríguez M, Braña AF, Méndez C, Salas JA, Olano C. New insights into paulomycin biosynthesis pathway in *Streptomyces* albus J1074 and generation of novel derivatives by combinatorial biosynthesis. Microbial Cell Factories. 2016:15

[127] Baltz RH. Synthetic biology, genome mining, and combinatorial biosynthesis of NRPS-derived antibiotics: A perspective. Journal of Industrial Microbiology & Biotechnology. 2017;**45**(7):635-649